T0226903

CONTINUED FRACTIONS
WITH APPLICATIONS

STUDIES IN
COMPUTATIONAL MATHEMATICS 3

Editors:

C. BREZINSKI
Univerity of Lille
Villeneuve d'Ascq, France

L. WUYTACK
University of Antwerp
Wilrijk, Belgium

NORTH-HOLLAND
AMSTERDAM · LONDON · NEW YORK · TOKYO

CONTINUED FRACTIONS
WITH APPLICATIONS

Lisa LORENTZEN
Division of Mathematical Sciences
University of Trondheim
Trondheim, Norway

Haakon WAADELAND
Department of Mathematics and Statistics
University of Trondheim
Trondheim, Norway

1992

NORTH-HOLLAND
AMSTERDAM · LONDON · NEW YORK · TOKYO

ELSEVIER SCIENCE PUBLISHERS B.V.
Sara Burgerhartstraat 25
P.O. Box 211, 1000 AE Amsterdam, The Netherlands

ISBN: 0 444 89265 6

This book is printed on acid-free paper.

Transferred to digital printing 2005

Preface

The name   Shortly before this book was finished we sent out a number of copies of Chapter I, under the name "A Taste of Continued Fractions". Now, in the process of working our way through the chapters on a last minute search for errors, unintended omissions and overlaps, or other unfortunate occurrences, we feel that this title might have been the right one even for the whole book. In most of the chapters, in particular in the applications, a lot of work has been put into the process of cutting, cancelling and "non-writing". In many cases we are just left with a "taste", or rather a glimpse of the role of the continued fractions within the topic of the chapter. We hope that we thereby can open some doors, but in most cases we are definitely not touring the rooms.

The chapters   Each chapter *starts* with some introductory information, "About this chapter". The purpose is not to tell about the contents in detail. That has been done elsewhere. What we want is to tell about the intention of the chapter, and thereby also to adjust the expectations to the right (moderate) level. Each chapter *ends* with a reference list, reflecting essentially literature used in preparing that particular chapter. As a result, books and papers will in many cases be referred to more than once in the book. On the other hand, those who look for a complete, updated bibliography on the field will look in vain. To present such a bibliography has not been one of the purposes of the book.

The authors   The two authors are different in style and approach. We have not made an effort to hide this, but to a certain extent the creative process of tearing up each other's drafts and telling him/her to glue it together in a better way (with additions and omissions) may have had a certain disguising effect on the differences. This struggling type of cooperation leaves us with a *joint* responsibility for the whole book. The way we then distribute blame and credit between us is an internal matter.

The treasure chest   Anybody who has lived with and loved continued fractions for a long time will also have lived with and loved the monographs by Perron, Wall and Jones/Thron. Actually the love for continued fractions most likely has been initiated by one or more of these books. This is at least the case for the authors of the present book, and more so: these three books have played an essential role in our lives. The present book is in no way an attempt to replace or compete with these books. To the contrary, we hope to urge the reader to go on to these sources for further information.

For whom?   We are aiming at two kinds of readers: On the one hand people *in or near* mathematics, who are curious about continued fractions; on the other hand senior-graduate level students who would like an introduction (and a little more) to the analytic theory of continued fractions. Some basic knowledge about functions of a complex variable, a little linear algebra, elementary differential equations and occasionally a little dash of measure theory is what is needed of mathematical background. Hopefully the students will appreciate the problems included and the examples. They may even appreciate that some examples precede a properly established theory. (Others may dislike it.)

Words of gratitude   We both owe a lot to Wolf Thron, for what we have learned from him, for inspiration and help, and for personal friendship. He has read most of this book, and his remarks, perhaps most of all his objections, have been of great help for us. Our gratitude also extends to Bill Jones, his closest coworker, to Arne Magnus, whose recent death struck us with sadness, and to all other members of the Colorado continued fraction community. Here in Trondheim Olav Njåstad has been a key person in the field, and we have on several occasions had a rewarding cooperation with him.

Many people, who had received our Chapter I, responded by sending friendly and encouraging letters, often with valuable suggestions. We thank them all for their interest and kind help.

The main person in the process of changing the hand-written drafts to a camera-ready copy was Leiv Arild Andenes Jacobsen. His able mastering of LaTeX, in combination with hard work, often at times when most people were in bed, has left us with a great debt of gratitude. We also want to thank Arild Skjølsvold and Irene Jacobsen for their part of the typing job. We finally thank Ruth Waadeland, who made all the drawings, except the LaTeX-made ones in Chapter XI.

The Department of Mathematics and Statistics, AVH, The University of Trondheim generously covered most of the typing expenses. The rest was covered by Elsevier Science Publishers.

We are most grateful to Claude Brezinski and Luc Wuytack for urging us to write this book, and to Elsevier Science Publishers for publishing it.

Trondheim, December 1991.

Lisa Lorentzen   Haakon Waadeland

Contents

Chapter I

Introductory examples

About this chapter

We have often been asked questions, by students as well as by established mathematicians, about continued fractions: what they are and what they can be used for. Sometimes the questions have been raised under circumstances where a *quick* answer is the only alternative to *no* answer: in the discussion after a talk or lecture, by a cup of coffee in a short break, in an airplane cabin or on a mountain hike. In responding to these questions we have very often been pleased by the sparks of interest we have seen, indicating that we had managed to transmit a glimpse of new and apparently appealing knowledge. In quite a few cases this led to a further contact and "follow-up activities".

This introductory chapter is to a large extent *inspired* by the questions we have received and *governed* by the answers we have given. There is of course a great danger: A quick answer is often a wrong answer. It may (and even ought to) tell the truth and nothing but the truth, but it definitely does not tell the whole truth. This may lead to false guesses. This danger is in particular great in cases where observations and experiments are used to create and support guesses, such as in Section 2 of the present chapter. But we still wanted to keep this (often

1

non-accepted, but highly necessary) aspect of mathematics as part of the introductory chapter. We have tried to reduce the danger partly by the way things are phrased, partly by indicating briefly how wrong such guesses can be, and finally by referring to a more careful treatment later in the book.

We decided to include, already in the introductory chapter, on the one hand three classical convergence theorems, on the other hand some of the newer thoughts on convergence and computation of continued fractions.

1 Definition and basic concepts. Convergence

1.1 Prelude to a definition

Let $\{t_n\}$ be a sequence of complex numbers. When we talk about the series

$$\sum_{n=1}^{\infty} t_n = t_1 + t_2 + \cdots + t_n + \cdots,\qquad(1.1.1)$$

we have in mind the sequence $\{T_n\}$ of *partial sums*

$$T_n = \sum_{k=1}^{n} t_k,$$

or *recursively defined*

$$T_{n+1} = T_n + t_{n+1}.$$

Convergence of the series (1.1.1) means convergence of $\{T_n\}$ to a complex number T, in which case we write

$$\sum_{n=1}^{\infty} t_n = T.\qquad(1.1.2)$$

Similarly we are familiar with infinite *products*

$$\prod_{n=1}^{\infty} p_n = p_1 \cdot p_2 \cdots p_n \cdots,\qquad(1.1.3)$$

where all p_n are complex numbers $\neq 0$. $\{P_n\}$ is the sequence of *partial products*

$$P_n = \prod_{k=1}^{n} p_k,$$

or *recursively defined*

$$P_{n+1} = P_n \cdot p_{n+1}.$$

Convergence of the infinite product (1.1.3) means convergence of the sequence $\{P_n\}$ to a complex number $P \neq 0$, in which case we write

$$\prod_{n=1}^{\infty} p_n = P.\qquad(1.1.4)$$

Let $\{a_n\}$ be a sequence of complex numbers $\neq 0$, and let $\{f_n\}$ be the sequence from $\hat{\mathbf{C}} = \mathbf{C} \cup \{\infty\}$ given by

$$f_1 = a_1, \qquad f_2 = \frac{a_1}{1 + a_2}, \qquad f_3 = \frac{a_1}{1 + \dfrac{a_2}{1 + a_3}},$$

and generally

$$f_n = \frac{a_1}{1 + \dfrac{a_2}{1 + \dfrac{a_3}{1 + \cdot_{\cdot_{\cdot + a_n}}}}}.$$

Similarly to what we have for sums and products this also leads to a concept, having to do with the nonterminating continuation of the process, in this case the concept of a *continued fraction*, constructed from a sequence $\{a_n\}$ of complex numbers, all $\neq 0$:

$$\overset{\infty}{\underset{n=1}{\mathbf{K}}} \, (a_n/1) = \frac{a_1}{1 + \dfrac{a_2}{1 + \dfrac{a_3}{1 + \cdot_{\cdot_\cdot}}}}. \tag{1.1.5}$$

Convergence of (1.1.5) means convergence of the sequence $\{f_n\}$ of *approximants*. We shall also accept convergence to ∞.

Example 1 For the continued fraction

$$\overset{\infty}{\underset{n=1}{\mathbf{K}}} \, (6/1) = \frac{6}{1 + \dfrac{6}{1 + \dfrac{6}{1 + \dfrac{6}{1 + \cdot_{\cdot_\cdot}}}}}$$

we find

$$f_1 = 6, \qquad f_2 = \frac{6}{7}, \qquad f_3 = \frac{42}{13}, \cdots.$$

It is easy to prove, by induction (see Problem 5, with $x = 2$, $y = -3$) that generally

$$f_n = -6\frac{(-3)^n - 2^n}{(-3)^{n+1} - 2^{n+1}}.$$

From this it follows that the continued fraction converges to 2.

$\diamondsuit$

Quite similarly we can construct, from *any* sequence $\{b_n\}$ of complex numbers, a continued fraction

$$\overset{\infty}{\underset{n=1}{\mathbf{K}}} (1/b_n) = \cfrac{1}{b_1 + \cfrac{1}{b_2 + \cfrac{1}{b_3 + \cfrac{1}{b_4 + \cdots}}}}, \qquad (1.1.6)$$

or from *two* sequences, $\{a_n\}$ and $\{b_n\}$ of complex numbers, where all $a_n \neq 0$, a continued fraction

$$\overset{\infty}{\underset{n=1}{\mathbf{K}}} (a_n/b_n) = \cfrac{a_1}{b_1 + \cfrac{a_2}{b_2 + \cfrac{a_3}{b_3 + \cfrac{a_4}{b_4 + \cdots}}}}. \qquad (1.1.7)$$

(1.1.5) and (1.1.6) are obviously special cases of (1.1.7). In the particular case when in (1.1.6) all b_n are natural numbers we get the *regular* continued fraction, well known in number theory, the one coming from the Euclidian algorithm. We shall look at regular continued fractions in Example 2 of the present chapter, and more seriously in Chapter IX.

Let us take a look at the common pattern in the three cases: series, products and continued fractions (and other constructions for that matter). In all three cases the construction can be described in the following way: We have a sequence $\{\phi_k\}$ of mappings from $\mathbf{C}$ into $\hat{\mathbf{C}}$. By composition we construct a new sequence $\{\Phi_n\}$ of mappings

$$\Phi_1 = \phi_1, \qquad \Phi_n = \Phi_{n-1} \circ \phi_n = \phi_1 \circ \phi_2 \circ \cdots \circ \phi_n. \qquad (1.1.8)$$

In all three cases there is a fixed complex number c, by means of which convergence is defined: as convergence of $\{\Phi_n(c)\}$ for that particular c. (There is a difference in the question of whether or not convergence to 0 or ∞ is counted as convergence or not.)

For *series* we have

$$\phi_k(w) = w + t_k \, ,$$

and the partial sums are

$$\Phi_n(0) = \phi_1 \circ \phi_2 \circ \cdots \circ \phi_n(0) = t_1 + t_2 + \cdots + t_n \, ,$$

i.e. here we have $c = 0$.

For *products* we have

$$\phi_k(w) = w \cdot p_k \, ,$$

and the partial products are

$$\Phi_n(1) = \phi_1 \circ \phi_2 \circ \cdots \circ \phi_n(1) = p_1 \cdot p_2 \cdots p_n \, ,$$

i.e. here we have $c = 1$.

For *continued fractions* (1.1.7) we have

$$\phi_k(w) = \frac{a_k}{b_k + w} \, ,$$

and the approximants are

$$\Phi_n(0) = \cfrac{a_1}{b_1 + \cfrac{a_2}{b_2 + \cfrac{a_3}{b_3 + \cfrac{\ddots}{\quad + \cfrac{a_n}{b_n}}}}} \, ,$$

i.e. here we have $c = 0$.

Remark: If all $a_k \neq 0$ the mappings ϕ_k are all non-singular linear fractional transformations. Hence ϕ_k^{-1} all exist. We shall later make much use of this property.

1.2 Formal definition. Convergence. Notation

The following definition is due to Henrici and Pfluger, see for instance [Henr77, p. 474] (for a slightly different version):

Definition *A continued fraction is an ordered pair*

$$(({\{a_n\}, \{b_n\}}), \{f_n\}) \, , \qquad (1.2.1)$$

where $\{a_n\}_1^\infty$ and $\{b_n\}_0^\infty$ are given sequences of complex numbers, $a_n \neq 0$, and where $\{f_n\}$ is the sequence of extended complex numbers, given by

$$f_n = S_n(0) \, , \qquad n = 0, 1, 2, 3, \cdots, \qquad (1.2.2)$$

where

$$S_0(w) = s_0(w) \, , \quad S_n(w) = S_{n-1}(s_n(w)) \, , \quad n = 1, 2, 3, \cdots, \quad (1.2.3a)$$

$$s_0(w) = b_0 + w \, , \qquad s_n(w) = \frac{a_n}{b_n + w} \, , \qquad n = 1, 2, 3, \cdots. \quad (1.2.3b)$$

See also [JoTh80, p. 17].

The *continued fraction algorithm* is the function **K** mapping a pair $(\{a_n\}, \{b_n\})$ onto the sequence $\{f_n\}$ defined by (1.2.2) and (1.2.3). Here the numbers a_n and b_n are called the nth *partial numerators* and *denominators*. A common name is *element*. The number

$$S_n(0) = b_0 + \cfrac{a_1}{b_1 + \cfrac{a_2}{b_2 + \cfrac{a_3}{b_3 + \cfrac{\ddots}{\quad + \cfrac{a_n}{b_n}}}}} \qquad (1.2.4)$$

is called the *n*th *approximant*. Several more convenient ways of writing the approximants are introduced in the literature on continued fractions. We shall here use:

$$S_n(0) = b_0 + \frac{a_1}{b_1 +} \frac{a_2}{b_2 +} \cdots + \frac{a_n}{b_n}, \qquad (1.2.4')$$

and more generally

$$S_n(w) = b_0 + \frac{a_1}{b_1} + \frac{a_2}{b_2} + \cdots + \frac{a_n}{b_n + w}.$$

Convergence of a continued fraction to an extended complex number f means convergence of $\{f_n\}$ to f, in which case we write

$$f = b_0 + \frac{a_1}{b_1 +} \frac{a_2}{b_2 +} \cdots + \frac{a_n}{b_n +} \cdots, \qquad (1.2.5)$$

or

$$f = b_0 + \mathop{\mathbf{K}}_{n=1}^{\infty} \frac{a_n}{b_n}. \qquad (1.2.5')$$

We even use the notation on the right-hand side when we discuss continued fractions more generally, regardless of convergence properties.

In computing numerically the value of a continued fraction the approximants, in particular those of high order, are essential. The first ones are:

$$f_0 = b_0, \qquad f_1 = \frac{b_0 b_1 + a_1}{b_1}, \qquad f_2 = \frac{b_0 b_1 b_2 + b_0 a_2 + a_1 b_2}{b_1 b_2 + a_2}.$$

A straightforward computation without cancellation leads to fractions for the approximants:

$$f_n = \frac{A_n}{B_n}, \qquad n = 0, 1, 2, \cdots. \qquad (1.2.6)$$

If we define

$$\begin{bmatrix} A_{-1} \\ B_{-1} \end{bmatrix} = \begin{bmatrix} 1 \\ 0 \end{bmatrix}, \qquad \begin{bmatrix} A_0 \\ B_0 \end{bmatrix} = \begin{bmatrix} b_0 \\ 1 \end{bmatrix}, \qquad (1.2.7)$$

the following is easily proved by induction:

$$S_n(w) = \frac{A_n + A_{n-1} w}{B_n + B_{n-1} w}, \qquad (1.2.8)$$

where the recurrence relation

$$\begin{bmatrix} A_n \\ B_n \end{bmatrix} = b_n \begin{bmatrix} A_{n-1} \\ B_{n-1} \end{bmatrix} + a_n \begin{bmatrix} A_{n-2} \\ B_{n-2} \end{bmatrix} \tag{1.2.9}$$

for $n = 1, 2, 3, \ldots$ holds. The proof is left as an exercise. Observe that S_n, being a composition of non-singular linear fractional transformations

$$s_k(w) = \frac{a_k}{b_k + w},$$

is itself a non-singular linear fractional transformation. We have in particular

$$S_n(0) = f_n = \frac{A_n}{B_n}, \qquad S_n(\infty) = f_{n-1} = \frac{A_{n-1}}{B_{n-1}}.$$

We shall here call A_n and B_n nth *canonical numerator* and *denominator* (sometimes just numerator and denominator). An important property of the numbers A_n and B_n, is the *determinant formula*

$$A_n B_{n-1} - A_{n-1} B_n = (-1)^{n-1} \prod_{k=1}^{n} a_k. \tag{1.2.10}$$

The proof is straight forward use of the recurrence relation (1.2.9), and is left as an exercise.

Let it finally be mentioned that we also may go "in the opposite direction", i.e. from given sequences $\{A_n\}$ and $\{B_n\}$ to a continued fraction (1.2.5'). Actually, any pair $(\{A_n\}, \{B_n\})$ with the initial condition (1.2.7) determines uniquely a continued fraction (1.2.5') with

$$f_n = \frac{A_n}{B_n},$$

provided that

$$A_n B_{n-1} - A_{n-1} B_n \neq 0, \qquad n = 0, 1, 2, \ldots$$

(Theorem 7 in Chapter II).

2   Some examples

2.1   The very best

Example 2 From the equality

$$\sqrt{2} - 1 = \frac{1}{2 + (\sqrt{2} - 1)}$$

follow the equalities

$$
\begin{aligned}
\sqrt{2} - 1 &= \frac{1}{2} + \frac{1}{2 + (\sqrt{2} - 1)} = \frac{1}{2} + \frac{1}{2} + \frac{1}{2 + (\sqrt{2} - 1)} \\
&= \frac{1}{2} + \frac{1}{2} + \frac{1}{2} + \frac{1}{2 + (\sqrt{2} - 1)},
\end{aligned}
$$

and so on, as long as we want to. Since obviously

$$\sqrt{2} = 1 + \frac{1}{2} + \frac{1}{2} + \cdots + \frac{1}{2} + \frac{1}{2 + (\sqrt{2} - 1)}$$

for any length of the row of dots, it seems to be a good idea to take a look at the approximants of the regular continued fraction

$$1 + \frac{1}{2} + \frac{1}{2} + \frac{1}{2} + \cdots + \frac{1}{2} + \cdots.$$

We find

$$
\begin{aligned}
1 + \frac{1}{2} &= 1.5 \\
1 + \frac{1}{2} + \frac{1}{2} &= \frac{7}{5} = 1.4 \\
1 + \frac{1}{2} + \frac{1}{2} + \frac{1}{2} &= \frac{17}{12} = 1.4166\ldots, \\
1 + \frac{1}{2} + \frac{1}{2} + \frac{1}{2} + \frac{1}{2} &= \frac{41}{29} = 1.41379\ldots, \\
1 + \frac{1}{2} + \frac{1}{2} + \frac{1}{2} + \frac{1}{2} + \frac{1}{2} &= \frac{99}{70} = 1.4142857\ldots.
\end{aligned}
$$

These fractions seem to approach $\sqrt{2}$ pretty quickly. Already the fifth one, the last one listed, has an error less than .00008. We phrase this observation in the following way: The fractions seem to be very good *rational approximations* to the irrational number $\sqrt{2}$.

$\diamond$

This is a good place for a warning: Identities such as the ones we have studied here, which led us into the temptation of studying $1 + K(1/2)$, in the hope of getting good approximations to $\sqrt{2}$, may just as well be a dead end. (Actually, the *normal* thing is that it goes wrong.) The equality

$$-\sqrt{2} - 1 = \frac{1}{2 + (-\sqrt{2} - 1)}$$

will for instance lead to

$$-\sqrt{2} = 1 + \frac{1}{2+}\frac{1}{2+}\cdots+\frac{1}{2 + (-\sqrt{2} - 1)},$$

but the continued fraction (still) seems to converge to $\sqrt{2}$. See also Subsection *3.4*. Such identities are special cases of more general *recurrence formulas*, which will play a crucial role later in this exposition.

But in the present case everything goes well. Our numerical observations are matched by mathematical reality: In fact, it can be proved, that the fractions obtained are not only very good approximations to $\sqrt{2}$, but the *very best*, in the following sense: If p/q is one of the fractions obtained in the way described above, and if m/n is a fraction such that

$$\left| \sqrt{2} - \frac{m}{n} \right| < \left| \sqrt{2} - \frac{p}{q} \right|,$$

i.e. m/n is a better approximation to $\sqrt{2}$ than p/q, then $n > q$. Slightly rephrased: In order to find a better approximation to $\sqrt{2}$ than p/q, we have to increase the denominator. (p, q, m, n are all positive integers.) We shall not go into the proof of this here, merely present a geometric illustration (also without proof) of this way of approximating an irrational number:

Let a be a positive irrational number (in our case $\sqrt{2}$). We shall let the number a be represented by the ray $y = ax$ from the origin and into the first quadrant of a cartesian coordinate system. Let furthermore

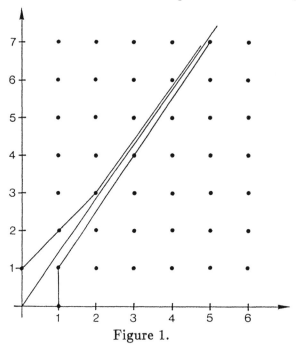

Figure 1.

the lattice point (n, m) represent the fraction m/n. Then the following, illustrated in Figure 1, can be proved: Assume that there is a nail in every lattice point, and a rubber band fastened in the points $(1,0)$ and $(0,1)$. Stretch the rubber band with a pencil following the ray $y = ax$ (or $y = \sqrt{2}x$ in our case). Then the corners of the polygon, in the order they turn up, are the lattice points corresponding to the rational approximants of the regular continued fraction for a. Observe e.g. on Figure 1. the points $(1,1)$, $(2,3)$, $(5,7)$, corresponding to the fractions $1/1$, $3/2$, $7/5$. Inside the polygon, i.e. where the ray is, there are no lattice points, showing the "bestness" of the fractions $1/1, 3/2, 7/5$ (and $17/12, 41/29, \cdots$) as approximants for $\sqrt{2}$.

This way of illustrating approximations was told to one of the authors by Viggo Brun in lectures at the University of Oslo [Brun50], but it goes way back, and different people have been given credit for it. Any positive irrational number has a unique representation by a continued fraction of this type. See remark 2 at the end of this chapter.

2.2 A differential equation

Example 3 From the differential equation

$$y = 2y' + y''$$

follow by differentiation the equalities

$$y' = 2y'' + y''',$$

$$\vdots$$

$$y^{(n)} = 2y^{(n+1)} + y^{(n+2)},$$

and hence, assuming that we do not divide by 0:

$$\frac{y}{y'} = 2 + \frac{1}{y'/y''},$$

$$\frac{y'}{y''} = 2 + \frac{1}{y''/y'''},$$

$$\vdots$$

$$\frac{y^{(n)}}{y^{(n+1)}} = 2 + \frac{1}{y^{(n+1)}/y^{(n+2)}}.$$

From this it follows that

$$\frac{y}{y'} = 2 + \underbrace{\frac{1}{2+}\frac{1}{2+}\cdots+\frac{1}{2}}_{n+1} + \frac{1}{y^{(n+1)}/y^{(n+2)}}.$$

This suggests to look at the continued fraction

$$2 + \frac{1}{2+}\frac{1}{2+}\cdots+\frac{1}{2+}\cdots.$$

We "know" from Example 2 that this converges to $\sqrt{2}+1$, which suggests that

$$\frac{y}{y'} = \sqrt{2} + 1,$$

or

$$\frac{y'}{y} = \sqrt{2} - 1,$$

from which it would follow that

$$y = C \exp \left[(\sqrt{2} - 1)x \right] .$$

This is actually a solution of the given differential equation.

$\diamond$

There is of course no good reason to use this "method" for the present differential equation, since there exists a perfectly good, simple method taught in elementary calculus classes. Furthermore, the continued fraction method (even after it is properly established) produces only a particular integral, not the general solution. But there are cases, where this idea leads to non-trivial results, and where the method may represent an alternative to (or at least a supplement to) existing power series methods for solving linear ODEs. See [Khov63], [Steen73], [Waad83] and the references there. See also Chapter XII of the present book.

2.3   An expansion of a function

Example 4 From the identity

$$\sqrt{1 + x} - 1 = \frac{x}{2 + (\sqrt{1 + x} - 1)}$$

follow, as it did in Example 2, the identities:

$$\sqrt{1 + x} - 1 = \frac{x}{2} + \frac{x}{2} + \cdots + \frac{x}{2} + \frac{x}{2 + (\sqrt{1 + x} - 1)} .$$

This suggests to look at the approximants of the continued fraction

$$\frac{x}{2} + \frac{x}{2} + \cdots + \frac{x}{2} + \cdots ,$$

i.e. at the rational functions

$$\frac{x}{2} , \qquad \frac{x}{2} + \frac{x}{2} = \frac{2x}{x + 4} , \qquad \frac{x}{2} + \frac{x}{2} + \frac{x}{2} = \frac{x^2 + 4x}{4x + 8} ,$$

and so on. We refer to the warning in Subsection *2.1*.

Continued fraction expansions are less known than power series expansions. In the present case the Taylor series expansion at 0 is

$$\sqrt{1+x} - 1 = \sum_{k=1}^{\infty} \binom{\frac{1}{2}}{k} x^k = \frac{1}{2}x - \frac{1}{8}x^2 + \frac{1}{16}x^3 - \frac{5}{128}x^4 + \cdots.$$

It converges for $|x| < 1$ and diverges for $|x| > 1$. The approximants of the series are the partial sums

$$s_1 = \frac{1}{2}x, \qquad s_2 = \frac{1}{2}x - \frac{1}{8}x^2, \qquad s_3 = \frac{1}{2}x - \frac{1}{8}x^2 + \frac{1}{16}x^3,$$

and so on. Observe that the series approximants are polynomials, whereas the continued fraction approximants are rational functions.

Let us make an experiment: We compute the two types of approximants for a certain x– value to see what happens. We choose $x = .96$, in which case the value of the function is exactly .4. In the table below some power series approximants (s_n) and some continued fraction approximants (f_n) are listed, all correctly rounded in the 4th decimal place:

n	1	2	3	4	5	6	7
s_n	.4800	.3648	.4201	.3869	.4092	.3932	.4053
f_n	.4800	.3871	.4022	.3996	.4001	.4000	.4000

This of course does not *prove* anything, but it suggests that in some cases the continued fraction may be better (converge faster) than the power series expansion. (It is, however, only fair to say, that such a comparison, based merely upon the order n of the approximant, does not always give the correct picture. Essential in the comparison is the resources needed, usually the time.)

Even more flattering for the continued fraction expansion is the choice $x = 3$. In this case it does not make sense to compute power series approximants, since we know that the power series diverges. In the next table the first seven continued fraction approximants are listed, correctly rounded in the 4th decimal place. Keep in mind that the value of the function is 1.

n	1	2	3	4	5	6	7
f_n	1.5000	.8571	1.0500	.9836	1.0055	.9982	1.0006

The table suggests that the continued fraction expansion converges to the right value for $x = 3$, i.e. for a value where the power series diverges, in which case the continued fraction is better than the power series also in that respect. (The next three approximants are .9998, 1.0001 and 1.0000.)

$\diamond$

In Chapter III it will be proved that the continued fraction in Example 4, with real x, converges for all $x \in [-1, \infty)$ and diverges for all $x \in (-\infty, -1)$. For complex x it will be proved, that the continued fraction expansion converges in the whole plane, except on the ray $(-\infty, -1)$ of the negative real axis, and to the right value. In Figure 2 we have convergence of the continued fraction expansion in the whole plane, except on the strongly indicated ray, whereas the power series expansion converges inside the dotted circle and diverges outside of it.

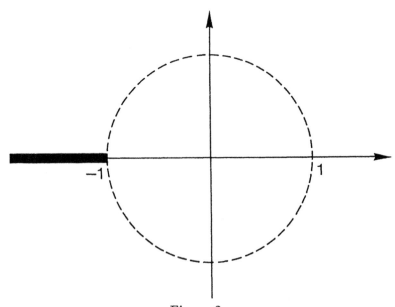

Figure 2.

For later use in the present chapter we rewrite the identities at the beginning of this example in the following form:

$$\frac{\sqrt{1+x}-1}{2} = \frac{x/4}{1+(\sqrt{1+x}-1)/2}$$

$$= \frac{x/4}{1} \; + \; \frac{x/4}{1} \; + \cdots + \frac{x/4}{1+(\sqrt{1+x}-1)/2} . \quad (2.3.1)$$

The approximants are the former ones, divided by 2.

2.4   A log-expansion

Example 5 An example, related to the previous one, but less trivial, is the expansion

$$\log(1+z) = \frac{z}{1} + \frac{z/2}{1} + \frac{z/6}{1} + \frac{2z/6}{1} + \frac{2z/10}{1} + \frac{3z/10}{1} + \cdots,$$

or more precisely

$$\log(1+z) = \frac{z}{1} + \frac{a_2 z}{1} + \frac{a_3 z}{1} + \cdots,$$

where log here shall mean the principal value of the natural logarithm, and where for all $k \geq 1$

$$a_{2k} = \frac{k}{2(2k-1)}, \qquad a_{2k+1} = \frac{k}{2(2k+1)} .$$

Right now we shall not worry about *how* one gets this, only use it for some experiments, to compare it to the power series expansion

$$\log(1+z) = z - \frac{z^2}{2} + \frac{z^3}{3} - \frac{z^4}{4} + \cdots .$$

Let us take the "worst" example, $z = 1$. The series then converges to $\log 2$, but very slowly. The first seven continued fraction approximants are listed in the table below, correctly rounded in the 5th place. The value is $\log 2 = .69314718$, correctly rounded in the 8th place.

n	1	2	3	4	5	6	7
f_n	1.00000	.66667	.70000	.69231	.69333	.69312	.69315

In order to get the series approximation s_n with the same accuracy we need $n \geq 10^5$.

Let us also here try a z-value where the series does not converge, for instance $z = 3$. In the next table the first 7 continued fraction approximants are listed, all correctly rounded in the 4th place. The value is $\log 4 = 2\log 2 = 1.38629436$, correctly rounded in the 8th place.

n	1	2	3	4	5	6	7
f_n	3.0000	1.2000	1.5000	1.3636	1.3973	1.3837	1.3874

This table also suggests convergence to the right value.

$\diamond$

We shall later see, that the present continued fraction expansion for $\log(1 + z)$ converges for all complex z, except on the ray $(-\infty, -1]$ of the negative axis, whereas the power series expansion converges for $|z| < 1$ and diverges for $|z| > 1$. An illustration would look like Figure 2.

3   More examples

3.1   *Hypergeometric functions*

Example 6 The hypergeometric functions

$$F(a, b; c; z) = 1 + \frac{ab}{c}\frac{z}{1!} + \frac{a(a + 1)b(b + 1)}{c(c + 1)}\frac{z^2}{2!} + \cdots, \qquad (3.1.1)$$

where a, b, c are complex numbers, and c not 0 or a negative integer, are of great importance in several applications. Many of the well known special functions are special cases of (3.1.1). If we assume that also a and b are different from 0 and the negative integers, the series (3.1.1) is an infinite power series whose radius of convergence is 1. The following formal identities can be established from (3.1.1) by comparing the power series on both sides term by term:

$$
\begin{aligned}
F(a,b;c;z) &= F(a,b+1;c+1;z) \\
&\quad - \frac{a(c-b)}{c(c+1)}zF(a+1,b+1;c+2;z) \\
F(a,b+1;c+1;z) &= F(a+1,b+1;c+2;z) \\
&\quad - \frac{(b+1)(c-a+1)}{(c+1)(c+2)}zF(a+1,b+2;c+3;z)
\end{aligned}
$$

Assuming that we avoid zeros in the denominators, this can be rewritten in the following way:

$$
\frac{F(a,b;c;z)}{F(a,b+1;c+1;z)} = 1 + \frac{\dfrac{-a(c-b)z}{c(c+1)}}{\dfrac{F(a,b+1;c+1;z)}{F(a+1,b+1;c+2;z)}}
$$

$$
\frac{F(a,b+1;c+1;z)}{F(a+1,b+1;c+2;z)} = 1 + \frac{\dfrac{-(b+1)(c-a+1)z}{(c+1)(c+2)}}{\dfrac{F(a+1,b+1;c+2;z)}{F(a+1,b+2;c+3;z)}}
$$

Observe that the denominator on the right-hand side of the first equality is equal to the left-hand side of the second one. Furthermore, the denominator on the right-hand side of the second equality coincides with the left-hand side of the first one, if in the former a is replaced by $a+1$, b is replaced by $b+1$ and c by $c+2$ in all places. Hence, by repeatedly increasing the first two parameters by 1 and the third one by 2 we are lead to a continued fraction in a similar way as we have seen in Subsection 2.2. The continued fraction, already studied by Gauss [Gauss13], [Gauss14], is of the form

$$
1 + \frac{a_1 z}{1} + \frac{a_2 z}{1} + \cdots + \frac{a_n z}{1} + \cdots, \tag{3.1.2}
$$

where

$$
a_{2n+1} = -\frac{(a+n)(c-b+n)}{(c+2n)(c+2n+1)}, \qquad n = 0,1,2,\cdots, \tag{3.1.2'}
$$

$$a_{2n} = -\frac{(b+n)(c-a+n)}{(c+2n-1)(c+2n)}, \qquad n = 1, 2, 3, \cdots. \qquad (3.1.2'')$$

We shall see later, in Chapter VI, that the sequence of approximants converges to a meromorphic function in the whole plane, except on the ray $z > 1$ of the positive real axis, and that this function is

$$\frac{F(a, b; c; z)}{F(a, b+1; c+1; z)}$$

(or its meromorphic continuation, if we regard the function as primarily defined by its power series).

Observe that

$$a_n \longrightarrow -\frac{1}{4}. \qquad (3.1.3)$$

The continued fraction (3.1.2) thus is an example of what is called a *limit periodic* continued fraction. The *tails* of the continued fraction "look more and more like" the continued fraction expansion for

$$\frac{\sqrt{1-z}-1}{2}, \qquad (3.1.4)$$

(see (2.3.1)). We shall later use this property to improve the procedure of computing the values of the continued fraction (3.1.2). The idea is to replace the actual tails of (3.1.2) by (3.1.4) in forming the approximants. This means, that we, instead of using $S_n(0)$ as approximants, use $S_n(w)$, with w as in (3.1.4). Recall that $S_n(w)$ is given by the formula in (1.2.8). $S_n(w)$ is called a *modified* approximant.

Later in this exposition we shall see more of hypergeometric functions and their expansions. The reasons for including this particular example *here* are: 1. To emphasize the connection between three term recurrence relations (such as the formal identities for the hypergeometric series or the differential equations in Subsection 2.2.) and continued fractions. 2. To put the example from Subsection 2.4 into a more general context: The log-function is a special hypergeometric function:

$$\log(1 + z) = z \cdot F(1, 1; 2; -z)$$

Three-term recurrence relations will be the topic of Chapter IV.

_____◇

3.2 From power series to continued fractions.

We have seen, that continued fractions of the form

$$\frac{a_1 z}{1} + \frac{a_2 z}{1} + \cdots + \frac{a_n z}{1} + \cdots$$

in some cases are of advantage compared to power series expansions both as far as *speed of convergence* and *domain of convergence* are concerned. Hence it is of interest, and sometimes useful, to go from a power series to this particular continued fraction. We shall illustrate the most primitive way of doing this (in [CuWu87] called the method of successive substitutions) by using the log-series as an example:

$$
\begin{aligned}
log(1 + z) = l_0(z) &= z - \frac{z^2}{2} + \frac{z^3}{3} - \frac{z^4}{4} + \cdots \\
&= z\left(1 - \frac{z}{2} + \frac{z^2}{3} - \frac{z^3}{4} + \cdots\right) = \frac{z}{1 + l_1(z)}, \\
l_1(z) &= \left(1 - \frac{z}{2} + \frac{z^2}{3} - \cdots\right)^{-1} - 1 \\
&= \frac{z}{2}\left(1 - \frac{z}{6} + \frac{z^2}{12} + \cdots\right) = \frac{z/2}{1 + l_2(z)}, \\
l_2(z) &= \left(1 - \frac{z}{6} + \frac{z^2}{12} + \cdots\right)^{-1} - 1 \\
&= \frac{z}{6}\left(1 - \frac{z}{3} + \cdots\right) = \frac{z/6}{1 + l_3(z)}, \\
l_3(z) &= \left(1 - \frac{z}{3} + \cdots\right)^{-1} - 1 = \frac{z/3}{1 + l_4(z)}.
\end{aligned}
$$

This leads to the identity

$$log(1 + z) = \frac{z}{1} + \frac{z/2}{1} + \frac{z/6}{1} + \frac{z/3}{1 + l_4(z)},$$

where $l_4(z)$ is a uniquely determined power series starting with a term cz. We recognize the start of the expansion from Subsection *2.4*, and the process could have been continued to any length (depending upon how many terms we start with in the log-series). It is of course not obvious, that we will get a continued fraction of the form

$$\frac{a_1 z}{1} + \frac{a_2 z}{1} + \cdots + \frac{a_n z}{1} + \cdots,$$

since we possibly may run into an $l_k(z)$, starting with a term cz^k with $k > 1$. It is known, however, that this will *not* happen in the present case. This procedure gives a partial answer to the question mentioned in Subsection *2.4, how* we can get the continued fraction expansion for $\log(1 + z)$. But by this procedure, as presented here, we do not get general formulas for a_n, let alone anything about convergence. In conclusion we may say, that the present method by far is not the best *practical* method, but it indicates a possible bridge from power series to a continued fraction expansions.

3.3   From continued fractions to power series

Sometimes we want to go in the opposite direction, i.e. from continued fraction to power series. We shall use our continued fraction expansion for $\log(1 + z)$ as an example. Here are the first (classical) approximants with their power series expansions at 0:

First approximant:

$$\frac{z}{1} = z + 0z^2 + 0z^3 + 0z^4 + \cdots$$

Second approximant:

$$\frac{z}{1 + z/2} = z - \frac{z^2}{2} + \frac{z^3}{4} - \frac{z^4}{8} + \cdots$$

Third approximant:

$$\frac{z}{1 + \dfrac{z/2}{1 + z/6}} = z - \frac{z^2}{2} + \frac{z^3}{3} - \frac{2}{9}z^4 + \cdots$$

Fourth approximant:

$$\cfrac{z}{1+\cfrac{z/2}{1+\cfrac{z/6}{1+z/3}}} = \underline{z - \frac{z^2}{2} + \frac{z^3}{3} - \frac{z^4}{4}} + \cdots$$

The underlined terms coincide with terms in the power series for $\log(1+z)$. Observe that the agreement increases with the order of the approximants. It can be proved (and it will, in Chapter V) that this continues. It is called *correspondence* between the power series and the continued fraction expansion. A proper definition will be given later.

3.4 One fraction, two series

We shall now look at a very special example, which however will prove very useful later.

Example 7 The identity

$$z = \frac{z}{1 - z + z}$$

used repeatedly leads to the identities

$$z = \frac{z}{1-z+}\frac{z}{1-z+z}, \qquad z = \frac{z}{1-z+}\frac{z}{1-z+}\frac{z}{1-z+z},$$

and generally

$$z = \frac{z}{1-z+}\frac{z}{1-z+}\cdots+\frac{z}{1-z+z}.$$

On the other hand, the identity

$$-1 = \frac{z}{1-z-1}, \qquad z \neq 0,$$

leads to

$$-1 = \frac{z}{1-z+}\frac{z}{1-z+}\cdots+\frac{z}{1-z-1}.$$

Inspired by these two identities we look at the continued fraction

$$\frac{z}{1-z+}\frac{z}{1-z+}\cdots+\frac{z}{1-z+}\cdots. \qquad (3.4.1)$$

For simplicity we assume z to be different from all roots of unity, in which case the approximants f_n are:

$$f_1 = \frac{z}{1-z} = \frac{z(1+z)}{1-z^2}$$

$$f_2 = \frac{z}{1-z+}\frac{z}{1-z} = \frac{z(1-z)}{1-z+z^2} = \frac{z(1-z^2)}{1+z^3}$$

$$f_3 = \frac{z}{1-z+}\frac{z}{1-z+}\frac{z}{1-z} = \frac{z(1+z^3)}{1-z^4}$$

By induction the following formula is easily established:

$$f_n = \frac{z(1-(-z)^n)}{1-(-z)^{n+1}}$$

See Problem 5, with $x = z$, $y = -1$.

We shall make *two* types of observations on the approximants, one on convergence, one on correspondence. We distinguish two cases:

$|z| < 1$:

$$\lim_{n \to \infty} f_n = z$$

$$f_n = z + (-z)^{n+1} + \text{ higher powers of } z$$

$|z| > 1$:

$$\lim_{n \to \infty} f_n = -1$$

$$f_n = -1 + (-z)^{-n} + \text{ higher powers of } z^{-1}$$

The way these observations will be phrased within the analytic theory of continued fractions is as follows:

The continued fraction

$$\frac{z}{1-z+}\frac{z}{1-z+}\cdots+\frac{z}{1-z+}\cdots \qquad (3.4.2)$$

a) converges in $|z| < 1$ to z, and corresponds at 0 to the series

$$z + 0z^2 + 0z^3 + \cdots,$$

b) converges in $|z| > 1$ to -1, and corresponds at ∞ to the series

$$-1 + 0z^{-1} + 0z^{-2} + \cdots.$$

$\diamond$

This example shows that one and the same continued fraction expansion may converge to two different functions in two different regions and correspond to two different series at two points (here 0 and ∞).

This trivial example has its non-trivial relatives, where one continued fraction simultaneously represents two different analytic or meromorphic functions by convergence and correspondence.

We conclude this section with another trivial remark (still in Example 7), which also has its non-trivial analogues. It has to do with replacing the classical approximants $f_n = S_n(0)$ by some *modified* approximant $S_n(w)$. The interesting w-values here are z and -1:

Case 1: If $S_n(0)$ is replaced by $S_n(z)$, all (classical) approximants will be replaced by z. This implies two things: The convergence to z in $|z| < 1$ is *accelerated* (bull's eye, the value is hit right away), and the convergence to z is extended also to the region $|z| \geq 1$, i. e. we have an *analytic continuation* of the limit function in $|z| < 1$ to the whole plane.

Case 2: If $S_n(0)$ is replaced by $S_n(-1)$, all classical approximants will be replaced by -1. This implies two things: Acceleration of convergence to -1 in $|z| > 1$ (again bull's eye), and the convergence to -1 is extended also to $0 < |z| \leq 1$, i.e. we have an analytic continuation of the limit function in $|z| > 1$ to the whole plane, minus the origin.

The continued fraction (3.4.1) is a special case of a *T-fraction*, named after W. J. Thron. We shall get back to T-fractions later in the book.

3.5   The length of an elliptic orbit

Example 8 We shall look at the computation of the circumference L of the ellipse

$$\frac{x^2}{a^2} + \frac{y^2}{b^2} = 1, \qquad a \geq b \geq 0, \quad a > 0. \qquad (3.5.1)$$

The well known arc length formula leads to an elliptic integral. One way of finding approximate values for it is to use power series. By using the arc length formula one easily proves the following, which is a well known formula, see for instance [Hütte55]:

$$L = \pi(a + b)\left(1 + \frac{t}{2^2} + \frac{t^2}{2^6} + \frac{t^3}{2^8} + \frac{25t^4}{2^{14}} + \cdots\right), \qquad (3.5.2)$$

where

$$t = \left(\frac{a - b}{a + b}\right)^2. \qquad (3.5.2')$$

For the general coefficient an explicit formula exists, and also the convergence properties are known. We shall leave out both here. Observe that the formula is exact for a circle, $a = b$, i.e. $t = 0$.

One way of finding approximate values for L is to truncate the series at different places. We shall, however, use a quite different approach:

We transform the series into a continued fraction the way it is shown in Subsection *3.2*. The start of the continued fraction is in this case

$$1 + \frac{t/4}{1} + \frac{-t/16}{1} + \frac{-3t/16}{1} + \frac{-3t/16}{1} + \cdots. \qquad (3.5.3)$$

The first approximants are

$$f_0(t) = 1,$$

$$f_1(t) = 1 + \frac{t}{4},$$

$$f_2(t) = 1 + \frac{t/4}{1 +} \frac{-t/16}{1} = \frac{16 + 3t}{16 - t},$$

$$f_3(t) = 1 + \frac{t/4}{1 +} \frac{-t/16}{1 +} \frac{-3t/16}{1} = \frac{64 - 3t^2}{64 - 16t},$$

$$f_4(t) = 1 + \frac{t/4}{1 +} \frac{-t/16}{1 +} \frac{-3t/16}{1 +} \frac{-3t/16}{1} = \frac{256 - 48t - 21t^2}{256 - 112t + 3t^2}.$$

These rational functions are what we later will learn to know as *Padé approximants* to the series in the *L*-formula.

The Padé approximants determined above give us a sequence of approximate formulas for *L*:

$$L \approx \pi(a + b)f_n(t)$$

We shall not include any discussion on the accuracy of the formulas, merely mention some points to indicate it: The formula with $n = 2$, simple as it is, has an error less than 3 mm for an ellipse with size and eccentricity as the orbit of the planet Mercury. The formula with $n = 3$ has for the same ellipse an error roughly $= 1/10$ of the wave length of blue light. In the "flat" case, which is likely to be the "worst" case $(t = 1)$, the exact value of *L* is $4a$. The approximate formulas give in this case:

$$\pi a f_0(1) = 3.1416a$$
$$\pi a f_1(1) = 3.9270a$$
$$\pi a f_2(1) = 3.9794a$$
$$\pi a f_3(1) = 3.9924a$$
$$\pi a f_4(1) = 3.9964a$$

The factors are all correctly rounded in the 4th decimal place.

We shall conclude this example by presenting two approximate formulas obtained in a different way, namely by using a *modified* approximant for the continued fraction (3.5.3). We have already touched upon the concept in the Subsections *3.1* and *3.4*.

In the continued fraction (3.5.3) we pretend that *all* partial fractions
from the second one are equal to

$$\frac{-t/16}{1},$$

i.e. we replace the continued fraction (3.5.3) by

$$1 + \frac{t/4}{1} \; \frac{-t/16}{+ \; 1} \; \frac{-t/16}{+ \; 1} \; \frac{-t/16}{+ \; 1} + \cdots.$$

(This is of course only an experiment, and we have no guarantee that
it will lead to a good approximation.) This continued fraction may be
written

$$1 + \frac{t/4}{1 + w}, \tag{3.5.4}$$

(which, in standard notation, is $S_1(w)$), where w is the value of the
continued fraction

$$\frac{-t/16}{1} \; \frac{-t/16}{+ \; 1} \; \frac{-t/16}{+ \; 1} + \cdots,$$

i.e. (see (2.3.1))

$$\frac{1}{2}\left(\sqrt{1 - t/4} - 1\right).$$

Since

$$1 + \frac{t/4}{1 + (\sqrt{1 - t/4} - 1)/2} = 3 - \sqrt{4 - t},$$

we get the formula

$$L \approx \pi(a + b)(3 - \sqrt{4 - t}). \tag{3.5.5}$$

This formula was first found by Ramanujan [Rama57]. For an ellipse of
size and eccentricity as the orbit of Mercury this formula has an error
less than 2 mm. For the degenerate case ("flat" case) it gives the value
$3.9834a$.

The next formula also uses a modified approximant (3.5.4), but with another w, based upon the observation that the continued fraction (3.5.3)
has two equal partial fractions

$$\frac{-3t/16}{1}$$

in a row, and in front of them

$$\frac{-t/16}{1}\,.$$

If we pretend that all subsequent partial fractions are

$$\frac{-3t/16}{1}\,,$$

(again merely an experiment) we have in (3.5.4)

$$w = \frac{1}{3}\left(\frac{-3t/16}{1} + \frac{-3t/16}{1} + \frac{-3t/16}{1} + \frac{-3t/16}{1} + \cdots\right),$$

which has the value

$$\frac{1}{6}\left(\sqrt{1 - 3t/4} - 1\right).$$

(See (2.3.1).) This suggests the approximate formula

$$L \approx \pi(a + b)\left(1 + \frac{t/4}{1 + (\sqrt{1 - 3t/4} - 1)/6}\right),$$

or, rewritten in a nicer form

$$L \approx \pi(a + b)\left(1 + \frac{3t}{10 + \sqrt{4 - 3t}}\right). \qquad (3.5.6)$$

This seems to be the best of the approximate formulas mentioned in this section. For the "flat" case it gives the value $3.9984a$.

This formula also is due to Ramanujan, although it is not known *how* he established it. It is assumed, though, that the method shown in the present section was the one he used [AlBe88].

$\diamondsuit$

4   Three classical convergence theorems

4.1   Śleszyński-Pringsheim's Theorem

Theorem 1 *The continued fraction* $\mathbf{K}(a_n/b_n)$ *converges if for all* n

$$|b_n| \geq |a_n| + 1 \,. \qquad (4.1.1)$$

Under the same condition

$$|f_n| < 1 \qquad (4.1.2)$$

holds for all approximants f_n, *and*

$$|f| \leq 1 \qquad (4.1.2')$$

for the value of the continued fraction.

Proof :   We first prove (4.1.2) by induction. For any $n \geq 1$ we have

$$\left| \frac{a_n}{b_n} \right| \leq \frac{|a_n|}{|a_n| + 1} < 1 \,,$$

which proves (4.1.2) for $n = 1$. Next, for any $n \geq 2$

$$\left| \frac{a_{n-1}}{b_{n-1} + a_n/b_n} \right| < \frac{|a_{n-1}|}{|a_{n-1}| + 1 - 1} = 1 \,,$$

which establishes (4.1.2) for $n = 2$.

Assume that for some k, $1 \leq k < n$

$$f_n^{(k)} = \frac{a_{k+1}}{b_{k+1} +} \cdots + \frac{a_n}{b_n}$$

has the property $\left| f_n^{(k)} \right| < 1$. Then

$$\left| f_n^{(k-1)} \right| = \left| \frac{a_k}{b_k + f_n^{(k)}} \right| \leq \frac{|a_k|}{|a_k| + 1 - \left| f_n^{(k)} \right|} < 1 \,.$$

Hence, by induction on k,

$$|f_n| = \left|f_n^{(o)}\right| < 1.$$

To prove the convergence of $\mathbf{K}(a_n/b_n)$ we observe that the determinant formula (1.2.10) gives

$$\frac{A_n}{B_n} - \frac{A_{n-1}}{B_{n-1}} = \frac{(-1)^{n-1} \prod_{k=1}^{n} a_k}{B_n B_{n-1}}.$$

Hence the convergence is established as soon as we have proved the convergence of the series

$$\sum_{n=1}^{\infty} \frac{(-1)^{n-1} \prod_{k=1}^{n} a_k}{B_n B_{n-1}}. \tag{4.1.3}$$

From the recurrence formulas (1.2.9) we have, for $n \geq 1$

$$|B_n| = |b_n B_{n-1} + a_n B_{n-2}| \geq |b_n||B_{n-1}| - |a_n||B_{n-2}|$$
$$\geq (|a_n| + 1)|B_{n-1}| - |a_n||B_{n-2}|,$$

and hence

$$|B_n| - |B_{n-1}| \geq |a_n|(|B_{n-1}| - |B_{n-2}|).$$

From this it follows that

$$|B_n| - |B_{n-1}| \geq \prod_{k=1}^{n} |a_k| \; (> 0),$$

and the general term in (4.1.3) thus satisfies

$$\left|\frac{(-1)^{n-1} \prod_{k=1}^{n} a_k}{B_n B_{n-1}}\right| \leq \frac{1}{|B_{n-1}|} - \frac{1}{|B_n|}.$$

We therefore find that (4.1.3) converges absolutely, and that the nth partial sum has absolute value less than or equal to

$$\frac{1}{|B_0|} - \frac{1}{|B_n|} = 1 - \frac{1}{|B_n|} < 1.$$

Hence the series (4.1.3) converges. (4.1.2') is now a simple consequence of (4.1.2). ∎

Remark: If (4.1.1) holds, then it follows from our proof that also $|S_n(w)| < 1$ for all $|w| < 1$ and $S_n(w) \to f$ locally uniformly for $|w| < 1$.

Example 9 Let z be a complex number, and assume that all $|a_n| \leq 1$. Then the continued fraction

$$\mathop{\mathbf{K}}_{n=1}^{\infty} \frac{a_n}{z}$$

converges for all $|z| \geq 2$. In the special case when $a_n = 1$ for all n we find for the value $f(z)$:

$$f(z) = \frac{1}{z + f(z)}$$

From this it follows that

$$f(z) = \frac{\sqrt{z^2 + 4} - z}{2} .$$

In Chapter III we will discuss periodic continued fractions more generally. The continued fraction $\mathbf{K}(1/z)$ is a special case. So are also the continued fractions in the Subsections *2.1* and *2.2*. Here in Example 9 the branch of the square root is to be chosen such that $f(z) \to 0$ when $z \to \infty$, i.e. such that

$$\sqrt{z^2 + 4} = z \left(1 + \frac{2}{z^2} + \cdots \right) = z + \frac{2}{z} + \cdots,$$

and hence

$$f(z) = \frac{1}{z} + \cdots .$$

(The $\cdots$ mean higher powers of z^{-1}.)

$\diamondsuit$

4.2   *Van Vleck's Theorem*

Theorem 2 (Van Vleck's Theorem) *Let $0 < \varepsilon < \pi/2$, and let b_n satisfy*

$$-\frac{\pi}{2} + \varepsilon < \arg b_n < \frac{\pi}{2} - \varepsilon \tag{4.2.1}$$

for all n. Then all approximants of $\mathbf{K}(1/b_n)$ *are finite and in the angular domain*

$$-\frac{\pi}{2} + \varepsilon < \arg f_n < \frac{\pi}{2} - \varepsilon. \qquad (4.2.2)$$

Furthermore, the sequences $\{f_{2m}\}$ *and* $\{f_{2m+1}\}$ *converge to finite values.*

If (and only if), in addition

$$\sum_{n=1}^{\infty} |b_n| = \infty, \qquad (4.2.3)$$

then $\mathbf{K}(1/b_n)$ *converges.*

Partial proof: We shall here restrict ourselves to a proof of the first part of the theorem, i.e. that the approximants all satisfy (4.2.2). The proof is closely related to the first part of the proof of Śleszyński-Pringsheim's Theorem, i.e. the proof of the statement on the location of the approximants f_n.

A crucial point in the present case is the following observation for the angular domain V_ε described in (4.2.1) and (4.2.2):

$$w \in V_\varepsilon \Rightarrow \frac{1}{b_k + w} \in V_\varepsilon \qquad (4.2.4)$$

for all $b_k \in V_\varepsilon$. This follows immediately from the fact that the sum of two elements in V_ε also is in V_ε, and that

$$\omega \in V_\varepsilon \Rightarrow \frac{1}{\omega} \in V_\varepsilon.$$

From (4.2.4) it follows by induction as in the previous section that (4.2.2) holds for all n. ∎

This argument is a special case of a basic type of argument in convergence theory for continued fractions. We shall return to this later, and also to the rest of the proof of Van Vleck's Theorem.

Remark: If (4.2.1) holds, then also the approximants $S_n(w)$ of $\mathbf{K}(1/b_n)$ are finite and $\in V_\varepsilon$ if $w \in V_\varepsilon$. Moreover $\{S_n(w)\}$ converges locally uniformly to the value of $\mathbf{K}(1/b_n)$ in V_ε. The convergence is also uniform with respect to the actual choice of b_n from compact subsets of V_ε.

Example 10 It follows immediately from Van Vleck's Theorem that a *regular continued fraction* always converges. (As defined in Subsection *1.1* and illustrated in Subsection *2.1* a regular continued fraction is a continued fraction $\mathbf{K}(1/b_n)$, where all b_n are natural numbers. Obviously then all b_n are in any V_ε with $0 < \varepsilon < \pi/2$, and also $\sum |b_n| = \sum b_n = \infty$.

$\diamond$

Example 11 It follows immediately from Van Vleck's Theorem that any *periodic* continued fraction $\mathbf{K}(1/b_n)$ where all b_n have positive real part will converge. We hope that the following *two-periodic* continued fraction will serve as an example. The continued fraction to be studied here is

$$\cfrac{1}{1+i+}\cfrac{1}{1-i+}\cfrac{1}{1+i+}\cfrac{1}{1-i+}\cfrac{1}{1+i+}\cfrac{1}{1-i+}\cdots .$$

Since we know that it converges, it is rather easy to find the value f it converges to. It must satisfy the equation

$$f = \cfrac{1}{1+i+\cfrac{1}{1-i+f}} ,$$

i.e.

$$(1+i)f^2 + 2f - (1-i) = 0 .$$

This quadratic equation has the two roots

$$\frac{(-1+\sqrt{3})(1-i)}{2} \quad \text{and} \quad \frac{(-1-\sqrt{3})(1-i)}{2} .$$

Since the real part of f has to be positive, we find

$$f = \frac{(\sqrt{3}-1)(1-i)}{2} .$$

$\diamond$

4.3 Worpitzky's Theorem

Theorem 3 (Worpitzky's Theorem) *Let for all* $n \geq 1$

$$|a_n| \leq \frac{1}{4}. \qquad (4.3.1)$$

Then $\mathbf{K}(a_n/1)$ *converges. All approximants* f_n *are in the disk*

$$|w| < \frac{1}{2}, \qquad (4.3.2)$$

and the value f *is in the disk* $|w| \leq \frac{1}{2}$.

Proof : Let

$$\frac{a_1}{1 +} \frac{a_2}{1 +} \cdots + \frac{a_n}{1 +} \cdots \qquad (4.3.3)$$

be such that $|a_n| \leq 1/4$ for all n. It is easily seen, that the sequence of approximants for the continued fraction

$$\frac{2a_1}{2 +} \frac{4a_2}{2 +} \frac{4a_3}{2 +} \cdots + \frac{4a_n}{2 +} \cdots \qquad (4.3.4)$$

is exactly the same sequence as the sequence of approximants for (4.3.3). Since $|a_n| \leq 1/4$ for all n, we have

$$2 \geq |4a_n| + 1,$$

and from Śleszyński-Pringsheim's Theorem it follows that the continued fraction (4.3.4), and hence (4.3.3) converges. If the continued fraction (4.3.4) is multiplied by two, which means to replace the first partial fraction by $4a_1/2$, we find from Śleszyński-Pringsheim's Theorem that all approximants have absolute value < 1, and hence all approximants of (4.3.3) have absolute value $< 1/2$. From the convergence it then follows, that the value of the continued fraction is in the disk $|w| \leq \frac{1}{2}$. This concludes the proof of Worpitzky's Theorem. (This is essentially Śleszyński's proof.) ∎

Remark: Again the convergence of $S_n(w)$ is uniform with respect to $\{a_n\}$ and w for $|a_n| \leq 1/4$ and $|w| < 1/2$.

We shall now, through an example, indicate how the knowledge of a set where the values must be, can be used in the computation of continued fraction values.

Example 12 Let
$$\frac{-1/4}{1} + \frac{1/8}{1} + \frac{a_3}{1} + \frac{a_4}{1} + \cdots \tag{4.3.5}$$
be a continued fraction where all a_n have absolute value $\leq 1/4$. What can be said about the value of the continued fraction?

From Worpitzky's Theorem it follows that the value of the tail
$$\frac{a_3}{1} + \frac{a_4}{1} + \cdots + \frac{a_n}{1} + \cdots$$
is in the disk $|w| \leq 1/2$. The linear fractional transformation (l.f.t.)
$$w \longrightarrow \frac{1/8}{1+w}$$
maps the disk $|w| \leq 1/2$ onto the disk
$$\left| w - \frac{1}{6} \right| \leq \frac{1}{12},$$
and the l.f.t.
$$w \longrightarrow \frac{-1/4}{1+w}$$
maps this disk onto the disk
$$\left| w + \frac{14}{65} \right| \leq \frac{1}{65}.$$

(This is established by using standard methods for mapping disks by linear fractional transformations, or simply by computing, in each case, the intersections with the real axis together with the knowledge that the images are disks.) Observe how quickly we reach good values. By taking $-14/65$ as an approximate value, the error is $\leq 1/65$, regardless of which continued fraction (4.3.5) we have, if it satisfies $|a_n| \leq 1/4$ for all $n \geq 3$.

$\diamond$

5 Convergence once again

5.1 Critical remarks on convergence

We return to some of the thoughts from the very first section in the present chapter. When a series

$$\sum_{n=1}^{\infty} t_n$$

converges, the nth tail goes to 0 when n goes to ∞:

$$\lim_{n \to \infty} \sum_{k=n+1}^{\infty} t_k = 0 \,. \tag{5.1.1}$$

The nth approximant of a series is obtained by removing the nth tail, or, phrased differently: by replacing it with its limit (which is 0), i.e. the nth approximant is

$$T_n = \sum_{k=1}^{n} t_k \,.$$

When a product

$$\prod_{n=1}^{\infty} p_n$$

converges, the nth tail goes to 1 when n goes to ∞:

$$\lim_{n \to \infty} \prod_{k=n+1}^{\infty} p_k = 1 \,. \tag{5.1.2}$$

The nth approximant of a product is obtained by removing the nth tail, or, phrased differently: by replacing it with its limit (which is 1), i.e. the nth approximant is

$$P_n = \prod_{k=1}^{n} p_k \,.$$

The nth approximant of a continued fraction is also obtained by cutting off the tail, i.e. the nth approximant is

$$\mathop{\mathbf{K}}_{k=1}^{n} \frac{a_k}{b_k} \,,$$

but for continued fractions this does *not* mean to replace the tail by its limit. Usually this limit does not exist *at all,* and if it exists, it is 0 only in very special cases. A continued fraction where the limit exists, is the one in Example 1,

$$\mathbf{K}(6/1),$$

where all the tails, including the continued fraction itself, have the value 2. This raises the questions:

1) In computing the value of a continued fraction other sequences $\{S_n(w_n)\}$ may be better than $\{S_n(0)\}$. (Look back to Subsections *2.3, 3.4* and *3.5.*)

2) Perhaps the concept of convergence of continued fractions should not have been tied to the sequence $\{S_n(0)\}$, but to some $\{S_n(w_n)\}$.

We shall consider these questions in the rest of the chapter.

5.2   Modified approximants

The *word* has been used earlier, and in the Subsections *3.4* and *3.5* we have seen examples indicating that in some cases sequences $\{S_n(w)\}$ or even $\{S_n(w_n)\}$ may be better than $\{S_n(0)\}$ in the computation of the value of a continued fraction. Here are two more examples:

Example 13 For the continued fraction

$$\mathop{\mathbf{K}}_{n=1}^{\infty} \frac{30 + 0.9^n}{1}, \tag{5.2.1}$$

the tails look more and more like

$$\mathop{\mathbf{K}}_{n=N+1}^{\infty} \frac{30}{1}, \tag{5.2.2}$$

if we let them start further and further out. One can prove, that (5.2.2) converges to the positive root of the quadratic equation

$$u = \frac{30}{1+u},$$

i.e. to 5. (Problem 2, the hint in Problem 6 and a little more.) This suggests, that in the computation of (5.2.1) the sequence $\{S_n(5)\}$ may turn out to be better than $\{S_n(0)\}$. (The convergence of (5.2.1) is not hard to prove directly, but after Chapter III it will be trivial. For the time being we take convergence for granted.) The following table indicates strongly that this is in fact true. In the table C stands for *classical* approximants, i.e. $S_n(0)$, whereas J stands for *modified* approximant, in the present case $S_n(5)$. See [JaWa84], [ThWa82].

n	C	J
1	30.9000	5.15000
2	0.97139	5.03667
3	15.6770	5.12176
4	1.85765	5.05762
35	5.10127	5.08507
36	5.07160	5.08506
37	5.09631	5.08507
38	5.07571	5.08506
39	5.09288	5.08507
40	5.07857	5.08507
41	5.09049	5.08507
85	5.08507	5.08507
86	5.08506	5.08507
87	5.08507	5.08507
88	5.08507	5.08507
89	5.08507	5.08507

$\diamond$

Example 14 For the continued fraction

$$\frac{3+1/1^2}{1} + \frac{4+3/2^2}{1} + \frac{3+1/3^2}{1} + \frac{4+3/4^2}{1} + \frac{3+1/5^2}{1} + \cdots \tag{5.2.3}$$

the tails "look more and more like"

$$\frac{3}{1} + \frac{4}{1} + \frac{3}{1} + \frac{4}{1} + \cdots \tag{5.2.4}$$

and

$$\frac{4}{1+} \frac{3}{1+} \frac{4}{1+} \frac{3}{1+} \cdots . \tag{5.2.5}$$

We take convergence for granted in all cases, since it will be obvious after Chapter III. The value of the continued fraction (5.2.4) is the positive root of the quadratic equation

$$u = \cfrac{3}{1 + \cfrac{4}{1 + u}} ,$$

which is 1. The value of the continued fraction (5.2.5) is the positive root of the quadratic equation

$$u = \cfrac{4}{1 + \cfrac{3}{1 + u}} ,$$

which is 2.

In the table below $\{S_n(0)\}$ is compared to $\{S_n(w_n)\}$, where $w_{2k} = 1$ and $w_{2k-1} = 2$.

n	C-app.	J-app.
1	4.00000	1.33333
2	0.69565	1.18519
3	1.85580	1.20055
4	1.00775	1.18752
5	1.38927	1.18941
6	1.12527	1.18745
7	1.25252	1.18777
8	1.16637	1.18740
9	1.20885	1.18746
10	1.18033	1.18738
11	1.19451	1.18740
12	1.18502	1.18738
13	1.18975	1.18738
23	1.18739	1.18738
24	1.18738	1.18738
25	1.18738	1.18738
26	1.18738	1.18738

The table indicates strongly that $\{S_n(w_n)\}$ converges to the same value as $\{S_n(0)\}$, and faster.

⸺⸺⸺⸺⸺⸺⸺⸺⸺⸺⸺⸺⸺◇

From Example 13 and Example 14 it seems (and it will later be proved) that with $\{w_n\}$ properly picked $\{S_n(w_n)\}$ converges to the value of the continued fraction, and faster than $\{S_n(0)\}$. What "properly picked" means will be discussed later. We easily can make "improper choices": Take *any* sequence $\{\beta_n\}$ of extended complex numbers, and choose

$$w_n = S_n^{-1}(\beta_n).$$

This is possible, since all S_n are non-singular. Then we have

$$\beta_n = S_n(w_n).$$

This shows, that we can make $\{S_n(w_n)\}$ converge to anything we want, or diverge, regardless of the convergence behavior of the continued fraction itself.

Let it finally be mentioned, that in some cases attempts to compute the value of some continued fraction $\mathbf{K}(a_n/b_n)$ lead to a sequence $\{S_n(w_n)\}$ which converges to something we have reason to believe is the value of the continued fraction, whereas $\{S_n(0)\}$ may be hard to handle, or even to get hold of. In such cases one needs results about going from convergence of $\{S_n(w_n)\}$ to convergence of the continued fraction, and to the same value (which, as we just have seen, is not true in general).

5.3 *Another concept of convergence*

We shall first look at an example [Jaco86]:

Example 15 For the 3-periodic continued fraction

$$\frac{2}{1+}\frac{1}{1+}\frac{-1}{1}\frac{2}{+1+}\frac{1}{1+}\frac{-1}{1}+\cdots$$

it is not hard to prove, by using the recurrence relations (1.2.9) and induction, that for all $n \geq 1$

$$
\begin{array}{llll}
A_{3n-2} = 2^n, & A_{3n-1} = 2^n, & A_{3n} = 0, \\
B_{3n-2} = 2^{n+1} - 3, & B_{3n-1} = 2^{n+1} - 2, & B_{3n} = 1.
\end{array}
$$

Hence the approximants A_n/B_n are

$$
f_{3n-2} = \frac{2^n}{2^{n+1} - 3}, \qquad f_{3n-1} = \frac{2^n}{2^{n+1} - 2}, \qquad f_{3n} = 0,
$$

from which

$$
\lim_{n \to \infty} f_{3n-2} = \frac{1}{2}, \qquad \lim_{n \to \infty} f_{3n-1} = \frac{1}{2}, \qquad \lim_{n \to \infty} f_{3n} = 0
$$

immediately follow. This shows that the continued fraction *diverges*.

For the modified approximants $S_n(w_n)$ we find from (1.2.8)

$$
S_{3n-2}(w_{3n-2}) = \frac{2^n + w_{3n-2} \cdot 0}{(2^{n+1} - 3) + w_{3n-2} \cdot 1} = \frac{1}{2 + (w_{3n-2} - 3)2^{-n}} \to \frac{1}{2}
$$

when $n \to \infty$, if the sequence $\{w_{3n-2}\}$ is bounded.

$$
\begin{aligned}
S_{3n-1}(w_{3n-1}) &= \frac{2^n + w_{3n-1} \cdot 2^n}{(2^{n+1} - 2) + w_{3n-1}(2^{n+1} - 3)} \\
&= \frac{1 + w_{3n-1}}{2(1 + w_{3n-1}) - (3w_{3n-1} + 2)2^{-n}} \to \frac{1}{2}
\end{aligned}
$$

when $n \to \infty$, if the sequence $\{w_{3n-1}\}$ is bounded away from -1.

$$
S_{3n}(w_{3n}) = \frac{0 + w_{3n} \cdot 2^n}{1 + w_{3n}(2^{n+1} - 2)} = \frac{w_{3n}}{2 \cdot w_{3n} + (1 - 2 \cdot w_{3n})2^{-n}} \to \frac{1}{2}
$$

when $n \to \infty$, if the sequence $\{w_{3n}\}$ is bounded away from 0.

Hence we have, in this example, that $\lim_{n \to \infty} S_n(0)$ does not exist, whereas

$$
\lim_{n \to \infty} S_n(w_n) = \frac{1}{2}
$$

for *all* $\{w_n\}$ bounded away from $0, -1, \infty$. This example suggests strongly, more so than earlier considerations, that the definition of convergence of continued fractions is "wrong", since the continued fraction in this example "ought to converge".

$\diamond$

This example (together with other observations) has led to a new definition of convergence [Jaco86]. In the definition we use the chordal distance $d(z, w)$, which is defined by

$$d(w, z) = \frac{|z - w|}{\sqrt{1 + |w|^2}\sqrt{1 + |z|^2}},$$

if w and z are both finite, whereas

$$d(w, \infty) = \frac{1}{\sqrt{1 + |w|^2}}.$$

This is a metric very much used in the theory of functions of one complex variable, in particular in cases where the point at infinity is not supposed to play a special role, different from the role of other points. The *name* comes from the fact, that $d(w, z)$ is the length of the chord between the images of w and z on a sphere by a suitable stereographic projection.

Definition $\mathbf{K}(a_n/b_n)$ *is said to converge generally to an* $f \in \hat{\mathbf{C}}$ *if there exist two sequences* $\{v_n\}$ *and* $\{w_n\}$ *in* $\hat{\mathbf{C}}$ *such that*

$$\liminf d(v_n, w_n) > 0 \qquad (5.3.1)$$

and

$$\lim_{n \to \infty} S_n(v_n) = \lim_{n \to \infty} S_n(w_n) = f. \qquad (5.3.2)$$

We shall see later, that f is unique. (If not, the definition would not make sense.) One property follows directly from the definition: If a continued fraction converges in the ordinary (classical) sense to f, we have

$$\lim_{n \to \infty} S_n(0) = \lim_{n \to \infty} S_n(\infty) = f.$$

Since $0, 0, 0, \ldots$ and $\infty, \infty, \infty, \ldots$ are two sequences satisfying (5.3.1), it follows that ordinary convergence to f implies general convergence to f. Thus the new concept includes the classical one, and picks up additional cases, for instance the one in Example 15.

More important is, that it in many cases is easier to apply. One advantage is, that we do not need to worry about $\{S_n(0)\}$. Once we have proved (5.3.2) for two sequences satisfying (5.3.1) we are through (as far as *general convergence* is concerned).

5.4 Another concept of continued fraction

The formal definition of a continued fraction was presented in Subsection *1.2*, and the sequence $\{S_n(0)\}$ played a crucial role in the definition. Recently there has been an increased use of and emphasis on modifications. Out of this has grown the concept of *modified continued fractions*, obtained by replacing the sequence $\{f_n\}$, $f_n = S_n(0)$, by $\{g_n\}$, $g_n = S_n(w_n)$. Of course the notation must then contain $\{w_n\}$. This appeared in print at first in [BaJo86], and the modified continued fraction was there written

$$((\{a_n\}, \{b_n\}, \{w_n\}), \{g_n\}).$$

An abbreviated notation is

$$\mathbf{K}(a_n, b_n, w_n).$$

We obviously have, with reference to notation introduced earlier:

$$
\begin{aligned}
((\{a_n\}, \{b_n\}, \{0\}), \{f_n\}) &= ((\{a_n\}, \{b_n\}), \{f_n\}), \\
\mathbf{K}(a_n, b_n, 0) &= \mathbf{K}(a_n/b_n), \\
\mathbf{K}(a_n, 1, 0) &= \mathbf{K}(a_n/1).
\end{aligned}
$$

In working with modified continued fractions the classical one is sometimes referred to as the *reference continued fraction*.

The earlier mentioned problem of going from convergence of $S_n(w_n)$ to convergence of $S_n(0)$ can now be expressed as a problem of going from convergence of modified continued fractions to convergence of ordinary continued fractions.

5.5 Computation of approximants

To compute

$$S_n(w_n) = \frac{a_1}{b_1} \frac{a_2}{+ \, b_2} + \cdots + \frac{a_n}{b_n + w_n} = \frac{A_n + A_{n-1}w_n}{B_n + B_{n-1}w_n}$$

there are several algorithms. We shall only mention the two obvious ones:

1. The forward recurrence algorithm consists of computing A_n and B_n by the recurrence relation (1.2.9).

2. The backward recurrence algorithm starts at the other end by setting

$$t_n = w_n$$

and then work backwards by setting

$$t_{k-1} = \frac{a_k}{b_k + t_k}$$

for $k = n, n-1, \ldots, 1$. Then $S_n(w_n) = t_0$ (or $b_0 + t_0$).

The first method has the advantage that if you have found $S_n(w_n)$, you can easily find $S_{n+1}(w_{n+1})$, whereas you must start again from scratch in the second method. On the other hand, the backward recurrence algorithm is in general more stable. (Why will become evident in Chapter IV.) The computations in this book are done by means of the backward recurrence algorithm.

Problems

(1) Use the identity
$$\frac{\sqrt{5}-1}{2} = \frac{1}{1 + \frac{\sqrt{5}-1}{2}}$$

to produce a continued fraction by the procedure of Example 2. Compute the first 7 approximants f_n and compare the values to $(\sqrt{5}-1)/2$. Prove that

$$f_n = \frac{F_{n-1}}{F_n},$$

where $F_0 = 1$, $F_1 = 1$, $F_2 = 2$, $F_3 = 3$, $F_4 = 5$, and generally

$$F_{n+1} = F_n + F_{n-1} \quad \text{for } n \geq 1.$$

(The sequence $\{F_n\}$ is the sequence of *Fibonacci* numbers, and the ratio $(\sqrt{5}-1)/2 = .61803\ldots$ is the *golden ratio*.)

(2) Prove the following: For any *real a, if* the continued fraction

$$\frac{a}{1+}\frac{a}{1+}\frac{a}{1+}\cdots$$

converges, then it converges to one of the roots of the equation

$$x = \frac{a}{1+x}.$$

Use this to prove that the continued fraction diverges for all $a < -1/4$.

(3) Assume that we know that the continued fraction

$$\frac{1}{1+}\frac{1}{1+}\frac{1}{1+}\cdots$$

converges. Prove that it then converges to $(\sqrt{5}-1)/2$.

(4) Find a particular integral of the differential equation

$$y = 2y' + 3y''$$

by using the "method" in Subsection *2.2*.

(5) Let x and y be complex numbers, $|x| \neq |y|$, and let f_n be the nth approximant of the continued fraction

$$\frac{-xy}{-(x+y)+}\frac{-xy}{-(x+y)+}\frac{-xy}{-(x+y)+}\cdots.$$

Find a formula for f_n in terms of x and y.

(6) Assume that we *know* that the continued fraction in Example 4,

$$\frac{x}{2+}\frac{x}{2+}\frac{x}{2+}\cdots$$

converges to $\sqrt{1+x}-1$ for positive x-values. Use this to find $\sqrt{5}$ with an error $< 5 \cdot 10^{-3}$. (Hint: Observe — and prove — that $\{f_{2n}\}$ and $\{f_{2n+1}\}$ are monotone. Take $x = 1/4$.)

(7) Compute for a suitable n the first n approximants of the continued fraction in Problem 6 for $x = 1 - 2i$. Try to *guess* in advance what the sequence will converge to, and compare to the computed approximants.

(8) Assume that we *know* that the $\log(1+z)$–expansion of Subsection 2.4 converges to $\log(1+z)$ for positive z-values. Use this to find $\log 5$ with an error $< 5 \cdot 10^{-3}$. (Hint as in Problem 6 with $z = 4$.)

(9) Compute for a suitable n the first n approximants of the continued fraction in Subsection 2.4 for $z = i - 1$. Try to *guess* what the sequence of approximants converges to.

(10) Use Example 6 in Subsection 3.1 to establish the expansion of $\log(1+z)$ of Subsection 2.4, assuming that the continued fraction expansion (3.1.2) of

$$\frac{F(a, b; c; z)}{F(a, b+1; c+1; z)}$$

is established. (Hint: Take $a = 1, b = 0, c = 1$, and replace z by $-z$.)

(11) Let α be real and not a positive integer. With F as in Subsection 3.1, prove that

$$F(-\alpha, 1; 1; -z) = (1+z)^\alpha.$$

Under the same assumptions as in Problem 10 find a continued fraction of the form

$$\frac{b_1 z}{1} + \frac{b_2 z}{1} + \frac{b_3 z}{1} + \cdots$$

for $z(1+z)^\alpha$ by using Subsection *3.1.*

(12) Let α be as in (11). Use the procedure of Subsection *3.2* to transform the power series expansion of $z(1+z)^\alpha$ at 0 to a continued fraction of the form

$$\frac{b_1 z}{1} + \frac{b_2 z}{1} + \frac{b_3 z}{1} + \cdots$$

(Of course only the start, for instance up to and including $b_4 z/1$.)

(13) Use the procedure of Subsection *3.2* to transform the power series expansion at 0 of $e^z - 1$ to a continued fraction of the form

$$\frac{b_1 z}{1} + \frac{b_2 z}{1} + \frac{b_3 z}{1} + \cdots$$

(Compute b_1, b_2, b_3, b_4.)

(14) Use the procedure of Subsection *3.3* to find the first terms of the power series expansion at 0 corresponding to the regular C-fraction

$$\frac{z}{1+} \, \frac{-z/2}{1} + \frac{z/6}{1} + \frac{-z/6}{1} + \cdots$$

(15) Let α be a positive number. For which values of α does the continued fraction $\mathbf{K}_{n=1}^{\infty}(1/n^{-\alpha})$ converge, and for which values does it diverge? (Hint: Use Van Vleck's Theorem.)

(16) Use Worpitzky's Theorem to prove the following:

a) The value of any continued fraction

$$\frac{1/4}{1} + \frac{-1/4}{1} + \frac{a_3}{1} + \frac{a_4}{1} + \cdots,$$

with $|a_n| \leq 1/4$ for all n must lie in the disk

$$\left| w - \frac{2}{5} \right| \leq \frac{1}{10}.$$

b) The value of any continued fraction

$$\frac{i/4}{1} + \frac{a_2}{1} + \frac{a_3}{1} + \cdots,$$

with $|a_n| \leq 1/4$ for all n must lie in the disk

$$\left| w - \frac{i}{3} \right| \leq \frac{1}{6}.$$

(17) For the continued fraction in Example 15 of Subsection *5.3* compute $S_n(w_n)$ for the following values of w_n. In which cases do we have convergence?

a) $w_n = 1$,

b) $w_n = n$,

c) $w_n = 2^{(n+2)/3}$,

d) $w_n = 1/n$,

e) $w_n = 2^{-n/3}$.

Remarks

1. For those who want to go deeper into the analytic theory of continued fractions we refer to the three standard monographs in
   the field: the classical text-book by Perron [Perr54], Wall's book
   [Wall48], with its introduction to some of the new ideas, upon
   which the modern theory is built, and finally the most modern
   exposition, by Jones and Thron from 1980 [JoTh80]. In Henrici's
   3 volume work on Applied and Computational Complex Analysis
   [Henr77] a large portion of Volume 2 is devoted to analytic theory
   of continued fraction. Khovanskii's book [Khov63] contains some
   interesting applications. References to further books and papers in
   the field are found in the texts above. As for the history of continued fractions we refer to a recent book by Brezinski [Brez91], but
   also to the texts mentioned, in particular to the book [JoTh80] by
   Jones and Thron, which contains many interesting comments on
   the historic development of concepts, methods and applications.

2. To most people (meaning mathematicians) continued fractions are
   most closely associated with Number Theory, for instance in connection with diophantine equations of degree 1 or 2. The continued
   fractions used there are mostly regular continued fractions. One
   very important thing in the Theory of Numbers is the connection
   between the Euclidean algorithm and the terminating regular continued fraction expansion of rational numbers. In the present exposition we shall not say much about continued fractions in number
   theory, except for a small chapter on some applications in number
   theory (Chapter IX). We also refer to [Perr54].

3. After Example 4 and Example 5 in Subsections *2.3* and *2.4*, indicating superiority of continued fractions over power series expansions, both as far as domain of convergence and speed of convergence are concerned, it is only fair to say, that this is not always
   the case. Sometimes it is the other way around. One trivial example, which has its non-trivial analogues, is Example 7 in Subsection
   3.4. There the power series $z + 0.z^2 + 0.z^3 + \cdots$ is transformed
   into the continued fraction (3.4.1). The series converges to z in
   the whole plane. The convergence is the "fastest possible", since
   all partial sums $= z$. The continued fraction converges to z in the

open unit disk $|z| < 1$, but more slowly, and for $|z| > 1$ it even converges to "something wrong", namely to -1. (It can be proved, that it diverges on the circle $|z| = 1$, except for $z = -1$, where it converges to -1.) On the other hand, we shall see in Chapter VI that the continued fraction in Example 6 always "wins over the hypergeometric series".

4. For references and comments connected to the three classical theorems in Section 4 we refer essentially to the book by Jones and Thron [JoTh80]. Let it be mentioned, though, that Worpitzky's Theorem was proved already in 1865, but remained unknown to workers in the field until Pringsheim rediscovered it more than 30 years later. It was not until 1905, through Van Vleck, that Worpitzky got credit for it. Part of the reason may be the way it was published, (in an annual report from the school where Worpitzky was teaching, [Worp65]), but there may be other more significant reasons, see [JaTW89]. Theorem 1 usually carries the name of Pringsheim. However, as pointed out to us by W. J. Thron, J. Śleszyński is the right one to give credit, since he already proved the theorem in 1888, see [Śles89].

5. Already Hamel [Hamel18] raised the question about the concept of convergence for continued fractions.

References

[AlBe88]   G. Almkvist and B. Berndt, *Gauss, Landen, Ramanujan, the Arithmetic-Geometric Mean, Ellipses, π, and the Ladies Diary.* Amer. Math. Monthly (1988), 585–608.

[BaJo86]   C. Baltus and W. B. Jones, *A Family of Best Value Regions for Modified Continued Fractions,* "Analytic Theory of Continued Fractions II", Lecture Notes in Mathematics **1199** (ed. W. J. Thron), Springer-Verlag, Berlin (1986), 1–20.

[Brez91]   C. Brezinski, "History of Continued Fractions and Padé Approximants", Springer Series in Computational Mathematics, **12**, Springer-Verlag, Berlin (1991).

[Brun50]   V. Brun, "Forelesninger over Kjedebrøk", Universitetet i Oslo (1950).

[CuWu87]   A. Cuyt and L. Wuytack, "Nonlinear Methods in Numerical Analysis", North-Holland Mathematics Studies, Amsterdam (1987).

[Gauss13]   C. F. Gauss, *Disquisitiones generales circa seriem infinitam* $1 + \frac{\alpha\beta}{1\cdot\gamma}x + \frac{\alpha(\alpha+1)\beta(\beta+1)}{1\cdot2\cdot\gamma\cdot(\gamma+1)}xx + \frac{\alpha(\alpha+1)(\alpha+2)\beta(\beta+1)(\beta+2)}{1\cdot2\cdot3\cdot\gamma\cdot(\gamma+1)(\gamma+2)}x^3$ *etc,* Commentationes Societatis Regiae Scientiarium Gottingensis Recentiores **2** (1813), 1–46; Werke, Vol. **3** Göttingen (1876), 134–138.

[Gauss14]   C. F. Gauss, *Methodus Nova Integralium Valores per Approximationem Inveniendi,* Commentationes Societatis Regiae Scientiarium Gottingensis Recentiores **15** (1814), 39–76; Werke **3** Göttingen (1876), 165–196.

[Hamel18]   G. Hamel, *Über einen limitärperiodischen Kettenbruch*, Archiv der Math. und Phys. **27** (1918), 37–43.

[Henr77]   P. Henrici, "Applied and Computational Complex Analysis", Vol. **2**, Wiley, New York (1977).

[Hütte55]   Hütte, "Des Ingenieurs Taschenbuch", 28. Aufl. **1**, Wilhelm Ernst & Sohn, Berlin (1955), Seite 139.

[Jaco86]   L. Jacobsen, *General Convergence for Continued Fractions*, Trans. Amer. Math. Soc. **281** (1986), 129–146.

[JaTW89]   L. Jacobsen, W. J. Thron and H. Waadeland, *Julius Worpitzky, his Contributions to the Analytic Theory of Continued Fractions and his Times*, "Analytic Theory of Continued Fractions III", (ed. L. Jacobsen), Lecture Notes in Mathematics **1406**, Springer-Verlag, Berlin (1989), 25–47.

[JaWa84]   L. Jacobsen and H. Waadeland, *Modification of Continued Fractions*, "Padé Approximation and its Applications, Bad Honnef 1983", (H. Werner and H. J. Bünger eds.) Lecture Notes in Mathematics, Springer-Verlag, Berlin **1071** (1984), 176–196.

[JoTh80]   W. B. Jones and W. J. Thron, "Continued Fractions: Analytic Theory and Applications", Encyclopedia of Mathematics and its Applications, **11**, Addison-Wesley Publishing Company, Reading, Mass. (1980). Now distributed by Cambridge University Press, New York.

[Khov63]   A. N. Khovanskii, "The Application of Continued Fractions and Their Generalizations to Problems in Approximation Theory", P. Noordhoff, Groningen (1963).

[Perr54,57]   O. Perron, "Die Lehre von den Kettenbrüchen", Vol. **1, 2** 3. Aufl., B. G. Teubner, Stuttgart (1954, 1957).

[Rama57]   S. Ramanujan, "Notebooks", Vol. **2** Tata Institute of Fundamental Research, Bombay (1957). Now distributed by Springer-Verlag.

[Śles89]   J. V. Śleszyński, *Zur Frage von der Konvergenz der Kettenbrüche* (in Russian), Mat. Sbornik **14** (1889), 337–343, 436–438.

[Steen73] A. Steen, *Integration af Lineære Differentialligninger af Anden Orden ved Hjælp af Kjædebrøker*, Köbenhavn (1873).

[ThWa82]   W. J. Thron and H. Waadeland, *Modifications of Continued Fractions, a Survey*, "Analytic Theory of Continued Fractions", Proceedings 1981, (W. B. Jones, W. J. Thron and H. Waadeland eds.), Lecture Notes in Mathematics **932**, Springer-Verlag, Berlin (1982), 38–66.

[Waad83]   H. Waadeland, *Differential Equations and Modifications of Continued Fractions, some Simple Observations*, "Padé Approximants and Continued Fractions", Proceedings 1982, (H. Waadeland and H. Wallin eds.) Det Kongelige Norske Videnskabers Selskabs Skrifter, Trondheim **1** (1983), 136–150.

[Wall48]   H. S. Wall, "Analytic Theory of Continued Fractions", Van Nostrand, New York (1948).

[Worp65]   J. Worpitzky, *Untersuchungen über die Entwickelung der monodromen und monogenen Funktionen durch Kettenbrüche*, Friedrichs-Gymnasium und Realschule Jahresbericht, Berlin (1865), 3–39.

Chapter II

More basics

About this chapter

Continued fractions are defined by means of linear fractional transformations. Such transformations have very nice properties. One way to take advantage of this is to introduce the concept of tail sequences. (The beautiful mapping properties of linear fractional transformations will be exploited in Chapter III.)

Continued fractions can also be described by means of linear recurrence relations. This will be treated at length in Chapter IV, but already now we observe the following from formulas (1.2.6)–(1.2.9) in Chapter I: If $\{A_n\}$ and $\{B_n\}$ are given, satisfying certain conditions, then the elements $\{a_n\}$ and $\{b_n\}$ of the continued fraction can be determined. This is the background for the useful transformations of continued fractions to be presented in this chapter, transformations that allow us to change the coefficients of a continued fraction without altering its approximants (too much).

1   Tails of continued fractions

1.1   *Tails*

The *N th tail* of the continued fraction $b_0 + \mathbf{K}(a_n/b_n)$ is the continued fraction

$$\underset{n=N+1}{\overset{\infty}{\mathbf{K}}} \frac{a_n}{b_n} = \frac{a_{N+1}}{b_{N+1}} + \frac{a_{N+2}}{b_{N+2}} + \frac{a_{N+3}}{b_{N+3}} + \cdots \qquad (1.1.1)$$

for $N \in \mathbf{N}_0$. Here and in the rest of the book $\mathbf{N}_0$ is the notation for the set of non-negative integers; i.e. $\mathbf{N}_0 = \mathbf{N} \cup \{0\}$. There are several reasons to study such tails, and one is described in the following theorem.

Theorem 1 *The following three statements are equivalent.*

A) $b_0 + \mathbf{K}(a_n/b_n)$ *converges/converges generally.*

B) *(1.1.1) converges/converges generally for an $N \in \mathbf{N}_0$.*

C) *(1.1.1) converges/converges generally for all $N \in \mathbf{N}_0$.*

Proof : $\mathbf{C} \Rightarrow \mathbf{B}$ and $\mathbf{C} \Rightarrow \mathbf{A}$ follow trivially.
$\mathbf{A} \Rightarrow \mathbf{C}$: Let $b_0 + \mathbf{K}(a_n/b_n)$ converge to f. That is, its approximants $f_n = S_n(0)$ converge to f. Let $N \in \mathbf{N}_0$ be chosen arbitrarily, and let $f_n^{(N)} = S_n^{(N)}(0)$ denote the approximants of (1.1.1). Then

$$f_{N+n} = b_0 + \frac{a_1}{b_1} + \frac{a_2}{b_2} + \cdots + \frac{a_N}{b_N + f_n^{(N)}} = S_N(f_n^{(N)}) \qquad (1.1.2)$$

so that $f_n^{(N)} = S_N^{-1}(f_{N+n})$. The convergence of $f_n^{(N)}$ as $n \to \infty$ follows therefore from the fact that S_N^{-1} is a linear fractional transformation, and therefore is a bijection of $\hat{\mathbf{C}}$ onto $\hat{\mathbf{C}}$, since all $a_n \neq 0$. (This will be discussed in Chapter III.)

The proof for general convergence uses the same idea, where we use the analogue

$$S_{N+n}(w) = S_N(S_n^{(N)}(w)) = S_N \circ S_n^{(N)}(w) \qquad (1.1.3)$$

to (1.1.2).

B $\Rightarrow$ C: This can also be proved in a related way. ∎

Example 1 We shall prove that the continued fraction $\mathbf{K}(a_n/1)$ converges if $a_n \to 0$.

Since $a_n \to 0$ there exists an $N \in \mathbf{N}_0$ such that $|a_n| \le 1/4$ for all $n > N$. Hence, the Nth tail of $\mathbf{K}(a_n/1)$ converges by Worpitzky's theorem, Theorem 3 in Chapter I.

Observe that we can no longer conclude that the value f of $\mathbf{K}(a_n/1)$ is in the closed disk $|f| \le 1/2$. This is only true for the value $f^{(N)}$ of its Nth tail, $|f^{(N)}| \le 1/2$ (and for $f^{(k)}$ for all $k > N$). The value f may even be infinite.

$\diamondsuit$

We have used A_n and B_n to denote the canonical numerators and denominators of $b_0 + \mathbf{K}(a_n/b_n)$, and $S_n(w_n)$ and $f_n = S_n(0)$ to denote its approximants. For its Nth tail we use $A_n^{(N)}$, $B_n^{(N)}$, $S_n^{(N)}(w_{N+n})$ and $f_n^{(N)} = S_n^{(N)}(0)$ to denote the corresponding quantities. This notation will be used throughout the book.

Just as for A_n, B_n and S_n we have

$$S_n^{(N)}(w) = \frac{A_n^{(N)} + A_{n-1}^{(N)} w}{B_n^{(N)} + B_{n-1}^{(N)} w} = s_{N+1} \circ s_{N+2} \circ \cdots \circ s_{N+n}(w), \quad (1.1.4)$$

where $A_{-1}^{(N)} = 1$, $A_0^{(N)} = 0$, $B_{-1}^{(N)} = 0$, $B_0^{(N)} = 1$ and

$$\begin{bmatrix} A_n^{(N)} \\ B_n^{(N)} \end{bmatrix} = b_{N+n} \begin{bmatrix} A_{n-1}^{(N)} \\ B_{n-1}^{(N)} \end{bmatrix} + a_{N+n} \begin{bmatrix} A_{n-2}^{(N)} \\ B_{n-2}^{(N)} \end{bmatrix} \quad (1.1.5)$$

for $n = 1, 2, 3, \ldots$, and

$$A_n^{(N)} B_{n-1}^{(N)} - A_{n-1}^{(N)} B_n^{(N)} = - \prod_{j=N+1}^{N+n} (-a_j). \quad (1.1.6)$$

In addition one can also prove:

Lemma 2 *With the notation just introduced the following equalities hold:*

$$A_n^{(N)} = a_{N+1} B_{n-1}^{(N+1)} \quad for\ N \geq 0, \quad n \geq 0, \quad (1.1.7)$$

$$B_n^{(N)} = b_{N+1} B_{n-1}^{(N+1)} + a_{N+2} B_{n-2}^{(N+2)} \quad for\ N \geq 0, \quad n \geq 1, \quad (1.1.8)$$

$$A_{n+k}^{(N)} B_{n-1}^{(N)} - A_{n-1}^{(N)} B_{n+k}^{(N)} = -B_k^{(N+n)} \prod_{j=N+1}^{N+n} (-a_j). \quad (1.1.9)$$

Proof :   (1.1.7) is trivially true for $n = 0$ and $n = 1$. Hence it holds for all n, since by (1.1.5)

$$A_n^{(N)} = b_{N+n} A_{n-1}^{(N)} + a_{N+n} A_{n-2}^{(N)} \quad \text{for all } n \geq 1$$

and

$$B_{n-1}^{(N+1)} = b_{N+n} B_{n-2}^{(N+1)} + a_{N+n} B_{n-3}^{(N+1)} \quad \text{for all } n \geq 2.$$

(1.1.8) holds trivially for $n = 1$ and $n = 2$. By induction on n we find that it holds for all n, since by (1.1.5)

$$
\begin{aligned}
B_n^{(N)} &= b_{N+n} B_{n-1}^{(N)} + a_{N+n} B_{n-2}^{(N)} \\
&= b_{N+n} \left(b_{N+1} B_{n-2}^{(N+1)} + a_{N+2} B_{n-3}^{(N+2)} \right) \\
&\quad + a_{N+n} \left(b_{N+1} B_{n-3}^{(N+1)} + a_{N+2} B_{n-4}^{(N+2)} \right) \\
&= b_{N+1} \left(b_{N+n} B_{n-2}^{(N+1)} + a_{N+n} B_{n-3}^{(N+1)} \right) \\
&\quad + a_{N+2} \left(b_{N+n} B_{n-3}^{(N+2)} + a_{N+n} B_{n-4}^{(N+2)} \right) \\
&= b_{N+1} B_{n-1}^{(N+1)} + a_{N+2} B_{n-2}^{(N+2)} \quad \text{for } n \geq 3.
\end{aligned}
$$

Finally, it follows from (1.1.6) that (1.1.9) holds for $k = 0$ since $B_0^{(N+n)} = 1$. For $k = 1$ we get by (1.1.5) that

$$
\begin{aligned}
A_{n+1}^{(N)} B_{n-1}^{(N)} - A_{n-1}^{(N)} B_{n+1}^{(N)} &= \left(b_{N+n+1} A_n^{(N)} + a_{N+n+1} A_{n-1}^{(N)} \right) B_{n-1}^{(N)} \\
&\quad - A_{n-1}^{(N)} \left(b_{N+n+1} B_n^{(N)} + a_{N+n+1} B_{n-1}^{(N)} \right) \\
&= b_{N+n+1} \left(A_n^{(N)} B_{n-1}^{(N)} - A_{n-1}^{(N)} B_n^{(N)} \right),
\end{aligned}
$$

which proves (1.1.9) for this value of k. By the same process we find that

$$A_{n+k}^{(N)} B_{n-1}^{(N)} - A_{n-1}^{(N)} B_{n+k}^{(N)}$$
$$= b_{N+n+k} \left(A_{n+k-1}^{(N)} B_{n-1}^{(N)} - A_{n-1}^{(N)} B_{n+k-1}^{(N)} \right)$$
$$+ a_{N+n+k} \left(A_{n+k-2}^{(N)} B_{n-1}^{(N)} - A_{n-1}^{(N)} B_{n+k-2}^{(N)} \right) \quad \text{for } k = 2, 3, 4, \ldots,$$

which is the same recurrence relation as (1.1.5) for $\left\{ B_k^{(N+n)} \right\}_{k=-1}^{\infty}$. Hence (1.1.9) follows. ∎

1.2   Tail sequences

Assume that $b_0 + \mathbf{K}(a_n/b_n)$ converges. Let $f^{(N)} \in \hat{\mathbf{C}}$ denote the value of its Nth tail (1.1.1) for $N = 0, 1, 2, \ldots$. Then we find from (1.1.1) that $f = b_0 + f^{(0)}$ is the value of $b_0 + \mathbf{K}(a_n/b_n)$ and

$$f^{(N)} = \frac{a_{N+1}}{b_{N+1} + f^{(N+1)}} \quad \text{for } N = 0, 1, 2, \ldots. \tag{1.2.1}$$

((1.2.1) is to be interpreted in the obvious way if $f^{(N)}$ or $f^{(N+1)}$ is infinite.) We say that $\{f^{(N)}\}_{N=0}^{\infty}$ is the *right tail sequence* for the convergent continued fraction $b_0 + \mathbf{K}(a_n/b_n)$.

More generally, we say that a sequence $\{t_n\}_{n=0}^{\infty}$ of elements from $\hat{\mathbf{C}}$ is a *tail sequence* for $b_0 + \mathbf{K}(a_n/b_n)$ if

$$t_{n-1} = \frac{a_n}{b_n + t_n} = s_n(t_n) \quad \text{for } n = 1, 2, 3, \ldots. \tag{1.2.2}$$

This means in particular that

$$t_0 = s_1(t_1) = s_1 \circ s_2(t_2) = \cdots = s_1 \circ \cdots \circ s_n(t_n) = S_n^{(0)}(t_n) \tag{1.2.3}$$

for all n, and thus that $\{S_n^{(0)^{-1}}(t_0)\}_{n=0}^{\infty}$ is a tail sequence (uniquely defined) for every $t_0 \in \hat{\mathbf{C}}$. Observe that if $\{t_n\}_{n=0}^{\infty}$ is a tail sequence for $b_0 + \mathbf{K}(a_n/b_n)$, then $\{t_n\}_{n=N}^{\infty}$ is a tail sequence for its Nth tail (1.1.1).

Theorem 3 *Let* $\{t_n\}$ *be a tail sequence for* $b_0 + \mathbf{K}(a_n/b_n)$. *Then*

$$\begin{aligned} t_n &= S_n^{-1}(b_0 + t_0) = s_n^{-1} \circ s_{n-1}^{-1} \circ \cdots \circ s_1^{-1}(t_0) \\ &= -\left\{ b_n + \cfrac{a_n}{b_{n-1}} \cfrac{a_{n-1}}{+\,b_{n-2}} + \cdots + \cfrac{a_2}{b_1} \cfrac{a_1}{+\,(-t_0)} \right\} \end{aligned} \qquad (1.2.4)$$

for all n.

Proof : Inverting (1.2.3) gives $t_n = s_n^{-1} \circ s_{n-1}^{-1} \circ \cdots \circ s_1^{-1}(t_0)$ which is equal to $s_n^{-1} \circ s_{n-1}^{-1} \circ \cdots \circ s_1^{-1} \circ s_0^{-1}(b_0 + t_0) = S_n^{-1}(b_0 + t_0)$. From (1.2.2) we find that

$$s_k^{-1}(w) = -b_k + \frac{a_k}{w} = -\left\{ b_k + \frac{a_k}{(-w)} \right\} \quad \text{for } k \geq 1 . \qquad (1.2.5)$$

This proves the last equality in (1.2.4).   ∎

Remark: Observe that if $\{t_n\}$ and $\{\tilde{t}_n\}$ are two tail sequences for $b_0 + \mathbf{K}(a_n/b_n)$ with $t_k = \tilde{t}_k$ for one index k, then $t_n = \tilde{t}_n$ for all n by (1.2.4) since all s_k are bijections of $\hat{\mathbf{C}}$ onto $\hat{\mathbf{C}}$.

We shall see later that the tail sequence $\{S_n^{-1}(\infty)\}$ plays a special role in our theory. We define

$$h_n = -S_n^{-1}(\infty) \quad \text{for } n = 0, 1, 2, \ldots , \qquad (1.2.6)$$

which gives $h_0 = \infty$, $h_1 = b_1$, and

$$h_n = \frac{B_n}{B_{n-1}} = b_n + \cfrac{a_n}{b_{n-1}} \cfrac{a_{n-1}}{+\,b_{n-2}} + \cdots + \cfrac{a_2}{b_1} \quad \text{for } n = 1, 2, \ldots . \qquad (1.2.7)$$

(Observe that $S_n^{-1}(\infty) = S_n^{(0)^{-1}}(\infty)$ since $s_0(\infty) = \infty$, regardless of the value of b_0.) This sequence $\{h_n\}$ is called the *critical tail sequence* of $b_0 + \mathbf{K}(a_n/b_n)$ (although it is strictly speaking $\{-h_n\}$ which is a tail sequence).

Example 2 In Example 15 in Chapter I we saw that the 3-periodic continued fraction

$$\mathbf{K}\frac{a_n}{1} = \cfrac{2}{1} \cfrac{1}{+\,1} \cfrac{-1}{+\,1} \cfrac{2}{+\,1} \cfrac{1}{+\,1} \cfrac{-1}{+\,1} \cfrac{2}{+\,1} + \cdots$$

converges generally to $f = 1/2$, but diverges in the classical sense since $S_{3n}(0) = 0$ for all n. We shall see that the tail sequence $\{t_n\}$ of $\mathbf{K}(a_n/1)$ with $t_0 = 0$ is 3-periodic, why this has any connection to the convergence behavior of $\mathbf{K}(a_n/1)$, and compute this tail sequence. We shall also see whether $\mathbf{K}(a_n/1)$ has any other 3-periodic tail sequences.

Let $t_0 = 0$. By (1.2.3) we have $t_n = S_n^{(0)^{-1}}(t_0) = S_n^{-1}(0)$ since $b_0 = 0$. Since $S_{3n}(0) = 0$ for all n, this means that $t_{3n} = 0$ for all n. From (1.2.2) with n replaced by $3n$, we find that t_{3n-1} has a value independent of n since $\{a_n\}$ is periodic and all $b_n = 1$. Similarly also t_{3n-2} has a value independent of n, so $\{t_n\}$ is 3-periodic.

We may say that $\{t_n\}$ reveals the "trouble" we have with the convergence of $f_n = S_n(0)$. For every tail sequence we have $b_0 + t_0 = S_n(t_n)$. So if $\{t_n\}$ is not a *right* tail sequence, we must stay away from t_n $(w_n \neq t_n)$ when we choose approximants $S_n(w_n)$ for our continued fraction, at least from some n on. Otherwise we destroy the convergence to f. Our $\{t_n\}$ is not the right tail sequence since $t_0 \neq f = 1/2$. And we did not stay away from $\{t_n\}$ when we chose our approximants $S_n(0)$. For every third index n we have $w_n = t_n = 0$.

With $t_0 = 0$ the terms of $\{t_n\}$ are

$$t_0 = t_{3n} = 0, \qquad t_1 = t_{3n+1} = s_1^{-1}(t_0) = -1 + \frac{2}{0} = \infty,$$

$$t_2 = t_{3n+2} = s_2^{-1}(t_1) = -1 + \frac{1}{\infty} = -1 \qquad \text{for } n = 1, 2, 3, \ldots.$$

The right tail sequence $\{f^{(n)}\}$ must also be periodic since every third tail of $\mathbf{K}(a_n/1)$ is identical to $\mathbf{K}(a_n/1)$ itself. It is given by

$$f^{(0)} \;=\; f^{(3n)} = \frac{1}{2},$$

$$f^{(1)} \;=\; f^{(3n+1)} = s_1^{-1}(f^{(0)}) = -1 + \frac{2}{1/2} = 3,$$

$$f^{(2)} \;=\; f^{(3n+2)} = s_2^{-1}(f^{(1)}) = -1 + \frac{1}{3} = -\frac{2}{3} \qquad \text{for } n = 1, 2, 3, \ldots.$$

These two sequences are the only 3-periodic tail sequences of $\mathbf{K}(a_n/1)$, since $\{t_n\}$ is a 3-periodic tail sequence if and only if $t_n = t_{n+3}$; that is, if and only if t_n is a solution of the equation

$$t_n = S_3^{(n)}(t_{n+3}) = S_3^{(n)}(t_n).$$

This is a quadratic equation which has at most two solutions.

$\diamond$

1.3   Some properties of linear fractional transformations

Both a continued fraction $b_0 + \mathbf{K}(a_n/b_n)$ and its tail sequences are closely tied to linear fractional transformations

$$t(w) = \frac{aw + b}{cw + d}, \quad ad - bc \neq 0. \tag{1.3.1}$$

(See for instance (1.1.4) and (1.2.2).) Such transformations $t(w)$ are bijective mappings of $\hat{\mathbf{C}}$ onto $\hat{\mathbf{C}}$ with very nice properties:

a) $t(w)$ maps (generalized) circles on the Riemann sphere $\hat{\mathbf{C}}$ onto (generalized) circles on $\hat{\mathbf{C}}$.

b) Let C be a (generalized) circle on $\hat{\mathbf{C}}$. Then t maps points symmetric with respect to C onto points symmetric with respect to $t(C)$. (The reflection property.)

c) The cross ratio is invariant under linear fractional transformations; that is, if w_1, w_2, w_3 and w_4 are four distinct points in $\hat{\mathbf{C}}$, then

$$\frac{(w_1 - w_2)(w_3 - w_4)}{(w_1 - w_3)(w_2 - w_4)} = \frac{(t(w_1) - t(w_2))(t(w_3) - t(w_4))}{(t(w_1) - t(w_3))(t(w_2) - t(w_4))}. \tag{1.3.2}$$

(If one of the points w_k or its image $t(w_k)$ is equal to infinity, then (1.3.2) has the standard meaning.)

Sometimes it is of advantage to use the chordal metric $d(w_1, w_2)$ as defined in Subsection 5.3 of Chapter I. The cross ratio is still invariant under t if we define it by the chordal metric:

$$\frac{d(w_1, w_2) \cdot d(w_3, w_4)}{d(w_1, w_3) \cdot d(w_2, w_4)} = \frac{d(t(w_1), t(w_2)) \cdot d(t(w_3), t(w_4))}{d(t(w_1), t(w_3)) \cdot d(t(w_2), t(w_4))} \tag{1.3.3}$$

for $w_1, w_2, w_3, w_4 \in \hat{\mathbf{C}}$ all distinct.

1.4 Speed of convergence. Truncation error bounds

Convergence properties are often important in applications of continued fractions. Not only the existence of a limit, but also how fast this limit is approached by the approximants f_n (or more generally $S_n(w_n)$). Hence it is important to have estimates for this speed of convergence. We distinguish between *a priori truncation error bounds*

$$|f - f_n| \le \lambda_n, \qquad (1.4.1)$$

where $\lambda_n > 0$ is a bound we can find in advance, before we start computing the approximants, and *a posteriori truncation error bounds*

$$|f - f_n| \le M_n |f_n - f_{n-1}|, \qquad (1.4.2)$$

where the bound $M_n |f_n - f_{n-1}|$ can be determined only after we have computed (at least) the approximants f_n and f_{n-1}. A priori bounds (1.4.1) can be used to determine, in advance, the index n we need in order to obtain a desired accuracy. This saves work in the sense that one only needs to compute f_n for this particular index. A posteriori bounds (1.4.2) work more like a stopping criterion. One computes approximants $f_1, f_2, f_3, \ldots$ until the right hand side of (1.4.2) is sufficiently small. Sometimes the a posteriori bounds are more accurate, so we can stop at a lower value of n than indicated by the a priori bound.

To find a posteriori truncation error bounds, and to compare speed of convergence of $\{S_n(w_n)\}$ for different sequences $\{w_n\}$, we shall use (1.3.2) with $t = S_n$. (This idea was presented by Thron [Thron89].) If we choose $w_1 = 0, w_2 = f_k^{(n)}, w_3 = \infty$, and $w_4 = -h_n$ (notation as in (1.1.2) and (1.2.6)), and we require that these four points in $\hat{C}$ are distinct, then (1.3.2) reduces to

$$-\frac{0 - f_k^{(n)}}{f_k^{(n)} - (-h_n)} = \frac{S_n(0) - S_n(f_k^{(n)})}{S_n(0) - S_n(\infty)} = \frac{f_n - f_{n+k}}{f_n - f_{n-1}}. \qquad (1.4.3)$$

Now, our four chosen points w_k are distinct if and only if the four points $S_n(w_k)$ for $k = 1, 2, 3, 4$ are distinct; i.e. if and only if

$$f_n, f_{n-1} \text{ and } f_{n+k} \text{ are distinct and finite.} \qquad (1.4.4)$$

Hence

$$f_{n+k} - f_n = -\frac{f_k^{(n)}}{f_k^{(n)} + h_n}(f_n - f_{n-1}) \quad \text{if (1.4.4) holds,} \qquad (1.4.5)$$

and thus

$$|f_{n+k} - f_n| \leq M_n |f_n - f_{n-1}|$$

if $f_k^{(n)}$ and h_n can be estimated properly to give a bound M_n. If $f_{n+k} \to f$ as $k \to \infty$ and M_n is independent of k, this gives us a posteriori truncation error bounds

$$|f - f_n| \leq M_n |f_n - f_{n-1}|. \qquad (1.4.6)$$

Example 3 We shall find a posteriori truncation error bounds for a continued fraction $\mathbf{K}(c_n/1)$ with all $|c_n| \leq g < 1/4$. (Tool: Worpitzky's theorem in Subsection *4.3* of Chapter I states that $\mathbf{K}(a_n/1)$ converges and has approximants $|f_n| < 1/2$ if all $|a_n| \leq 1/4$.)

All tails of $\mathbf{K}(c_n/1)$ satisfy Worpitzky's theorem, so $|f_n^{(N)}| < 1/2$ for all n and N. However, since $f_n^{(N)} = s_{N+1}(f_{n-1}^{(N+1)}) = c_{N+1}/(1 + f_{n-1}^{(N+1)})$, we really have

$$\left| f_n^{(N)} \right| = \frac{|c_{N+1}|}{\left| 1 + f_{n-1}^{(N+1)} \right|} < \frac{g}{1 - 1/2} = 2g$$

for all $N \geq 0$ and $n = 1, 2, 3, \ldots$. The critical tail sequence

$$h_n = 1 + \frac{c_n}{1 +} \frac{c_{n-1}}{1 +} \cdots + \frac{c_2}{1}$$

can be regarded as $1 +$ (approximant of continued fraction $\mathbf{K}(d_n/1)$ with all $|d_n| \leq g$). Hence, $|h_n| > 1 - 2g$ for $n = 1, 2, 3, \ldots$. Assume that f_n, f_{n-1} and f_{n+k} are distinct. (We shall see later that this is really so.) Then by combining the above we find from (1.4.5) that

$$|f_{n+k} - f_n| < \frac{2g}{1 - 4g}|f_n - f_{n-1}|,$$

and thus

$$|f - f_n| \leq \frac{2g}{1 - 4g}|f_n - f_{n-1}|.$$

This is a useful bound. To find the value of $K(0.2e^{in}/1)$ with an error less than 0.05, we compute approximants f_n for $n = 1, 2, 3, \ldots$ until

$$\frac{2 \cdot 0.2}{1 - 4 \cdot 0.2} |f_n - f_{n-1}| = 2|f_n - f_{n-1}| < 0.05$$

for some index n, and then we use $f \approx f_n$ for this n.

n	f_n	$2|f_n - f_{n-1}|$
1	$0.1081 + i0.1683$	
2	$0.1484 + i0.1541$	0.0856
3	$0.1567 + i0.1462$	0.0229

Hence we can use $f \approx f_3 \approx 0.16 + i0.15$. We shall return to the question of a priori truncation error bounds in Chapter III.

$\diamond$

Next we choose $w_1 = f_k^{(n)}, w_2 = w_n, w_3 = 0$ and $w_4 = -h_n$ in (1.3.2). If these are distinct; i.e. if $f_{n+k}, S_n(w_n)$ and f_n are distinct and finite, then

$$\frac{(f_k^{(n)} - w_n)(0 + h_n)}{(f_k^{(n)} - 0)(w_n + h_n)} = \frac{(f_k^{(n)} - w_n)h_n}{f_k^{(n)}(w_n + h_n)} = \frac{f_{n+k} - S_n(w_n)}{f_{n+k} - f_n}. \tag{1.4.7}$$

This formula is useful for comparing modified approximants $S_n(w_n)$ to the approximants $f_n = S_n(0)$. If $f_{n+k} \to f \neq \infty$ as $k \to \infty$, and $|h_n/f_k^{(n)}(w_n + h_n)|$ is bounded by some constant M, then we get in particular that

$$\left| \frac{f - S_n(w_n)}{f - f_n} \right| \leq M|f^{(n)} - w_n| \to 0 \quad \text{if} \quad (f^{(n)} - w_n) \to 0.$$

That is, $S_n(w_n)$ converges faster to f than f_n. This idea was used in Example 13 and 14 in Chapter I.

To compare various approximants $S_n(u_n), S_n(v_n)$ and $S_n(w_n)$ of $b_0 + K(a_n/b_n)$ one can choose the four points v_n, u_n, w_n and $-h_n$ in (1.3.2) (if these are distinct) to get

$$\frac{(v_n - u_n)(w_n + h_n)}{(v_n - w_n)(u_n + h_n)} = \frac{S_n(v_n) - S_n(u_n)}{S_n(v_n) - S_n(w_n)}. \tag{1.4.8}$$

For later use we record that the chordal metric version of (1.4.8) is

$$\frac{d(v_n, u_n)d(w_n, -h_n)}{d(v_n, w_n)d(u_n, -h_n)} = \sqrt{\frac{1 + |S_n(u_n)|^2}{1 + |S_n(w_n)|^2}} \cdot \frac{d(S_n(v_n), S_n(u_n))}{d(S_n(v_n), S_n(w_n))}. \quad (1.4.9)$$

But when are the approximants f_n, f_{n+k} distinct? Assume that $f_n = f_{n+k}$; i.e. that $S_n(0) = S_n(f_k^{(n)})$. Then we have $f_k^{(n)} = 0$. So, if we can ascertain that $f_k^{(n)} \neq 0$, then $f_n \neq f_{n+k}$. Since $f_1^{(n)} = a_{n+1}/b_{n+1} \neq 0$ we always have $f_n \neq f_{n+1}$. We shall return to this later.

1.5   *More about general convergence*

In Subsection *5.3* of Chapter I we introduced the concept of general convergence of continued fractions $b_0 + \mathbf{K}(a_n/b_n)$ due to Jacobsen [Jaco86]. (See (5.3.1)–(5.3.2) in Chapter I.) We also found that if $b_0 + \mathbf{K}(a_n/b_n)$ converges to a value f in the classical sense, then it converges generally to f, and we saw an example of a continued fraction which converges generally but not in the classical sense. (Example 15 in Chapter I.) But what about the following questions?

A. Why do we require a common limit for *two* sequences of modifying factors $\{v_n\}$ and $\{w_n\}$ in the definition of general convergence?

B. If $b_0 + \mathbf{K}(a_n/b_n)$ converges generally to f, for which sequences $\{u_n\}$ will $\lim S_n(u_n) = f$?

C. Is the value f of a generally convergent continued fraction really unique?

We shall look at some answers.

A. Assume that we have the following information about a continued fraction $b_0 + \mathbf{K}(a_n/b_n)$: $\lim S_n(w_n) = f$ for some sequence $\{w_n\}$ from $\hat{\mathbf{C}}$. What can we then say about $b_0 + \mathbf{K}(a_n/b_n)$? If we do not have any additional information we can say nothing! In fact, let $b_0 + \mathbf{K}(a_n/b_n)$ be an arbitrarily chosen continued fraction, and let $\{q_n\}$ be an arbitrary sequence of numbers from $\hat{\mathbf{C}}$, converging to f. Then the choice $w_n = S_n^{-1}(q_n)$ for all n gives the approximants $S_n(w_n) = q_n \to f$ for $b_0 + \mathbf{K}(a_n/b_n)$.

Hence, it would not suffice to require convergence of just *one* sequence $\{S_n(w_n)\}$ in the definition. We need more. Common limit for *two* sequences is one way of doing this. Another possibility is demonstrated in Theorem 4 to come.

B. Let $b_0 + \mathbf{K}(a_n/b_n)$ converge generally to f. When will $S_n(u_n) \to f$? It is easy to see that if $\{u_n\}$ is a tail sequence for $b_0 + \mathbf{K}(a_n/b_n)$ with $u_0 \neq f - b_0$, then $S_n(u_n) = b_0 + u_0 \neq f$. A deeper result is that $\lim S_n(u_n) = f$ if $\{u_n\}$ "stays far enough away asymptotically" from *one* such tail sequence:

Theorem 4 *The continued fraction* $b_0 + \mathbf{K}(a_n/b_n)$ *converges generally to* f *if and only if* $\lim S_n(u_n) = f$ *for every sequence* $\{u_n\}$ *from* $\hat{\mathbf{C}}$ *such that*

$$
\begin{aligned}
&\liminf_{n \to \infty} d(u_n, -h_n) > 0 && \text{if } f \neq \infty, \\
&\liminf_{n \to \infty} d(u_n, -A_n/A_{n-1}) > 0 && \text{if } f = \infty.
\end{aligned} \tag{1.5.1}
$$

Proof : The if-part follows from the definition of general convergence. To prove the only if-part we let

$$
\lim S_n(v_n) = \lim S_n(w_n) = f \tag{1.5.2}
$$

where

$$
\liminf d(v_n, w_n) > 0; \tag{1.5.3}
$$

i.e. $b_0 + \mathbf{K}(a_n/b_n)$ converges generally to f. Let $\{u_n\}$ satisfy (1.5.1), and assume first that $f \neq \infty$. We then know that from some n on, say $n \geq N$, we have $v_n \neq w_n$, $u_n \neq -h_n$, $S_n(v_n) \neq \infty$ and $S_n(w_n) \neq \infty$, where the two last statements are equivalent to $v_n \neq -h_n$ and $w_n \neq -h_n$. It suffices to prove that if $\{n_k\}_{k=1}^{\infty}$ is the subsequence of $\mathbf{N}$ where $n \geq N$ and $u_n \neq v_n$, $u_n \neq w_n$ for all $n = n_k$, then

$$
\lim_{k \to \infty} S_{n_k}(u_{n_k}) = f. \tag{1.5.4}
$$

For $n = n_k$ we can use (1.4.9). The left side of (1.4.9) stays bounded as $k \to \infty$. Since $d(S_n(v_n), S_n(w_n))$ in the denominator of the right side of (1.4.9) approaches 0 as $k \to \infty$ ($n = n_k$) and $S_n(w_n) \to f \neq \infty$, we therefore need that $d(S_n(u_n), S_n(v_n))$ in the numerator also approaches 0 when $n = n_k$ and $k \to \infty$. This proves (1.5.4).

The case $f = \infty$ is not much different. $\{-h_n\}$ is no longer dangerous, but $-A_n/A_{n-1} = S_n^{-1}(0)$ is!   ■

C. The uniqueness of f is a simple corollary of Theorem 4:

Corollary 5 *Let $b_0 + \mathbf{K}(a_n/b_n)$ converge generally to f and to g. Then $f = g$.*

Proof :   Let $\{v_n\}$ and $\{w_n\}$ be such that (1.5.2) and (1.5.3) hold, and let

$$\lim_{n \to \infty} S_n(p_n) = \lim_{n \to \infty} S_n(q_n) = g \qquad (1.5.5)$$

where

$$\liminf d(p_n, q_n) > 0 .$$

Assume first that $f \neq \infty$. For each n define

$$u_n = \begin{cases} p_n & \text{if } d(p_n, -h_n) \geq d(q_n, -h_n), \\ q_n & \text{otherwise .} \end{cases}$$

Then (1.5.1) holds, and thus $S_n(u_n) \to f$. On the other hand $S_n(u_n) \to g$ by (1.5.5). Hence $f = g$.

If $f = \infty$ we repeat the argument with h_n replaced by A_n/A_{n-1}.   ■

Thus having answered our three questions A, B and C, a new question springs to the mind: How can it be that we only have to stay "sufficiently far away" from one particular tail sequence when all tail sequences $\{t_n\}$ with $t_0 \neq f - b_0$ are dangerous choices for w_n? There can only be one answer to this:

Theorem 6 *Let $b_0 + \mathbf{K}(a_n/b_n)$ converge generally to f, and let $\{t_n\}$ and $\{\tilde{t}_n\}$ be tail sequences for $b_0 + \mathbf{K}(a_n/b_n)$ with $t_0 \neq f - b_0$ and $\tilde{t}_0 \neq f - b_0$. Then*

$$\lim d(t_n, \tilde{t}_n) = 0 \,. \tag{1.5.6}$$

2 Transformations of continued fractions

2.1 Generating a continued fraction from a sequence

In the previous chapter we saw examples of how a continued fraction $b_0 + \mathbf{K}(a_n/b_n)$ can be derived to represent a number f or a function $f(z)$. The hope was that the continued fraction would converge to f, i.e. that its sequence of approximants would converge to f. In fact, our interest was in the behaviour of the approximant sequence, not in the continued fraction itself. The continued fraction was just an intermediate step, as it is for any such limiting process.

Still we prefer to study the continued fractions because we have something to gain by doing so. For instance, the convergence criteria for $\{f_n\}$ in Chapter I were all based on the elements of the continued fraction. And in Example 13 and 14 in Chapter I, these elements helped us to choose favorable approximants.

Indeed, one might ask the question: Given a sequence $\{f_n\}$, which continued fraction $b_0 + \mathbf{K}(a_n/b_n)$ has this as an approximant sequence, if such one exists at all?

Theorem 7 *The sequences $\{A_n\}_{n=-1}^{\infty}$ and $\{B_n\}_{n=-1}^{\infty}$ of complex numbers are the canonical numerators and denominators of some continued fraction $b_0 + \mathbf{K}(a_n/b_n)$ if and only if*

$$A_{-1} = B_0 = 1\,, \quad B_{-1} = 0\,, \quad \Delta_n = A_n B_{n-1} - B_n A_{n-1} \neq 0 \tag{2.1.1}$$

for all $n \in \mathbf{N}$. If (2.1.1) holds, then $b_0 + \mathbf{K}(a_n/b_n)$ is uniquely determined by

$$b_0 = A_0, \qquad b_1 = B_1, \qquad a_1 = A_1 - A_0 B_1,$$

$$a_n = -\frac{\Delta_n}{\Delta_{n-1}}, \qquad b_n = \frac{A_n B_{n-2} - B_n A_{n-2}}{\Delta_{n-1}} \quad \text{for } n \geq 2. \quad (2.1.2)$$

Proof : If $b_0 + \mathbf{K}(a_n/b_n)$ is given, then $(2.1.1)$ holds by the determinant formula $(1.2.10)$ and the initial conditions $(1.2.7)$ in Chapter I. If $\{A_n\}$ and $\{B_n\}$ are given, satisfying $(2.1.1)$, then a_n and b_n are solutions of the system

$$\begin{aligned} b_n A_{n-1} + a_n A_{n-2} &= A_n, \\ b_n B_{n-1} + a_n B_{n-2} &= B_n \end{aligned} \qquad (2.1.3)$$

of linear equations. The determinant of this system is $\Delta_{n-1} \neq 0$. Hence the solution $(2.1.2)$ is unique. ∎

Example 4 We shall find the continued fraction $b_0 + \mathbf{K}(a_n/b_n)$ which has $A_n = n^2$ and $B_n = n^2 + 1$ for $n = 0, 1, 2, \ldots$.

Using Theorem 7 we get

$$\Delta_n = A_n B_{n-1} - B_n A_{n-1} = 2n - 1$$

and

$$A_n B_{n-2} - B_n A_{n-2} = 4n - 4,$$

which means that

$$b_0 = 0, \quad b_1 = 2, \quad a_1 = 1,$$

$$a_n = -\frac{2n-1}{2n-3}, \quad b_n = \frac{4n-4}{2n-3} \quad \text{for } n = 2, 3, 4, \ldots.$$

Hence, the continued fraction

$$\frac{1}{2+} \; \frac{-3/1}{4/1 +} \; \frac{-5/3}{8/3 +} \; \frac{-7/5}{12/5 +} \; \frac{-9/7}{16/7 +} \; \frac{-11/9}{20/9 +} \cdots$$

has canonical approximants $n^2/(n^2 + 1)$, and converges to 1.

◇

If we only have given the sequence $\{f_n\}_{n=0}^{\infty}$ of approximants, the continued fraction $b_0 + \mathbf{K}(a_n/b_n)$ is no longer unique. One way to use Theorem 7 is then to choose

$$
\begin{array}{llll}
B_n = 1, & A_n = f_n & \text{if } f_n \neq \infty, \\
B_n = 0, & A_n = 1 & \text{if } f_n = \infty
\end{array}
\tag{2.1.4}
$$

for $n = 0, 1, 2, \ldots$. To emphasize that $\{f_n\}$ shall be approximants of the type $f_n = S_n(0)$, we shall call them *classical approximants* in contrast to the (modified) approximants $S_n(w_n)$.

Corollary 8 *The sequence $\{f_n\}_{n=0}^{\infty}$ from $\hat{\mathbf{C}}$ is a sequence of classical approximants for some continued fraction $b_0 + \mathbf{K}(a_n/b_n)$ if and only if*

$$
f_0 \neq \infty \quad \text{and} \quad f_n \neq f_{n-1} \quad \text{for } n = 1, 2, 3, \ldots.
\tag{2.1.5}
$$

Proof : Let $b_0 + \mathbf{K}(a_n/b_n)$ be given. Then, by (2.1.1)

$$
f_n - f_{n-1} = \frac{A_n}{B_n} - \frac{A_{n-1}}{B_{n-1}} = \frac{\Delta_n}{B_n B_{n-1}} \neq 0.
\tag{2.1.6}
$$

(This holds also if B_n or B_{n-1} are equal to 0, since two consecutive B_n's can not be equal to 0 by the difference equations (2.1.3).)

If $\{f_n\}$ is given, we define $\{A_n\}$ and $\{B_n\}$ by (2.1.4), and the result follows from Theorem 7. ∎

Example 5 We shall find a continued fraction $b_0 + \mathbf{K}(a_n/b_n)$ which has classical approximants $f_n = n^2/(n^2 + 1)$ for $n = 0, 1, 2, \ldots$

With the choice (2.1.4) for A_n and B_n we find from Theorem 7 that $b_0 + \mathbf{K}(a_n/b_n)$ has classical approximants f_n if $b_0 = f_0 = 0, b_1 = 1, a_1 = f_1 - f_0 = 1/2$ and

$$
\begin{aligned}
a_n &= -\frac{f_n - f_{n-1}}{f_{n-1} - f_{n-2}} = -\frac{(2n-1)(n^2 - 4n + 5)}{(2n-3)(n^2 + 1)} \quad \text{for } n = 2, 3, 4, \ldots, \\
b_n &= \frac{f_n - f_{n-2}}{f_{n-1} - f_{n-2}} = \frac{(4n-4)(n^2 - 2n + 2)}{(2n-3)(n^2 + 1)} \quad \text{for } n = 2, 3, 4, \ldots.
\end{aligned}
$$

◇

Example 6 The infinite product $\prod_{k=0}^{\infty} \rho_k$ has partial products $f_n = \prod_{k=0}^{n} \rho_k$. Let all $\rho_k \neq 1, 0, \infty$. We shall find a continued fraction $b_0 + \mathbf{K}(a_n/b_n)$ which has classical approximants $\{f_n\}$.

With the choice (2.1.4) for A_n and B_n we find by Theorem 7 that $b_0 = f_0 = \rho_0, b_1 = 1, a_1 = f_1 - f_0 = \rho_0(\rho_1 - 1)$ and

$$a_n = -\frac{f_n - f_{n-1}}{f_{n-1} - f_{n-2}} = -\frac{\rho_{n-1}(\rho_n - 1)}{\rho_{n-1} - 1} \quad \text{for } n = 2, 3, 4, \dots,$$

$$b_n = \frac{f_n - f_{n-2}}{f_{n-1} - f_{n-2}} = \frac{\rho_n \rho_{n-1} - 1}{\rho_{n-1} - 1} \quad \text{for } n = 2, 3, 4, \dots.$$

Hence the continued fraction

$$\rho_0 + \frac{\rho_0(\rho_1 - 1)}{1} \ \frac{\rho_1(\rho_2 - 1)/(\rho_1 - 1)}{-(\rho_1\rho_2 - 1)/(\rho_1 - 1)} \ \frac{\rho_2(\rho_3 - 1)/(\rho_2 - 1)}{-(\rho_2\rho_3 - 1)/(\rho_2 - 1)} -\cdots$$

has classical approximants $f_n = \prod_{k=0}^{n} \rho_k$, and all its canonical denominators B_n are equal to 1.

$\diamond$

2.2   Equivalence transformations

Definition   *We say that two continued fractions are equivalent if they have the same sequence of classical approximants.*

We write $b_0 + \mathbf{K}(a_n/b_n) \approx d_0 + \mathbf{K}(c_n/d_n)$ to express that $b_0 + \mathbf{K}(a_n/b_n)$ and $d_0 + \mathbf{K}(c_n/d_n)$ are equivalent. Let the canonical numerators and denominators be denoted by A_n and B_n for $b_0 + \mathbf{K}(a_n/b_n)$ and by C_n and D_n for $d_0 + \mathbf{K}(c_n/d_n)$. If we require that all $C_n = A_n$ and $D_n = B_n$, then it follows from Theorem 7 that the two continued fractions are identical; that is, $c_n = a_n$ and $d_n = b_n$ for all n. So that has no point. We have required too much. What we can do, and shall do, is to require that $A_n/B_n = C_n/D_n$ for all n.

The idea of equivalent continued fractions is due to Seidel [Seid55] who also proved:

Theorem 9 $b_0 + \mathbf{K}(a_n/b_n) \approx d_0 + \mathbf{K}(c_n/d_n)$ *if and only if there exists a sequence $\{r_n\}$ of complex numbers with $r_0 = 1, r_n \neq 0$ for all $n \in \mathbf{N}$, such that*

$$d_0 = b_0, \qquad c_n = r_n r_{n-1} a_n, \qquad d_n = r_n b_n \quad \text{for all } n \in \mathbf{N}. \quad (2.2.1)$$

Proof : Let A_n, B_n be the canonical numerators and denominators of $b_0 + \mathbf{K}(a_n/b_n)$. Then $b_0 + \mathbf{K}(a_n/b_n) \approx d_0 + \mathbf{K}(c_n/d_n)$ if and only if there exist numbers $r_n \neq 0$ such that the canonical numerators C_n and denominators D_n of $d_0 + \mathbf{K}(c_n/d_n)$ can be written

$$C_{-1} = 1, \quad D_{-1} = 0, \quad C_n = A_n \prod_{k=0}^{n} r_k, \quad D_n = B_n \prod_{k=0}^{n} r_k \quad (2.2.2)$$

for all n. Since $D_0 = B_0 = 1$ we need $r_0 = 1$. From Theorem 7 it follows then that $d_0 + \mathbf{K}(c_n/d_n)$ is given by (2.2.1). ■

Remarks:

1. The concept of equivalence is tied to the classical approximants. If $b_0 + \mathbf{K}(a_n/b_n) \approx d_0 + \mathbf{K}(c_n/d_n)$ by the relations (2.2.1), then

$$S_n(w_n) = T_n(r_n w_n) \quad \text{for } n = 0, 1, 2, \dots, \quad (2.2.3)$$

   where $S_n(w)$ are approximants of $b_0 + \mathbf{K}(a_n/b_n)$, and $T_n(w)$ are approximants of $d_0 + \mathbf{K}(c_n/d_n)$.

2. If $\{t_n\}$ is a tail sequence for $b_0 + \mathbf{K}(a_n/b_n)$, then $\{t_n r_n\}$ is a tail sequence for $d_0 + \mathbf{K}(c_n/d_n)$, where $b_0 + \mathbf{K}(a_n/b_n)$ and $d_0 + \mathbf{K}(c_n/d_n)$ are as in Theorem 9.

Example 7 The continued fractions in Example 4 and 5 in the previous subsection are equivalent since they have the same sequence of approximants. To derive the one in Example 4 from the one in Example 5 we use

$$r_0 = 1, \qquad r_1 = 2, \qquad r_n = \frac{n^2 + 1}{(n-1)^2 + 1} \quad \text{for } n = 2, 3, 4, \dots.$$

An even simpler equivalent continued fraction can be obtained from the one in Example 4 by using $r_0 = 1, r_1 = 1, r_n = 2n - 3$ for $n = 2, 3, 4, \ldots$. We get

$$\frac{1}{2+} \; \frac{-3}{4+} \; \frac{-5 \cdot 1}{8} \; \frac{-7 \cdot 3}{+ \; 12} \; \frac{-9 \cdot 5}{+ \; 16} \; \frac{-11 \cdot 7}{+ \; 20} \; + \cdots$$

which therefore also has approximants $n^2/(n^2 + 1) \to 1$.

$\diamond$

Example 8 The continued fraction in Example 6 also has a simpler, equivalent form. The choice $r_0 = 1$, $r_1 = 1$, $r_n = \rho_{n-1} - 1$ for $n \geq 2$ leads to

$$\rho_0 + \frac{\rho_0(\rho_1 - 1)}{1} \; \frac{\rho_1(\rho_2 - 1)}{- \; \rho_1\rho_2 - 1 \;-} \; \frac{\rho_2(\rho_3 - 1)(\rho_1 - 1)}{\rho_2\rho_3 - 1} \; \frac{\rho_3(\rho_4 - 1)(\rho_2 - 1)}{- \; \rho_3\rho_4 - 1} \; - \cdots$$

which therefore also has approximants $f_n = \prod_{k=0}^{n} \rho_k$.

$\diamond$

Example 9 We shall prove that $\mathbf{K}(n^2/3n)$ converges.

An equivalence transformation with $r_n = 1/n$ for $n = 1, 2, 3, \ldots$ brings $\mathbf{K}(n^2/3n)$ over to the form

$$\mathbf{K} \frac{n^2}{3n} \approx \frac{1}{3+} \; \frac{2/1}{3} \; \frac{3/2}{+ \; 3} \; \frac{4/3}{+ \; 3} \; \frac{5/4}{+ \; 3} \; \frac{6/5}{+ \; 3} \; + \cdots$$

which converges by the Śleszyński-Pringsheim theorem, Theorem 1 in Chapter I.

$\diamond$

The following two equivalence transformations are of particular interest:

Corollary 10

A. $b_0 + \mathbf{K}(a_n/b_n) \approx b_0 + \mathbf{K}(1/d_n)$ *where*

$$d_n = b_n \prod_{k=1}^{n} a_k^{(-1)^{n+1-k}} \qquad \text{for } n = 1, 2, 3, \ldots. \tag{2.2.4}$$

B. *If $b_n \neq 0$ for all $n \geq 1$, then $b_0 + \mathbf{K}(a_n/b_n) \approx b_0 + \mathbf{K}(c_n/1)$, where*

$$c_1 = \frac{a_1}{b_1}, \qquad c_n = \frac{a_n}{b_n b_{n-1}} \quad \textit{for } n = 2, 3, 4, \ldots . \qquad (2.2.5)$$

Remarks:

1. The transformation in A can always be performed. The elements d_n have the structure

$$d_1 = b_1 \cdot \frac{1}{a_1}, \quad d_2 = b_2 \frac{a_1}{a_2}, \quad d_3 = b_3 \frac{a_2}{a_1 a_3}, \quad d_4 = b_4 \frac{a_1 a_3}{a_2 a_4}, \ldots .$$

2. The transformation in B can only be applied if all $b_n \neq 0$, since otherwise c_n would not be a well defined complex number. Combined with Worpitzky's theorem in Subsection 4.3 in Chapter I, it shows for instance that every continued fraction $b_0 + \mathbf{K}(a_n/b_n)$ with $|a_1/b_1| \leq 1/4$ and $|a_n/b_n b_{n-1}| \leq 1/4$ for all $n \geq 2$ converges to a finite value.

Proof :
A: Use Theorem 9 with

$$r_n = \prod_{k=1}^{n} a_k^{(-1)^{n-k+1}} \qquad \text{for all } n \geq 1 .$$

B: Use Theorem 9 with $r_n = 1/b_n$ for all $n \geq 1$. ∎

Example 10 We shall see that the continued fraction $\mathbf{K}((30+(0.9)^n)/1)$ in Example 13 of Chapter I converges.

By Corollary 10A it follows that $\mathbf{K}((30 + (0.9)^n)/1) \approx \mathbf{K}(1/d_n)$ where

$$
\begin{aligned}
d_1 &= \frac{1}{30 + 0.9} > \frac{1}{31} \\[2mm]
d_2 &= \frac{30 + 0.9}{30 + (0.9)^2} > 1 \\[2mm]
d_3 &= \frac{30 + (0.9)^2}{(30 + 0.9)(30 + (0.9)^3)} > \frac{1}{31} \\[2mm]
d_4 &= \frac{(30 + 0.9)(30 + (0.9)^3)}{(30 + (0.9)^2)(30 + (0.9)^4)} > 1, \quad \text{etc.}
\end{aligned}
$$

In fact, we find $d_{2n} > 1$ and $d_{2n+1} > 1/31$ for all n, so $\sum d_n = \infty$. Hence the convergence follows by Van Vleck's theorem, Theorem 2 in Chapter I. We also get that the value of the continued fraction is finite.

$\diamond$

2.3   The Bauer-Muir transformation

Definition   *The Bauer-Muir transform of a continued fraction $b_0 + \mathbf{K}(a_n/b_n)$ with respect to a sequence $\{w_n\}$ from $\mathbf{C}$ is the continued fraction $d_0 + \mathbf{K}(c_n/d_n)$ whose canonical numerators C_n and denominators D_n are given by*

$$
\begin{aligned}
C_{-1} &= 1, & D_{-1} &= 0, \\
C_n &= A_n + A_{n-1}w_n, & D_n &= B_n + B_{n-1}w_n
\end{aligned} \tag{2.3.1}
$$

for $n = 0, 1, 2, \ldots$, where $\{A_n\}$ and $\{B_n\}$ are the canonical numerators and denominators of $b_0 + \mathbf{K}(a_n/b_n)$.

This transformation dates back to the 1870's [Bauer72], [Muir77]. What the Bauer-Muir transformation does, is to give a continued fraction $d_0 + \mathbf{K}(c_n/d_n)$ whose classical approximants $T_n(0)$ are equal to the modified approximants $S_n(w_n)$ of $b_0 + \mathbf{K}(a_n/b_n)$. With this notation we have:

Theorem 11   *The Bauer-Muir transform of $b_0 + \mathbf{K}(a_n/b_n)$ with respect to $\{w_n\}$ from $\mathbf{C}$ exists if and only if*

$$
\lambda_n = a_n - w_{n-1}(b_n + w_n) \neq 0 \quad \text{for } n = 1, 2, 3, \ldots . \tag{2.3.2}
$$

If it exists, then it is given by

$$b_0 + w_0 + \cfrac{\lambda_1}{b_1 + w_1} \; \cfrac{c_2}{+ \, d_2} \; \cfrac{c_3}{+ \, d_3 +} \cdots \qquad (2.3.3)$$

where

$$c_n = a_{n-1}q_{n-1}, \quad d_n = b_n + w_n - w_{n-2}q_{n-1}, \quad q_n = \lambda_{n+1}/\lambda_n. \quad (2.3.4)$$

Proof : Let $\{C_n\}$ and $\{D_n\}$ be given by (2.3.1). Then $\{C_n\}$ and $\{D_n\}$ are canonical numerators and denominators of a continued fraction $d_0 + \mathbf{K}(c_n/d_n)$ if and only if $C_{-1} = D_0 = 1, D_{-1} = 0$ and $\Delta_n = C_n D_{n-1} - D_n C_{n-1} \neq 0$ for all $n \geq 1$. (See Theorem 7.) The initial conditions for C_n and D_n are satisfied. Moreover

$$
\begin{aligned}
\Delta_n &= C_n D_{n-1} - D_n C_{n-1} \\
&= (A_n + A_{n-1}w_n)(B_{n-1} + B_{n-2}w_{n-1}) \\
&\quad - (B_n + B_{n-1}w_n)(A_{n-1} + A_{n-2}w_{n-1}) \\
&= (b_n A_{n-1} + a_n A_{n-2} + A_{n-1}w_n)(B_{n-1} + B_{n-2}w_{n-1}) \\
&\quad - (b_n B_{n-1} + a_n B_{n-2} + B_{n-1}w_n)(A_{n-1} + A_{n-2}w_{n-1}) \\
&= (A_{n-2}B_{n-1} - A_{n-1}B_{n-2})(a_n - w_{n-1}b_n - w_{n-1}w_n)
\end{aligned}
$$

where the first factor is different from 0 by the determinant formula (1.2.10) in Chapter I and the second factor is equal to λ_n in (2.3.2). This proves the existence part of Theorem 11. The elements of $d_0 + \mathbf{K}(c_n/d_n)$ follows now from (2.1.2). ∎

We shall see examples of three different applications of the Bauer-Muir transformation.

Example 11 We shall see later (Theorem 28 in Chapter III) that the limit periodic continued fraction $\mathbf{K}((30 + (0.9)^n)/1)$ has critical tail sequence $h_n \to 6$ and right tail sequence $f^{(n)} \to 5$ as $n \to \infty$. Using this here, we shall prove that $S_n(5)$ converges faster to the value f of $\mathbf{K}((30 + (0.9)^n)/1)$ than $S_n(0)$, in the sense that

$$(f - S_n(5))/(f - S_n(0)) \to 0 \quad \text{as } n \to \infty, \qquad (2.3.5)$$

and find a Bauer-Muir transform of $\mathbf{K}((30 + (0.9)^n)/1)$ with respect to $w_n = 5$.

We proved in Example 10 that $\mathbf{K}((30 + (0.9)^n)/1)$ converges to a finite value f. Indeed, by the same type of argument we find that all $f_k^{(n)} \neq \infty$ and all $f^{(n)} \neq \infty$ for this continued fraction. Since

$$f_k^{(n)} = \frac{a_{n+1}}{b_{n+1} + f_{k-1}^{(n+1)}} \quad \text{and} \quad f^{(n)} = \frac{a_{n+1}}{b_{n+1} + f^{(n+1)}},$$

we therefore also have that all $f_k^{(n)} \neq 0$, $f^{(n)} \neq 0$. This means that all approximants f_n, f_{n+k} are distinct and finite (see remark at the end of Subsection *1.4*), and (2.3.5) follows from (1.4.7). (This is consistent with our observations in Example 13 of Chapter I.)

From Theorem 11 we get

$$\lambda_n = 30 + (0.9)^n - 5(5 + 1) = (0.9)^n,$$

which means that the Bauer-Muir transform is

$$
\begin{aligned}
&d_0 + \mathbf{K}(c_n/d_n) \\
&= \; 5 + \frac{0.9}{6 +} \; \frac{(30 + 0.9)0.9}{6 - 0.9 \cdot 5 +} \; \frac{(30 + (0.9)^2)0.9}{6 - 0.9 \cdot 5} + \cdots \\
&= \; 5 + \frac{0.9}{6 +} \; \frac{27 + (0.9)^2}{1.5 +} \; \frac{27 + (0.9)^3}{1.5 +} \; \frac{27 + (0.9)^4}{1.5} + \cdots \quad (2.3.6)
\end{aligned}
$$

Idea: This continued fraction has the same structure as $\mathbf{K}((30+(0.9)^n)/1)$, and for the same reasons we know that the modified approximants $T_n(w)$ of $d_0 + \mathbf{K}(c_n/d_n)$ converge faster to f than $T_n(0)$ if w is the positive root of the quadratic equation

$$w = \frac{27}{1.5 + w} \quad \text{i.e. } w = 4.5 = 5 \cdot 0.9.$$

Hence, let us replace the first tail of $d_0 + \mathbf{K}(c_n/d_n)$ by its Bauer-Muir transform with respect to $w_n = 4.5$. This time we get

$$\lambda_n = 27 + (0.9)^{n+1} - 4.5(1.5 + 4.5) = (0.9)^{n+1},$$

so the result is

$$5 + \cfrac{0.9}{6 + 4.5+} \cfrac{(0.9)^2}{1.5 + 4.5+} \cfrac{(27 + (0.9)^2) \cdot 0.9}{1.5 + 4.5 - 0.9 \cdot 4.5+} \cfrac{(27 + (0.9)^3) \cdot 0.9}{1.5 + 4.5 - 0.9 \cdot 4.5+} \cdots$$

$$= 5 + \cfrac{0.9}{10.5+} \cfrac{(0.9)^2}{6 +} \cfrac{24.3 + (0.9)^3}{1.95 +} \cfrac{24.3 + (0.9)^4}{1.95 +} \cfrac{24.3 + (0.9)^5}{1.95 +} \cdots$$

$$(2.3.7)$$

Again we can repeat the process, this time with $w_n = 4.5 \cdot 0.9 = 4.05$ and so on. Each time we get a continued fraction converging faster than the previous one. It can be proved that this leads to the continued fraction

$$5 + \cfrac{0.9}{10.5+} \cfrac{(0.9)^2}{10.05 +} \cfrac{(0.9)^3}{9.645 +} \cfrac{(0.9)^4}{9.2805+} \cfrac{(0.9)^5}{\tilde{b}_5} \cdots \qquad (2.3.8)$$

where

$$\tilde{b}_n = 6 + 5 \cdot (0.9)^n \quad \text{for } n \geq 1.$$

The table below gives the first classical approximants for the given continued fraction and for the continued fractions (2.3.6), (2.3.7) and (2.3.8).

n	$K((30 + (0.9)^n)/1)$	(2.3.6)	(2.3.7)	(2.3.8)
1	30.90000	5.15000	5.08571	5.08571
2	0.97139	5.03667	5.08463	5.08506
3	1.56770	5.12176	5.08536	5.08507
4	1.85765	5.05762	5.08486	5.08507
⋮	⋮	⋮	⋮	⋮
16	4.59286	5.08418	5.08506	
17	5.53149	5.08573	5.08507	
18	4.73830	5.08457	5.08507	
⋮	⋮	⋮	⋮	
38	5.07571	5.08506		
39	5.09288	5.08507		
40	5.07857	5.08507		
⋮	⋮	⋮		
86	5.08506			
87	5.08507			
88	5.08507			

In every column we have stopped when the computed value has reached the accuracy of the table. (To determine this accuracy, we have used a theorem for continued fractions with positive elements which will be proved in Chapter III, Theorem 2.)

Let it finally be mentioned, that the continued fraction (2.3.8) is even better than the table shows. With more figures in the approximants of order 3 and 4 we actually find for the value f:

$$f_4 \approx 5.085066164 < f < 5.085066199 \approx f_3 .$$

—————————————————————————————————◇

Example 12 One can prove that the approximants $S_n(-6)$ of $\mathbf{K}((30 + (0.5)^n)/1)$ converge, but not to the value f of the continued fraction. But if we try to compute $S_n(-6)$ from the continued fraction, we have a problem. Small inaccuracies in the input or computation will have the effect that our computed sequence still converges to f. The computation of $S_n(-6)$ in this way is unstable. (See the table.) How can we find a more stable method to compute $S_n(-6)$? This problem is important for (for instance) analytic continuation by the method to be described in Chapter III.

We find the Bauer-Muir transform of $\mathbf{K}((30 + (0.5)^n)/1)$ with respect to $w_n = -6$. Since

$$\lambda_n = 30 + (0.5)^n - (-6)(1 - 6) = (0.5)^n ,$$

this transform $d_0 + \mathbf{K}(c_n/d_n)$ is given by

$$-6 + \frac{0.5}{-5+} \frac{(30 + 0.5) \cdot 0.5}{-5 - (-6) \cdot 0.5+} \frac{(30 + (0.5)^2) \cdot 0.5}{-5 - (-6) \cdot 0.5} +\cdots$$

$$= -6 + \frac{0.5}{-5+} \frac{15 + (0.5)^2}{-2} + \frac{15 + (0.5)^3}{-2} +\cdots ,$$

and its classical approximants $T_n(0)$ (which can be computed stably) are exactly $S_n(-6)$.

We observe that the correct "table-value", -6.06220, of $\lim S_n(-6)$ is taken on for $T_n(0)$ for all $n \geq 19$.

n	$S_n(-6)$	$T_n(0)$
1	-6.09999	-6.03960
2	-6.03962	-6.07582
3	-6.07580	-6.05404
⋮	⋮	⋮
12	-6.06215	-6.06228
13	-6.06218	-6.06215
14	-6.06229	-6.06223
15	-6.06208	-6.06218
⋮	⋮	⋮
18	-6.06247	-6.06221
19	-6.06189	-6.06220
20	-6.06257	-6.06220
⋮	⋮	⋮
161	5.05858	-6.06220
162	5.05859	-6.06220
163	5.05859	-6.06220

Observe also, that the computed values of $S_n(-6)$ for $n = 13$ and 14 are pretty close to the value -6.06220. By increasing n, however, the values "take off" and in the long run ($n \geq 162$) approach the value 5.05859 of $\lim S_n(0)$.

$\diamond$

Example 13 The continued fraction $\mathbf{K}((z + n)/n)$ is equivalent to $\mathbf{K}(a_n(z)/1)$ where

$$a_n(z) = \frac{z + n}{n(n - 1)} \to 0 \quad \text{as } n \to \infty \,.$$

Hence, $\mathbf{K}((z + n)/n)$ converges to some value $f(z)$ for all $z \in \mathbf{C}$ by the argument of Example 1. What does this function $f(z)$ look like? What is the value of $\mathbf{K}((z + n)/n)$?

Let us assume that the modified approximants $S_n(1)$ of $\mathbf{K}((z + n)/n)$ also converge to the same value $f(z)$. (It is possible to prove this, as we shall see later. It is for instance a consequence of the parabola theorem, Theorem 20 in Chapter III.) Then its Bauer-Muir transform $d_0 + \mathbf{K}(c_n/d_n)$ with respect to $w_n = 1$ must also converge to $f(z)$. To determine $d_0 + \mathbf{K}(c_n/d_n)$ we find that

$$\lambda_n = z + n - 1(n+1) = z - 1 \quad \text{for all } n.$$

Hence

$$d_0 + \mathbf{K}\,\frac{c_n}{d_n} = 1 + \frac{z-1}{2} + \frac{z+1}{2} + \frac{z+2}{3} + \frac{z+3}{4} + \cdots,$$

which looks similar to $\mathbf{K}((z+n)/n)$. Indeed, the first tail $g^{(1)}(z)$ of $d_0 + \mathbf{K}(c_n/d_n)$ is such that

$$\frac{z}{1+g^{(1)}(z)} = \frac{z}{1+} \frac{z+1}{2} + \frac{z+2}{3} + \cdots = \mathbf{K}\,\frac{z-1+n}{n} = f(z-1).$$

Since $d_0 + \mathbf{K}(c_n/d_n)$ also converges to $f(z)$, this means that

$$f(z) = 1 + \frac{z-1}{2+g^{(1)}(z)} = 1 + \frac{z-1}{1 + \dfrac{z}{f(z-1)}} = z\,\frac{f(z-1)+1}{f(z-1)+z},$$

which is a functional equation for $f(z)$. If we let $z = 1$ in this equation, we find that

$$f(1) = 1 \cdot \frac{f(0)+1}{f(0)+1} = 1$$

since $f(0)$ must be a positive number. Hence, we can at least obtain the value of $f(k)$ for all $k \in \mathbf{N}$. For instance

$$f(2) = 2\frac{1+1}{1+2} = \frac{4}{3}, \qquad f(3) = 3\frac{\dfrac{4}{3}+1}{\dfrac{4}{3}+3} = \frac{21}{13},$$

$$f(4) = 4\frac{\dfrac{21}{13}+1}{\dfrac{21}{13}+4} = \frac{136}{73}, \qquad f(5) = 5\frac{\dfrac{136}{73}+1}{\dfrac{136}{73}+5} = \frac{1045}{501}, \dots$$

$\diamond$

2.4 Contractions and extensions

We shall call $d_0 + \mathbf{K}(c_n/d_n)$ a *contraction* of $b_0 + \mathbf{K}(a_n/b_n)$ if its classical approximants $\{g_n\}$ form a subsequence of the classical approximants $\{f_n\}$ of $b_0 + \mathbf{K}(a_n/b_n)$. We call $b_0 + \mathbf{K}(a_n/b_n)$ an *extension* of $d_0 + \mathbf{K}(c_n/d_n)$ in this case. Also this idea is due to Seidel [Seid55] although Lagrange had some special cases already in 1774–76 [Lagr74], [Lagr76].

We call in particular $d_0 + \mathbf{K}(c_n/d_n)$ a *canonical contraction* of $b_0 + \mathbf{K}(a_n/b_n)$ if

$$C_k = A_{n_k}, \qquad D_k = B_{n_k} \quad \text{for } k = 0, 1, 2, \ldots, \tag{2.4.1}$$

where C_n, D_n, A_n and B_n are canonical numerators and denominators of $d_0 + \mathbf{K}(c_n/d_n)$ and $b_0 + \mathbf{K}(a_n/b_n)$ respectively. To derive a general expression for a canonical contraction we can use Theorem 7 combined with formula (1.1.9) in Lemma 2 (with $N = 0$). Rather than considering the general case we shall restrict ourselves to some important special cases.

Theorem 12 *The canonical contraction of* $b_0 + \mathbf{K}(a_n/b_n)$ *with*

$$C_k = A_{2k}, \qquad D_k = B_{2k} \quad \text{for } k = 0, 1, 2, \ldots$$

exists if and only if $b_{2k} \neq 0$ *for* $k = 1, 2, 3, \ldots,$ *and is then given by*

$$b_0 + \cfrac{b_2 a_1}{b_2 b_1 + a_2} - \cfrac{a_2 a_3 b_4/b_2}{a_4 + b_3 b_4 + a_3 b_4/b_2} - \cfrac{a_4 a_5 b_6/b_4}{a_6 + b_5 b_6 + a_5 b_6/b_4} - \cdots \tag{2.4.2}$$

If $\{t_n\}_{n=0}^{\infty}$ *is a tail sequence for* $b_0 + \mathbf{K}(a_n/b_n)$ *with all* $t_n \neq \infty$, *then* $t_0, -t_1 t_2, -t_3 t_4, \ldots$ *is a tail sequence for (2.4.2).*

Proof : From Theorem 7 we find that the canonical contraction has elements

$$
\begin{aligned}
d_0 &= C_0 = A_0 = b_0, \\
d_1 &= D_1 = B_2 = b_2 b_1 + a_2, \\
c_1 &= C_1 - C_0 D_1 = A_2 - A_0 B_2 = b_2 a_1,
\end{aligned}
$$

$c_n = -\Delta_n/\Delta_{n-1}$, where

$$
\begin{aligned}
\Delta_n &= C_n D_{n-1} - D_n C_{n-1} = A_{2n} B_{2n-2} - B_{2n} A_{2n-2} \\
&= -B_1^{(2n-1)} \prod_{j=1}^{2n-1} (-a_j) = -b_{2n} \prod_{j=1}^{2n-1} (-a_j)
\end{aligned}
$$

by formula (1.1.9), and finally $d_n = P_n/\Delta_{n-1}$, where

$$
\begin{aligned}
P_n &= C_n D_{n-2} - D_n C_{n-2} = A_{2n} B_{2n-4} - B_{2n} A_{2n-4} \\
&= -B_3^{(2n-3)} \prod_{j=1}^{2n-3} (-a_j) \\
&= -[b_{2n}(b_{2n-1}b_{2n-2} + a_{2n-1}) + a_{2n}b_{2n-2}] \prod_{j=1}^{2n-3} (-a_j).
\end{aligned}
$$

This proves (2.4.2). Let $\{t_n\}$ be a tail sequence for $b_0 + \mathbf{K}(a_n/b_n)$ with all $t_n \neq \infty$. Then $a_n = t_{n-1}(b_n + t_n)$ for all n. To see that $t_0, -t_1 t_2, -t_3 t_4, \ldots$ is a tail sequence for $d_0 + \mathbf{K}(c_n/d_n)$ given by (2.4.2) it suffices to prove that

$$
c_1 = t_0(d_1 - t_1 t_2), \quad c_n = -t_{2n-3}t_{2n-2}(d_n - t_{2n-1}t_{2n})
$$

for $n = 2, 3, \ldots$. This follows by straight forward computation using the substitution $a_n = t_{n-1}(b_n + t_n)$.   ∎

A contraction of this kind, where the approximants are the even-numbered approximants of $b_0 + \mathbf{K}(a_n/b_n)$ is often called the *even part* of $b_0 + \mathbf{K}(a_n/b_n)$. By an equivalence transformation, (2.4.2) can be written in the form

$$
b_0 + \cfrac{b_2 a_1}{b_2 b_1 + a_2} \underset{-}{} \cfrac{a_2 a_3 b_4}{b_2(a_4 + b_3 b_4) + a_3 b_4}
$$
$$
\underset{-}{} \cfrac{a_4 a_5 b_6 b_2}{b_4(a_6 + b_5 b_6) + a_5 b_6} \underset{-}{} \cfrac{a_6 a_7 b_8 b_4}{b_6(a_8 + b_7 b_8) + a_7 b_8} - \cdots \tag{2.4.3}
$$

which is more widely used (but which is no longer canonical). If $\{t_n\}$; $t_n \neq \infty$ is a tail sequence for $b_0 + \mathbf{K}(a_n/b_n)$, then $t_0, -t_1 t_2, -t_2 t_3 t_4,$ $-b_4 t_5 t_6, -b_6 t_7 t_8, \ldots$ is a tail sequence for (2.4.3). This follows from Remark 2 to Theorem 9.

Theorem 13 *The canonical contraction of $b_0 + \mathbf{K}(a_n/b_n)$ with $C_0 = A_1/B_1$, $D_0 = 1$ and*

$$C_k = A_{2k+1}, \qquad D_k = B_{2k+1} \quad for \; k = 1, 2, 3, \ldots$$

exists if and only if $b_{2k+1} \neq 0$ for $k = 0, 1, 2, \ldots$, and is then given by

$$\frac{b_0 b_1 + a_1}{b_1} - \frac{a_1 a_2 b_3 / b_1}{b_1(a_3 + b_2 b_3) + a_2 b_3} - \frac{a_3 a_4 b_5 b_1 / b_3}{a_5 + b_4 b_5 + a_4 b_5 / b_3}$$

$$\frac{a_5 a_6 b_7 / b_5}{- a_7 + b_6 b_7 + a_6 b_7 / b_5} - \frac{a_7 a_8 b_9 / b_7}{a_9 + b_8 b_9 + a_8 b_9 / b_7} - \cdots \qquad (2.4.4)$$

If $\{t_n\}$ is a tail sequence for $b_0 + \mathbf{K}(a_n/b_n)$ with all $t_n \neq \infty$, then $-t_0 t_1 / b_1$, $-b_1 t_2 t_3$, $-t_4 t_5$, $-t_6 t_7, \ldots$ is a tail sequence for (2.4.4).

The proof follows the same lines as the proof of Theorem 12 and is omitted. Contractions such as (2.4.4) which have the odd-numbered approximants of $b_0 + \mathbf{K}(a_n/b_n)$ as classical approximants, are called *odd parts* of $b_0 + \mathbf{K}(a_n/b_n)$. An equivalence transformation changes (2.4.4) to the form

$$\frac{b_0 b_1 + a_1}{b_1} - \frac{a_1 a_2 b_3 / b_1}{b_1(a_3 + b_2 b_3) + a_2 b_3}$$

$$\frac{a_3 a_4 b_5 b_1}{- b_3(a_5 + b_4 b_5) + a_4 b_5} - \frac{a_5 a_6 b_7 b_3}{b_5(a_7 + b_6 b_7) + a_6 b_7} - \cdots . \qquad (2.4.5)$$

By Remark 2 to Theorem 9, the tail sequence $-t_0 t_1 / b_1$, $-b_1 t_2 t_3$, $-t_4 t_5, \ldots$ of (2.4.4) transforms into the tail sequence $-t_0 t_1 / b_1$, $-b_1 t_2 t_3$, $-b_3 t_4 t_5, \ldots$ of (2.4.5).

Problems

(1) Prove that $\mathbf{K}(a_n/1)$ converges if $a_n \to -0.2$.

(2) Prove that $\mathbf{K}(a_n/b_n)$ converges if $a_n \to 2$ and all $|b_n| \geq 4$.

(3) Let $\{t_n\}$ be a tail sequence for $\mathbf{K}(a_n/b_n)$ with all $t_n \neq \infty$, and let A_n and B_n denote its canonical numerators and denominators. Show that then

$$B_n + B_{n-1}t_n \;=\; \prod_{k=1}^{n}(b_k + t_k), \tag{1}$$

$$A_n - B_n t_0 \;=\; \prod_{k=0}^{n}(-t_k), \tag{2}$$

and

$$f_n - t_0 = -t_0\left(1 + \frac{t_n}{h_n}\right)\prod_{k=1}^{n}\frac{-t_k}{b_k + t_k} \quad \text{if } B_n \neq 0, \tag{3}$$

where $f_n = A_n/B_n$ and $h_n = B_n/B_{n-1}$.

(4) Given the periodic continued fraction

$$\mathbf{K}\frac{a_n}{b_n} \;=\; \frac{2}{4} - \frac{4}{1} + \frac{2}{4} - \frac{4}{1} + \frac{2}{4} - \frac{4}{1} + \cdots$$
$$\left(\;=\; \frac{2}{4+} \frac{-4}{1} + \frac{2}{4+} \frac{-4}{1} + \frac{2}{4+} \frac{-4}{1} + \cdots\right).$$

(a) Find the periodic tail sequences of $\mathbf{K}(a_n/b_n)$.

(b) Find the first ten terms of its critical tail sequence $\{h_n\}$.

(c) Show that

$$\frac{1}{3} \leq h_{2n} \leq 1 \quad \text{and} \quad h_{2n+1} \geq 6 \quad \text{for all } n \geq 2.$$

(d) Use the results from a) and c) and the formula (3) in Problem 3 to prove that $\mathbf{K}(a_n/b_n)$ converges to 1.

(e) Use formula (1.4.5) to find a posteriori truncation error bounds for $\mathbf{K}(a_n/b_n)$.

(f) Use Theorem 6 to determine the asymptotic behavior of $\{h_n\}$.

(g) Compute the 10 first approximants of the types $S_n(0)$, $S_n(t_n)$ and $S_n(\tilde{t}_n)$ for the continued fraction

$$\frac{2+0.5}{4} - \frac{4+(0.5)^2}{1} + \frac{2+(0.5)^3}{4} - \frac{4+(0.5)^4}{1} + \frac{2+(0.5)^5}{4} - \cdots$$

where $\{t_n\}$ and $\{\tilde{t}_n\}$ are the periodic tail sequences of $\mathbf{K}(a_n/b_n)$ from a). Compare these sequences of approximants.

(5) We return to the 3-periodic continued fraction

$$\mathbf{K}\frac{a_n}{1} = \frac{2}{1} + \frac{1}{1} - \frac{1}{1} + \frac{2}{1} + \frac{1}{1} - \frac{1}{1} + \frac{2}{1} + \cdots$$

from Example 15 in Chapter I and Example 2 in *this* chapter.

(a) Find the first 5 terms of its critical tail sequence.

(b) Determine the critical tail sequence of its first tail

$$\frac{1}{1} - \frac{1}{1} + \frac{2}{1} + \frac{1}{1} - \frac{1}{1} + \frac{2}{1} + \cdots$$

Compare this to the results in Example 2.

(c) Determine the asymptotic behavior of the critical tail sequence in (a).

(d) Explain the convergence result

$$\lim_{n \to \infty} S_n(w_n) = \frac{1}{2} \text{ for all } \{w_n\} \text{ bounded away from } 0, -1, \infty$$

for $\mathbf{K}(a_n/1)$ by means of Theorem 4. (This result was proved by another method in Example 15 in Chapter I.)

(6) (a) Use the method of Example 3 to compute the value of $\mathbf{K}((-0.2 + (0.4)^n)/1)$ with an absolute error less than 0.05.

(b) Which approximants $S_n(w_n)$ would you choose to compute $\mathbf{K}((-0.2 + (0.4)^n)/1)$?

(7) (a) Show that if $\mathbf{K}(a/b)$ is a 1-periodic continued fraction which converges generally to f, then it converges to f also in the classical sense.

(b) Show that if $\mathbf{K}(a_n/1)$ is a 2-periodic continued fraction which converges generally to f, then it converges to f also in the classical sense.

(c) Give an example (other than the one in Problem (5)) of a periodic continued fraction which converges in the general sense but not in the classical sense.

(8) Prove that the continued fraction

$$\rho_0 + \frac{\rho_1}{1} \ \frac{\rho_2}{-1+\rho_2} \ \frac{\rho_3}{-1+\rho_3} - \cdots \qquad \text{where all } \rho_k \neq 0$$

has canonical approximants A_n/B_n with

$$A_n = \rho_0 + \sum_{k=1}^{n} \left(\prod_{j=1}^{k} \rho_j \right), \qquad B_n = 1 \quad \text{for } n = 0, 1, 2, \ldots .$$

(9) Let N be a fixed natural number. Show that the canonical contraction of $b_0 + \mathbf{K}(a_n/b_n)$ with

$$C_n = A_n, D_n = B_n \quad \text{for } n = 0, 1, 2, \ldots, N-1$$

and

$$C_n = A_{n+1}, D_n = B_{n+1} \quad \text{for } n = N, N+1, N+2, \ldots$$

exists if and only if $b_{N+1} \neq 0$, and show that then it is given by

$$b_0 + \frac{a_1}{b_1 + \cdots + } \ \frac{a_{N-1}}{b_{N-1} + } \ \frac{b_{N+1}a_N}{b_{N+1}b_N + a_{N+1}}$$
$$+ \ \frac{-a_{N+1}a_{N+2}/b_{N+1}}{(b_{N+2}b_{N+1} + a_{N+2})/b_{N+1} + } \ \frac{a_{N+3}}{b_{N+3} + \cdots} .$$

Show further that if $\{t_n\}_{n=0}^{\infty}$ is a tail sequence for $b_0 + \mathbf{K}(a_n/b_n)$ with $t_N \neq \infty$ and $t_{N+1} \neq \infty$, then
$t_0, t_1, \ldots, t_{N-1}, -t_N t_{N+1}, t_{N+2}, t_{N+3}, \ldots$ is a tail sequence for this contraction.

(10) Given $b_0 + \mathbf{K}(a_n/b_n)$ with critical tail sequence $\{h_n\}$ such that all $h_n \neq 0$. Prove that its equivalent continued fraction

$$b_0 + \frac{a_1/h_1}{1} \ + \ \frac{a_2/h_1 h_2}{b_2/h_2} \ + \ \frac{a_3/h_2 h_3}{b_3/h_3} + \cdots$$

has all canonical denominators equal to 1 for $n \geq 0$.

(11) Let $b_0 + \mathbf{K}(a_n/b_n)$ have classical approximants $S_n(0) = f_n$, let $N \in \mathbf{N}, N \geq 2$ and let $g \in \hat{\mathbf{C}}$ be chosen such that

$$\rho = -S_N^{-1}(g) = \frac{A_N - B_N g}{A_{N-1} - B_{N-1}g} \neq 0, \infty.$$

Prove that

$$b_0 + \frac{a_1}{b_1} + \cdots + \frac{a_{N-1}}{b_{N-1}} + \frac{a_N}{b_N - \rho} + \frac{\rho}{1 - b_{N+1}} + \frac{a_{N+1}/\rho}{a_{N+1}/\rho + b_{N+2}} + \frac{a_{N+2}}{} + \cdots$$

is an extension of $b_0 + \mathbf{K}(a_n/b_n)$ with classical approximants

$$f_n^* = \begin{cases} f_n & \text{for } n = 0, 1, \ldots, N-1, \\ g & \text{for } n = N, \\ f_{n-1} & \text{for } n = N+1, N+2, \ldots. \end{cases}$$

(This idea can be found in [Perr57, p. 15] and [JoTh80, p. 43].)

(12) The Khovanskii transform of $\mathbf{K}(a_n/1)$ is given by

$$\frac{a_1}{1 + 2a_2} - \frac{a_2}{1} - \frac{a_3}{1 + 2a_3 + 2a_4} - \frac{a_4}{1} - \frac{a_5}{1 + 2a_5 + 2a_6}$$
$$- \cdots - \frac{a_{2n}}{1} - \frac{a_{2n+1}}{1 + 2a_{2n+1} + 2a_{2n+2}} - \cdots$$

See also [Khov63, p. 22]. Prove that if both $\mathbf{K}(a_n/1)$ and its Khovanskii transform converge (in the classical sense), then they converge to the same value.

(13) Prove that the following continued fractions converge.

(a) $\mathbf{K}_{n=1}^{\infty}(n/n)$
(b) $\mathbf{K}_{n=1}^{\infty}(n^2/n)$
(c) $\mathbf{K}_{n=1}^{\infty}((1 + \frac{1}{n})/2)$

(14) (a) Find the Bauer-Muir transform of

$$z^2 - 1 + \frac{2^2}{1 + z^2 - 1} + \frac{2^2}{1} + \frac{4^2}{z^2 - 1} + \frac{4^2}{1} + \frac{6^2}{z^2 - 1} + \frac{6^2}{1} + \cdots$$

with respect to

$$w_n = \begin{cases} \dfrac{n+1}{z+1} - \dfrac{1}{2} & \text{if } n \text{ is odd,} \\ n(z+1) + \frac{1}{2}(3 + 2z - z^2) & \text{if } n \text{ is even.} \end{cases}$$

(b) Assume that the continued fraction in (a) and its Bauer-Muir transform converge to the same value $f(z)$ for $z > 1$. Find a functional equation for $f(z)$.

References

[Bauer72]   G. Bauer, *Von einem Kettenbruch von Euler und einem Theorem von Wallis*, Abh. der Kgl. Bayr. Akad. der Wiss., München, Zweite Klasse, **11** (1872), 99–116.

[Jaco86]   L. Jacobsen, *General Convergence of Continued Fractions*, Trans. Amer. Math. Soc. **294**, no. 2 (1986), 477–485.

[JoTh80]   W. B. Jones and W. J. Thron, "Continued Fractions: Analytic Theory and Applications", Encyclopedia of Mathematics and its Applications, **11**, Addison-Wesley Publishing Company, Reading, Mass. (1980). Now distributed by Cambridge University Press, New York.

[Khov63]   A. N. Khovanskii, "The Application of Continued Fractions and their Generalizations to Problems in Approximation Theory", P. Noordhoff N. V., Groningen, The Netherlands (1963).

[Lagr74]   J. L. Lagrange, *Additions aux Eléments d'Algèbre d'Euler*, Lyon (1774).

[Lagr76]   J. L. Lagrange, *Sur l'usage des fractions continues dans le calcul intégral*, Nouveaux Mém. Acad. Sci. Berlin **7** (1776), 236–264; Oeuvres, **4** (J. A. Serret, ed.), Gauthier Villars, Paris (1869), 301–322.

[Muir77]   T. Muir, *A Theorem in Continuants*, Phil. Mag., (5) **3** (1877), 137–138.

[Perr57]   O. Perron, "Die Lehre von den Kettenbrüchen" Band **2**, B.
             G. Teubner, Stuttgart (1957).

[Seid55]   L. Seidel, *Bemerkungen über den Zusammenhang zwischen
             dem Bildungsgesetze eines Kettenbruches und der Art des
             Fortgangs seiner Näherungsbrüche*, Abh. der Kgl. Bayr.
             Akad. der Wiss., München, Zweite Klasse, **7**:3 (1855), 559.

[Thron89]   W. J. Thron, *Continued Fraction Identities Derived from
             the Invariance of the Crossratio under Linear Fractional
             Transformations*, "Analytic Theory of Continued Fractions
             III", Proceedings, Redstone 1988, (L. Jacobsen ed.), Lec-
             ture Notes in Mathematics **1406**, Springer-Verlag, Berlin
             (1989), 124–134.

Chapter III

Convergence criteria

About this chapter

Applications of continued fractions are often tied to their possible convergence. It is therefore important to have convergence criteria which are easy to check and which cover large classes of continued fractions.

Rather than discussing a large variety of such criteria we shall emphasize *how* one can derive them. This means in particular that only the best known and/or the widest applicable convergence theorems will be presented here. For a more complete list we refer to [JoTh80].

The methods we use are based upon some very nice mapping properties of linear fractional transformations. Some of these methods will also lead to truncation error estimates. In this respect we have taken the attitude that relatively simple and easy to use bounds are often to be preferred to more complicated but slightly tighter ones.

1 Two classical results

1.1 *The Stern-Stolz divergence theorem*

In Chapter I we presented three classical convergence theorems. We shall now see two more. The first one is in fact a divergence theorem. It dates back at least to the 1860's [Stern60], [Stolz86]. We state it in a slightly more general form:

Theorem 1 (The Stern-Stolz Theorem) *The continued fraction* $b_0 + \mathbf{K}(1/b_n)$ *diverges generally if* $\sum |b_n| < \infty$. *In fact*

$$\lim_{n \to \infty} A_{2n+p} = P_p \neq \infty, \qquad \lim_{n \to \infty} B_{2n+p} = Q_p \neq \infty \qquad (1.1.1)$$

for $p = 0, 1$, *where*

$$P_1 Q_0 - P_0 Q_1 = 1. \qquad (1.1.2)$$

Remarks:

1. We say that $b_0 + \mathbf{K}(a_n/b_n)$ diverges (generally) if it fails to converge (generally) in $\hat{\mathbb{C}}$. Since general convergence is a slightly wider concept than classical convergence, it follows that general divergence is a slightly stronger property than divergence in the classical sense.

2. From (1.1.1) it follows that

$$\lim_{n \to \infty} S_{2n}(w) = \frac{P_0 + P_1 w}{Q_0 + Q_1 w},$$

$$\lim_{n \to \infty} S_{2n+1}(w) = \frac{P_1 + P_0 w}{Q_1 + Q_0 w}. \qquad (1.1.3)$$

That is, the even and odd parts of $\mathbf{K}(1/b_n)$ converge in the classical sense to P_0/Q_0 and P_1/Q_1. However, by (1.1.2) these limits are distinct, so the continued fraction itself diverges. It even diverges generally.

Here is the place for a little reflection. The even part of $K(1/b_n)$ converges and thus converges generally. Still it follows from $(1.1.3)$ that the limit of $S_{2n}(w)$ is totally dependent on the choice of w. How can this be? Does not this violate Theorem 4 in Chapter II? Please note that $\{S_{2n}(0)\}$ is the sequence of classical approximants $\{T_n(0)\}$ for the even part of $\mathbf{K}(1/b_n)$, whereas $S_{2n}(w) \neq T_n(w)$ for $w \neq 0$. In fact, if we consider the canonical even part of $\mathbf{K}(1/b_n)$ as described in Theorem 12 in Chapter II, then $S_{2n}(w) = T_n(w_n)$ if and only if

$$\frac{A_{2n} + A_{2n-1}w}{B_{2n} + B_{2n-1}w} = \frac{A_{2n} + A_{2n-2}w_n}{B_{2n} + B_{2n-2}w_n},$$

that is, if and only if

$$w_n = \frac{-a_{2n}w}{b_{2n} + w} \quad \text{for all } n.$$

Hence, the classical convergence of the even part of $\mathbf{K}(1/b_n)$ implies general convergence of the even part, but not convergence of $S_{2n}(w)$ to a value independent of w.

3. An equivalence transformation does not change the classical approximants of a continued fraction. Hence, from Corollary 10A in Chapter II it follows that $\mathbf{K}(a_n/b_n)$ diverges in the classical sense if

$$\sum_{n=1}^{\infty} \left| b_n \prod_{k=1}^{n} a_k^{(-1)^{n-k+1}} \right| < \infty. \tag{1.1.4}$$

The series in $(1.1.4)$ is called the *Stern-Stolz series* of $\mathbf{K}(a_n/b_n)$. It is invariant under equivalence transformations of $\mathbf{K}(a_n/b_n)$. The *general* divergence of $\mathbf{K}(a_n/b_n)$ also follows easily.

Proof of Theorem 1:   It suffices to prove $(1.1.1)$–$(1.1.2)$. $\{A_n\}$ and $\{B_n\}$ are solutions of the recurrence relation

$$X_n = b_n X_{n-1} + X_{n-2} \quad \text{for } n = 1, 2, 3, \ldots. \tag{1.1.5}$$

By induction it follows that any such solution satisfies

$$|X_n| \leq \max\{|X_{-1}|, |X_0|\} \cdot (|b_1| + 1)(|b_2| + 1) \cdots (|b_n| + 1).$$

Hence $\{A_n\}$ and $\{B_n\}$ are bounded under our conditions. This means that $\sum b_n A_{n-1}$ and $\sum b_n B_{n-1}$ converge absolutely. Since $X_n - X_{n-2} = b_n X_{n-1}$ by (1.1.5), we get for instance

$$A_{2n} = \sum_{m=1}^{n} (A_{2m} - A_{2m-2}) = \sum_{m=1}^{n} b_{2m} A_{2m-1},$$

and similar expressions for A_{2n+1}, B_{2n} and B_{2n+1}. This proves (1.1.1). (1.1.2) follows then since by the determinant formula (see formula (1.2.10) in Chapter I)

$$A_{2n+1} B_{2n} - A_{2n} B_{2n+1} = 1 \quad \text{for all } n.$$

■

1.2   Continued fractions with positive elements

Let $\mathbf{K}(a_n/b_n)$ have all $a_n > 0$ and $b_n > 0$. Then

$$S_1(0) \;=\; \frac{a_1}{b_1} > 0,$$

$$0 < S_2(0) \;=\; \cfrac{a_1}{b_1 + \cfrac{a_2}{b_2}} < \frac{a_1}{b_1} = S_1(0)$$

since $(a_2/b_2) > 0$. Furthermore

$$S_3(0) = \cfrac{a_1}{b_1 + \cfrac{a_2}{b_2 + \cfrac{a_3}{b_3}}} > \cfrac{a_1}{b_1 + \cfrac{a_2}{b_2}} = S_2(0)$$

since

$$\cfrac{a_2}{b_2 + \cfrac{a_3}{b_3}} < \frac{a_2}{b_2}.$$

Moreover

$$S_3(0) = \cfrac{a_1}{b_1 + \cfrac{a_2}{b_2 + \cfrac{a_3}{b_3}}} < \frac{a_1}{b_1} = S_1(0),$$

and so on. We get:

Theorem 2 *Let all the elements* a_n *and* b_n *of* $\mathbf{K}(a_n/b_n)$ *be positive. Then*

$$S_2(0) < S_4(0) < S_6(0) < \cdots < S_5(0) < S_3(0) < S_1(0). \qquad (1.2.1)$$

Remarks:

1. We find from (1.2.1) that $\{S_{2n}(0)\}$ is a bounded, monotonely increasing sequence. This implies that $\{S_{2n}(0)\}$ converges to a finite value L_0. Similarly $\{S_{2n+1}(0)\}$ decreases monotonely to a finite value L_1, and $L_0 \leq L_1$. Hence, both the even and odd parts of $\mathbf{K}(a_n/b_n)$ converge to finite values.

2. If we know that $\mathbf{K}(a_n/b_n)$ itself converges, then (1.2.1) can be used to estimate its value f. By setting

$$f \approx f_n^* = \frac{1}{2}(S_{2n+1}(0) + S_{2n}(0)) \qquad (1.2.2)$$

(the average value of the two approximants), we know that the error is bounded by

$$|f - f_n^*| < \frac{1}{2}(S_{2n+1}(0) - S_{2n}(0)). \qquad (1.2.3)$$

Example 1 In Example 5 in Chapter I we used a continued fraction to estimate the value of $\log 2$:

$$\log 2 = \frac{1}{1+} \; \frac{1/2}{1 +} \; \frac{1/(2 \cdot 3)}{1 +} \; \frac{2/(2 \cdot 3)}{1 +} \; \frac{2/(2 \cdot 5)}{1 +} \; \frac{3/(2 \cdot 5)}{1 +} + \cdots .$$

The first seven approximants $f_n = S_n(0)$ were given in a table. The oscillatory character of $\{S_n(0)\}$ is consistent with (1.2.1). In particular we get

$$\log 2 \approx \frac{1}{2}(f_7 + f_6) \pm \frac{1}{2}(f_7 - f_6) \approx 0.693135 \pm 0.000015$$

which agrees with the correct value of $\log 2$.

$\diamond$

Knowing Theorem 2 makes it easy to prove the second classical result due to Seidel [Seid46] and Stern [Ster48]:

Theorem 3 (The Seidel-Stern Theorem) *Let all the elements b_n of* $\mathbf{K}(1/b_n)$ *be positive. Then* $\mathbf{K}(1/b_n)$ *converges if and only if* $\sum b_n = \infty$.

Remarks:

1. From Remark 1 to Theorem 2, we know that if $\mathbf{K}(1/b_n)$ converges, then it converges to a finite value, and if it diverges, then its even and odd parts still converge to finite values.

2. An equivalent formulation of Theorem 3 is that $\mathbf{K}(a_n/b_n)$ with all $a_n > 0, b_n > 0$ converges if and only if its Stern-Stolz series (1.1.4) diverges to ∞.

Proof : If $\sum b_n < \infty$, then $\mathbf{K}(1/b_n)$ diverges by Theorem 1. Let $\sum b_n = \infty$. To prove that then $\mathbf{K}(1/b_n)$ converges, it suffices to prove that

$$S_{2n+1}(0) - S_{2n}(0) = \frac{A_{2n+1}}{B_{2n+1}} - \frac{A_{2n}}{B_{2n}} = \frac{1}{B_{2n+1}B_{2n}} \to 0. \qquad (1.2.4)$$

Since

$$B_n = b_n B_{n-1} + B_{n-2} \quad \text{for } n = 1, 2, \ldots \quad \text{and } B_{-1} = 0, \quad B_0 = 1,$$

it follows that all $B_{2n} > B_{2n-2} > \cdots > B_0 = 1$ and $B_{2n+1} > B_{2n-1} > \cdots > B_1 = b_1$, so that

$$B_{2n} > b_{2n}b_1 + B_{2n-2} > \cdots > (b_{2n} + b_{2n-2} + \cdots + b_2)b_1 + 1$$

and

$$B_{2n+1} > b_{2n+1} \cdot 1 + B_{2n-1} > \cdots > b_{2n+1} + b_{2n-1} + \cdots + b_1.$$

The divergence of $\sum b_n$ now proves (1.2.4).   ∎

Example 2 In Example 2 in Chapter I we used the continued fraction

$$1 + \frac{1}{2} + \frac{1}{2} + \frac{1}{2} + \cdots$$

to compute approximations to $\sqrt{2}$. This can be done since the continued fraction converges by Theorem 3, and its value must be the positive solution of the equation $f = 1 + 1/(1+f)$; i.e. $f = \sqrt{2}$. The approximants $f_n = S_n(0)$ were computed for $n = 1, 2, \ldots, 5$. They show the oscillation property described in (1.2.1). In particular we get $f_4 = 41/29 < \sqrt{2} < f_5 = 99/70$, i.e.

$$\begin{aligned}
\sqrt{2} &= \frac{1}{2}(f_4 + f_5) \pm \frac{1}{2}(f_5 - f_4) \\
&= \frac{5741}{4060} \pm \frac{1}{4060} \approx 1.41404 \pm 0.00025
\end{aligned}$$

which agrees with the value $\sqrt{2} = 1.41421356\ldots$.

$\diamond$

If we use modified approximants $S_n(w_n)$, this useful oscillation property sometimes gets lost. When is it preserved?

Example 3 In Example 14 in Chapter I we used the classical approximants $S_n(0)$ and the modified approximants $S_n(w_n)$ where $w_{2k} = 1$, $w_{2k+1} = 2$, for the continued fraction

$$\frac{3 + 1/1^2}{1} + \frac{4 + 3/2^2}{1} + \frac{3 + 1/3^2}{1} + \frac{4 + 3/4^2}{1} + \frac{3 + 1/5^2}{1} + \cdots.$$

The table for $f_n = S_n(0)$ displays the oscillation property (1.2.1). Also the table for $S_n(w_n)$ shows the same property, but only up to and including $n = 5$. In this case we have

$$S_1(w_1) = \frac{3 + 1/1^2}{1 + 2} > S_2(w_2) = \frac{3 + 1/1^2}{1 + \dfrac{4 + 3/2^2}{1 + 1}}$$

because

$$w_1 = 2 < \frac{4 + 3/2^2}{1 + 1} = \frac{a_2}{1 + w_2},$$

and

$$S_1(w_1) = \frac{3 + 1/1^2}{1 + 2} > S_3(w_3) = \cfrac{3 + 1/1^2}{1 + \cfrac{4 + 3/2^2}{1 + \cfrac{3 + 1/3^2}{1 + 2}}}$$

because

$$w_1 = 2 < \cfrac{4 + 3/2^2}{1 + \cfrac{3 + 1/3^2}{1 + 2}} = \cfrac{a_2}{1 + \cfrac{a_3}{1 + w_3}},$$

and so on. But $S_4(w_4) > S_6(w_6)$.

$\diamond$

If we follow the same line of argument as in Example 3 we find:

Theorem 4 *Let all the elements a_n and b_n of $\mathbf{K}(a_n/b_n)$ be positive and let all $w_n \geq 0$. Then*

$$S_2(w_2) < S_4(w_4) < S_6(w_6) < \cdots < S_5(w_5) < S_3(w_3) < S_1(w_1) \quad (1.2.5)$$

if

$$w_n < \frac{a_{n+1}}{b_{n+1} + w_{n+1}} \quad \text{and} \quad w_n < \cfrac{a_{n+1}}{b_{n+1} + \cfrac{a_{n+2}}{b_{n+2} + w_{n+2}}} \quad (1.2.6)$$

for all $n \in \mathbf{N}$.

Remarks:

1. $(1.2.6)$ holds trivially if all $w_n = 0$.

2. If $(1.2.6)$ holds with the opposite inequality signs, then $(1.2.5)$ holds with the opposite inequality signs.

3. In Example 3 the second inequality in $(1.2.6)$ fails to hold for $n = 4$, since $w_4 = 1$ and the righthand side $= (3 + 1/25)/(3 + 1/24)$.

2 Periodic continued fractions

2.1 Introduction

A continued fraction $b_0 + \mathbf{K}(a_n/b_n)$ is called *periodic* with *period length* $k \in \mathbf{N}$, or k-periodic for short, if the sequences $\{a_n\}_{n=1}^\infty$ and $\{b_n\}_{n=1}^\infty$ of its elements are all k-periodic; i.e. if

$$
\begin{aligned}
a_{N+kn+p} &= a_p^*, \\
b_{N+kn+p} &= b_p^* \quad \text{for all } n \in \mathbf{N}_0, \quad p \in \{1, \ldots, k\}
\end{aligned}
\tag{2.1.1}
$$

for some $N \in \mathbf{N}_0$. We say that the period begins at $n = N + 1$. We have already seen several examples of periodic continued fractions.

The approximants of $b_0 + \mathbf{K}(a_n/b_n)$ with property (2.1.1) can be written

$$
S_{N+kn+p}(w) = S_N \circ T_k^n \circ T_p(w)
\tag{2.1.2}
$$

where S_N is as used earlier, and

$$
T_m(w) = \frac{a_1^*}{b_1^*} \frac{a_2^*}{+\, b_2^*} + \cdots + \frac{a_m^*}{b_m^* + w} \quad \text{for } m = 1, 2, \ldots, k\,.
\tag{2.1.3}
$$

The convergence behavior of $b_0 + \mathbf{K}(a_n/b_n)$ depends therefore on how the linear fractional transformation T_k behaves under iterations.

2.2 Classification of linear fractional transformations

Linear fractional transformations

$$
t(w) = \frac{aw + b}{cw + d}, \quad ad - bc \neq 0
\tag{2.2.1}
$$

are often classified according to how iterations behave asymptotically; i.e. to what happens to

$$
t^n(w) = t \circ t \circ \cdots \circ t(w) \qquad n \text{ times}
\tag{2.2.2}
$$

as $n \to \infty$. If $t^n(w) \to x$, then clearly x must be a *fixed point* of t; i.e. $t(x) = x$. Unless t is the identity function $t(w) = w$ (i.e. $a = d \neq 0, b =$

$c = 0$), it can have at most two distinct fixed points, since they have to
be solutions of the equation

$$cx^2 + (d - a)x - b = 0 \,. \tag{2.2.3}$$

We allow $x = \infty$ as a fixed point. From (2.2.1) we see that $x = \infty$ is a
fixed point for t if and only if $c = 0$.

Basis for the classification is the following result:

Theorem 5 *Let t be a linear fractional transformation (2.2.1) with at
most two fixed points x and y.*

A. *If $x = y$ (only one fixed point) then*

$$\lim_{n \to \infty} t^n(w) = x \quad \text{for all } w \in \hat{\mathbf{C}} \,. \tag{2.2.4}$$

B. *If $x \neq y$ and*

$$\begin{aligned} |cx + d| &= |cy + d| & \text{if } c \neq 0 \,, \\ |a| &= |d| & \text{if } c = 0 \,, \end{aligned} \tag{2.2.5}$$

then $t^n(w)$ diverges (by oscillation) for all $w \neq x, y$.

C. *If $x \neq y$ and*

$$\begin{aligned} |cx + d| &> |cy + d| & \text{if } c \neq 0 \,, \\ |a| &\neq |d| & \text{if } c = 0 \,, \end{aligned} \tag{2.2.6}$$

then

$$\lim_{n \to \infty} t^n(w) = x \quad \text{for all } w \neq y \,. \tag{2.2.7}$$

(If $c = 0$ then $x = \infty$ if $|d| < |a|$ and $y = \infty$ if $|d| > |a|$.)

For the proof of Theorem 5 we refer to text-books on complex analysis.
These three different types of linear fractional transformations are given

special names. We say that t is *parabolic* if it has only one fixed point as in Theorem 5A, *elliptic* if it is as in Theorem 5B and *loxodromic* if it is as in Theorem 5C. It is also common to say that t is *hyperbolic* if it is loxodromic with

$$
\begin{aligned}
(cx + d)/(cy + d) > 0 \qquad &\text{if } c \neq 0 \,, \\
a/d > 0 \qquad &\text{if } c = 0 \,.
\end{aligned}
\tag{2.2.8}
$$

(In some books one has chosen to say that t is not loxodromic if it is hyperbolic.) Note that these three possibilities: parabolic, elliptic and loxodromic, are the only ones we have in addition to the identity transformation. Note also that t is parabolic (and not elliptic) if $cx + d = cy + d$ and $c \neq 0$ or if $a = d \neq 0$, $c = 0$ and $b \neq 0$.

We also say that the fixed point x of t is *attractive* if $t^n(w) \to x$ for all w different from a point y, and *repulsive* if $t^n(w) \to p \neq x$ for all $w \neq x$.

Properties

1. Let x be a fixed point of t. Then x is also a fixed point of t^{-1}. That is, t and t^{-1} have the same fixed points. However, if x is an attractive (repulsive) fixed point of t, then x is a repulsive (attractive) fixed point of t^{-1}. In particular this means that the classification (parabolic/elliptic/loxodromic) of t is invariant under inversion.

2. We say that a linear fractional transformation t_c is conjugate (or similar) to t if there exists a linear fractional transformation p such that

$$
t_c = p \circ t \circ p^{-1} \,.
\tag{2.2.9}
$$

Since then $t_c^n = p \circ t^n \circ p^{-1}$, it follows that our classification is invariant under conjugation.

If x is a fixed point for t, then $p(x)$ is a fixed point for t_c in (2.2.9). And if x is attractive (repulsive) for t, then $p(x)$ is attractive (repulsive) for t_c.

2.3   Convergence of periodic continued fractions

By combining (2.1.2) and Theorem 5, we find our main result in this section. We refer to (2.1.1)–(2.1.3) for notation.

Theorem 6 *Let $b_0 + \mathbf{K}(a_n/b_n)$ be a k-periodic continued fraction satisfying (2.1.1).*

A. *If T_k is parabolic, then $b_0 + \mathbf{K}(a_n/b_n)$ converges to $S_N(x)$, where x is the fixed point of T_k.*

B. *If T_k is elliptic or the identity transformation, then $b_0 + \mathbf{K}(a_n/b_n)$ diverges generally.*

C. *If T_k is loxodromic, then $b_0 + \mathbf{K}(a_n/b_n)$ converges generally to $S_N(x)$, where x is the attractive fixed point of T_k. It also converges to $S_N(x)$ in the classical sense if*

$$T_p(0) \neq y \quad \text{for } p = 1, 2, \ldots, k, \tag{2.3.1}$$

where y is the repulsive fixed point of T_k.

Example 4 In Example 2 we proved that the 1-periodic continued fraction

$$1 + \frac{1}{2} \frac{1}{+2} \frac{1}{+2} +\cdots$$

converges to $f = \sqrt{2}$. This agrees with Theorem 6 since $T_1(w) = 1/(2 + w)$ is loxodromic with attractive fixed point $x = \sqrt{2} - 1$.

                                                                                    ◇

Example 5 In Chapter I, Subsection *2.3*, we promised to return to the continued fraction

$$\frac{x}{2} \frac{x}{+2} \frac{x}{+2} \frac{x}{+2} +\cdots$$

and prove that it converges to $f = \sqrt{1 + x} - 1$ for all x in the cut plane $\mathbf{C} \setminus L$, where the cut L is the real ray $(-\infty, -1)$. This follows now easily from Theorem 6 since $T_1(w) = x/(2+w)$ is parabolic for $x = -1$, elliptic for $x \in \mathbf{L}$ and loxodromic otherwise, and since f is an attractive fixed point for T_1 when T_1 is parabolic or loxodromic when we choose the principal branch for the square root; i.e. $\Re\left(\sqrt{1+x}\right) \geq 0$.

$\diamond$

2.4 Thiele oscillation

Let us return to condition (2.3.1) in Theorem 6C. If T_k is loxodromic and $T_p(0) = y$ for some $p \in \{1, 2, \ldots, k\}$, then $b_0 + \mathbf{K}(a_n/b_n)$ diverges in the classical sense although it converges generally. This follows easily since then

$$S_{N+kn+p}(0) = S_N \circ T_k^n \circ T_p(0) = S_N(y) \quad \text{for all } n \in \mathbf{N}, \qquad (2.4.1)$$

whereas $S_{N+kn+m}(0) \to S_N(x)$ as $n \to \infty$ for all m such that $T_m(0) \neq y$. This phenomenon is called *Thiele oscillation*, due to Thiele [Thie79] who was the first one to point out that this thing could happen.

In Example 2 in Chapter II this phenomenon was connected to properties of tail sequences for $b_0 + \mathbf{K}(a_n/b_n)$. Let us do so here too. Since T_k is loxodromic, it has two distinct fixed points $x = x^{(0)}$ and $y = y^{(0)}$. Moreover, since

$$
\begin{aligned}
T_k^{(j)}(w) &= \frac{a_{j+1}^*}{b_{j+1}^*} + \frac{a_{j+2}^*}{b_{j+2}^*} + \cdots + \frac{a_k^*}{b_k^*} + \frac{a_1^*}{b_1^*} + \cdots + \frac{a_j^*}{b_j^* + w} \\
&= T_j^{-1} \circ T_k \circ T_j(w),
\end{aligned}
\qquad (2.4.2)
$$

and thus is a conjugate of T_k, we know by Property 2 in Subsection 2.2 that also $T_k^{(j)}$ is loxodromic. Let $x^{(j)}$ and $y^{(j)}$ be the attractive and repulsive fixed points of $T_k^{(j)}$. Then the right tail sequence $\{f^{(n)}\}$ of $b_0 + \mathbf{K}(a_n/b_n)$ is periodic, looking like

$$x^{(0)}, x^{(1)}, \ldots, x^{(k-1)}, x^{(k)} = x^{(0)}, x^{(1)}, \ldots, x^{(k-1)}, x^{(0)}, \ldots \qquad (2.4.3)$$

from some n on. The second periodic tail sequence $\{t_n\}$ of $b_0 + \mathbf{K}(a_n/b_n)$ looks like

$$y^{(0)}, y^{(1)}, \ldots, y^{(k-1)}, y^{(k)} = y^{(0)}, y^{(1)}, \ldots, y^{(k-1)}, y^{(0)}, \ldots \qquad (2.4.4)$$

from some n on. If $T_p(0) = y = y^{(0)}$, then

$$T_k^{(p)}(0) = T_p^{-1} \circ T_k \circ T_p(0) = T_p^{-1}(y) = 0 \,;$$

i.e. $y^{(p)} = 0$. Hence we have the following alternative characterization of Thiele oscillation: T_k loxodromic and $y^{(p)} = 0$ for some $p \in \{0, 1, \ldots, k - 1\}$.

Or, since $y^{(p)} = 0$ if and only if $y^{(p+1)} = \infty$: $b_0 + \mathbf{K}(a_n/b_n)$ *diverges by Thiele oscillation if and only if T_k is loxodromic and $y^{(p+1)} = \infty$ for some $p \in \{0, 1, \ldots, k - 1\}$.*

It is worth noticing that Thiele oscillation can never occur if $k = 1$ or if $k = 2$ and $b_1^* = b_2^* = 1$. (See Problem 7 in Chapter II.)

Example 6 The periodic continued fraction

$$\frac{2}{1} + \frac{1}{1} - \frac{1}{1} + \frac{2}{1} + \frac{1}{1} - \frac{1}{1} + \frac{2}{1} + \cdots$$

in Example 15 in Chapter I and Example 2 in Chapter II, diverges by Thiele oscillation. Its periodic tail sequences are

$$\frac{1}{2}, 3, -\frac{2}{3}, \frac{1}{2}, 3, -\frac{2}{3}, \frac{1}{2}, \ldots$$

and

$$0, \infty, -1, 0, \infty, -1, 0, \infty, -1, 0, \ldots .$$

The first one is the right tail sequence. The second one reveals the Thiele oscillation.

$\diamond$

2.5   Tail sequences

For simplicity we let $\mathbf{K}(a_n/b_n)$ be a k-periodic continued fraction where the period begins at $n = 1$. That is, (2.1.1) holds with $N = 0$. Then $\mathbf{K}(a_n/b_n)$ has the periodic tail sequences (2.4.3) and (2.4.4). Combining this knowledge with Theorem 6 in this chapter, we immediately find:

Theorem 7 *Let* $\mathbf{K}(a_n/b_n)$ *be as described above. Let* $\{t_n\} \subseteq \hat{\mathbf{C}}$ *be a tail sequence for* $\mathbf{K}(a_n/b_n)$.

A. *If* T_k $(= S_k)$ *is parabolic with fixed point* x, *then*

$$\lim_{n \to \infty} t_{kn+p} = T_p^{-1}(x) = x^{(p)} \quad \text{for } p = 0, 1, \ldots, k-1. \quad (2.5.1)$$

B. *If* T_k *is loxodromic with attractive fixed point* x *and repulsive fixed point* y, *and* $t_0 \neq x$, *then*

$$\lim_{n \to \infty} t_{kn+p} = T_p^{-1}(y) = y^{(p)} \quad \text{for } p = 0, 1, \ldots, k-1. \quad (2.5.2)$$

Example 7 The 4-periodic continued fraction

$$\frac{1}{1} + \frac{1}{1} - \frac{3}{1} - \frac{2}{1} + \frac{1}{1} + \frac{1}{1} - \frac{3}{1} - \frac{2}{1} + \frac{1}{1} + \cdots \quad (2.5.3)$$

converges generally to the attractive fixed point $x = 1$ of the loxodromic transformation

$$T_4(w) = \frac{1}{1} + \frac{1}{1} - \frac{3}{1} - \frac{2}{1+w} = \frac{4+2w}{5+w}.$$

The repulsive fixed point of T_4 is $y = -4$. Hence

$$y^{(0)} = -4, \qquad y^{(1)} = s_1^{-1}(-4) = -\frac{5}{4},$$

$$y^{(2)} = s_2^{-1}\left(-\frac{5}{4}\right) = -\frac{9}{5} \quad \text{and} \quad y^{(3)} = s_3^{-1}\left(-\frac{9}{5}\right) = \frac{2}{3}.$$

Therefore, every tail sequence $\{t_n\}$ with $t_0 \neq 1$ satisfies

$$\lim_{n \to \infty} t_{4n+p} = \begin{cases} -4 & \text{if } p = 0, \\ -\frac{5}{4} & \text{if } p = 1, \\ -\frac{9}{5} & \text{if } p = 2, \\ \frac{2}{3} & \text{if } p = 3. \end{cases}$$

In particular this is the case for $t_n = -h_n$ where $\{h_n\}$ is the critical tail sequence for (2.5.3).

Note also that since all $y^{(p)} \neq \infty$, we have no Thiele oscillation, so (2.5.3) converges to $x = 1$ also in the classical sense.

$\diamond$

3   Techniques to prove convergence

3.1   Convergence sets

Convergence criteria for continued fractions are often given in terms of *convergence sets* $\Omega \subseteq \mathbf{C} \times \mathbf{C}$: If $(a_n, b_n) \in \Omega$ for all n, then $\mathbf{K}(a_n/b_n)$ converges. Examples of such sets are the Śleszyński-Pringsheim set

$$\Omega = \{(a,b) \in \mathbf{C} \times \mathbf{C};\ |b| \geq |a| + 1\} \qquad (3.1.1)$$

(Theorem 1 in Chapter I) and the Worpitzky disk

$$\Omega = E \times \{1\} \quad \text{where } E = \{a \in \mathbf{C};\ |a| \leq 1/4\}, \qquad (3.1.2)$$

(Theorem 3 in Chapter I). Observe that here $\{1\}$ means the one-point set consisting of only the element 1. This use of symbols is not consistent with the use of $\{\cdot\}$ for sequences, but will be used in a few places where the context prevents confusion. A *conditional convergence set* Ω is a set $\Omega \subseteq \mathbf{C} \times \mathbf{C}$ such that: If $(a_n, b_n) \in \Omega$ for all n, then $\mathbf{K}(a_n/b_n)$ converges if and only if its Stern-Stolz series diverges to ∞, i.e.

$$\sum_{n=1}^{\infty} \left| b_n \prod_{k=1}^{n} a_k^{(-1)^{n-k+1}} \right| = \infty. \qquad (3.1.3)$$

The Van Vleck sector

$$\Omega = \{1\} \times G_\epsilon \quad \text{where } G_\epsilon = V_\epsilon = \left\{ b \in \mathbf{C};\ |\arg b| < \frac{\pi}{2} - \epsilon \right\} \qquad (3.1.4)$$

for an $\epsilon > 0$ is an example of a conditional convergence set, (Theorem 2 in Chapter I). Another example is $\Omega = \{(a,b) \in \mathbf{C} \times \mathbf{C};\ a > 0, b > 0\}$, (Theorem 3 and the subsequent Remark 2 in this chapter).

A *uniform convergence set* Ω is a convergence set to which there corresponds a sequence $\{\lambda_n\}$ of positive numbers converging to 0 such that

$$|S_{n+m}(0) - S_n(0)| \leq \lambda_n \quad \text{for all } m, n \in \mathbf{N} \qquad (3.1.5)$$

for every continued fraction $\mathbf{K}(a_n/b_n)$ from Ω; i.e. for every continued fraction $\mathbf{K}(a_n/b_n)$ with all $(a_n, b_n) \in \Omega$.

All these types of convergence sets refer to classical convergence. For general convergence we shall use the terms *general convergence sets, conditional general convergence sets,* and *uniform general convergence sets with respect to some set $W \subseteq \hat{\mathbf{C}}$,* where

$$|S_{n+m}(w_{n+m}) - S_n(w_n)| \leq \lambda_n \qquad \text{for all } m, n \in \mathbf{N} \quad \text{and} \quad w_k \in W.$$
(3.1.6)

For continued fractions $\mathbf{K}(a_n/1)$, a convergence set Ω can always be described as $\Omega = E \times \{1\}$. For short we say that $E \subseteq \mathbf{C}$ is a convergence set for continued fractions $\mathbf{K}(a_n/1)$ if $\Omega = E \times \{1\}$ is a convergence set.

Similarly, we say that $G \subseteq \mathbf{C}$ is a convergence set for continued fractions $\mathbf{K}(1/b_n)$ if $\Omega = \{1\} \times G$ is a convergence set. For instance, the Van Vleck sector G_ϵ in (3.1.4) is a conditional convergence set for continued fractions $\mathbf{K}(1/b_n)$.

Sometimes we need the more general notion of a *sequence* $\{\Omega_n\}_{n=1}^\infty \subseteq \mathbf{C} \times \mathbf{C}$ *of convergence sets*: If $(a_n, b_n) \in \Omega_n$ for all n, then $\mathbf{K}(a_n/b_n)$ converges.

Example 8 The 3-periodic continued fraction

$$\frac{2}{1+}\frac{1}{1-}\frac{1}{1+}\frac{2}{1+}\frac{1}{1-}\frac{1}{1+}\cdots = \mathbf{K}\frac{a_n}{1}$$

from Example 6 converges generally, but diverges in the classical sense. Hence $\{\Omega_n\}$ where $\Omega_n = \{a_n\} \times \{1\}$ is a 3-periodic sequence of general convergence sets, but not a sequence of convergence sets.

── ◇

A 2-periodic sequence $\{\Omega_n\}$ of convergence sets is determined by the pair (Ω_1, Ω_2). We say that (Ω_1, Ω_2) is a pair of *twin convergence sets*, or, for short, that Ω_1, Ω_2 are twin convergence sets.

Without loss of generality we shall restrict ourselves to continued fractions $b_0 + \mathbf{K}(a_n/b_n)$ with $b_0 = 0$ in this section.

3.2   Value sets

To determine whether a continued fraction $\mathbf{K}(a_n/b_n)$ converges or not, we have the following tool:

Definition. *We say that $\{V_n\}_{n=0}^{\infty}$ is a sequence of value sets for $\mathbf{K}(a_n/b_n)$ if all $V_n \subseteq \hat{\mathbf{C}}$, $V_n \neq \emptyset$ and*

$$s_n(V_n) = \frac{a_n}{b_n + V_n} \subseteq V_{n-1} \quad \text{for } n = 1, 2, 3, \ldots . \tag{3.2.1}$$

(In the literature $\{V_n\}$ is often referred to as pre value sets.) The importance of value sets lies mainly in the fact that they contain values of approximants of $\mathbf{K}(a_n/b_n)$:

Theorem 8 *Let $\{V_n\}$ be a sequence of value sets for $\mathbf{K}(a_n/b_n)$. Then*

$$S_k^{(n)}(w_{n+k}) = \frac{a_{n+1}}{b_{n+1}} \frac{a_{n+2}}{+ b_{n+2}} + \cdots + \frac{a_{n+k}}{b_{n+k} + w_{n+k}} \in V_n \tag{3.2.2}$$

for all $w_{n+k} \in V_{n+k}$, for $n = 0, 1, 2, \ldots$ and $k = 1, 2, 3, \ldots$.

In particular $S_k(w_k) = S_k^{(0)}(w_k) \in V_0$ if $w_k \in V_k$. Theorem 8 is a simple consequence of (3.2.1) and the fact that $S_k^{(n)} = s_{n+1} \circ s_{n+2} \circ \cdots \circ s_{n+k}$.

If $0 \in V_n$ for all n, then $S_k^{(n)}(0) \in V_n$ for all k by Theorem 8. That is, V_n contains all the classical approximants $f_k^{(n)}$ of $\mathbf{K}(a_n/b_n)$. In this case we say that $\{V_n\}$ is a sequence of *classical value sets* for $\mathbf{K}(a_n/b_n)$. The reason for this is that historically the emphasis has mostly been on classical approximants, and thus, when one referred to value sets or value regions, one always meant sets containing the classical approximants. See for instance [JoTh80, p. 64]. (For information on the historical development of this concept we refer to the section of remarks at the end of this chapter.)

Following the classical ideas, we say that V is a value set for $\mathbf{K}(a_n/b_n)$ if $\{V_n\}$, where all $V_n = V$, is a sequence of values sets for $\mathbf{K}(a_n/b_n)$. If

$\{V_n\}$ is 2-periodic such that $V_{2n} = V_0$ and $V_{2n+1} = V_1$ for all n, we say that V_0, V_1 are twin value sets for $\mathbf{K}(a_n/b_n)$.

Assume that $\mathbf{K}(a_n/b_n)$ converges generally to some value $f \in \hat{\mathbf{C}}$. Will then $f \in \bar{V}_0$? Here and in the rest of the book $\bar{A}$ denotes the closure of the set A in $\hat{\mathbf{C}}$. We shall try to avoid confusion with the complex conjugate $\bar{z}$ of a complex number z.) The answer is of course YES if $S_n(u_n) \to f$ for some sequence $\{u_n\}$ from $\{V_n\}$; i.e. $u_n \in V_n$ for all n. According to Theorem 4 in Chapter II it suffices that we can find two sequences $\{p_n\}$ and $\{q_n\}$ from $\{V_n\}$ such that

$$\liminf d(p_n, q_n) > 0\,,$$

where $d(u, v)$ denotes the chordal metric on the Riemann sphere. In that case we can always construct a sequence $\{u_n\}$ from $\{V_n\}$ such that $S_n(u_n) \to f$. We just use the same technique as in the proof of Corollary 5 in Chapter II. This proves the first two statements in the following theorem. We shall return to the proof of Theorem 9C after having seen some examples.

Theorem 9 *Let $\{V_n\}$ be a sequence of value sets for the generally convergent continued fraction $\mathbf{K}(a_n/b_n)$ such that $\bar{V}_0 \neq \hat{\mathbf{C}}$ and*

$$\liminf_{n \to \infty} \operatorname{diam}_d(V_n) > 0, \quad \text{where}$$
$$\operatorname{diam}_d(V_n) = \sup\{d(u, v); u, v \in V_n\}\,. \tag{3.2.3}$$

Then:

A. *The value f of $\mathbf{K}(a_n/b_n)$ is contained in $\bar{V}_0$.*

B. *The value $f^{(n)}$ of the nth tail of $\mathbf{K}(a_n/b_n)$ is contained in $\bar{V}_n$.*

C. $\lim_{n \to \infty} S_n(w_n) = f$ *for every sequence $\{w_n\}$ such that $w_n \in V_n$ and*

$$\liminf_{n \to \infty} \operatorname{dist}_d(w_n, \partial V_n) > 0, \quad \text{where}$$
$$\operatorname{dist}_d(w_n, \partial V_n) = \inf\{d(w_n, v); v \in \partial V_n\}\,. \tag{3.2.4}$$

Here ∂V_n denotes the boundary of the set V_n, (and $\bar{V}_n$ denotes the closure of V_n in $\hat{\mathbf{C}}$).

Example 9 The unit disk

$$V = U = \{w \in \mathbf{C}; \, |w| < 1\} \tag{3.2.5}$$

is a (classical) value set for the Śleszyński-Pringsheim convergence set (3.1.1). This can be seen from the proof of Theorem 1 in Chapter I (or by direct verification).

We know that every continued fraction from the Śleszyński-Pringsheim set (3.1.1) converges to some value $f \in \bar{V}$ since $0 \in V$ and thus $S_n(0) \in V$ for all n. So in this example Theorem 9A, B does not bring anything new. From Theorem 9C we can conclude that not only the classical approximants of such a continued fraction from (3.1.1) converge to the value f, but

$$\lim_{n \to \infty} S_n(w) = f \quad \text{for all } w \in V. \tag{3.2.6}$$

From Stieltjes-Vitali's theorem, to be presented later (Subsection *3.6*), we can conclude that the convergence in (3.2.6) is uniform on compact subsets of V. This was mentioned in a remark to the Śleszyński/Pringsheim theorem in Chapter I.

The Worpitzky disk (3.1.2) has the value set

$$V = \{w \in \mathbf{C}; \, |w| < 1/2\}. \tag{3.2.7}$$

Hence, every continued fraction $\mathbf{K}(a_n/1)$ from the Worpitzky disk has approximants $S_n(w)$ converging locally uniformly to a constant function in this V. This was also mentioned in Chapter I.

_____◇

Example 10 A value set for the Van Vleck sector (3.1.4) is

$$V_\epsilon = G_\epsilon = \left\{w \in \mathbf{C}; \, |\arg w| < \frac{\pi}{2} - \epsilon\right\}. \tag{3.2.8}$$

(See the proof of Theorem 2 in Chapter I.) Since $b_{n+1} \in G_\epsilon \Rightarrow 1/b_{n+1} \in V_\epsilon$, we find from Theorem 8 that $S_k^{(n)}(1/b_{n+k+1}) = S_{k+1}^{(n)}(0) = f_{k+1}^{(n)} \in V_\epsilon$

for all $n \geq 0$ and $k \geq 0$. Hence every convergent continued fraction $\mathbf{K}(1/b_n)$ from G_ϵ converges to a value in $\bar{V}_\epsilon$, and A and B in Theorem 9 are obvious. From Theorem 9C we find that if $\mathbf{K}(1/b_n)$ is a convergent continued fraction from G_ϵ, then (3.2.6) holds with $V = V_\epsilon$.

$\diamond$

Other examples where Part A and B are no longer obvious will come later.

Proof of Theorem 9C: Let $t_0 \in \hat{\mathbf{C}} \setminus \bar{V}_0$. Then $t_0 \neq f$, and the tail sequence $\{t_n\}$; $t_n = S_n^{-1}(t_0)$ is not a right tail sequence. Since $S_n(V_n) \subseteq V_0$ it follows that $V_n \subseteq S_n^{-1}(V_0)$ and thus that

$$S_n^{-1}(\hat{\mathbf{C}} \setminus \bar{V}_0) \subseteq \hat{\mathbf{C}} \setminus \bar{V}_n . \qquad (3.2.9)$$

Hence $t_n \in \hat{\mathbf{C}} \setminus \bar{V}_n$ for all n. All tail sequences $\{\tilde{t}_n\}$ which are not the right tail sequence of $\mathbf{K}(a_n/b_n)$, have the same asymptotic behavior in the sense that $d(t_n, \tilde{t}_n) \to 0$. (See Theorem 6 in Chapter II). Hence $\{w_n\}$ stays "sufficiently far away" from all such "dangerous" tail sequences in the sense of Theorem 4 in Chapter II, and thus $S_n(w_n) \to f$. ∎

Value sets are in no way unique. This is illustrated by the following two examples:

Example 11 Let $\{t_n\}_{n=0}^\infty$ be a tail sequence for $\mathbf{K}(a_n/b_n)$. Then $\{V_n\}_{n=0}^\infty$, where V_n is the one-point set containing t_n, is a sequence of value sets for $\mathbf{K}(a_n/b_n)$.

$\diamond$

Example 12 We shall find some value sets for the 3-periodic continued fraction in Example 8. Combining results in Example 2 in Chapter II with results in Example 11 above, we find that

$$V_{3n} = \left\{\frac{1}{2}\right\} , \quad V_{3n+1} = \{3\} , \quad V_{3n+2} = \left\{-\frac{2}{3}\right\} , \quad \text{for } n = 0, 1, 2, \ldots$$

is a sequence of value sets. Similarly

$$W_{3n} = \{0\} , \quad W_{3n+1} = \{\infty\} , \quad W_{3n+2} = \{-1\} , \quad \text{for } n = 0, 1, 2, \ldots$$

is another such sequence. Still another one is given by

$$U_{3n} = \left\{ w \in \mathbf{C}; \left| w - \frac{1}{2} \right| < \frac{1}{6} \right\}$$

$$U_{3n+1} = \{ w \in \mathbf{C}; |w - 3| < 1 \}$$

$$U_{3n+2} = \left\{ w \in \mathbf{C}; \left| w + \frac{2}{3} \right| < \frac{1}{12} \right\} .$$

——◇

3.3 *Value set techniques I. A posteriori truncation error bounds*

In Formula (1.4.5) of Chapter II we proved that if f_n, f_{n-1} and f_{n+k} are distinct and finite, then

$$f_{n+k} - f_n = \frac{-f_k^{(n)}}{h_n + f_k^{(n)}}(f_n - f_{n-1}) \tag{3.3.1}$$

for $n, k \in \mathbf{N}$. But when are f_n, f_{n-1} and f_{n+k} distinct and finite? The following theorem gives a very simple criterion:

Theorem 10 *Let $\{V_n\}$ be a sequence of classical value sets for $\mathbf{K}(a_n/b_n)$ such that $\infty \notin V_n$ for all n. Then all the approximants f_n of $\mathbf{K}(a_n/b_n)$ are distinct and finite.*

Proof : Since $f_n \in V_0$ and $\infty \notin V_0$, it follows that $f_n \neq \infty$ for all n. That $f_n \neq f_{n+k}$ is a consequence of the determinant formula: Assume that $f_n = f_{n+k}$ for some $k \geq 1$; i.e. that $S_n(0) = S_n(f_k^{(n)})$. Then $0 = f_k^{(n)}$ (which rules out the case $k = 1$), and thus $f_{k-1}^{(n+1)} = s_{n+1}^{-1}(f_k^{(n)}) = -b_{n+1} + a_{n+1}/f_k^{(n)} = \infty$ which is impossible since $f_{k-1}^{(n+1)} \in V_{n+1}$ whereas $\infty \notin V_{n+1}$. ∎

To derive useful a posteriori truncation error bounds from (3.3.1) we want to estimate the factor $f_k^{(n)}/(h_n + f_k^{(n)})$. We shall show how this can be done by means of value sets in the special example where the continued fraction has the form $\mathbf{K}(a_n/1)$ and has a bounded classical value set V. Then $f_k^{(n)} \in V$ for all n and k, and by (1.2.7) in Chapter II

$$h_n = 1 + \frac{a_n}{1} + \frac{a_{n-1}}{1} + \cdots + \frac{a_2}{1} \in 1 + V \quad \text{for all } n \in \mathbf{N}. \qquad (3.3.2)$$

Example 13 Let $E = \{w \in \mathbf{C};\ |w| \le g(1-g)\}$ and $V = \{w \in \mathbf{C};\ |w| < g\}$ for a positive number $g \le 1/2$. Then V is a classical value set for E since $0 \in V$ and

$$\left| \frac{a}{1+w} \right| < \frac{g(1-g)}{1-g} = g \quad \text{if } a \in E, \quad w \in V.$$

For $g = 1/2$ this is exactly the Worpitzky situation.

Let $g < 1/2$. Since $h_n \in 1 + V$ by (3.3.2) and $f_k^{(n)} \in V$, we then have that $|f_k^{(n)}| \le g$ and $|h_n + f_k^{(n)}| \ge 1 - 2g$, and thus by (3.3.1)

$$|f - f_n| \le \frac{g}{1 - 2g}|f_n - f_{n-1}| \quad \text{for } n = 1, 2, 3, \ldots.$$

In Example 3 in Chapter II we also considered continued fractions $\mathbf{K}(c_n/1)$ with $|c_n| \le M < 1/4$. (The notation differed slightly.) We found that

$$|f - f_n| < \frac{2M}{1 - 4M}|f_n - f_{n-1}|.$$

Our new truncation error bound is slightly better since $M = g(1-g)$ implies that $g = (1 - \sqrt{1 - 4M})/2$ so that

$$\frac{g}{1-g} = \frac{1 - \sqrt{1 - 4M}}{2\sqrt{1 - 4M}} = \frac{4M}{2\sqrt{1 - 4M}(1 + \sqrt{1 - 4M})} < \frac{2M}{1 - 4M}.$$

For $M = 0.2$ we have for instance

$$\frac{g}{1-g} = 0.382, \qquad \frac{2M}{1 - 4M} = 2.$$

Numerically this improvement is not much worth, of course. As is often the case, we can find reasonable truncation error bounds by using rough estimates, and we gain very little by careful refinements (unless we can pull in some new factors going to zero. We shall return to this point in Section 5).

————————————————————————————————◇

The crucial points are really that $f^{(n)} = \lim_{k \to \infty} f_k^{(n)} \in \bar{V}$ and that $h_n \in 1 + V$. If we turn to the more general situation where $\{V_n\}_{n=0}^{\infty}$ is a sequence of value sets for $\mathbf{K}(a_n/b_n)$, we still have that $f^{(n)} \in \bar{V}_n$ if the V_ns are "large enough" in the sense of Theorem 9. But we loose control over

$$h_n = b_n + \cfrac{a_n}{b_{n-1} + \cfrac{a_{n-1}}{b_{n-2} + \cdots + \cfrac{a_2}{b_1}}}. \tag{3.3.3}$$

We find, though, that if (V_0, V_1) is a pair of classical twin value sets for $\mathbf{K}(a_n/1)$, then $h_{2n} \in 1 + V_1$ and $h_{2n+1} \in 1 + V_0$ for $n \geq 1$.

The same idea also works if V is a classical value set (or V_0, V_1 are classical twin value sets) for a continued fraction of the form $\mathbf{K}(1/b_n)$. Then $s_n(V) = 1/(b_n + V) \subseteq V$ and $(1/b_n) \in V$ for all n and therefore $h_n \in b_n + V \subseteq 1/V$. Hence, if V is bounded, say $|w| \leq M$ for all $w \in V$, and

$$\inf\{|x + y|;\ x \in V, y \in 1/V\} = d > 0, \tag{3.3.4}$$

then

$$|f_{n+m} - f_n| \leq \frac{M}{d}|f_n - f_{n-1}|. \tag{3.3.5}$$

3.4 Value set techniques II. A priori truncation error bounds

Let $\{V_n\}_{n=0}^{\infty}$ be a sequence of value sets for $\mathbf{K}(a_n/b_n)$. Then it follows from the definition (3.2.1) that if $w_n \in \bar{V}_n$ then

$$S_n(w_n) \in S_n(\bar{V}_n) = S_{n-1}(s_n(\bar{V}_n)) \subseteq S_{n-1}(\bar{V}_{n-1}), \tag{3.4.1}$$

so that $K_n = S_n(\bar{V}_n)$ forms a sequence of nested closed sets, $\bar{V}_0 \supseteq K_1 \supseteq K_2 \supseteq \cdots$. This sequence will therefore converge to a non-empty set K.

Assume first that K contains only one point, $K = \{f\}$. (The limit point case.) Then $f \in K_n$ for all n, and

$$|f - S_n(w_n)| \leq \operatorname{diam}(K_n) \to 0 \quad \text{if } w_n \in \bar{V}_n, \tag{3.4.2}$$

so that $S_n(w_n) \to f$. If now $\liminf \operatorname{diam}_d(V_n) > 0$, where the diameter is measured by the chordal metric as in (3.2.3), then $\mathbf{K}(a_n/b_n)$ converges generally to f. Bounds for $\operatorname{diam}(K_n)$ can be used as a priori bounds for the truncation error $|f - S_n(w_n)|$.

To keep the computation simple one often chooses V_n to be circular disks on the Riemann sphere $\hat{\mathbf{C}}$. The following lemma may then be of help:

Lemma 11 *Let D be a circular disk with center at c and radius r, and let $-b \notin \bar{D}$. Then*

$$s(D) = \frac{a}{b+D}; \qquad a \neq 0$$

is a circular disk with center c_a and radius r_a given by

$$c_a = \frac{(\bar{b} + \bar{c})a}{|b+c|^2 - r^2}, \qquad r_a = \frac{r|a|}{|b+c|^2 - r^2}. \qquad (3.4.3)$$

The proof is a simple exercise in mapping theory for linear fractional transformations and is left out here.

Example 14 Let E and V be as in Example 13 for a fixed g, $0 < g \leq 1/2$. Then $K_n = S_n(\bar{V}) = s_1 \circ s_1 \circ \cdots \circ s_n(\bar{V}) \subseteq \bar{V}$ when all $s_k(w) = a_k/(1+w)$ and $a_k \in E$. By Lemma 11, $D_{n,n} = s_n(\bar{V})$ is a circular disk with center and radius given by

$$c_{n,n} = \frac{a}{1-g^2}, \qquad r_{n,n} = \frac{g|a|}{1-g^2} \leq \frac{g^2}{1+g} \leq \frac{1}{6}.$$

Furthermore $D_{n,n-1} = s_{n-1}(D_{n,n})$ is a circular disk with center $c_{n,n-1}$ and radius

$$r_{n,n-1} \leq \frac{r_{n,n} g(1-g)}{|1+c_{n,n}|^2 - r_{n,n}^2},$$

and so on. Observe that since $D_{n,k} \subseteq \bar{V}$ it follows that $|c_{n,k}| + r_{n,k} \leq g$, and thus that $|1 + c_{n,k}| \geq 1 - |c_{n,k}| \geq 1 - g + r_{n,k}$, which again means that

$$|1 + c_{n,k}|^2 - r_{n,k}^2 \geq (1 - g + r_{n,k})^2 - r_{n,k}^2 = (1-g)^2 + 2(1-g)r_{n,k}.$$

Therefore

$$r_{n,k-1} \leq \frac{r_{n,k}g(1-g)}{|1+c_{n,k}|^2 - r_{n,k}^2} \leq \frac{r_{n,k}g(1-g)}{(1-g)^2 + 2(1-g)r_{n,k}}$$

$$= \frac{g}{\dfrac{1-g}{r_{n,k}} + 2} \leq \frac{g}{\dfrac{1-g}{(r_{n,k})_{max}} + 2} = \frac{g(r_{n,k})_{max}}{1-g+2(r_{n,k})_{max}},$$

where $(r_{n,k})_{max}$ is a positive number such that $r_{n,k} \leq (r_{n,k})_{max}$. This means that

$$r_{n,n-1} \leq \frac{1}{6}\frac{g}{1-g+2/6} \leq \frac{1}{6}\frac{1/2}{1/2+2/6} = \frac{1}{10},$$

$$r_{n,n-2} \leq \frac{1}{10}\frac{g}{1-g+2/10} \leq \frac{1}{10}\frac{1/2}{1/2+2/10} = \frac{1}{14},$$

and so on, and thus the radius R_n of $K_n = S_n(\bar{V})$ is bounded by

$$R_n = r_{n,1} \leq \frac{1}{4n+2} \quad \text{for } n = 1, 2, 3, \ldots.$$

Since therefore $R_n \to 0$, we have proved that E is a convergence set (which we already knew from Worpitzky's theorem), and we have proved that

$$|f - S_n(w)| \leq \text{diam}(K_n) = 2R_n \leq \frac{1}{2n+1} \quad \text{for } w \in \bar{V}. \qquad (3.4.4)$$

If $g < 1/2$ we can do even better. Then

$$r_{n,k-1} \leq \frac{r_{n,k}g(1-g)}{(1-g)^2 + 2(1-g)r_{n,k}} < \frac{g}{1-g}r_{n,k}$$

where $g/(1-g) < 1$. By the same argument we get $R_1 \leq g^2/(1+g)$ and

$$|f - S_n(w)| \leq 2R_n \leq \frac{2g^2}{1+g}\left(\frac{g}{1-g}\right)^{n-1} \quad \text{for } w \in \bar{V}. \qquad (3.4.5)$$

If $g < 1/2$, the bound (3.4.5) approaches 0 faster than (3.4.4) and is thus better.

$\diamond$

Example 15 We want to compute the value of

$$\mathbf{K}\, \frac{0.2e^{ni}}{1} = \frac{0.2e^{i}}{1} + \frac{0.2e^{2i}}{1} + \frac{0.2e^{3i}}{1} + \cdots$$

with an error less than 0.05, as we did in Example 13, but this time we want to determine the index n for $S_n(0)$ in advance by means of the a priori truncation error bound (3.4.5). We use $g(1-g) = 0.2$ so that $g = (1 - \sqrt{0.2})/2 < 0.28$, and we require

$$\frac{2g^2}{1+g}\left(\frac{g}{1-g}\right)^{n-1} < 0.1225 \cdot 0.39^{n-1} \le 0.05$$

which holds already for $n = 2$. ◇

3.5   Value set techniques III. The Hillam-Thron theorem

If the sequence of nested sets $\{K_n\}$ in Subsection 3.4 converges to a larger set K instead of a one-point-set $\{f\}$, the question about convergence of $\{S_n(w_n)\}$ is a little more tricky. But the very fact that $\lim \text{diam}(K_n) = d > 0$ can have implications which lead to convergence of the continued fraction. This is demonstrated in the following theorem due to Hillam and Thron, [HiTh65]:

Theorem 12 (The Hillam-Thron theorem) *Let $V \subseteq \mathbf{C}$ be an open circular disk with $0 \in V$. If V is a value set for the continued fraction $\mathbf{K}(a_n/b_n)$, then $\mathbf{K}(a_n/b_n)$ converges.*

Notice that both the Śleszyński-Pringsheim theorem and the Worpitzky theorem are simple consequences of the Hillam-Thron theorem.

To prove Theorem 12 we shall use a value set technique, but not directly on V. We shall rather prove the following lemma which has wider applications:

Lemma 13 *Let $U = \{z \in \mathbf{C}; |z| < 1\}$ be the unit disk, and let $0 \le k < 1$. Let $\{t_n\}$ be a sequence of linear fractional transformations such that*

$$t_n(U) \subseteq U \quad and \quad |t_n(\infty)| \le k \quad for \ all \ n, \qquad (3.5.1)$$

and let $T_1 = t_1, T_n = T_{n-1} \circ t_n = t_1 \circ t_2 \circ \cdots \circ t_n$ for all n. Then $\{T_n(w)\}$ converges locally uniformly in U to a constant function $T(w) \equiv c \in \bar{U}$.

Also this can be found (in a slightly weaker form) in [HiTh65]. In this setting U plays the role of the value set and $\{T_n\}$ the role of the continued fraction. We still have

$$K_n = T_n(\bar{U}) \subseteq T_{n-1}(\bar{U}) = K_{n-1} \subseteq \cdots \subseteq \bar{U} \qquad (3.5.2)$$

where K_n now are circular disks, and we distinguish between the limit point case where $\operatorname{diam}(K_n) \to 0$ and what we can call the limit circle case where $\operatorname{diam}(K_n) \to d > 0$.

Proof of Lemma 13: Let C_n and R_n be center and radius of K_n for each n. If $R_n \to 0$ then the lemma is trivial, so assume that $R_n \to R > 0$. The nestedness (3.5.2) then ensures that $C_n \to C$, the center of K. The most general linear fractional transformation which maps U onto U is given by

$$S(w) = e^{\iota \omega} \frac{w - \alpha}{1 - \bar{\alpha} w} \quad where \ \omega \in \mathbf{R}, \qquad \alpha \in U.$$

(Here $\bar{\alpha}$ denotes the complex conjugate of α.) Hence, we can write

$$T_n(w) = C_n + R_n e^{\iota \omega_n} \frac{w - \alpha_n}{1 - \bar{\alpha}_n w} \quad for \ n = 1, 2, 3, \ldots, \qquad (3.5.3)$$

where all $\omega_n \in \mathbf{R}$ and $\alpha_n \in U$. The plan now is to develop two inequalities which together will prove our lemma. The first one is a result of the condition $t_n(\infty) \in \bar{U}$, which leads to $T_n(\infty) = T_{n-1}(t_n(\infty)) \in T_{n-1}(\bar{U}) = K_{n-1}$; that is, $T_n(\infty) = C_{n-1} + R_{n-1} z_n$ for some $z_n \in \bar{U}$. Since therefore

$$T_n(\infty) - C_n = -\frac{R_n e^{\iota \omega_n}}{\bar{\alpha}_n} = C_{n-1} + R_{n-1} z_n - C_n,$$

we have

$$\frac{R_n}{|\alpha_n|} = |(C_{n-1} - C_n) + R_{n-1}z_n| \le (R_{n-1} - R_n) + R_{n-1}, \qquad (3.5.4)$$

and thus

$$\frac{R_n}{R_{n-1}} \le \frac{2|\alpha_n|}{1 + |\alpha_n|} = 1 - \frac{1 - |\alpha_n|}{1 + |\alpha_n|},$$

which gives us the inequality

$$R_n \le R_1 \prod_{j=2}^{n}(1 - \delta_j) \quad \text{where } \delta_j = \frac{1 - |\alpha_j|}{1 + |\alpha_j|} < 1. \qquad (3.5.5)$$

The second inequality depends on the stronger condition that if $k_n = t_n(\infty)$, then $|k_n| \le k$ for all n. We get

$$\begin{aligned}
T_{n+1}(\infty) - T_n(\infty) &= T_n(t_{n+1}(\infty)) - T_n(\infty) = T_n(k_{n+1}) - T_n(\infty) \\
&= R_n e^{i\omega_n}\left(\frac{k_{n+1} - \alpha_n}{1 - \bar{\alpha}_n k_{n+1}} + \frac{1}{\bar{\alpha}_n}\right) \\
&= R_n e^{i\omega_n}\frac{1 - |\alpha_n|^2}{\bar{\alpha}_n(1 - \bar{\alpha}_n k_{n+1})} \qquad (3.5.6)
\end{aligned}$$

so that

$$|T_{n+1}(\infty) - T_n(\infty)| \le \frac{R_n}{|\alpha_n|}\frac{1 - |\alpha_n|^2}{1 - k} \le \frac{2R_1}{1 - k}(1 - |\alpha_n|^2),$$

by use of (3.5.4), and thus the inequality

$$|T_{n+m}(\infty) - T_n(\infty)| \le \frac{2R_1}{1 - k}\sum_{j=0}^{m-1}(1 - |\alpha_{n+j}|^2) \le \frac{8R_1}{1 - k}\sum_{j=0}^{m-1}\delta_{n+j}. \qquad (3.5.7)$$

Since $R_n \to R > 0$, it follows by (3.5.5) that $\sum \delta_j < \infty$. Hence, by (3.5.7) $\{T_n(\infty)\}$ is a Cauchy sequence and thus converges to a value $c \in K$. Since by (3.5.3)

$$|T_n(w) - T_n(\infty)| = R_n\left|\frac{w - \alpha_n}{1 - \bar{\alpha}_n w} + \frac{1}{\bar{\alpha}_n}\right| = \frac{R_n}{|\alpha_n|}\cdot\frac{1 - |\alpha_n|^2}{|1 - \bar{\alpha}_n w|}, \qquad (3.5.8)$$

where $|\alpha_n| \to 1$ (since $\delta_n \to 0$), this actually proves that $T_n(w) \to c$ locally uniformly in $\mathbf{C} \setminus \partial U$ in the limit circle case. (∂U is the boundary of U; i.e. the unit circle.) ∎

Proof of Theorem 12:   Let γ and ρ be the center and the radius of ∂V, and let $t(w) = \rho w + \gamma$ so that $t(U) = V$. Then $t_n = t^{-1} \circ s_n \circ t$ maps U into U and

$$t_n(\infty) = t^{-1} \circ s_n(\infty) = t^{-1}(0) = -\frac{\gamma}{\rho}$$

for all n, where $|\gamma/\rho| = k < 1$ since $0 \in V$. Hence $T_n(w)$ converges locally uniformly in U to a constant function $T(w) \equiv c \in \bar{U}$. Now

$$
\begin{aligned}
T_n &= t_1 \circ t_2 \circ \cdots \circ t_n = t^{-1} \circ s_1 \circ t \circ t^{-1} \circ s_2 \circ t \circ \cdots \circ t^{-1} \circ s_n \circ t \\
&= t^{-1} \circ s_1 \circ s_2 \circ \cdots \circ s_n \circ t = t^{-1} \circ S_n \circ t. \qquad (3.5.9)
\end{aligned}
$$

Hence $S_n(w)$ converges for all $w \in t(U) = V$ to the constant function $S(w) = f$ where $t^{-1}(f) = c$; i.e. $f = t(c) \in \bar{V}$.   ∎

We can also prove a slightly more general result by means of Lemma 13:

Theorem 14 *Let $\{V_n\}_{n=0}^{\infty}$ be a sequence of value sets for $\mathbf{K}(a_n/b_n)$ consisting of open circular disks with centers γ_n and radii ρ_n such that $|\gamma_n/\rho_n| \leq k$ for all n for some $k < 1$. Then $\mathbf{K}(a_n/b_n)$ converges to a value $f \in \bar{V}_0$.*

Remark: Since $0 \in V_n$ for all n, it follows that $\{V_n\}$ are indeed classical value sets for $\mathbf{K}(a_n/b_n)$.

Proof :   Let $\tau_n(w) = \gamma_n + \rho_n w$ so that $\tau_n(U) = V_n$ for all n. Then $t_n = \tau_{n-1}^{-1} \circ s_n \circ \tau_n$ maps U into U and $t_n(\infty) = \tau_{n-1}^{-1}(0) = -\gamma_{n-1}/\rho_{n-1}$. The result follows by the same line of argument as Theorem 12 since

$$
\begin{aligned}
T_n &= t_1 \circ t_2 \circ \cdots \circ t_n \\
&= \tau_0^{-1} \circ s_1 \circ \tau_1 \circ \tau_1^{-1} \circ s_2 \circ \tau_2 \circ \cdots \circ \tau_{n-1}^{-1} \circ s_n \circ \tau_n \quad (3.5.10) \\
&= \tau_0^{-1} \circ S_n \circ \tau_n.
\end{aligned}
$$

                                                                                    ∎

Theorem 14 and the Hillam-Thron theorem illustrate an important point in the value set techniques to derive convergence theorems. One picks "nice" value sets $\{V_n\}$. And then $\mathbf{K}(a_n/b_n)$ converges if all s_n have the "right mapping properties" $s_n(V_n) \subseteq V_{n-1}$. Collections of such continued fractions $\mathbf{K}(a_n/b_n)$ can then be described by convergence sets $\{\Omega_n\}$.

We shall return to this in Section 4.

3.6   Value set techniques IV. The Stieltjes-Vitali theorem

Theorem 15 (Stieltjes-Vitali's theorem) *Let* $\{f_n\}$ *be a sequence of holomorphic functions in a region* D, *such that*

(i) *there exist two points* $a, b \in \mathbf{C}$ *such that* $f_n(z) \neq a, f_n(z) \neq b$ *for all* n *and all* $z \in D$, *and*

(ii) $\{f_n(z)\}$ *converges to finite values for every* z *in an infinite set* $\Delta \subseteq D$ *which has at least one point of accumulation in* D.

Then $\{f_n(z)\}$ *converges locally uniformly in* D *to a holomorphic function.*

We recall that the word "region" here (as it is generally in this book) is used in the *strict* sense: open, connected set. This theorem is a consequence of Montel's theorem for normal families. For the proof we refer to text books on functions of a complex variable, for instance [Hille62, p. 248–251].

The idea of application of this theorem is best explained if V is a classical value set for an element region Ω; i.e. $\Omega \subseteq \mathbf{C} \times \mathbf{C}$ is an open, connected set and $a/(b+V) \subseteq V$ for all $(a,b) \in \Omega$. Assume that we know that every continued fraction $\mathbf{K}(a_n/b_n)$ from a subset $\Omega_0 \subseteq \Omega$ converges. We can then introduce an auxiliary variable z such that

$$
\begin{aligned}
(a_n(z), b_n(z)) \in \Omega_0 \qquad &\text{if } z \in \Delta, \\
(a_n(z), b_n(z)) \in \Omega \qquad &\text{if } z \in D,
\end{aligned}
\tag{3.6.1}
$$

to get $f_n(z) \in V$ for $z \in \Omega$, where f_n is assumed to be holomorphic. So if there exist two points $a, b \in \mathbf{C}$ which are not contained in V, then $\lim_{n \to \infty} f_n(z)$ exists for all $z \in D$.

We can also follow the same idea if V is a non–classical value set for Ω. Then we use modified approximants $f_n^*(z) = S_n(w, z)$ for some $w \in V$ and consider general convergence.

In Chapter I we presented Van Vleck's convergence theorem in Theorem 2. We promised to return to the proof. We shall prove this result now by means of the Stieltjes-Vitali theorem, by extending the convergence result in Theorem 3 for continued fractions with positive elements. The proof is taken from [JoTh80, p. 89].

Proof of Van Vleck's theorem: We want to prove that G_ϵ given by (3.1.4) is a conditional convergence set. We know that $V_\epsilon = G_\epsilon$ is a value set for G_ϵ. (See (3.2.8).)

Let $\mathbf{K}(1/b_n)$ be an arbitrarily chosen continued fraction from G_ϵ. If $\sum |b_n| < \infty$, then $\mathbf{K}(1/b_n)$ diverges, so assume that $\sum |b_n| = \infty$. Let

$$\beta_n = \arg(b_n) \quad \text{and} \quad d_n(z) = |b_n| e^{i\beta_n z} \quad \text{for all } n . \tag{3.6.2}$$

Then $d_n(z) \in G_{\epsilon/2}$ if $|\arg(|b_n| e^{i\beta_n z})| < \pi/2 - \epsilon/2$; i.e. if $|\beta_n \Re(z)| < \pi/2 - \epsilon/2$ where $|\beta_n| < \pi/2 - \epsilon$. Hence

$$d_n(z) \in G_{\epsilon/2} \quad \text{if } z \in D = \left\{ z \in \mathbf{C}; \, |\Re(z)| < \frac{\pi - \epsilon}{\pi - 2\epsilon} \right\} . \tag{3.6.3}$$

Moreover
$$d_n(z) > 0 \quad \text{if } z \in \Delta = \{ z \in \mathbf{C}; \, \Re(z) = 0 \} . \tag{3.6.4}$$

Since
$$\sum d_n(z) = \sum |b_n| e^{-\beta_n \Im(z)} > e^{-|\Im(z)| \pi/2} \sum |b_n| = \infty; \qquad z \in \Delta ,$$

it follows from Seidel-Stern's theorem, Theorem 3, that $\mathbf{K}(1/d_n(z))$ converges for $z \in \Delta$; i.e. that the classical approximants $f_n(z)$ of $\mathbf{K}(1/d_n(z))$ converge for $z \in \Delta$. By Stieltjes-Vitali's theorem it follows therefore that $\mathbf{K}(1/d_n(z))$ converges for all $z \in D$. In particular it converges for $z = 1$. Hence $\mathbf{K}(1/b_n)$ converges since $b_n = d_n(1)$.   ∎

The idea in this subsection can be extended to sequences $\{V_n\}$ of value sets for sequences $\{\Omega_n\}$ of element regions.

3.7 Smaller value sets for truncation error bounds

If we want to use value sets to estimate truncation error bounds, as we did in Subsection *3.3* and *3.4*, we really want these sets to be "as small as possible" to obtain best possible bounds. In some cases we can then use:

Lemma 16 *Let $\{U_n\}$ and $\{W_n\}$ be two sequences of value sets for $\mathbf{K}(a_n/b_n)$. If $V_n = U_n \cap W_n \neq \emptyset$ for all n, then $\{V_n\}$ is also a sequence of value sets for $\mathbf{K}(a_n/b_n)$.*

Proof : If $a_n/(b_n + U_n) \subseteq U_{n-1}$ and $a_n/(b_n + W_n) \subseteq W_{n-1}$, then $a_n/(b_n + U_n \cap W_n) \subseteq U_{n-1} \cap W_{n-1}$. ∎

Lemma 17 *Let W_0, W_1 be twin value sets for $\mathbf{K}(a_n/b_n)$. Then:*

A. *If all $b_n = 1$ and*

$$V_0 = W_0 \setminus (-1 - \bar{W}_1) \neq \emptyset, \quad V_1 = W_1 \setminus (-1 - \bar{W}_0) \neq \emptyset, \quad (3.7.1)$$

then V_0, V_1 are also twin value sets for $\mathbf{K}(a_n/1)$.

B. *If all $a_n = 1$ and*

$$V_0 = W_0 \setminus (-1/\bar{W}_1) \neq \emptyset, \quad V_1 = W_1 \setminus (-1/\bar{W}_0) \neq \emptyset, \quad (3.7.2)$$

then V_0, V_1 are also twin value sets for $\mathbf{K}(1/b_n)$.

Proof : **A:** In view of Lemma 16 it suffices to prove that $U_0 = \hat{\mathbf{C}} \setminus (-1 - \bar{W}_1)$ and $U_1 = \hat{\mathbf{C}} \setminus (-1 - \bar{W}_0)$ are value sets for $\mathbf{K}(a_n/1)$; i.e. that

$$s_{2n}(U_0) = \frac{a_{2n}}{1 + U_0} \subseteq U_1 \quad \text{and} \quad s_{2n+1}(U_1) = \frac{a_{2n+1}}{1 + U_1} \subseteq U_0,$$

or, since s_n is bijective, that

$$s_{2n}(\hat{\mathbf{C}} \setminus U_0) \supseteq \hat{\mathbf{C}} \setminus U_1 \quad \text{and} \quad s_{2n+1}(\hat{\mathbf{C}} \setminus U_1) \supseteq \hat{\mathbf{C}} \setminus U_0,$$

i.e.

$$s_{2n}(-1 - \bar{W}_1) \supseteq -1 - \bar{W}_0 \quad \text{and} \quad s_{2n+1}(-1 - \bar{W}_0) \supseteq -1 - \bar{W}_1,$$

i.e.

$$-1 - \bar{W}_1 \supseteq s_{2n}^{-1}(-1 - \bar{W}_0) = -1 + \frac{a_{2n}}{-1 - W_0}$$

and

$$-1 - \bar{W}_0 \supseteq s_{2n+1}^{-1}(-1 - \bar{W}_1) = -1 + \frac{a_{2n+1}}{-1 - W_1}.$$

But this follows directly from the fact that W_0, W_1 are twin value sets for $\mathbf{K}(a_n/1)$.

B: This part can be proved in a similar way, observing that if $s(w) = 1/(b + w)$, then $s^{-1}(w) = -b + 1/w$. ∎

4   Convergence results

4.1   Two useful lemmas

Value set techniques give results on general convergence for continued fractions. If the value sets do not contain the classical approximants from some n on, it may be difficult or even impossible to prove classical convergence. This is not so if $\mathbf{K}(a_n/b_n)$ has the form $\mathbf{K}(a_n/1)$ or $\mathbf{K}(1/b_n)$ and $\{V_n\}$ is either 1-periodic or 2-periodic and bounded.

Lemma 18 *Let (V_0, V_1) be a pair of bounded twin value sets for the generally convergent continued fraction $\mathbf{K}(a_n/b_n)$, and let V_0 (or V_1) contain at least two points. Then $\mathbf{K}(a_n/b_n)$ converges in the classical sense if either all $a_n = 1$ or all $b_n = 1$.*

Proof : Let f be the value of $\mathbf{K}(a_n/b_n)$, and observe that both V_0 and V_1 must contain at least two elements since $s_1(V_1) \subseteq V_0$ and $s_2(V_0) \subseteq V_1$. It follows therefore from Theorem 9A that $f \in \bar{V}_0$ and thus $f \neq \infty$. For a given $t_0 \in \hat{\mathbf{C}}$ we recall from Theorem 3 in Chapter II that the tail sequence $\{t_n\}_{n=0}^{\infty}$ for $\mathbf{K}(a_n/b_n)$ is given by

$$t_n = S_n^{-1}(t_0) = -\left\{b_n + \frac{a_n}{b_{n-1}} \frac{a_{n-1}}{+b_{n-2}} + \cdots + \frac{a_2}{b_1} \frac{a_1}{+(-t_0)}\right\}. \qquad (4.1.1)$$

Assume first that all $b_n = 1$ and choose $t_0 \in -1 - V_1, t_0 \neq f$. Then, by (4.1.1), $t_n \in -1 - V_{(n+1)mod2}$ for all n, and is thus bounded. From properties of tail sequences of generally convergent continued fractions (Theorem 6 in Chapter II) it follows therefore that $\limsup |h_n| < \infty$. This in turn means that $\liminf d(\infty, -h_n) > 0$ so that $\lim S_n(\infty) = f$ by Theorem 4 in Chapter II. This proves the classical convergence since $S_n(\infty) = S_{n-1}(0)$.

Assume next that all $a_n = 1$, and choose $t_0 \in -1/V_1, t_0 \neq f$. Then $t_n \in -b_n - V_{(n)mod2} \subseteq -1/V_{(n+1)mod2}$ by (4.1.1). Hence $\liminf |t_n| > 0$. By the same argument as above we now get that $\liminf |h_n| > 0$, so that $\liminf d(0, -h_n) > 0$ and thus $\lim S_n(0) = f$. ∎

This first lemma gave a method to conclude classical convergence from general convergence. The next lemma shows a method to conclude classical convergence from the convergence of the even and the odd part of a continued fraction. It is due to Lane and Wall, [LaWa49].

Lemma 19 *Let $\mathbf{K}(a_n/b_n)$ have finite approximants $f_n = S_n(0)$ satisfying*

$$\sum_{n=1}^{\infty} |f_{2n+2} - f_{2n}| < \infty \quad and \quad \sum_{n=1}^{\infty} |f_{2n+1} - f_{2n-1}| < \infty. \qquad (4.1.2)$$

Then $\mathbf{K}(a_n/b_n)$ converges if and only if its Stern-Stolz series

$$\sum_{n=1}^{\infty} \left| b_n \prod_{k=1}^{n} a_k^{(-1)^{n-k+1}} \right| \qquad (4.1.3)$$

diverges to ∞.

Remarks:

1. The condition (4.1.2) implies that $\sum(f_{2n+2} - f_{2n})$ and $\sum(f_{2n+1} - f_{2n-1})$ converge to finite values; i.e. that $f_{2n} \to L_0 \neq \infty$ and $f_{2n+1} \to L_1 \neq \infty$. When (4.1.2) holds, we say that the even and odd parts of $\mathbf{K}(a_n/b_n)$ converge *absolutely*.

2. From the proof of the Śleszyński-Pringsheim theorem in Chapter I, we find that

$$\sum_{n=1}^{\infty} \left| \frac{A_n}{B_n} - \frac{A_{n-1}}{B_{n-1}} \right| = \sum_{n=1}^{\infty} \left| \frac{\prod_{k=1}^{n} a_k}{B_n B_{n-1}} \right| \leq \sum_{n=1}^{\infty} \left(\frac{1}{|B_{n-1}|} - \frac{1}{|B_n|} \right) = 1,$$

that is, if the unit disk is a value region for $\mathbf{K}(a_n/b_n)$, then $\mathbf{K}(a_n/b_n)$ converges absolutely.

Proof of Lemma 19: Since the approximants f_n and the Stern-Stolz series (4.1.3) are invariant under equivalence transformations, we may assume that $\mathbf{K}(a_n/b_n)$ has the form $\mathbf{K}(1/b_n)$. We use the standard notation $f_n = A_n/B_n$ and $h_n = -S_n^{-1}(\infty) = B_n/B_{n-1}$. Since $f_n \neq \infty$, it follows that $B_n \neq 0$ and thus $h_n \neq \infty, 0$. We shall see later (Formula (3.3.4) in Chapter IV combined with Formula (1.1.7) in Chapter II) that then A_n can be written

$$A_n = \sum_{k=1}^{n} \left(\prod_{j=2}^{k} (b_j - h_j) \prod_{j=k+1}^{n} h_j \right) \tag{4.1.4}$$

for all $n \geq 1$, where the empty sum is 0 and the empty product is 1. Since $B_1 = b_1$ we have $B_n = B_1 \prod_{j=2}^{n} (B_j/B_{j-1}) = b_1 \prod_{j=2}^{n} h_j$, and therefore

$$\frac{A_n}{B_n} = \frac{1}{b_1} \sum_{k=1}^{n} P_k \quad \text{where } P_k = \prod_{j=2}^{k} \frac{b_j - h_j}{h_j} \tag{4.1.5}$$

and

$$\frac{b_j - h_j}{h_j} = \frac{-1/h_{j-1}}{b_j + 1/h_{j-1}} = \frac{-1}{\delta_j + 1} \quad \text{for } \delta_j = b_j h_{j-1}. \tag{4.1.6}$$

By Theorem 1, the Stern-Stolz theorem, we know that $\mathbf{K}(1/b_n)$ diverges if $\sum |b_n| < \infty$. So, assume that $\sum |b_n| = \infty$. We want to prove that $\{f_{2n}\}$ and $\{f_{2n+1}\}$ have a common limit $L_0 = L_1$; i.e. that

$$f_{n+1} - f_n = \frac{P_{n+1}}{b_1} \to 0. \tag{4.1.7}$$

By (4.1.2) and (4.1.5) we find, using (4.1.6), that

$$
\begin{aligned}
\sum_{n=1}^{\infty} |f_{2n+2} - f_{2n}| &= \sum_{n=1}^{\infty} \frac{|P_{2n+1} + P_{2n+2}|}{|b_1|} \\
&= \sum_{n=1}^{\infty} \left| \frac{P_{2n+2}}{b_1} \right| |1 - (\delta_{2n+2} + 1) + 1| \\
&= \sum_{n=1}^{\infty} \left| \frac{P_{2n+2} \delta_{2n+2}}{b_1} \right| < \infty,
\end{aligned}
$$

and similarly

$$\sum_{n=1}^{\infty} |f_{2n+1} - f_{2n-1}| = \sum_{n=1}^{\infty} \left| \frac{P_{2n+1} \delta_{2n+1}}{b_1} \right| < \infty.$$

Now, if $\sum |\delta_n| = \infty$ and $\sum |P_n \delta_n| < \infty$, then $P_n \to 0$ which proves (4.1.7). Hence we only need to prove that $\sum |\delta_n| = \infty$. We have

$$h_k = b_k + \frac{1}{h_{k-1}} = \frac{\delta_k + 1}{h_{k-1}},$$

so that

$$
\begin{aligned}
b_{2n} &= \frac{\delta_{2n}}{h_{2n-1}} = \frac{\delta_{2n}}{\delta_{2n-1} + 1} h_{2n-2} = \cdots \\
&= \frac{\delta_{2n}(\delta_{2n-2} + 1)(\delta_{2n-4} + 1) \cdots (\delta_2 + 1)}{(\delta_{2n-1} + 1)(\delta_{2n-3} + 1) \cdots (\delta_3 + 1) b_1},
\end{aligned}
\tag{4.1.8}
$$

and similarly

$$b_{2n+1} = \frac{\delta_{2n+1}(\delta_{2n-1} + 1)(\delta_{2n-3} + 1) \cdots (\delta_3 + 1) b_1}{(\delta_{2n} + 1)(\delta_{2n-2} + 1) \cdots (\delta_2 + 1)}. \tag{4.1.9}$$

If $\sum |\delta_n| < \infty$, then $\sum |b_n| < \infty$ by (4.1.8)-(4.1.9), a contradiction. Hence $\sum |\delta_n| = \infty$. ∎

Example 16 In Remark 2 to the Seidel-Stern theorem, Theorem 3, we noted that $\mathbf{K}(a_n/b_n)$ converges if and only if its Stern-Stolz series (4.1.3) diverges to ∞ if all $a_n > 0$ and $b_n > 0$. How does this relate to Lemma 19?

We shall first prove that the even and odd parts of $\mathbf{K}(a_n/b_n)$ converge absolutely when all $a_n > 0$ and $b_n > 0$. From Theorem 2 we know that $\{f_{2n}\}$ is a bounded, monotonely increasing sequence. Hence $f_{2n} \to L_0 \neq \infty$ and

$$\sum_{n=1}^{\infty} |f_{2n+2} - f_{2n}| = \sum_{n=1}^{\infty} (f_{2n+2} - f_{2n}) = L_0.$$

Similarly, $\{f_{2n+1}\}$ is bounded and monotonely decreasing, so $f_{2n+1} \to L_1 \neq \infty$ and

$$\sum_{n=1}^{\infty} |f_{2n+1} - f_{2n-1}| = -\sum_{n=1}^{\infty} (f_{2n+1} - f_{2n-1}) = -L_1 + f_1.$$

Hence, by Lemma 19, $\mathbf{K}(a_n/b_n)$ converges if and only if (4.1.3) diverges to ∞, i.e. $\Omega = \{(a,b) \in \mathbf{C} \times \mathbf{C}; a > 0 \text{ and } b > 0\}$ is a conditional convergence set.

_____◇

4.2   Parabola Theorems

The value set techniques described in this chapter are all based on mapping properties of the linear fractional transformations s_n and S_n. The convergence set or the element set is the set of all (a_n, b_n) which give s_n the wanted mapping properties. Since linear fractional transformations map circles and lines into circles and lines, it is nice to work with value sets that are circular disks, the exterior of circular disks or halfplanes. In this subsection we shall work with value sets that are halfplanes.

The parabola theorem is a convergence theorem for continued fractions $\mathbf{K}(a_n/1)$:

Theorem 20 (The parabola theorem) *Let α be a fixed, real number, $-\pi/2 < \alpha < \pi/2$. Then:*

A. *The parabolic region*

$$P_\alpha = \left\{ a \in \mathbf{C}; \ |a| - \Re(ae^{-i2\alpha}) \leq \frac{1}{2} \cos^2 \alpha \right\}$$

$$= \left\{ re^{i(\theta + 2\alpha)}; \ 0 \leq r \leq \frac{\frac{1}{2} \cos^2 \alpha}{1 - \cos \theta} \right\} \qquad (4.2.1)$$

$$= \left\{ c^2 \in \mathbf{C}; \ |\Im(ce^{-i\alpha})| \leq \frac{1}{2} \cos \alpha \right\}$$

is a conditional convergence set for continued fractions $\mathbf{K}(a_n/1)$.

B. *The even and odd parts of continued fractions* $\mathbf{K}(a_n/1)$ *from* P_α *converge to finite values.*

C. *The half plane*

$$V_\alpha = \left\{ w \in \mathbf{C}; \ \Re(we^{-i\alpha}) > -\frac{1}{2} \cos \alpha \right\} \qquad (4.2.2)$$

is a value set for P_α.

D. *If* $a_n \in P_\alpha$ *for all* n *and* $\sum(n|a_n|)^{-1} = \infty$, *then* $\{S_n(w)\}$ *converges uniformly in* $\bar{V}_\alpha$ *to a value* f. *In fact,*

$$|f - S_n(w)| \leq \frac{2|a_1|/\cos \alpha}{\displaystyle\prod_{j=2}^{n} \left(1 + \frac{\cos^2 \alpha}{4(j-1)|a_j|} \right)} \qquad \text{for all } w \in \bar{V}_\alpha. \quad (4.2.3)$$

E. *The set* $E_{\alpha,M} = \{a \in P_\alpha; \ |a| \leq M\}$ *is a uniform convergence set for continued fractions* $\mathbf{K}(a_n/1)$.

Remarks:

1. The boundary of P_α is a parabola with axis along the ray $\arg z = 2\alpha$, focus at the origin and vertex at $-(1/4)e^{i2\alpha}\cos^2 \alpha$. It intersects the real axis at $z = -1/4$. (See Figure 1.)

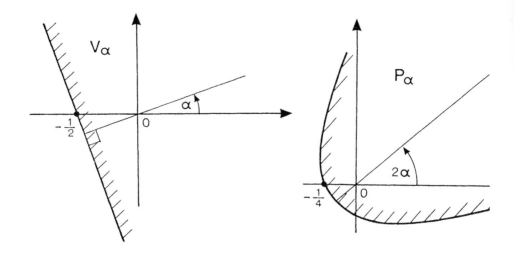

Figure 1.

2. Since $\infty \notin V_\alpha$ and $0 \in V_\alpha$, it follows that all approximants $S_n(0) = A_n/B_n$ of a continued fraction from P_α are finite. In particular all $B_n \neq 0$.

3. It follows from Part B that if $\mathbf{K}(a_n/1)$ from P_α converges, then it converges to a finite value.

4. In Part C we actually prove the stronger result that $a/(1 + V_\alpha) \subseteq V_\alpha$ if and only if $a \in P_\alpha$.

5. The truncation error bound (4.2.3) is also valid in Part E.

6. The parabola theorem generalizes the Worpitzky theorem, since the Worpitzky disk $E = \{a \in \mathbf{C}; |a| \leq 1/4\} \subseteq P_0$.

Sketch of proof: We shall first prove part C and B, and then use these to prove part A. Finally we look at part D and E.

C: The mapping $s(w) = a/(1+w)$ maps $\bar{V}_\alpha$ onto the closed circular disk

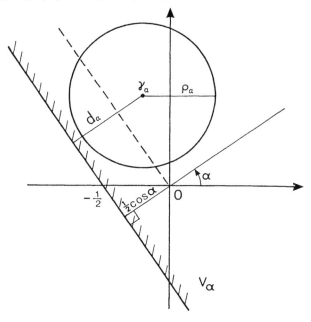

Figure 2.

with center at $\gamma_a = (ae^{-i\alpha})/\cos\alpha$ and radius $\rho_a = |a|/\cos\alpha$. This disk is contained in $\bar{V}_\alpha$ if and only if $\gamma_a \in V_\alpha$ and γ_a has a distance d_a to ∂V_α such that $d_a \geq \rho_a$. Now

$$d_a = \frac{1}{2}\cos\alpha + \Re(\gamma_a e^{-i\alpha}),$$

where $d_a > 0$ if and only if $\gamma_a \in V_\alpha$. (See Figure 2.) So $d_a \geq \rho_a$ if and only if

$$\frac{1}{2}\cos\alpha + \Re\left(\frac{ae^{-i\alpha}}{\cos\alpha}e^{-i\alpha}\right) \geq \frac{|a|}{\cos\alpha},$$

i.e. if and only if $a \in P_\alpha$.

B: We define the linear fractional transformation

$$t(w) = \frac{-1 + e^{i\alpha}\cos\alpha - w}{1+w}. \tag{4.2.4}$$

Then t maps the closed unit disk $\bar{U}$ onto $\bar{V}_\alpha$ with $t(\infty) = -1$ and $t(-1) = \infty$. Let $\mathbf{K}(a_n/1)$ be from P_α and let

$$t_n(w) = t^{-1} \circ s_{2n-1} \circ s_{2n} \circ t(w) \quad \text{for } n = 1, 2, 3, \ldots. \qquad (4.2.5)$$

Then, $t_n(\bar{U}) = t^{-1} \circ s_{2n-1} \circ s_{2n}(\bar{V}_\alpha) \subseteq t^{-1} \circ s_{2n-1}(\bar{V}_\alpha) \subseteq t^{-1}(\bar{V}_\alpha) = \bar{U}$, and $t_n(\infty) = t^{-1} \circ s_{2n-1} \circ s_{2n}(-1) = t^{-1} \circ s_{2n-1}(\infty) = t^{-1}(0) = -1 + e^{i\alpha} \cos \alpha = k \in U$. It follows therefore by Lemma 13 that $T_n(w) = t_1 \circ t_2 \circ \cdots \circ t_n(w)$ converges locally uniformly in U to a constant function $T(w) \equiv c \in \bar{U}$. Since

$$\begin{aligned}
T_n &= t_1 \circ t_2 \circ \cdots \circ t_n \\
&= \left(t^{-1} \circ s_1 \circ s_2 \circ t\right) \circ \left(t^{-1} \circ s_3 \circ s_4 \circ t\right) \circ \cdots \qquad (4.2.6) \\
&\quad \circ \left(t^{-1} \circ s_{2n-1} \circ s_{2n} \circ t\right) \\
&= t^{-1} \circ S_{2n} \circ t,
\end{aligned}$$

we therefore have that $S_{2n}(w) \to t(c) = L_0$ for all $w \in t^{-1}(U) = V_\alpha$, and in particular $S_{2n}(0) \to L_0$. To see that $L_0 \neq \infty$, we observe that $S_{2n}(0) \in S_{2n}(\bar{V}_\alpha) = K_{2n} \subseteq s_1(\bar{V}_\alpha)$ for all n, so that $L_0 \in s_1(\bar{V}_\alpha)$. Further, $\infty \notin s_1(\bar{V}_\alpha)$ since $s_1^{-1}(\infty) = -1 \notin \bar{V}_\alpha$.

Similarly one can also prove that $S_{2n+1}(0) \to L_1 \neq \infty$.

A: It follows by Theorem 1 that if $\mathbf{K}(a_n/1)$ converges, then its Stern-Stolz series diverges. We need to prove that if $\mathbf{K}(a_n/1)$ diverges, then the Stern-Stolz series converges. As before we let $K_n = S_n(\bar{V}_\alpha)$ so that $K_{n+1} \subseteq K_n \subseteq \cdots \subseteq s_1(\bar{V}_\alpha)$, where $s_1(\bar{V}_\alpha)$ is bounded. If $\operatorname{diam}(K_n) \to 0$, then $\mathbf{K}(a_n/1)$ converges, so assume that $\operatorname{diam}(K) > 0$ where $K = \lim K_n$. Then the radius R_n of $T_n(\bar{U}) = t^{-1} \circ S_{2n} \circ t(\bar{U})$ must converge to an $R > 0$; i.e. we have the limit circle case for $\{T_n\}$ in (4.2.6). We shall prove that

$$\sum_{n=1}^\infty |S_{2n+2}(0) - S_{2n}(0)| < \infty. \qquad (4.2.7)$$

Since $\{T_n\}$ satisfies the conditions of Lemma 13, it follows by (3.5.6) and (3.5.8) that $\sum |T_{n+1}(\infty) - T_n(\infty)| < \infty$, and thus by (3.5.8) that $\sum |T_{n+1}(w) - T_n(w)| < \infty$ for $|w| \neq 1$. In particular $\sum |T_{n+1}(w) - T_n(w)| < \infty$ for $w = -1 + e^{i\alpha} \cos \alpha$. (4.2.7) follows therefore, since $S_n(0) = t \circ T_n(-1 + e^{i\alpha} \cos \alpha) \in s_1(\bar{V}_\alpha)$ which is bounded.

In a similar way we prove the second condition in (4.1.2), and the result follows from Lemma 19.

D: It is possible to prove by elementary methods that the radius R_n of the circular disk $K_n = S_n(\bar{V}_\alpha)$ satisfies the inequality

$$R_n \leq \frac{|a_1|/\cos\alpha}{\prod\limits_{j=2}^{n}\left(1 + \dfrac{\cos^2\alpha}{4(j-1)|a_j|}\right)} \qquad \text{for } n = 1, 2, 3, \ldots. \qquad (4.2.8)$$

([Thron58].) This proves Part D.

E: This follows directly from D. ∎

The parabola theorem is in several ways the queen among the convergence theorems for continued fractions $\mathbf{K}(a_n/1)$. It is best in several respects. In particular one can not enlarge the conditional convergence set P_α, not even by adding just one point, without destroying that property; i.e. if $a \notin P_\alpha$, then $\{a\} \cup P_\alpha$ is not a conditional convergence set, [Lore].

Example 17 The continued fraction

$$\mathbf{K}\frac{x}{2} = \frac{x}{2} \frac{x}{+2} \frac{x}{+2} + \cdots \approx \frac{x/2}{1} \frac{x/4}{+1} \frac{x/4}{+1} \frac{x/4}{+1} + \cdots$$

was discussed in Chapter I, Example 4. Let

$$\alpha = \begin{cases} \frac{1}{2}\arg(x) & \text{if } |\arg(x)| < \pi, \\ 0 & \text{if } |\arg(x)| = \pi. \end{cases}$$

Then $x/4 \in P_\alpha$ if $|\arg(x)| < \pi$, and thus $\mathbf{K}(x/2)$ converges. If $x < 0$, then $x/4 \in P_0$ if and only if $|x| \leq 1$. Hence $\mathbf{K}(x/2)$ converges in the cut plane $D = \{x \in \mathbf{C}; |\arg(1+x)| < \pi\} \cup \{-1\}$, just as proved in Example 5 in Subsection 2.3, and illustrated in Figure 2 in Chapter I.

From (4.2.3) we further get the a priori truncation error bound

$$|f - S_n(w)| \leq \frac{|x|}{\cos\alpha} \bigg/ \prod_{j=2}^{n}\left(1 + \frac{\cos^2\alpha}{(j-1)|x|}\right) \qquad \text{for } w \in \bar{V}_\alpha$$

which shows that $\mathbf{K}(x/2)$ converges locally uniformly in D. We can not expect that this bound is the best possible, since we have not taken into account the special periodic character of $\mathbf{K}(x/2)$.

───◇

Example 18 In Example 5 of Chapter I we presented the continued fraction expansion

$$\mathbf{K}\frac{a_n z}{1} = \frac{z}{1} + \frac{z/2}{1} + \frac{z/6}{1} + \frac{2z/6}{1} + \frac{2z/10}{1} + \frac{3z/10}{1} +\cdots$$

of $\log(1+z)$, where

$$a_{2n} = \frac{n}{2(2n-1)}, \quad a_{2n+1} = \frac{n}{2(2n+1)} \quad \text{for } n = 1,2,3,\ldots.$$

Let

$$\alpha = \begin{cases} \frac{1}{2}\arg(z) & \text{if } |\arg(z)| < \pi, \\ 0 & \text{if } |\arg(z)| = \pi. \end{cases}$$

Then $a_n z \in P_\alpha$ if $|\arg(z)| < \pi$. If $|\arg(z)| = \pi$, then $a_n z \in P_0$ if and only if $|a_n z| \leq 1/4$. Hence, $a_n z \in P_\alpha$ for all $n \geq 2$ (with our choice of α) if z is in the cut plane $D_1 = \{z \in \mathbf{C}; |\arg(z+1/2)| < \pi\}$, and thus $\mathbf{K}(a_n z/1)$ converges for all $z \in D_1$. We can even conclude that this convergence is uniform on compact subsets of D_1, since for every $z \in D_1$ there exist a neighborhood B_z and a permissible α such that $B_z \subseteq P_\alpha$, and the result follows therefore from Theorem 20E. (Keep in mind that every compact subset of D_1 can be covered by a finite number of such neighborhoods B_z.)

───◇

We can improve the result in Example 18 if we let V_α vary with n as in the following theorem by Jones and Thron, [Thron58], [JoTh68]:

Theorem 21 (The parabola sequence theorem) *Let α be a fixed number, $-\pi/2 < \alpha < \pi/2$, $0 < g_0 \leq 1$ and $0 < g_n < 1$ for $n = 1,2,3,\ldots.$ Then:*

A. The parabolic regions

$$P_{\alpha,n} = \{a \in \mathbf{C};\ |a| - \Re(ae^{-i2\alpha}) \le 2g_{n-1}(1 - g_n)\cos^2\alpha\} \quad (4.2.9)$$

for $n = 1, 2, 3, \ldots$ form a sequence of conditional convergence regions for continued fractions $\mathbf{K}(a_n/1)$.

B. The even and odd parts of continued fractions $\mathbf{K}(a_n/1)$ from $\{P_{\alpha,n}\}$ converge to finite values.

C. The half planes

$$V_{\alpha,n} = \{w \in \mathbf{C};\ \Re(we^{-i\alpha}) > -g_n\cos\alpha\} \qquad (4.2.10)$$

for $n = 0, 1, 2, \ldots$ form a sequence of value sets for $\{P_{\alpha,n}\}$.

D. If $a_n \in P_{\alpha,n}$ for all n, then

$$|f - S_n(w)| \le \frac{|a_1|/((1 - g_1)\cos\alpha)}{\displaystyle\prod_{j=2}^{n}\left(1 + \frac{d_{j-1}g_{j-1}(1 - g_j)\cos^2\alpha}{|a_j|}\right)}, \qquad (4.2.11)$$

for all $w \in \bar{V}_{\alpha,n}$ and $n \in \mathbf{N}$, where

$$d_n = \left(\prod_{k=1}^{n}\frac{1 - g_k}{g_k}\right)\Big/\left(\sum_{m=0}^{n-1}\prod_{k=1}^{m}\frac{1 - g_k}{g_k}\right) \quad \text{for } n \in \mathbf{N}. \quad (4.2.12)$$

E. If

$$\sum_{m=0}^{\infty}\prod_{k=1}^{m}\frac{1 - g_k}{g_k} = \infty, \qquad (4.2.13)$$

then $E_n = \{a \in P_{\alpha,n};\ |a| \le M\}$ is a uniform sequence of convergence sets for continued fractions $\mathbf{K}(a_n/1)$.

(Part A, B and C follow from [JoTh68, Theorem 5.1] with $\psi_n = \alpha$ and $p_n = g_n\cos\alpha$ for all n. Part D and E follow from [Thron58].) The proof follows the same pattern as the proof of Theorem 20. Notice that also here all canonical denominators $B_n \ne 0$ for continued fractions $\mathbf{K}(a_n/1)$ from $\{P_{\alpha,n}\}$ since $\infty \notin V_{\alpha,n}$.

Example 19 Let $\mathbf{K}(a_n z/1)$ and α be as in Examle 18. If z is real and negative, we now have that $a_n z \in P_{0,n}$ if and only if $|a_n z| \leq g_{n-1}(1 - g_n)$. Let us choose

$$g_{2n-1} = n/(2n - 1), \quad g_{2n} = 1/2 \quad \text{for } n = 1, 2, 3, \dots .$$

Then $a_n = g_{n-1}(1 - g_n)$ for all $n \geq 2$. Hence $\mathbf{K}(a_n z/1)$ converges locally uniformly in the cut plane $D = \{z \in \mathbf{C}; \, |\arg(z + 1)| < \pi\}$ to a holomorphic function.

———————————————————————————————◇

4.3   S-fractions

S-fractions, or Stieltjes fractions, are continued fractions of the form $\mathbf{K}(a_n z/1)$, where all $a_n > 0$ and z is a complex variable. Example 17 and 18 were studies of some special S-fractions. The parabola theorems are well suited to prove convergence of such continued fractions. But we also have:

Theorem 22 *The S-fraction $\mathbf{K}(a_n z/1)$ where all $a_n > 0$, has the following properties.*

A. *Its even and odd parts converge locally uniformly in $D = \{z \in \mathbf{C}; \, |\arg(z)| < \pi\}$ to holomorphic functions.*

B. *It converges to a holomorphic function in D if and only if the Stern-Stolz series*

$$\sum_{n=1}^{\infty} \prod_{k=1}^{n} |a_k|^{(-1)^{n-k+1}} \tag{4.3.1}$$

of $\mathbf{K}(a_n/1)$ diverges to ∞.

C. *It diverges for all $z \in D$ if the series (4.3.1) converges.*

This result is due to Stieltjes [Stie94], but it is also a corollary of Theorem 21. In some cases we obtain better results by using the parabola theorems, such as in Example 17, 18 and 19.

A very nice consequence of Theorem 22 is that an S-fraction converges to a holomorphic function in D if and only if it converges at a single point $z \in D$.

Henrici and Pfluger [HePf66] have proved the following a posteriori truncation error bounds for S-fractions:

Theorem 23 (The Henrici-Pfluger truncation error bounds)
Let $-\pi < \theta < \pi$. Then the sector

$$V = \{w \in \mathbf{C};\; 0 \leq \operatorname{sgn}(\theta) \cdot \arg(w) \leq |\theta|\} \tag{4.3.2}$$

is a value set containing the classical approximants of the S-fraction $\mathbf{K}(a_n z/1)$, *where all $a_n > 0$ and $\arg(z) = \theta$. If this S-fraction converges to $f(z)$, then*

$$|f(z) - f_n(z)| \leq \begin{cases} |f_n(z) - f_{n-1}(z)| & \text{if } |\theta| \leq \pi/2, \\[2mm] \dfrac{|f_n(z) - f_{n-1}(z)|}{|\sin \theta|} & \text{if } \frac{\pi}{2} < |\theta| < \pi, \end{cases} \tag{4.3.3}$$

where $f_n(z)$ denotes the nth classical approximant of $\mathbf{K}(a_n z/1)$.

Proof : $a_n z/(1 + V) \subseteq V$, since for $w \in V$

$$\arg\left(\frac{a_n z}{1 + w}\right) = \arg(z) - \arg(1 + w)$$

which lies between θ and 0. Since $a_n z \in V$, it follows therefore by Theorem 8 that $f_n(z) \in V$ for all $n \geq 1$.

Let $K_n = S_n(\bar{V})$ where $S_n = s_1 \circ s_2 \circ \cdots \circ s_n$ and $s_k(w) = a_k z/(1 + w)$. Then K_n is bounded by the circular arcs $S_n(\mathbf{R}^+)$ and $S_n(R_\theta)$, where R_θ is the ray $\arg(w) = \theta$. The truncation error bound (4.3.3) is derived by a careful study of these convex, lens-shaped, closed regions K_n. For details we refer to [HePf66]. ∎

The value set V in (4.3.2) can also be used to derive a priori truncation error bounds for S-fractions $\mathbf{K}(a_n z/1)$, [GrWa83]:

Theorem 24 (The Gragg-Warner bounds) *Let* $\mathbf{K}(a_n z/1)$ *be an S-fraction with* $z = r e^{i2\alpha}$ *for* $|\alpha| < \pi/2$ *and all* $a_n > 0$. *Then*

$$|f_{n+m}(z) - f_n(z)| \leq 2 \frac{a_1 r}{\cos \alpha} \prod_{k=2}^{n} \frac{\sqrt{1 + 4a_k r / \cos^2 \alpha} - 1}{\sqrt{1 + 4a_k r / \cos^2 \alpha} + 1} \qquad (4.3.4)$$

for $n \geq 2$ *and* $m \geq 1$.

Example 20 Let us once again turn to the continued fraction expansion

$$\mathbf{K} \frac{a_n z}{1} = \frac{z}{1+} \frac{z/2}{1+} \frac{z/6}{1+} \frac{2z/6}{1+} \frac{2z/10}{1+} \frac{3z/10}{1+} \cdots$$

of $\log(1 + z)$. (See Example 18 and 19.) This is an S-fraction, and it converges, not only for $|\arg(z)| < \pi$, but for $|\arg(z+1)| < \pi$ by Example 19.

We want to compute $\log(1 + z)$ for $z > 0$. Then, by (4.3.3),

$$|f(z) - f_n(z)| \leq |f_n(z) - f_{n-1}(z)| \quad \text{for } z > 0$$

which is consistent with results from Subsection *1.2*. From (4.3.4) we get the a priori bounds

$$|f(z) - f_n(z)| \leq 2z \prod_{k=2}^{n} \frac{\sqrt{1 + 4a_k z} - 1}{\sqrt{1 + 4a_k z} + 1} \quad \text{for } z > 0,$$

which for $z = 1$ is

$$2 \frac{\sqrt{1 + 2} - 1}{\sqrt{1 + 2} + 1} \cdot \frac{\sqrt{1 + 2/3} - 1}{\sqrt{1 + 2/3} + 1} \cdot \frac{\sqrt{1 + 4/3} - 1}{\sqrt{1 + 4/3} + 1}$$
$$\cdot \frac{\sqrt{1 + 4/5} - 1}{\sqrt{1 + 4/5} + 1} \cdots \frac{\sqrt{1 + 4a_n} - 1}{\sqrt{1 + 4a_n} + 1}.$$

Hence $|f(1) - f_3(1)| \leq 0.0681$, $|f(1) - f_4(1)| \leq 0.0142$, $|f(1) - f_5(1)| \leq 0.0021$ and so on. Compare these estimates to the table of the first approximants $f_n(1)$ of $\mathbf{K}(a_n z/1)$ in Chapter I, Example 5.

———————————————————————————————◇

4.4   Oval theorems

As pointed out in Subsection 3.7, we want as "small" value sets as possible for a given continued fraction to derive as good truncation error bounds as possible. For this reason, although we are happy with the parabola theorem, we also want other value set results for continued fractions $\mathbf{K}(a_n/1)$.

In the first theorem in this subsection, we let

$$V = \{w \in \mathbf{C}; \ |w - C| < R\}, \quad C \in \mathbf{C}, \quad R > 0 \tag{4.4.1}$$

be the value set, and we consider continued fractions $\mathbf{K}(a_n/1)$ where $a_n/(1 + V) \subseteq V$ for all n. Since V is bounded, we can not allow $-1 \in \bar{V}$, so a necessary condition is $|1 + C| > R$.

Theorem 25 (The oval theorem) *Let $C \in \mathbf{C}$ with $\Re(C) > -1/2$ and $0 < R < |1 + C|$ be given. Then*

$$E = \{a \in \mathbf{C}; \ \left|a(1 + \bar{C}) - C(|1 + C|^2 - R^2)\right| + R|a| \leq R(|1 + C|^2 - R^2)\} \tag{4.4.2}$$

is a convergence set for continued fractions $\mathbf{K}(a_n/1)$, and

$$V = \{w \in \mathbf{C}; \ |w - C| < R\}$$

is a value set for E. Moreover

$$|f - S_n(w)| \leq 2R \frac{|C| + R}{|1 + C| - R} M^{n-1} \tag{4.4.3}$$

where

$$M = \max\left\{\left|\frac{w}{1 + w}\right|; \ w \in \bar{V}\right\} \tag{4.4.4}$$

for every continued fraction $\mathbf{K}(a_n/1)$ from E and $w \in \bar{V}$, where f is the value of $\mathbf{K}(a_n/1)$.

Remarks:

1. We shall see from the proof of Theorem 25 that the condition $\Re(C) \geq -1/2$ is necessary for the existence of an $a \in \mathbf{C}$, $a \neq 0$, such that $a/(1 + V) \subseteq V$.

2. We shall also see that $a/(1 + V) \subseteq V$ if and only if $a \in E$.

3. The oval theorem also gives some new convergence criteria in the sense that not every E is contained in some parabolic region P_α from the parabola theorem. (See [JaTh86].)

4. If $M < 1$, then (4.4.3) implies that E is a uniform general convergence set with respect to V. M is always less than 1 if $R < \Re(C + 1/2)$.

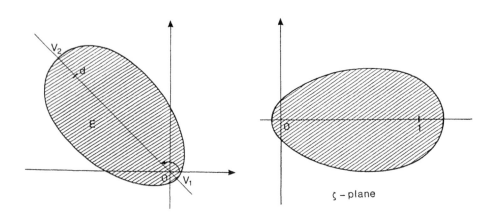

Figure 3.

5. The boundary ∂E of E is called a Cartesian oval. If $C = 0$ it is the circle which bounds

$$E = \{a \in \mathbf{C};\ |a| \leq R(1 - R)\}, \quad 0 < R < 1,$$

a convergence set known from Example 13 in Subsection *3.3*. If $C \neq 0$ we define

$$d = C(1 + C)\left(1 - \frac{R^2}{|1 + C|^2}\right). \qquad (4.4.5)$$

Then E can be written

$$E = \left\{a = d\zeta \in \mathbf{C}; \; |\zeta - 1| + \frac{R}{|1 + C|}|\zeta| \leq \frac{R}{|C|}\right\}. \qquad (4.4.6)$$

From this we see that the Cartesian oval ∂E is symmetric about its axis through 0 and d, and that it is some kind of a "weighted ellipse" with foci at $a = d$ and $a = 0$ and vertices at

$$v_1 = \begin{cases} d\left(1 - \frac{R}{|C|}\right)\Big/\left(1 - \frac{R}{|C+1|}\right) & \text{if } R \leq |C|, \\[2mm] d\left(1 - \frac{R}{|C|}\right)\Big/\left(1 + \frac{R}{|C+1|}\right) & \text{if } R > |C|, \end{cases} \qquad (4.4.7)$$

and

$$v_2 = d\left(1 + \frac{R}{|C|}\right)\Big/\left(1 + \frac{R}{|C + 1|}\right). \qquad (4.4.8)$$

One can also prove that E contains the circular disk with center at d and radius $|v_2 - d|$ and that E is contained in the closed circular disk with center at d and radius $|v_1 - d|$. (See Figure 3 and [JaTh86].)

6. The oval theorem generalizes the Hillam-Thron theorem, Theorem 12, for continued fractions $\mathbf{K}(a_n/1)$, since $0 \in V$ is no longer required.

We shall not prove the fact that E is a convergence set in general. For that we refer to [LoRu]. We shall prove the rest of Theorem 25, though, in addition to the points mentioned in Remark 1 and 2.

Proof of Theorem 25: By Lemma 11 it follows that $a/(1 + V)$ is a circular disk with center and radius given by

$$c_a = \frac{(1 + \bar{C})a}{|1 + C|^2 - R^2}, \quad r_a = \frac{R|a|}{|1 + C|^2 - R^2}. \qquad (4.4.9)$$

This disk is contained in V if and only if $|C - c_a| + r_a \leq R$; i.e. if and only if $a \in E$.

Next we shall prove that $E \neq \emptyset$ if and only if $\Re(C) \geq -1/2$. From Remark 5 it follows that $E = \emptyset$ if and only if $C \neq 0$ and R and C are chosen such that

$$|\zeta - 1| + \frac{R}{|1 + C|}|\zeta| \leq \frac{R}{|C|} \tag{4.4.10}$$

is impossible for every $\zeta \in \mathbf{C}$. The left side of (4.4.10) attains it minimum for $\zeta = 1$. So (4.4.10) is impossible if and only if $R/|1+C| > R/|C|$; i.e. $|1 + C| < |C|$ i.e. $\Re(C) < -1/2$.

Finally, to prove (4.4.3), we observe that the circumference of $K_n = S_n(\bar{V})$ is given by

$$2\pi R_n = \int_{\partial V} |S_n'(w)|\,dw, \tag{4.4.11}$$

where $S_n'(w)$ denotes the derivative of S_n and R_n is the radius of K_n. Since $S_n = s_1 \circ s_2 \circ \cdots \circ s_n$ we get by the chain rule that

$$S_n'(w) = s_1'(w_{n,1}) \cdot s_2'(w_{n,2}) \cdots s_n'(w_{n,n}) \tag{4.4.12}$$

where $w_{n,n} = w \in \bar{V}$, $w_{n,k} = s_{k+1}(w_{n,k+1}) \in \bar{V}$ for $k < n$ and

$$s_k'(w_{n,k}) = \left.\frac{d}{dw}\left(\frac{a_k}{1 + w}\right)\right|_{w=w_{n,k}} = -\frac{a_k}{(1 + w_{n,k})^2} = \frac{-w_{n,k-1}}{1 + w_{n,k}}. \tag{4.4.13}$$

Combining (4.4.12)–(4.4.13) we get (4.4.3) since

$$
\begin{aligned}
|S_n'(w)| &= \prod_{k=1}^{n}\left|\frac{w_{n,k-1}}{1 + w_{n,k}}\right| = \left|\frac{w_{n,0}}{1 + w_{n,n}}\right|\prod_{k=1}^{n-1}\left|\frac{w_{n,k}}{1 + w_{n,k}}\right| \\
&\leq \frac{|C| + R}{|1 + C| - R}M^{n-1},
\end{aligned}
$$

so that by (4.4.11)

$$R_n \leq \frac{1}{2\pi}\frac{|C| + R}{|1 + C| - R}M^{n-1} \cdot 2\pi R.$$

Example 21 How do the truncation error bounds (4.4.3) turn out for the S-fraction expansion of $\log(1 + z)$ that we studied in the previous examples? Again we choose $z = 1$. Then $\mathbf{K}(a_n z/1)$ is a continued fraction with positive elements such that $1/6 \leq a_n z \leq 1$.

Of course we can not expect the bounds (4.4.3) to be as good as the Gragg-Warner bounds in Example 20 or the practical bounds we get from Theorem 2 for continued fractions with positive elements. Those types of upper bounds are "designed" for the particular S-fraction (or continued fraction with positive elements) in question. The bounds (4.4.3) hold for all continued fractions from E.

On the other hand it is useful to get an idea of how well (4.4.3) is doing compared to these more specialized bounds. Since $1/6 \leq a_n \leq 1$, we want to choose C and R such that the element set E is symmetric about the positive real axis with vertices $v_1 = 1/6$ and $v_2 = 1$. From (4.4.7)–(4.4.8) we find, if $0 < R < C$,

$$v_1 = (1 + C + R)(C - R) = \frac{1}{6}, \qquad v_2 = (1 + C - R)(C + R) = 1$$

i.e. $R = 5/12$ and $C = (\sqrt{145} - 6)/12$, and

$$M = \frac{C + R}{1 + C + R} = \frac{\sqrt{145} - 1}{\sqrt{145} + 11} \approx 0.479 .$$

Hence

$$|f(1) - f_n(1)| \leq \frac{5}{6} \left(\frac{\sqrt{145} - 1}{\sqrt{145} + 1} \right) (0.479)^{n-1} \approx 0.706 \cdot (0.479)^{n-1} .$$

That is, $|f(1) - f_3(1)| \leq 0.16$, $|f(1) - f_4(1)| \leq 0.078$, $|f(1) - f_5(1)| \leq 0.037$, and so on.

$\diamond$

Again we can do better if we allow V (and thus E) to vary with n:

Theorem 26 (The oval sequence theorem) *Let $C_n \in \mathbf{C}$ and $0 < R_n < |1 + C_n|$ be given for $n = 0, 1, 2, \ldots$ such that*

$$|C_{n-1}| R_n < |1 + C_n| R_{n-1} \quad \text{for } n = 1, 2, 3, \ldots . \tag{4.4.14}$$

Then

$$V_n = \{w \in \mathbf{C}; |w - C_n| < R_n\} \quad for\ n = 0, 1, 2, \ldots \qquad (4.4.15)$$

is a sequence of value sets for

$$\begin{aligned} E_n &= \{a \in \mathbf{C}; |a(1 + \bar{C}_n) - C_{n-1}(|1 + C_n|^2 - R_n^2)| + R_n|a| \\ &\leq R_{n-1}(|1 + C_n|^2 - R_n^2)\} \quad for\ n = 1, 2, 3, \ldots. \qquad (4.4.16) \end{aligned}$$

If $\mathbf{K}(a_n/1)$ *is a continued fraction from* $\{E_n\}$ *and all* $w_k \in \bar{V}_k$, *then*

$$|S_{n+m}(w_{n+m}) - S_n(w_n)| \leq 2R_0 \frac{|C_0| + R_0}{|1 + C_n| - R_n} \prod_{k=1}^{n-1} M_k \qquad (4.4.17)$$

where

$$M_k = \max\left\{\left|\frac{w}{1 + w}\right|; w \in \bar{V}_k\right\}. \qquad (4.4.18)$$

The proof follows the same lines as the proof of Theorem 25 and will be left out.

Example 22 How can we find better truncation error estimates for the S-fraction expansion $\mathbf{K}(a_n z/1)$ of $\log(1 + z)$ by using (4.4.17)? Again we let $z = 1$. To keep things simple, we want to choose all $C_n = C$ and let R_n vary. Since

$$a_1 = 1, \quad a_{2k} = \frac{k}{4k - 2} > \frac{1}{4}, \quad a_{2k+1} = \frac{k}{4k + 2} < \frac{1}{4}$$

and $a_n \to 1/4$, it seems reasonable to choose C such that $C(1+C) = 1/4$; i.e. $C = (\sqrt{2} - 1)/2$, and to choose R_n such that, except for a_1, the elements a_{2k} and a_{2k+1} are alternatingly close to the right vertex

$$v_{2,n} = (|C_{n-1}| + R_{n-1})(|1 + C_n| - R_n)e^{i2\alpha_n}, \qquad (4.4.19)$$

where $2\alpha_n = \arg(C_{n-1}(1 + C_n)) = 0$, and the left vertex

$$v_{1,n} = \begin{cases} (|C_{n-1}| - R_{n-1})(|1 + C_n| + R_n)e^{i2\alpha_n} & \text{if } R_{n-1} \leq |C_{n-1}|, \\ (|C_{n-1}| - R_{n-1})(|1 + C_n| - R_n)e^{i2\alpha_n} & \text{if } R_{n-1} > |C_{n-1}|. \end{cases} \qquad (4.4.20)$$

A simple (but not the best) choice is for instance

$$R_0 = 1.06 \quad \text{and} \quad R_n = \frac{1}{4n - 0.2} \quad \text{for } n = 1, 2, 3, \ldots .$$

We then get

$$M_k = \frac{C + R_k}{1 + C + R_k} \approx \frac{(\sqrt{2} - 1)k + 0.5}{(\sqrt{2} + 1)k + 0.4} \quad \text{for } k = 1, 2, 3, \ldots ,$$

and thus

$$|S_{n+m}(w_{n+m}) - S_n(w_n)| \leq 2.12 \frac{\sqrt{2} + 1.12}{\sqrt{2} + 1 - \dfrac{1}{2n - 0.1}} \cdot \prod_{k=1}^{n-1} \frac{(\sqrt{2} - 1)k + 0.5}{(\sqrt{2} + 1)k + 0.4} .$$

That is, $|f - S_3(w_3)| \leq 0.20$, $|f - S_4(w_4)| \leq 0.044$, $|f - S_5(w_5)| \leq 0.009$, and so on.

$\diamondsuit$

Remarks:

1. If $\{C_n\}$ and $\{R_n\}$ are chosen such that $\liminf_{n \to \infty} R_n/(1+|C_n|) > 0$ and the right side of (4.4.17) diverges to 0 as $n \to \infty$, then (4.4.17) shows that $\{E_n\}$ is a uniform sequence of general convergence sets with respect to $\{V_n\}$.

2. If all $C_n = C$ and $R_n = R$, then Theorem 26 reduces to parts of the oval theorem. The condition (4.4.14) is then equivalent to the condition $\Re(C) > -1/2$ in the oval theorem.

3. If all $C_{2n} = C_0, C_{2n+1} = C_1, R_{2n} = R_0, R_{2n+1} = R_1$, then (V_0, V_1) is a pair of twin value sets for the twin element sets (E_0, E_1). Condition (4.4.14) now implies that the centers C_0, C_1 must satisfy the inequality

$$\Delta = |1 + C_0||1 + C_1| - |C_0 C_1| > 0 . \tag{4.4.21}$$

It is no longer sufficient that $0 < R_n < |1 + C_n|$. (4.4.14) imposes a second condition on the radii. The choice

$$R_0 = \frac{\Delta - \mu}{|1 + C_1| + |C_1|}, \qquad R_1 = \frac{\Delta - \mu}{|1 + C_0| + |C_0|} \tag{4.4.22}$$

where $0 \leq \mu < \Delta$ is one possibility.

4. For period lengths $k > 2$, the corresponding expressions for Δ and R_n are

$$\Delta = \prod_{j=0}^{k-1} |1 + C_j| - \prod_{j=0}^{k-1} |C_j| > 0, \qquad (4.4.21)$$

and

$$R_n = \frac{\Delta - \mu}{\Delta_n} \quad \text{where } \Delta_{k-1} = \sum_{m=-1}^{k-2} \prod_{j=0}^{m} |1 + C_j| \prod_{j=m+1}^{k-2} |C_j|$$
$$(4.4.22)$$

and the other Δ_ns are determined from Δ_{k-1} by cyclic shifts. For instance, for $k = 3$ the three Δ_n are given by

$$
\begin{aligned}
\Delta_0 &= |C_1 C_2| + |1 + C_1||C_2| + |1 + C_1||1 + C_2|, \\
\Delta_1 &= |C_2 C_0| + |1 + C_2||C_0| + |1 + C_2||1 + C_0|, \\
\Delta_2 &= |C_0 C_1| + |1 + C_0||C_1| + |1 + C_0||1 + C_1|.
\end{aligned}
$$

Example 23 We want to find a priori truncation error bounds for the continued fraction $\mathbf{K}(a_n/1)$ given by

$$\frac{3 + 1/1^2}{1} + \frac{4 + 3/2^2}{1} + \frac{3 + 1/3^2}{1} + \frac{4 + 3/4^2}{1} + \frac{3 + 1/5^2}{1} + \cdots$$

in Example 14 of Chapter I. To keep things simple we choose $C_{2n} = 1$ and $C_{2n+1} = 2$ for all n, since then $C_{2n-1}(1 + C_{2n}) = 4$ and $C_{2n}(1 + C_{2n+1}) = 3$. Since we want a_n to the left of the vertex $v_{2,n}$ given by

$$v_{2,n} = (C_{n-1} + R_{n-1})(1 + C_n - R_n),$$

(see (4.4.19)) it follows that

$$R_{2n} = \frac{1}{2n + 1}, \quad R_{2n+1} = \frac{1}{2n + 1} \quad \text{for } n = 0, 1, 2 \ldots$$

is a possible choice. We get

$$M_{2k-1} = \frac{C_1 + R_{2k-1}}{1 + C_1 + R_{2k-1}} = \frac{4k - 1}{6k - 2} \quad \text{for } k = 1, 2, 3, \ldots$$

and
$$M_{2k} = \frac{C_0 + R_{2k}}{1 + C_0 + R_{2k}} = \frac{2k+2}{4k+3} \quad \text{for } k = 1, 2, 3, \ldots,$$

which lead to the truncation error bounds

$$|f - S_{2n}(w_{2n})| \leq 2\frac{1+1}{2 - \dfrac{1}{2n+1}} \prod_{k=1}^{2n-1} M_k,$$

$$|f - S_{2n+1}(w_{2n+1})| \leq 2\frac{1+1}{4 - \dfrac{1}{2n+1}} \prod_{k=1}^{2n} M_k$$

for $w_n \in \bar{V}_n$. For instance

$$|f - S_6(w_6)| \leq 0.24, \qquad |f - S_7(w_7)| \leq 0.062,$$

which is not so very impressive compared to the table in Example 14 in Chapter I, of course. Still this type of bound is useful.

$\diamond$

The oval theorem and the oval sequence theorem are results on continued fractions of the form $\mathbf{K}(a_n/1)$. The only reason for this restriction was to keep expressions like (4.4.16) and (4.4.17) as simple as possible. For continued fractions $\mathbf{K}(a_n/b_n)$ the oval sequence theorem takes the form

Theorem 27 *Let $C_n \in \mathbf{C}$ and $R_n > 0$ be given for $n = 0, 1, 2, \ldots$. Then $\{V_n\}$ given by (4.4.15) is a sequence of value sets for*

$$\begin{aligned}
\Omega_n = \ & \{(a, b) \in \mathbf{C} \times \mathbf{C}; \ |a(\bar{b} + \bar{C}_n) - C_{n-1}(|b + C_n|^2 - R_n^2)| \\
& + R_n|a| \leq R_{n-1}(|b + C_n|^2 - R_n^2)\} \quad \text{for } n = 1, 2, 3, \ldots.
\end{aligned}$$

$$(4.4.23)$$

If $\mathbf{K}(a_n/b_n)$ is a continued fraction from $\{\Omega_n\}$ and all $w_k \in \bar{V}_k$, then

$$|S_{n+m}(w_{n+m}) - S_n(w_n)| \leq 2R_0 \frac{|C_0| + R_0}{|b_n + C_n| - R_n} \prod_{k=1}^{n-1} M_k, \qquad (4.4.24)$$

where

$$M_k = \max\left\{ \left|\frac{w}{b_k + w}\right|; \ w \in \bar{V}_k \right\}. \qquad (4.4.25)$$

5 Limit periodic continued fractions

5.1 Definition

A continued fraction $\mathbf{K}(a_n/b_n)$ is said to be *limit periodic* with period length k, or limit k-periodic for short, if its sequences $\{a_n\}$ and $\{b_n\}$ of elements are limit k-periodic, i.e. the limits

$$\lim_{n \to \infty} a_{kn+p} = a_p^*, \qquad \lim_{n \to \infty} b_{kn+p} = b_p^* \quad \text{for } p = 1, 2, \ldots, k \quad (5.1.1)$$

exist in $\hat{\mathbf{C}}$. One only has to cast a glimpse at the appendix of this book to see that most of the continued fraction expansions of special functions that are in use are indeed limit periodic.

Limit periodic continued fractions have been widely studied. Not only for their importance, but also for their nice properties. They resemble periodic continued fractions, both by appearance and behavior in several aspects. This connection to the well understood periodic continued fraction is very useful. We shall use it to prove properties of limit periodic continued fractions and to choose approximants for them.

5.2 Finite limits, loxodromic case

We assume in this subsection that all the limits in (5.1.1) are finite.

Consider first the case where also all $a_p^* \neq 0$. Since convergence of a continued fraction is a property which always depends on terms far out in the continued fraction, and since a limit periodic continued fraction looks more and more like the *corresponding periodic* continued fraction

$$\mathbf{K}\frac{a_n^*}{b_n^*} = \frac{a_1^*}{b_1^*} + \frac{a_2^*}{b_2^*} + \cdots + \frac{a_k^*}{b_k^*} + \frac{a_1^*}{b_1^*} + \frac{a_2^*}{b_2^*} + \cdots, \qquad (5.2.1)$$

it is to be expected that the convergence/divergence of $\mathbf{K}(a_n/b_n)$ is closely tied to the convergence/divergence of (5.2.1). And, as we saw in Subsection 2.3, Theorem 6, this convergence is determined by the classification of the linear fractional transformation

$$T_k(w) = \frac{a_1^*}{b_1^*} + \frac{a_2^*}{b_2^*} + \cdots + \frac{a_k^*}{b_k^* + w} \tag{5.2.2}$$

and by the fixed points $x^{(p)}$ and $y^{(p)}$ of

$$T_k^{(p)}(w) = \frac{a_{p+1}^*}{b_{p+1}^*} + \frac{a_{p+2}^*}{b_{p+2}^*} + \cdots + \frac{a_{p+k}^*}{b_{p+k}^* + w}. \tag{5.2.3}$$

If $a_p^* = 0$ for some p, then (5.2.1) is no longer a continued fraction by our own definition. But if T_k is well defined, then $\mathbf{K}(a_n/b_n)$ may still be a convergent continued fraction.

Let us first study limit 1-periodic continued fractions $\mathbf{K}(a_n/b_n)$ of *loxodromic type*: that is, $a_n \to a_1^* \in \mathbf{C}$, $b_n \to b_1^* \in \mathbf{C}$ and $T_1(w) = a_1^*/(b_1^*+w)$ is loxodromic if $a_1^* \neq 0$, and $b_1^* \neq 0$ if $a_1^* = 0$. In the latter case T_1 is a singular transformation, $T_1(w) \equiv 0$ for $w \neq -b_1^*$. We say that $x = 0$ is the attractive fixed point of (the singular transformation) T_1 in this case, and that $y = -b_1^* \neq x$ is the repulsive fixed point of T_1.

Theorem 28 *Let* $\mathbf{K}(a_n/b_n)$ *be limit 1-periodic of loxodromic type, where* T_1 *has attractive fixed point* x *and repulsive fixed point* y. *Then:*

A. $\mathbf{K}(a_n/b_n)$ *converges to a value* $f \in \hat{\mathbf{C}}$.

B.

$$\lim_{n \to \infty} t_n = \begin{cases} x & \text{if } t_0 = f \\ y & \text{if } t_0 \neq f \end{cases} \tag{5.2.4}$$

for every tail sequence $t_n = S_n^{-1}(t_0)$ *of* $\mathbf{K}(a_n/b_n)$.

C. *Let all* $b_n = 1$ *and* $d_n = \sup\{|a_m - a_1^*|; \ m \geq n\}$ *for* $n = 1, 2, 3, \ldots$. *If* $d_2 < \Delta^2/4$ *where* $\Delta = |1 + x| - |x|$, *then*

$$|f - S_n(w)| \leq 2R_0 \frac{|x| + R_0}{|1 + x| + |x| + \sqrt{\Delta^2 - 4d_{n+1}}} \prod_{k=1}^{n-1} M_k \tag{5.2.5}$$

for $|w - x| \leq R_n$, where

$$R_0 \quad = \quad \frac{2d_1 + |x|(\Delta - \sqrt{\Delta^2 - 4d_2})}{|1 + x| + |x| + \sqrt{\Delta^2 - 4d_2}}, \qquad (5.2.6)$$

$$R_n \quad = \quad (\Delta - \sqrt{\Delta^2 - 4d_{n+1}})/2 \qquad (5.2.7)$$

and

$$M_k \quad = \quad \max\left\{\left|\frac{w}{1 + w}\right|; |w - x| \leq R_k\right\} \qquad (5.2.8)$$

$$\leq \quad \frac{|1 + x| + |x| - \sqrt{\Delta^2 - 4d_{n+1}}}{|1 + x| + |x| + \sqrt{\Delta^2 - 4d_{n+1}}}. \qquad (5.2.9)$$

Proof :   **A:** We want to use Theorem 27 with all $V_n = V$ and all $\Omega_n = \Omega$. As center for V we choose the attractive fixed point x. (This is possible since $x \neq \infty$ under our conditions.)   To make sure that $(a_n, b_n) \in \Omega$, at least from some n on, we require that (a_1^*, b_1^*) be contained in the interior of Ω; i.e.

$$|a_1^*(\bar{b}_1^* + \bar{x}) - x(|b_1^* + x|^2 - R^2)| + R|a_1^*| < R(|b_1^* + x|^2 - R^2),$$

where R is the radius of V. Since $a_1^* = x(b_1^* + x)$ this can be written

$$|x| < |b_1^* + x| - R.$$

Hence, the choice $R = \Delta - \mu$ where $\Delta = |b_1^* + x| - |x|$ and $\Delta/2 < \mu < \Delta$ is fine. By Theorem 27 we then find that

$$|S_{n+m}^{(N)}(w_{N+n+m}) - S_n^{(N)}(w_{N+n})| \leq 2R\frac{|x| + R}{|b_{N+n} + x| - R}\prod_{k=1}^{n-1} M_{N+k}$$

for $w_j \in \bar{V}$ for all j, where $N \in \mathbf{N}_0$ is chosen so large that $(a_n, b_n) \in \Omega$ for all $n > N$ and

$$
\begin{aligned}
M_{N+k} \quad &\leq \quad \frac{|x| + R}{|b_{N+k} + x| - R} \\
&\longrightarrow \quad \frac{|x| + R}{|b_1^* + x| - R} = \frac{|b_1^* + x| - \mu}{|x| + \mu} < \frac{|b_1^* + x| - \Delta/2}{|x| + \Delta/2} = 1
\end{aligned}
$$

as $k \to \infty$. This proves that $\{S_n^{(N)}(w_{N+n})\}$ is a Cauchy sequence and thus converges to a value $f^{(N)} \in \bar{V}$. This value is independent of the actual choice of $w_{N+n} \in \bar{V}$. Hence the Nth tail of $\mathbf{K}(a_n/b_n)$ converges generally to $f^{(N)}$. The classical convergence of this tail follows then by Theorem 9. Finally Theorem 1 in Chapter II gives the classical convergence of the continued fraction $\mathbf{K}(a_n/b_n)$ itself.

B: Let first $t_0 = f$. Then t_n is the value of nth tail of $\mathbf{K}(a_n/b_n)$ for every $n \in \mathbf{N}_0$. Therefore $t_n \in \bar{V}$ for all $n \geq N$ by the argument in the proof of part A. Since μ can be chosen arbitrarily close to Δ (such that R is arbitrarily close to 0), this proves that $t_n \to x$.

Next we consider the case $t_0 \neq f$. We shall use that

$$t_n = - \left\{ b_n + \frac{a_n}{b_{n-1}} + \frac{a_{n-1}}{b_{n-2}} + \cdots + \frac{a_2}{b_1} + \frac{a_1}{(-t_0)} \right\} \qquad (5.2.10)$$

(see Theorem 3 in Chapter II). Let μ, $\Delta/2 < \mu < \Delta$, be arbitrarily chosen, and let $N = N(\mu)$ be an integer such that $(a_n, b_{n-1}) \in \Omega$ for all $n \geq N$. This is possible since also $(a_n, b_{n-1}) \to (a_1^*, b_1^*)$ in the interior of Ω. Further, let t_N be chosen such that $-(b_N + t_N) \in V$ and $t_N \neq f^{(N)}$. Since by (5.2.10)

$$- (b_{N+n} + t_{N+n}) = \frac{a_{N+n}}{b_{N+n-1}} + \frac{a_{N+n-1}}{b_{N+n-1}} + \cdots + \frac{a_{N+1}}{b_N - (b_N + t_N)}, \qquad (5.2.11)$$

we then have that $-(b_n + t_n) \in V$ for all $n \geq N$. Hence $\{t_n\}$ has all its limit points $\in -(b_1^* + \bar{V})$. By Theorem 6 in Chapter II we know that all tail sequences of a generally convergent continued fraction have the same asymptotic behavior, except the right tail sequence $\{f^{(N)}\}$. Hence every tail sequence $\{t_n\}$ with $t_0 \neq f$ has all its limit points in $-(b_1^* + \bar{V})$. Since μ can be chosen arbitrarily close to Δ so that the radius R of V is arbitrarily close to 0, we find that $\lim_{n \to \infty} t_n = -(b_1^* + x)$ if $t_0 \neq f$.

Now x and y are the two solutions of the quadratic equation $T_1(w) = w$. Hence $x + y = -b_1^*$, and so $t_n \to -(b_1^* + x) = y$.

C: As in Part A we let V_n be circular disks with centers at $C_n = x$, but this time we shall vary the radii R_n. Moreover, since we want to derive truncation error bounds for $\mathbf{K}(a_n/1)$, we want $(a_n, 1) \in \Omega$ for all $n \geq 1$; that is

$$|a_n(1 + \bar{x}) - x(|1 + x|^2 - R_n^2)| + R_n|a_n| \leq R_{n-1}(|1 + x|^2 - R_n^2) \quad (5.2.12)$$

for $n = 1, 2, 3, \ldots$. Since $a_n = a_1^* + \epsilon_n$ where $|\epsilon_n| \leq d_n$ and $a_1^* = x(1+x)$, the choice for R_0 and R_n as given in the theorem works fine. Inserted into (4.4.24) this gives (5.2.5).   ∎

An alternative way to prove Theorem 28A is as follows: Since $b_1^* \neq 0$ under our conditions, we can transform some tail of $\mathbf{K}(a_n/b_n)$ to a continued fraction of the form $\mathbf{K}(a_n/1)$ where $c_n \to a_1^*/b_1^{*2}$. (For such equivalence transformations, see Corollary 10B in Chapter II.) The convergence then follows from the parabola theorem, Theorem 20. We chose to use Theorem 27 since then the proof can be generalized to limit k-periodic continued fractions with period lengths $k > 1$.

Theorem 29 *Let* $\mathbf{K}(a_n/b_n)$ *be a limit k-periodic continued fraction of loxodromic type with finite limits (5.1.1). Let $x^{(p)}$ and $y^{(p)}$ denote the attractive and repulsive fixed points of $T_k^{(p)}$ for $p = 0, 1, \ldots, k$. If all $x^{(p)} \neq \infty$, then:*

A. $\mathbf{K}(a_n/b_n)$ *converges generally to a value* $f \in \hat{\mathbf{C}}$.

B. $\mathbf{K}(a_n/b_n)$ *converges in the classical sense if all* $y^{(p)} \neq \infty$.

C.

$$\lim_{n \to \infty} S_{kn+p}^{-1}(t_0) = \begin{cases} x^{(p)} & \text{if } t_0 = f \\ y^{(p)} & \text{if } t_0 \neq f \end{cases} \quad \text{for } p = 1, \ldots, k \,.$$

$$(5.2.13)$$

Remark: $\mathbf{K}(a_n/b_n)$ is limit k-periodic of loxodromic type if either T_k is a loxodromic linear fractional transformation or a singular, well defined transformation $T_k(w) \equiv c$ for all $w \neq w_0$ for a $w_0 \neq c$. We then say that $x = c$ is the attractive fixed point of T_k and that $y = w_0$ is the repulsive fixed point of T_k.

We shall not prove Theorem 29 in detail, but the idea is to use Theorem

27 with $C_{kn+p} = C_p = x^{(p)}$ as centers of V_{kn+p}. Hence we need that

$$\Delta = \prod_{j=0}^{k-1} |b_j^* + x^{(j)}| - \prod_{j=0}^{k-1} |x^{(j)}| > 0 \,. \tag{5.2.14}$$

Fortunately (5.2.14) holds in our case:

Lemma 30 *Let* $x^{(p)} \neq \infty$ *be a fixed point of* $T_k^{(p)}$ *(as given in (5.2.3))* *for* $p = 0, 1, \ldots, k$, *chosen such that* $x^{(p-1)} = a_p^*/(b_p^* + x^{(p)})$ *for* $p = 1, 2, \ldots, k$.

A. *If* $x^{(p)}$ *is the attractive fixed point of* $T_k^{(p)}$ *and* $y^{(p)} \neq x^{(p)}$ *is the repulsive fixed point, then (5.2.14) holds.*

B. *If* T_k *is parabolic, then*

$$\prod_{j=0}^{k-1} (b_j^* + x^{(j)}) = \prod_{j=0}^{k-1} (-x^{(j)}) \,. \tag{5.2.15}$$

C. *If* T_k *is elliptic, then* $\Delta = 0$ *(Δ as given in (5.2.14)) but (5.2.15) does not hold.*

Proof : A: Assume first that T_k is non-singular (i.e. all $a_p^* \neq 0$). Then T_k is a loxodromic linear fractional transformation, which can be written

$$T_k(w) = \frac{A_k^* + A_{k-1}^* w}{B_k^* + B_{k-1}^* w} \,.$$

According to Theorem 5C we have

$$\begin{aligned} |B_k^* + B_{k-1}^* x^{(0)}| &> |B_k^* + B_{k-1}^* y| &&\text{if } B_{k-1}^* \neq 0 \,, \\ |A_{k-1}^*| &\neq |B_k^*| &&\text{if } B_{k-1}^* = 0 \,, \end{aligned} \tag{5.2.16}$$

and if $B_{k-1}^* = 0$ then $x^{(0)} = \infty$ if $|B_k^*| < |A_{k-1}^*|$ and $y = \infty$ if $|B_k^*| > |A_{k-1}^*|$. We have assumed that $x^{(0)} \neq \infty$.

Let first $B^*_{k-1} \neq 0$. Then $x^{(0)}$ and y are finite solutions of the quadratic equation $T_k(w) = w$. Hence $x^{(0)} + y = (A^*_{k-1} - B^*_k)/B^*_{k-1}$ which means that

$$B^*_k + B^*_{k-1} y = A^*_{k-1} - B^*_{k-1} x^{(0)} . \tag{5.2.17}$$

In Problem 3 in Chapter II you were asked to prove that if $\{t_n\}$ is a tail sequence for a continued fraction $\mathbf{K}(a_n/b_n)$ and all $t_n \neq \infty$, then

$$B_n + B_{n-1} t_n = \prod_{k=1}^{n} (b_k + t_k), \qquad A_n - B_n t_0 = \prod_{k=0}^{n} (-t_k). \tag{5.2.18}$$

The periodic sequence $x^{(0)}, x^{(1)}, \ldots, x^{(k-1)}, x^{(k)} = x^{(0)}, x^{(1)}$ is a tail sequence with all $x^{(p)} \neq \infty$ for the periodic continued fraction $\mathbf{K}(a^*_n/b^*_n)$ given by (5.2.1). Hence, by (5.2.17),

$$B^*_k + B^*_{k-1} y = A^*_{k-1} - B^*_{k-1} x^{(0)} = \prod_{p=0}^{k-1} \left(-x^{(p)} \right)$$

and

$$B^*_k + B^*_{k-1} x^{(0)} = B^*_k + B^*_{k-1} x^{(k)} = \prod_{p=0}^{k-1} \left(b^*_p + x^{(p)} \right) .$$

Hence (5.2.16) is equivalent to (5.2.14) in this case.

Now let $B^*_{k-1} = 0$. Then $|B^*_k| > |A^*_{k-1}|$ and $x^{(0)} = A^*_k/(B^*_k - A^*_{k-1})$. By (5.2.18) with $t_n = x^{(n)}$ we thus have

$$B^*_k = \prod_{j=0}^{k} (b^*_j + x^{(j)}), \qquad A^*_{k-1} = \prod_{j=0}^{k-1} (-x^{(j)})$$

and (5.2.14) follows since $|B^*_k| > |A^*_{k-1}|$.

Finally assume that $a^*_p = 0$. Then $x^{(p-1)} = 0$. Hence (5.2.14) follows since $b^*_n + x^{(n)} \neq 0$ for all n. (If $b^*_n + x^{(n)} = 0$ for an $n \in \{1, \ldots, k\}$, then $x^{(n-1)} = a^*_n/(b^*_n + x^{(n)}) = \infty$, since T_k is well defined and thus, in particular, $a^*_n \neq 0$.)

The results B and C follow by the same kind of arguments.   ■

To prove Theorem 29C we can use that the classification of linear fractional transformations T_k are invariant under inversion. (See Property

1 in Subsection *2.2.*) Theorem 29B follows then easily from part A and C, using Theorem 4 in Chapter II.

In applications it is often functions of a complex variable z which are expanded in limit periodic continued fractions $\mathbf{K}(a_n(z)/b_n(z))$ with polynomial elements $a_n(z)$ and $b_n(z)$. Theorem 29 gives a domain D where $\mathbf{K}(a_n(z)/b_n(z))$ converges, but we can only conclude pointwise convergence. However, Theorem 29 was based on Theorem 27, where the bound (4.4.24) can be used to prove uniform convergence.

Theorem 31 *Let* $\mathbf{K}(a_n(z)/b_n(z))$ *with polynomial elements* $a_n(z)$ *and* $b_n(z)$ *be limit* k-*periodic. Let* $D \subseteq \mathbf{C}$ *be an open set where* $\mathbf{K}(a_n(z)/b_n(z))$ *satisfies the hypotheses of Theorem 29 for all* $z \in D$. *Let* $F(z)$ *be the value of* $\mathbf{K}(a_n(z)/b_n(z))$ *in* D. *Then the (general) convergence of* $\mathbf{K}(a_n(z)/b_n(z))$ *in* D *is uniform on compact subsets* $C \subseteq D$ *where* $\infty \neq F(C)$.

The proof is based on the fact that a compact set C which is contained in a union $\bigcup_{n=1}^{\infty} D_n$ of open sets is also contained in a union of a finite number of these sets D_n.

5.3 Finite limits, parabolic case

Also in this section we consider limit k-periodic continued fractions $\mathbf{K}(a_n/b_n)$ where the limits (5.1.1) are finite, but now we assume that either T_k given by (5.2.2) is a (non-singular) linear fractional transformation of parabolic type, or T_k is singular ($a_p^* = 0$ for some $p \in \{1, \ldots, k\}$) and well defined for all $w \neq c$ for a $c \in \hat{\mathbf{C}}$, with $T(w) \equiv c$ for all $w \neq c$. We say that $\mathbf{K}(a_n/b_n)$ is limit k-periodic of parabolic type.

The situation is then substantially different from the loxodromic case. Periodic continued fractions of parabolic type converge generally (see Theorem 6), but they are in a way on the border line between the periodic continued fractions of loxodromic type which converge generally and the ones of elliptic type which diverge generally. This is reflected by the limit k-periodic continued fractions of parabolic type. They may converge or diverge generally depending on *how* the elements (a_n, b_n) approach their limit points (5.1.1).

Example 24 The continued fraction $\mathbf{K}(a_n/1)$ is limit 1-periodic of parabolic type if and only if $a_1^* = \lim_{n \to \infty} a_n = -1/4$. Let first $a_n = -1/4 + \epsilon_n$ where all $\epsilon_n > 0, \epsilon_n \to 0$. Then $a_n \in P_0$ in the parabola theorem, and $\mathbf{K}(a_n/1)$ converges.

If $a_n = (-1/4) - 1/(4n^2 - 1)$ for all n, then one can prove that $\mathbf{K}(a_n/1)$ diverges generally.

On the other hand, if

$$a_n = -\frac{1}{4} - \frac{1}{4(4n^2 - 1)} \quad \text{for all } n, \tag{5.3.1}$$

then $a_n \in P_{0,n}$ in the parabola sequence theorem with $g_n = (n+1)/(2n+1)$, and thus $\mathbf{K}(a_n/1)$ converges. In fact, $f^{(n)} = -g_n = -(n+1)/(2n+1)$ is the right tail sequence of $\mathbf{K}(a_n/1)$ in this case, so the continued fraction converges to $-g_0 = -1$.

$\diamond$

The parabola theorems, the oval theorems etc. are useful tools for determining whether a given limit k-periodic continued fraction of parabolic type converges. From these we can find "safe directions" in which (a_n, b_n) may approach the limit points without disturbing the convergence (such as $a_n > -1/4$, $a_n \to -1/4$ in Example 24). We can also find conditions for "safe speed in unsafe directions" (such as in (5.3.1)). The next result belongs to this latter category. It describes the borderline between the convergent and divergent continued fractions $\mathbf{K}(a_n/1)$ where $a_n \to -1/4$ monotonely from the "unsafe direction" $a_n < -1/4$. It is due to Jacobsen and Masson, [JaMa90]. We use the notation

$$\log_0 n = n, \quad \log_m n = \log(\log_{m-1} n) \quad \text{for } m = 1, 2, \ldots \tag{5.3.2}$$

for the natural logarithms, and let $L_k(n)$ denote the product

$$L_k(n) = \prod_{m=0}^{k} \log_m n = n(\log n)(\log_2 n) \cdots (\log_k n). \tag{5.3.3}$$

Theorem 32 *Let $p \in \mathbf{N}_0$ be a fixed number. The limit 1-periodic continued fraction $\mathbf{K}(a_n/1)$ converges if there exists an $N \in \mathbf{N}$ such that*

$$\left| a_{N+n} - \left(-\frac{1}{4} \right) \right| \leq \sum_{j=0}^{p} (4L_j(n))^{-2} \qquad (5.3.4)$$

from some n on. All tail sequences of $\mathbf{K}(a_n/1)$ then converge to $x = -1/2$. The limit 1-periodic continued fraction $\mathbf{K}(a_n/1)$ diverges generally if

$$a_n = -\frac{1}{4} - \sum_{j=0}^{p-1} (4L_j(n))^{-2} - d \big/ (4L_p(n))^2 \qquad (5.3.5)$$

from some n on with $d > 1$.

Sketch of proof: $\mathbf{K}(a_n/1)$ converges if a tail of $\mathbf{K}(a_n/1)$ converges. A tail of $\mathbf{K}(a_n/1)$ converges if $a_{N+n} \in P_{0,n}$ from some n on, where $P_{0,n}$ are the parabolic element regions in the parabola sequence theorem. $a_{N+n} \in P_{0,n}$ from some n on if

$$g_n = \frac{1}{2} + \frac{1}{4n} + \frac{1}{4n \log n} + \cdots + \frac{1}{4L_p(n)} \qquad (5.3.6)$$

from some n on. ∎

Observe that the case (5.3.1) is covered by (5.3.4) with $p = 0$, and the case $a_n = (-1/4) - 1/(4n^2 - 1)$ is covered by (5.3.5) with $p = 0$ and $d = 4$. Similar results can be obtained for limit k-periodic continued fractions of parabolic type with period lengths $k > 1$.

5.4   Finite limits, elliptic case

If T_k in (5.2.2) is an elliptic transformation, we say that $\mathbf{K}(a_n/b_n)$ in (5.1.1) is limit k-periodic continued fraction of elliptic type. Such continued fractions may also converge or diverge generally depending on how (a_n, b_n) approach their limit points. In [Gill73] it is proved that $\mathbf{K}(a_n/1)$, where $\lim_n \to \infty \, a_n = a_1^* < -1/4$, may converge or diverge. It always diverges if $a_n \to a_1^*$ fast enough; i.e. if $|a_n - a_1^*| \leq Cr^n$ for some positive $r < 1$.

5.5   Choice of approximants

Let $\mathbf{K}(a_n/b_n)$ be limit k-periodic of loxodromic type. Since

$$\lim_{n \to \infty} f^{(kn+p)} = x^{(p)} \quad \text{for } p = 0, 1, \ldots, k-1 \qquad (5.5.1)$$

by Theorem 29C, it seems reasonable to use the approximants $S_{kn+p}(x^{(p)})$. Actually, we find that if $f \neq \infty$ and all $x^{(p)} \neq 0, \infty$, then

$$\lim_{n \to \infty} \frac{f - S_{(kn+p)}(x^{(p)})}{f - S_{kn+p}(0)} = \lim_{n \to \infty} \frac{(f^{(kn+p)} - x^{(p)})h_{kn+p}}{f^{(kn+p)}(h_{kn+p} + x^{(p)})} = 0 \quad (5.5.2)$$

since $h_{kn+p} \to -y^{(p)}$ as $n \to \infty$ and $y^{(p)} \neq x^{(p)}$. Here we have used formula (1.4.7) in Chapter II and Theorem 29C.) We say that $S_{kn+p}(x^{(p)})$ converges faster to f than $S_{kn+p}(0)$, or that the modification $w_{kn+p} = x^{(p)}$ accelerates the convergence of $\mathbf{K}(a_n/b_n)$.

In (5.5.2) we used our information (5.1.1) to choose approximants for $\mathbf{K}(a_n/b_n)$. What if we also know that the limits

$$\lim_{n \to \infty} \frac{\delta_{kn+p+1}}{\delta_{kn+p}} = r_p \quad \text{for } p = 1, 2, \ldots, k \qquad (5.5.3)$$

exist, where

$$\delta_n = a_n - w_{n-1}(b_n + w_n), \qquad (5.5.4)$$

what then? Can one use this information to find even better approximants $S_n(w_n^{(1)})$ for $\mathbf{K}(a_n/b_n)$? The answer is yes, under mild conditions. Let us demonstrate how this works for the case where the period length is $k = 1$. We get:

Theorem 33 *Let $\mathbf{K}(a_n/b_n)$ be a limit 1-periodic continued fraction of loxodromic type, with finite limits $a_n \to a_1^* \neq 0$ and $b_n \to b_1^*$ and with finite value f. Further let $\{w_n\}$ be a sequence from $\hat{\mathbf{C}}$ such that $\epsilon_n = w_n - f^{(n)} \to 0$ as $n \to \infty$, and let δ_n be given by (5.5.4). Then:*

A.

$$\lim_{n \to \infty} \delta_{n+1}/\delta_n = r \quad \Longleftrightarrow \quad \lim_{n \to \infty} \epsilon_{n+1}/\epsilon_n = r. \qquad (5.5.5)$$

B. If $\lim_{n \to \infty} \delta_{n+1}/\delta_n = r$, then $\lim_{n \to \infty} \delta_n/\epsilon_{n-1} = b_1^* + x + rx$.

C. If $\lim_{n \to \infty} \delta_{n+1}/\delta_n = r \neq 0, \infty$, then

$$\lim_{n \to \infty} \frac{f - S_n(w_n^{(1)})}{f - S_n(w_n)} = 0,$$

where $w_n^{(1)} = w_n + \delta_{n+1}/(b_1^* + x + rx)$.

Proof : **A:** We know from Theorem 28B that $f^{(n)} \to x \neq \infty$. Since $a_1^* \neq 0$ we also have $x \neq 0$. Assume first that $\epsilon_{n+1}/\epsilon_n \to r$. Then

$$
\begin{aligned}
\delta_n &= a_n - w_{n-1}(b_n + w_n) \\
&= f^{(n-1)}\left(b_n + f^{(n)}\right) - \left(f^{(n-1)} + \epsilon_{n-1}\right)\left(b_n + f^{(n)} + \epsilon_n\right) \\
&= -\epsilon_{n-1}\left(b_n + f^{(n)}\right) - f^{(n-1)}\epsilon_n - \epsilon_n\epsilon_{n-1} \\
&= -\epsilon_{n-1}\left\{b_n + f^{(n)} + f^{(n-1)}\epsilon_n/\epsilon_{n-1} + \epsilon_n\right\}, \qquad (5.5.6)
\end{aligned}
$$

where the expression in the parentheses { } converges to $b_1^* + x + xr =: d$ as $n \to \infty$. This expression d is $\neq 0$ since $|b_1^* + x| > |x|$ by Lemma 30 and $|r| \leq 1$. Hence $\lim \delta_{n+1}/\delta_n = \lim \epsilon_n/\epsilon_{n-1} = r$.

Assume next that $\delta_{n+1}/\delta_n \to r$. From (5.5.6) we obtain the equality

$$\frac{\delta_{n+1}}{\delta_n} = \frac{\epsilon_n}{\epsilon_{n-1}}\left\{\frac{b_{n+1} + f^{(n+1)} + f^{(n)}\epsilon_{n+1}/\epsilon_n + \epsilon_{n+1}}{b_n + f^{(n)} + f^{(n-1)}\epsilon_n/\epsilon_{n-1} + \epsilon_n}\right\}, \qquad (5.5.7)$$

that is

$$f^{(n-1)}\frac{\epsilon_n}{\epsilon_{n-1}} = \frac{\frac{\delta_{n+1}}{\delta_n}(b_n + w_n)f^{(n-1)}}{b_{n+1} + w_{n+1} - f^{(n-1)}\frac{\delta_{n+1}}{\delta_n} + f^{(n)}\frac{\epsilon_{n+1}}{\epsilon_n}},$$

which means that $\{f^{(n)}\epsilon_{n+1}/\epsilon_n\}$ is a tail sequence for a continued fraction $\mathbf{K}(c_n/d_n)$ where

$$c_n = \frac{\delta_{n+1}}{\delta_n}(b_n + w_n)f^{(n-1)} \to r(b_1^* + x)x = ra_1^* = c_1^*$$

and

$$d_n = b_{n+1} + w_{n+1} - \frac{\delta_{n+1}}{\delta_n} f^{(n-1)} \to b_1^* + x - xr =: d_1^* .$$

Since $T_1^*(w) = c_1^*/(d_1^* + w)$ is loxodromic or singular with attractive fixed point rx and repulsive fixed point $-(b_1^* + x)$, where $|b_1^* + x| > |rx|$, it follows from Theorem 28 that $f^{(n-1)}\epsilon_n/\epsilon_{n-1} \to rx$. This proves that $\epsilon_n/\epsilon_{n-1} \to r$ since $x \neq 0, \infty$.

B: This is a direct result of (5.5.6) and the subsequent remark.

C: We have as in (5.5.2) that

$$\frac{f - S_n(w_n^{(1)})}{f - S_n(w_n)} = \frac{f^{(n)} - w_n^{(1)}}{f^{(n)} - w_n} \cdot \frac{h_n + w_n}{h_n + w_n^{(1)}}$$

where $(h_n + w_n) \to (-y + x) \neq 0, \infty$ and where

$$\frac{f^{(n)} - w_n^{(1)}}{f^{(n)} - w_n} = \frac{\epsilon_n - \delta_{n+1}/(b_1^* + x + rx)}{\epsilon_n} \to 0$$

by Part B. ∎

In a way one might say that Theorem 33 describes a device to improve approximants in the sense that the new approximants converge faster to the value f of the continued fraction $\mathbf{K}(a_n/b_n)$ than the old ones. As a starting point we need to have approximants $\{S_n(w_n)\}$ working better than $\{S_n(0)\}$ in the sense of (5.5.2).The process works under mild conditions if the asymptotic side condition (5.5.3) holds.

Example 25 The Stieltjes fraction

$$\mathbf{K}\frac{a_n z}{1} = \frac{\frac{2^2 z}{3 \cdot 5}}{1} + \frac{\frac{3^2 z}{5 \cdot 7}}{1} + \frac{\frac{4^2 z}{7 \cdot 9}}{1} + \frac{\frac{5^2 z}{9 \cdot 11}}{1} + \cdots$$

converges to a holomorphic function $f(z)$ for $|\arg(z + 1)| < \pi$ since $a_n \to 1/4$. For such values of z this S-fraction is limit 1-periodic of loxodromic type. We have $a_1^*(z) = z/4$ and $b_1^* = 1$, and thus $x = x(z) = (\sqrt{1 + z} - 1)/2$ and $y = y(z) = (-\sqrt{1 + z} - 1)/2$, where $\Re(\sqrt{1 + z}) > 0$. Hence we can choose the approximants

$$S_n(w_n^{(0)}) \quad \text{where } w_n^{(0)}(z) = x(z) = (\sqrt{1 + z} - 1)/2 .$$

These approximants converge faster to $f(z)$ than $S_n(0)$ in the sense of (5.5.2). With this choice we find that

$$\delta_n^{(0)}(z) = a_n z - w_{n-1}^{(0)}(z)\left(1 + w_n^{(0)}(z)\right) = a_n z - \frac{z}{4} = \frac{z}{(2n+1)(2n+3)}.$$

Hence $\delta_{n+1}^{(0)}/\delta_n^{(0)} \to 1$, so by Theorem 33C the approximants $S_n(w_n^{(1)}(z))$ where

$$w_n^{(1)}(z) = w_n^{(0)}(z) + \frac{\delta_{n+1}^{(0)}(z)}{1 + 2w_n^{(0)}(z)}$$

converge even faster to $f(z)$. Suppressing the variable z we have

$$w_n^{(1)} = x + \frac{z}{(1+2x)(2n+3)(2n+5)},$$

and thus

$$
\begin{aligned}
\delta_n^{(1)} &= a_n z - w_{n-1}^{(1)}\left(1 + w_n^{(1)}\right) \\
&= a_n z - \left(x + \frac{z/(1+2x)}{(2n+1)(2n+3)}\right)\left(1 + x + \frac{z/(1+2x)}{(2n+3)(2n+5)}\right) \\
&= \frac{z/(1+2x)}{(2n+1)(2n+3)(2n+5)}\left(4x - \frac{z}{(2n+3)(1+2x)}\right),
\end{aligned}
$$

so also $\delta_{n+1}^{(1)}/\delta_n^{(1)} \to 1$. Hence $S_n(w_n^{(2)})$ where

$$w_n^{(2)} = w_n^{(1)} + \frac{\delta_{n+1}^{(1)}}{1 + 2x}$$

converge even faster again. We can continue the process. In this example it is possible to prove that $\delta_{n+1}^{(m)}/\delta_n^{(m)} \to 1$ at every step $m \in \mathbf{N}_0$, and therefore

$$w_n^{(m+1)} = w_n^{(m)} + \frac{\delta_{n+1}^{(m)}}{1 + 2x}$$

at every step m. In Table 1 we show the first approximants $S_n(0)$, $S_n(x)$, $S_n(w_n^{(1)})$ and $S_n(w_n^{(2)})$ for $\mathbf{K}(a_n/1)$ (i.e. $z = 1$).

The table stops when we have reached the correct value with 8 digits, and this value is repeated for all larger indices. Of course, in this example we would always choose the approximants $S_n(0)$ or $S_n(x)$ since they

n	$S_n(0)$	$S_n(x)$	$S_n(w_n^{(1)})$	$S_n(w_n^{(2)})$
1	1.267	1.221	1.219993	1.2199356
2	1.2121	1.21984	1.2199258	1.2199305
3	1.2213	1.219941	1.21993125	1.2199308
4	1.21970	1.2199296	1.2199307	$\vdots$
5	1.219971	1.2199309	1.2199308	
6	1.219924	1.2199308	$\vdots$	
7	1.2199320	$\vdots$		
8	1.2199306			
9	1.2199308			
$\vdots$				

Table 1: Example 25.

converge so fast anyway. But for values of z close to the ray $|\arg(z+1)| = \pi$ where the continued fraction diverges, or with $|z|$ large, there is much to be gained by this method.

$\diamond$

Example 26 Let us once more consider the S-fraction

$$\mathbf{K}\,\frac{a_n z}{1} = \frac{z}{1+}\,\frac{z/2}{1}\,+\,\frac{z/6}{1}\,+\,\frac{2z/6}{1}\,+\,\frac{2z/10}{1}\,+\,\frac{3z/10}{1}\,+\cdots$$

which converges to $\log(1+z)$ for $|\arg(1+z)| < \pi$. Also here $a_1^*(z) = z/4$ and $b_1^* = 1$, so the approximants $S_n(x)$ converge faster than $S_n(0)$ to the value $f(z) = \log(1+z)$ in the sense of (5.5.2). With this choice we find for $n \geq 2$

$$\delta_n^{(0)} \;=\; a_n z - w_{n-1}^{(0)}\left(1 + w_n^{(0)}\right) = a_n z - x(1+x)$$

$$=\; a_n z - z/4 = \begin{cases} z/4(n-1) & \text{if } n \text{ is even,} \\ -z/4n & \text{if } n \text{ is odd.} \end{cases}$$

This means that $\lim_{n \to \infty} \delta_{n+1}^{(0)}/\delta_n^{(0)} = -1$, so by Theorem 33C the

n	$S_n(0)$	$S_n(x)$	$S_n(w_n^{(1)})$
1	0.667	0.707	0.6921
2	0.700	0.6948	0.69309
3	0.6923	0.69336	0.693137
4	0.69333	0.693177	0.6931464
5	0.693122	0.693152	0.69314703
6	0.693152	0.69314783	0.693147168
7	0.6931464	0.69314728	0.693147178
8	0.69314733	0.693147196	0.693147180
9	0.693147158	0.693147183	0.693147181
10	0.693147185	0.693147181	0.693147181

Table 2: Example 26.

approximants $S_n(w_n^{(1)})$ where

$$w_n^{(1)} = w_n + \delta_{n+1}^{(0)} = \begin{cases} x - z/4(n+1) & \text{if } n \text{ is even,} \\ x + z/4n & \text{if } n \text{ is odd,} \end{cases}$$

converge even faster to $\log(1 + z)$. Now we find

$$
\begin{aligned}
\delta_n^{(1)} &= a_n z - w_{n-1}^{(1)}(1 + w_n^{(1)}) = a_n z - (x + \delta_n^{(0)})(1 + x + \delta_{n+1}^{(0)}) \\
&= \delta_n^{(0)} - \delta_n^{(0)}(1 + x) - x\delta_{n+1}^{(0)} - \delta_n^{(0)}\delta_{n+1}^{(0)} \\
&= -x(\delta_n^{(0)} + \delta_{n+1}^{(0)}) - \delta_n^{(0)}\delta_{n+1}^{(0)} \\
&= \begin{cases} \dfrac{z(z - 8x)}{16(n^2 - 1)} & \text{if } n \text{ is even,} \\ z^2/16n^2 & \text{if } n \text{ is odd.} \end{cases}
\end{aligned}
$$

Hence

$$\lim_{n \to \infty} \frac{\delta_{n+1}^{(1)}}{\delta_n^{(1)}} = \begin{cases} \dfrac{z}{z - 8x} & \text{if } n \text{ is even,} \\ \\ \dfrac{z - 8x}{z} & \text{if } n \text{ is odd.} \end{cases}$$

To continue the process we therefore need results similar to Theorem 33 for limit 2-periodic situations. We shall return to this later. In Table 2 we have compared the values of $S_n(0)$, $S_n(x)$ and $S_n(w_n^{(1)})$ for $z = 1$.

$\diamond$

For the limit k-periodic situations one can prove a similar result.

Theorem 34 *Let* $\mathbf{K}(a_n/b_n)$ *be a limit k-periodic continued fraction of loxodromic type with finite limits (5.1.1), finite value f and finite and nonzero attractive fixed points $x^{(p)}$ for $T_k^{(p)}$. Further let $\{w_n\}$ be a sequence from $\hat{\mathbf{C}}$ such that $\epsilon_n = (w_n - f^{(n)}) \to 0$ as $n \to \infty$, and let δ_n be given by (5.5.4). Then:*

A. *If for an* $m \in \{1, \ldots, k\}$

$$\lim_{n \to \infty} \frac{\epsilon_{kn+p+1}}{\epsilon_{kn+p}} = s_p \in \mathbf{C} \quad \text{for } p = m, m-1, \qquad (5.5.8)$$

and $s_p \neq -(b_{p+1}^* + x^{(p+1)})/x^{(p)}$ *for at least one of the indices* $p = m, m-1$, *then*

$$\lim_{n \to \infty} \frac{\delta_{kn+m+1}}{\delta_{kn+m}} = t_m = s_{m-1} \frac{b_{m+1}^* + x^{(m+1)} + x^{(m)} s_m}{b_m^* + x^{(m)} + x^{(m-1)} s_{m-1}}. \qquad (5.5.9)$$

B. *If*

$$\lim_{n \to \infty} \frac{\delta_{kn+p+1}}{\delta_{kn+p}} = t_p \in \mathbf{C} \quad \text{for } p = 0, 1, 2, \ldots, k-1, \qquad (5.5.10)$$

then

$$\lim_{n \to \infty} \frac{\epsilon_{kn+p+1}}{\epsilon_{kn+p}} = s_p \neq -\frac{b_{p+1}^* + x^{(p+1)}}{x^{(p)}} \quad \text{for } p = 0, 1, 2, \ldots, k-1.$$
$$(5.5.11)$$

C. *If (5.5.10) holds, then*

$$\lim_{n \to \infty} \frac{f - S_n(w_n^{(1)})}{f - S_n(w_n)} = 0, \qquad (5.5.12)$$

where

$$w_{kn+p}^{(1)} = w_{kn+p} + \frac{\delta_{kn+p+1}}{b_{p+1}^* + x^{(p+1)} + x^{(p)}s_p} \qquad (5.5.13)$$

and s_p is given by the equations (5.5.9) for $m = 1, 2, \ldots, k$.

To simplify the notation we have used that $b_{m+k}^* = b_m^*$, $x^{(m+k)} = x^{(m)}$, $t_{m+k} = t_m$ and $s_{m+k} = s_m$. The proof of this result follows the same lines as the proof of Theorem 33. (It can be found for the special cases all $a_n = 1$ or all $b_n = 1$ in [JaWa88] and [JaWa90].)

Example 26 continued. If we regard $\mathbf{K}(a_n z/1)$ as a limit 2-periodic continued fraction, then our observation

$$\lim_{n \to \infty} \frac{\delta_{2n+1}^{(1)}}{\delta_{2n}^{(1)}} = t_0 = t_2 = \frac{z}{z - 8x}, \qquad \lim_{n \to \infty} \frac{\delta_{2n+2}^{(1)}}{\delta_{2n+1}^{(1)}} = t_1 = \frac{z - 8x}{z}$$

n	$S_n(w_n^{(2)})$
1	0.693170
2	0.693159
3	0.69314740
4	0.69314728
5	0.693147183
6	0.693147182
7	0.693147181

Table 3: Example 26.

gives by (5.5.9)

$$s_0 = t_1 = \frac{b_1^* + x^{(1)} - x^{(1)}t_0}{b_2^* + x^{(2)} - x^{(2)}t_1} = \frac{z - 8x - 8x^2}{z + 8x^2},$$

$$s_1 = \frac{1}{s_0} = \frac{z + 8x^2}{z - 8x - 8x^2}$$

and thus

$$\lim_{n \to \infty} \frac{f - S_n(w_n^{(2)})}{f - S_n(w_n^{(1)})} = 0$$

when

$$w_{2n+1}^{(2)} = w_{2n+1}^{(1)} + \frac{\delta_{2n+2}^{(1)}}{1 + x + \dfrac{x(z + 8x^2)}{z - 8x - 8x^2}},$$

$$w_{2n+2}^{(2)} = w_{2n+2}^{(1)} + \frac{\delta_{2n+3}^{(1)}}{1 + x + \dfrac{x(z - 8x - 8x^2)}{z + 8x^2}}.$$

The first approximants $S_n(w_n^{(2)})$ are given in Table 3. The value is repeated for all $n \geq 7$.

$\diamondsuit$

Another question is: How can we improve the convergence of limit periodic continued fractions of parabolic type? These continued fractions often converge very slowly, if they converge at all, so the question is very relevant. Let us look at an example:

Example 27 The continued fraction $\mathbf{K}(a_n/1)$ where

$$a_n = -\frac{1}{4} + \frac{1}{8n} \quad \text{for } n = 1, 2, 3, \ldots$$

is limit 1-periodic of parabolic type. It converges to a finite value f by virtue of the parabola theorem, Theorem 20, since all $a_n \in P_0$. This value is in fact $f = -0.172160$, correctly rounded to 6 decimal places. Two other properties can also be derived from this fact that all a_n are real and $\in P_0$. The first one is that $f_k^{(n)}$ is real and $\in V_0$ for all n and k; i.e. $f_k^{(n)} > -\frac{1}{2}$ for all n and k. Hence

$$S_n(0) = \frac{a_1}{1 + f_{n-1}^{(1)}} < 0 \quad \text{for all } n,$$

and

$$S_n(0) > S_{n+1}(0) \quad \text{for all } n.$$

Hence $\{S_n(0)\}$ is a decreasing sequence of negative numbers converging to f. The other property is that the right tail sequence of $\mathbf{K}(a_n/1)$, $\{f^{(n)}\}$, converges to $-\frac{1}{2}$. (Clearly, all $f^{(n)}$ are negative and $\geq -1/2$ by the arguments above. If f^* is a limit point of $\{f^{(n)}\}$, then so is f_1^*, where $f^* = (-1/4)/(1 + f_1^*)$. Hence, the set of limit points for $\{f^{(n)}\}$ must contain the tail sequence $\{t_n^*\}$ of the periodic continued fraction $K((-1/4)/1)$ which begins with $t_0^* = f^*$. The only such tail sequence which is contained in $\bar{V}_0$ is the right tail sequence, where all $t_n^* = -1/2$.) This suggests that $S_n(-1/2)$ is a good choice for the approximants of $K(a_n/1)$. Since $f^{(n)} \geq -1/2$ for all n, this in turn means that

$$f = S_n(f^{(n)}) \geq S_n(-1/2)$$

for all n, so $S_n(0) > f \geq S_n(-1/2)$, which is a useful truncation error bound. On the other hand, since also $h_n \to 1/2$ (by a similar argument), we do not have high expectations to the speed of convergence of $S_n(-1/2)$ as compared to $S_n(0)$. (See the formula in (5.5.2).)

We shall estimate $f^{(n)}$ a little better than just using $f^{(n)} \approx -1/2$. The idea was suggested by Gill in another context, [Gill80]. We have

$$
\begin{aligned}
f^{(n)} &= \frac{a_{n+1}}{1} + \frac{a_{n+2}}{1} + \frac{a_{n+3}}{1} + \cdots \\
&\approx \frac{a_{n+1}}{1} + \frac{a_{n+1}}{1} + \frac{a_{n+1}}{1} + \cdots = \frac{\sqrt{1 + 4a_{n+1}} - 1}{2} \\
&= -\frac{1}{2} + \frac{1}{2\sqrt{2(n+1)}} =: w_n.
\end{aligned}
$$

Therefore $S_n(w_n)$ ought to converge faster to f than $S_n(0)$ and $S_n(-1/2)$. Table 4 gives the first approximants of the types $S_n(0)$, $S_n(-1/2)$ and $S_n(w_n)$ for $\mathbf{K}(a_n/1)$.

$\diamond$

5.6   *Continued fractions* $\mathbf{K}(a_n/1)$ *where* $a_n \to \infty$

Let $\mathbf{K}(a_n/1)$ have elements $a_n \to \infty$. Then $\mathbf{K}(a_n/1)$ may converge or diverge depending on how $\{a_n\}$ approaches infinity. For instance, if

n	$S_n(0)$	$S_n(-1/2)$	$S_n(w_n)$
1	-0.125	-0.25	-0.167
2	-0.154	-0.20	-0.1704
3	-0.164	-0.184	-0.1714
4	-0.168	-0.1780	-0.1718
$\vdots$	$\vdots$	$\vdots$	$\vdots$
17	-0.172147	-0.172176	-0.172160
18	-0.172150	-0.172171	-0.172160
19	-0.172153	-0.172169	-0.172160
20	-0.172155	-0.172166	-0.172160

Table 4: Example 27.

all $a_n > 0$, then $\mathbf{K}(a_n/1)$ converges if and only if its Stern-Stolz series diverges. (See Theorem 3 and the subsequent Remark 2.)

Evidently the techniques of Subsections 5.2– 5.5 can not be applied here. So, how can we find good approximants and truncation error bounds? We shall illustrate some ideas in the following examples.

Example 28 Does the continued fraction $\mathbf{K}(a_n/1)$, where $a_n = i^n n$, converge or diverge? The even part of $\mathbf{K}(a_n/1)$ is

$$
\begin{aligned}
K_e &= \cfrac{a_1}{1 + a_2 -} \cfrac{a_2 a_3}{1 + a_3 + a_4 -} \cfrac{a_4 a_5}{1 + a_5 + a_6 -} \cdots \\
&= \cfrac{i}{1 - 2 -} \cfrac{(-2)(-3i)}{1 - 3i + 4 -} \cfrac{4 \cdot 5i}{1 + 5i - 6 -} \cfrac{(-6)(-7i)}{1 - 7i + 8 -} \cdots \\
&= \cfrac{i}{-1 -} \cfrac{2 \cdot 3i}{1 + (4 - 3i) -} \cfrac{4 \cdot 5i}{1 - (6 - 5i) -} \cfrac{6 \cdot 7i}{1 + (8 - 7i) -} \cdots
\end{aligned}
$$

(See Theorem 12 and formula (2.4.3) in Chapter II.) This continued fraction K_e is equivalent to a continued fraction of the form $\mathbf{K}(c_n/1)$, where

$$
c_n = \frac{-(2n - 2)(2n - 1)i}{\{1 + (-1)^n(2n - 2 - (2n - 3)i)\}\{1 + (-1)^{n-1}(2n - (2n - 1)i)\}}
$$

for $n \geq 3$. (See Corollary 10B in Chapter II.) Since

$$\lim_{n \to \infty} c_n = \frac{-4i}{-(2-2i)^2} = \frac{i}{(1-i)^2} = -\frac{1}{2},$$

we find that $\mathbf{K}(c_n/1)$ is limit 1-periodic of elliptic type. One can prove that $\mathbf{K}(c_n/1)$ diverges by methods presented in Chapter IV.) Hence also $\mathbf{K}(a_n/1)$ diverges.

$\diamond$

Example 29 Let $a_{2n-1} = in^2$ and $a_{2n} = n$ for all $n \in \mathbf{N}$. Will then $\mathbf{K}(a_n/1)$ converge or diverge? The even part of $\mathbf{K}(a_n/1)$ is

$$K_e = \frac{i1^2}{1+1} \frac{1 \cdot i2^2}{-1+i2^2+2} \frac{2 \cdot i3^2}{-1+i3^2+3} \frac{3 \cdot i4^2}{-1+i4^2+4} \cdots \approx \mathbf{K}\frac{c_n}{1}$$

where

$$c_n = \frac{-i(n-1)n^2}{(n+i(n-1)^2)(n+1+in^2)} \quad \text{for } n = 3, 4, 5, \dots .$$

Hence $c_n \to 0$ as $n \to 0$ and thus $\mathbf{K}(c_n/1)$ converges. By the same type of argument we find that the odd part of $\mathbf{K}(a_n/1)$ converges. But we can not say whether $\mathbf{K}(a_n/1)$ itself converges or not, unless we can prove that $(f_{n+1} - f_n) \to 0$ or not.

$\diamond$

Example 30 Let $a_n = n$ for all $n \in \mathbf{N}$. Then we know by Theorem 3 that $\mathbf{K}(a_n/1)$ converges. But the convergence of $\{S_n(0)\}$ is slow. Can we find better approximants?

The even part of $\mathbf{K}(n/1)$ is

$$K_e = \frac{1}{1+2} \frac{2 \cdot 3}{-1+3+4} \frac{4 \cdot 5}{-1+5+6} \cdots \approx \mathbf{K}\frac{c_n}{1}$$

where $c_1 = 1/3$, $c_2 = -1/4$ and

$$\begin{aligned}
c_n &= \frac{-(2n-2)(2n-1)}{(1+2n-3+2n-2)(1+2n-1+2n)} \\
&= -\frac{n-1/2}{4n} = -\frac{1}{4} + \frac{1}{8n} \to -\frac{1}{4} \quad \text{as } n \to \infty .
\end{aligned}$$

We recognize the continued fraction from Example 27. The second tail of that one is identical to the second tail of $\mathbf{K}(c_n/1)$. Hence we can use the same modification $w_n = -(1/2) + 1/(2\sqrt{2(n+1)})$ to compute approximants $S_n(w_n)$ of $\mathbf{K}(c_n/1)$.

$\diamond$

Example 31 Another approach for finding good approximants for $\mathbf{K}(n/1)$ is the following: Since $a_n = n$ is not so very different from a_{n+1}, we guess that the value of the nth tail

$$f^{(n)} = \frac{a_{n+1}}{1} + \frac{a_{n+2}}{1} + \frac{a_{n+3}}{1} + \cdots = \frac{n+1}{1} + \frac{n+2}{1} + \frac{n+3}{1} + \cdots$$

of $\mathbf{K}(n/1)$ is not so very different from

$$w_n = \frac{a_{n+1}}{1} + \frac{a_{n+1}}{1} + \frac{a_{n+1}}{1} + \cdots = \frac{\sqrt{1+4a_{n+1}}-1}{2} = \frac{\sqrt{4n+5}-1}{2}.$$

Hence we want to try the approximants

$$S_n(w_n) = \frac{1}{1+} \frac{2}{1+} \cdots + \frac{n}{1+w_n} = \frac{1}{1+} \frac{2}{1+} \cdots + \frac{n}{(1+\sqrt{4n+5})/2}.$$

Table 5 compares the approximants $S_n(0)$ and $S_n(w_n)$. The last column contains the approximants suggested in Example 30.

We can in fact prove that $S_n(w_n)$ converges faster than $S_n(0)$ to the same value f; i.e. that

$$\lim_{n \to \infty} \frac{f - S_n(w_n)}{f - S_n(0)} = 0.$$

To do this we can use the oval sequence theorem, Theorem 26, with centers $C_n = w_n = (\sqrt{4n+5}-1)/2$ and radii $R_n = R = 1/2$. This choice satisfies the conditions of Theorem 26. Moreover $a \in E_n$ as given by (4.4.16) if and only if

$$\left| \frac{a}{2}(1+\sqrt{4n+5}) - \frac{1}{2}(\sqrt{4n+1}-1)\left(\frac{1}{4}(1+\sqrt{4n+5})^2 - \frac{1}{4}\right) \right| + \frac{|a|}{2}$$

$$\leq \frac{1}{2}\left(\frac{1}{4}(1+\sqrt{4n+5})^2 - \frac{1}{4}\right).$$

n	$S_n(0)$	$S_n(w_n)$	Ex.30	
1	1.0	0.5		
2	0.33	0.535		
3	0.67	0.5205		
4	0.44	0.5275	n=2:	0.5168
5	0.583	0.5238		
6	0.487	0.52592	n=3:	0.5218
7	0.553	0.5247		
8	0.506	0.52544	n=4:	0.5236
9	0.540	0.5249		
10	0.515	0.52527	n=5:	0.5244

Table 5: Example 31.

Hence $a_n = n \in E_n$ if and only if

$$\left| 4n(1 + \sqrt{4n+5}) - (\sqrt{4n+1} - 1)(2\sqrt{4n+5} + 4n + 5) \right|$$
$$+4n \leq 2\sqrt{4n+5} + 4n + 5$$

that is if

$$-4n(1 + \sqrt{4n+5}) + (\sqrt{4n+1} - 1)(2\sqrt{4n+5} + 4n + 5) + 4n$$
$$\leq 2\sqrt{4n+5} + 4n + 5.$$

Straight forward computation shows that this holds for all n. Hence $V_n = \{w; |w - w_n| < 1/2\}$ is a sequence of value regions for $\mathbf{K}(n/1)$. Since V_n is contained in the right half plane $\Re(w) \geq -1/2$, at least from some n on, it follows from the parabola theorem with $\alpha = 0$ that $f^{(n)} \in \bar{V}_n$ for all n; that is, $|f^{(n)} - w_n| \leq 1/2$. Hence, by formula (1.4.7) in Chapter II

$$\left| \frac{f - S_n(w_n)}{f - S_n(0)} \right| = \left| \frac{h_n(f^{(n)} - w_n)}{(h_n + w_n)f^{(n)}} \right| \leq \frac{1/2}{w_n - 1/2} \to 0$$

since $h_n > 0$ and $w_n > 0$ and thus $|h_n/(h_n + w_n)| < 1$.

We can also derive truncation error estimates. From Theorem 26 it follows that

$$|f - S_n(w_n)| \leq 2R\frac{w_0 + R}{1 + w_n - R} \prod_{k=1}^{n-1} \frac{w_k + R}{1 + w_k + R}$$

$$= \frac{\sqrt{5}}{\sqrt{4n + 5}} \prod_{k=1}^{n-1} \frac{\sqrt{4k + 5}}{\sqrt{4k + 5} + 2}.$$

(A slightly smaller R would have given better bounds.)

$\diamondsuit$

5.7 Analytic continuation

Let us illustrate the idea by a trivial example.

Example 32 The periodic C-fraction

$$\mathbf{K}\frac{az}{1} = \frac{az}{1} + \frac{az}{1} + \frac{az}{1} + \cdots; \quad a \in \mathbf{C} \setminus \{0\} \tag{5.7.1}$$

converges in the cut plane $D = \{z \in \mathbf{C}; \, |\arg(1 + 4az)| < \pi\}$ to the holomorphic function

$$w(z) = (\sqrt{1 + 4az} - 1)/2 \quad \text{where } \Re(\sqrt{1 + 4az}) > 0. \tag{5.7.2}$$

(See Theorem 28.) This function $w(z)$ can be extended analytically to a 2-sheeted Riemann surface D^* with branch points of order 1 at $z = -1/4a$ and at $z = \infty$. Let $W(z)$ denote this function,

$$W(z) = ((1 + 4az)^{1/2} - 1)/2 \quad \text{for } z \in D^*. \tag{5.7.3}$$

Then $\mathbf{K}(a_n z/1)$ converges to $W(z)$ for $z \in D \subseteq D^*$. The classical approximants of $\mathbf{K}(a_n z/1)$ are rational functions of z, and thus have no branch points. Hence there is no way that these approximants can converge to $W(z)$ for $z \in D^* \setminus \bar{D}$. For the modified approximants

$$S_n(W(z)) = \frac{az}{1} + \frac{az}{1} + \cdots + \frac{az}{1 + W(z)} \quad (n \text{ terms})$$

however, the picture is totally different. For these we have $S_n(W(z)) = W(z)$ for all n and all $z \in D^*$. That is, they "converge" to $W(z)$ for all

$z \in D^*$. So this choice of approximants lead to analytic continuation of the value $w(z)$ of the continued fraction $\mathbf{K}(az/1)$.

$\diamond$

As already mentioned, this example was trivial. But what if we try to use modified approximants

$$S_n(W(z)) = \frac{a_1 z}{1} + \frac{a_2 z}{1} + \frac{a_3 z}{1} + \cdots + \frac{a_n z}{1 + W(z)} \qquad (5.7.4)$$

for a continued fraction $\mathbf{K}(a_n z/1)$ where $a_n \to a$? Can we also then obtain convergence of $S_n(W(z))$ to a holomorphic or meromorphic function in a larger domain $\subseteq D^*$ than D where $\mathbf{K}(a_n z/1)$ is known to converge? The answer is yes under proper conditions:

Theorem 35 *Let* $a \in \mathbf{C} \setminus \{0\}, c > 0, 0 \leq r < 1$ *and*

$$D_r^* = \left\{ z \in D^*; \left| \frac{W(z)}{1 + W(z)} \right| < \frac{1}{r} \right\}, \qquad (5.7.5)$$

where $W(z)$ *is as given in (5.7.3). If* $\mathbf{K}(a_n z/1)$ *satisfies*

$$|a_n - a| \leq Cr^n \quad for \ n = 2, 3, 4, \ldots, \qquad (5.7.6)$$

then $\{S_n(W(z))\}$ *converges to a meromorphic function* $F(z)$ *in* D_r^*. *The convergence is uniform on compact subsets of* D_r^* *where* $F(z) \neq \infty$.

Remarks:

1. D_r^* is a domain in D^*. $D_r^* \subseteq D_t^*$ if $r \geq t$. $D_1^* = D$ and D_0^* is equal to D^* where the point 0 is removed from the sheet $D^* \setminus \bar{D}$.

2. D_r^* is all of D^* except for a bounded hole $H = D^* \setminus D_r^*$. This hole is contained in the sheet $D^* \setminus \bar{D}$, and it is symmetric about the axis $\arg(-az) = 2\pi$. It is bounded away from the branch points $z = -1/4a$ and $z = \infty$.

3. The observations in the previous remark imply that the limit function $F(z)$ also has branch points of order 1 at $z = -1/4a$ and $z = \infty$.

4. The computation of the approximants $S_n(W(z))$ is unstable for $z \in D_r^* \setminus \bar{D}$. Small inaccuracies lead to approximants which converge to $f(\check{z}) \neq F(z)$, where $\check{z}$ is the projection of z onto D. This instability can be avoided by using the Bauer–Muir transformation as described in Chapter II. The new continued fraction thus obtained (the Bauer–Muir transform of $K(a_n z/1)$ with respect to $W(z)$) can often be accelerated by methods described in the present section, or be extended analyticly even further.

For the proof of Theorem 35 we refer to [Lore]. We shall rather show an example.

Example 33 The limit 1-periodic S-fraction $\mathbf{K}(a_n z/1)$ where

$$a_n = 0.25 + (0.3)^n \quad \text{for } n = 1, 2, 3, \ldots$$

converges to a meromorphic function in the cut plane $D = \{z \in \mathbf{C}; |\arg(1+z)| < \pi\}$. According to Theorem 35, however, its approximants

$$S_n(W(z)) = \frac{a_1 z}{1} + \frac{a_2 z}{1} + \cdots + \frac{a_n z}{1 + W(z)}; \qquad W(z) = \frac{(1+z)^{1/2} - 1}{2}$$

for $z \in D^*$ converges to a meromorphic function for $z \in D_{0.3}^*$. Here D^* is the 2-sheeted Riemann surface with branch points of order 1 at $z = -1$ and at $z = \infty$, and $D_{0.3}^*$ is the subset of D^* where

$$\left| \frac{W(z)}{1 + W(z)} \right| = \left| \frac{(1+z)^{1/2} - 1}{(1+z)^{1/2} + 1} \right| < \frac{1}{0.3} = \frac{10}{3}.$$

Observe that $z \in D_{0.3}^*$ if $\Re(1+z)^{1/2} > -7/13$ whereas $z \in D$ if and only if $\Re(1+z)^{1/2} > 0$. Moreover $z \in D_{0.3}^*$ if $|1+z|^{1/2} > 13/7$. Hence $D_{0.3}^*$ contains D and neighborhoods of the branchpoints $z = -1$ and $z = \infty$.

$\diamondsuit$

Problems

(1) Determine whether $\mathbf{K}(a_n/b_n)$ converges or diverges and whether its even and odd parts converge or diverge in each of the following cases.

    (a) All $a_n = 1$ and $b_n = z/n^2$ for $z \in \mathbf{C}$.

    (b) All $a_n = z/n^2$ and $b_n = 1$ for $z \in \mathbf{C} \setminus \{0\}$.

    (c) All $a_n = zn^2$ and $b_n = 1$ for $z \in \mathbf{C} \setminus \{0\}$.

(2) Given the continued fraction

$$b + \mathbf{K}\,\frac{a}{2b} = b + \frac{a}{2b +}\ \frac{a}{2b +}\ \frac{a}{2b +}\cdots$$

    (a) Prove that if $a > 0$ and $b > 0$ then $b + \mathbf{K}(a/2b)$ converges to $\sqrt{a + b^2}$. Use this to find a rational approximation to $\sqrt{13}$ with an error less than 10^{-4}.

    (b) For which values of $(a, b) \in \mathbf{C} \times \mathbf{C}$ does $b + \mathbf{K}(a/2b)$ converge/diverge? Find its value if it converges.

(3) In Example 26 we suggested approximants $S_n(w_n^{(m)})$ for the continued fraction

$$\mathbf{K}\,\frac{a_n}{1} = \frac{1}{1 +}\ \frac{1/2}{1 +}\ \frac{1/(2 \cdot 3)}{1} \ + \ \frac{2/(2 \cdot 3)}{1}\ + \ \frac{2/(2 \cdot 5)}{1} \ + \ \frac{3/(2 \cdot 5)}{1}\ +\cdots$$

$$+ \ \frac{n/(2(2n - 1))}{1} \ + \ \frac{n/(2(2n + 1))}{1} \ +\cdots$$

for $\log 2$ (natural logarithm).

Does $\{S_n(w_n^{(0)})\}$ or $\{S_n(w_n^{(1)})\}$ have an oscillating character which can be used to obtain upper bounds for the truncation error $|\log 2 - S_n(w_n^{(m)})|$?

(4) We want to improve the speed of convergence of the continued fraction

$$\mathbf{K}\,\frac{a_n}{1} = \frac{6 + (0.9)}{1}\ + \ \frac{6 + (0.9)^2}{1} \ + \ \frac{6 + (0.9)^3}{1}\ + \ \frac{6 + (0.9)^4}{1}\ +\cdots.$$

    (a) Prove that $\mathbf{K}(a_n/1)$ converges to a finite value f.

(b) Suggest a value $w_n = w$ for all n such that

$$\lim_{n \to \infty} \frac{f - S_n(w)}{f - S_n(0)} = 0.$$

Does the sequence $\{S_n(w)\}$ have an oscillating character which can be used to obtain upper bounds for the truncation error $|f - S_n(w)|$?

(c) Suggest values for w_n such that

$$\lim_{n \to \infty} \frac{f - S_n(w_n)}{f - S_n(w)} = 0$$

where w is the value from a). Does $\{S_n(w_n)\}$ have such an oscillating character?

(5) For which values of z does the 4-periodic continued fraction

$$\underset{}{\mathbf{K}} \frac{a_n}{1} = \frac{1}{1+} \frac{1}{1-} \frac{2}{1-} \frac{2z}{1} + \frac{1}{1+} \frac{1}{1+} \frac{2}{1-} \frac{2z}{1} + \cdots$$

(a) converge/diverge generally?

(b) oscillate by Thiele oscillation?

(c) converge in the classical sense?

What is the value of this continued fraction when it converges generally?

(6) Let z be chosen such that the continued fraction $\mathbf{K}(a_n/1)$ in Problem 5 converges generally. Determine the asymptotic behavior of the tail sequences of $\mathbf{K}(a_n/1)$ in the sense of Theorem 7.

(7) For which $b_n > 0$ is $V = \{w \in \mathbf{C}; |w - 1| < 1\}$ a value set for $\mathbf{K}(1/b_n)$? For which $a_n > 0$ is V a value set for $\mathbf{K}(a_n/1)$?

(8) Let V be the half plane $V = \{w \in \mathbf{C}; \Re(w) > p\}$ where $p \geq 0$ or $p \leq -1$. Do there exist continued fractions $\mathbf{K}(a_n/1)$ or $\mathbf{K}(1/b_n)$ (with complex elements a_n, b_n) such that V is a value set for this continued fraction?

(9) Let $\mathbf{K}(1/b_n)$ be the continued fraction where $b_n = 4 + (0.9)^n$ for all n.

(a) Find a connected value set V for $\mathbf{K}(1/b_n)$. (Try to make V small.)

(b) Does $\mathbf{K}(1/b_n)$ converge to a finite value f?

(c) Are the classical approximants f_n of $\mathbf{K}(1/b_n)$ all distinct; i.e. $f_n \neq f_m$ if $n \neq m$?

(d) Use the value set V found in (a) to derive upper bounds for the truncation error $|f - S_n(w)|$ for suitably chosen $w \in \mathbf{C}$.

(10) Let $p, q, r \in \mathbf{N}_0$, and

$$a_{2n-1}(z) = \sum_{k=0}^{p} \alpha_k(z) n^k, \quad a_{2n}(z) = \sum_{k=0}^{p} \gamma_k(z) n^k$$

$$b_{2n-1}(z) = \sum_{k=0}^{q} \beta_k(z) n^k, \quad b_{2n}(z) = \sum_{k=0}^{r} \delta_k(z) n^k$$

be polynomials in n for $n = 1, 2, 3, \ldots$, where all α_k, γ_k, β_k and δ_k are entire functions of z and $\alpha_p(z)\gamma_p(z)\beta_q(z)\delta_r(z) \not\equiv 0$. Further let $\tilde{D} = \{z \in \mathbf{C}; \beta_q(z)\delta_r(z) \neq 0$ and all $a_n(z) \neq 0\}$.

Prove that $\mathbf{K}(a_n(z)/b_n(z))$ converges in D if D is open and connected and

(a) $q + r > p$ and $D \subseteq \tilde{D}$

(b) $q + r = p$ and

$$D \subseteq \left\{ z \in \tilde{D}; \; \frac{\alpha_p(z)\gamma_p(z)}{(\alpha_p(z) + \gamma_p(z) + \beta_q(z)\delta_r(z))^2} \notin \left[\frac{1}{4}, \infty\right] \right\}$$

(c) $q + r = p - 1$, $\alpha_p(z) \equiv \gamma_p(z)$ and

$$D \subseteq \{z \in \tilde{D}; \beta_q(z)\delta_r(z)/\alpha_p(z) \notin [-\infty, 0]\}$$

(d) $q + r = p - 2$, $\alpha_p(z) \equiv \gamma_p(z)$ and

$$D \subseteq \left\{ z \in \tilde{D}; \; 4\frac{\beta_q(z)\delta_r(z)}{\alpha_p(z)} \right.$$

$$\left. + \left(r + 1 + \frac{\alpha_{p-1}(z) - \gamma_{p-1}(z)}{\alpha_p(z)}\right)^2 \notin [-\infty, 0] \right\}.$$

(Hint: To prove (b)–(d) one can first prove that the even and odd parts of $\mathbf{K}(a_n(z)/b_n(z))$ converge to meromorphic functions or functions identically ∞ in D. Then use the parabola theorem to prove that these functions are identical on some (large enough) subset of D. Finally one can use Stieltjes-Vitali's theorem to prove that they are identical in D.)

(11) Use the parabola theorem to derive a priori truncation error bounds for

$$\log(1 + i) = \mathbf{K}\,\frac{a_n i}{1} = \frac{i}{1+}\ \frac{i/2}{1+}\ \frac{i/(2 \cdot 3)}{1}\ +\ \frac{2i/(2 \cdot 3)}{1}\ +\ \frac{2i/(2 \cdot 5)}{1}\ +\cdots$$

where $a_{2n} = n/(2(2n-1))$ and $a_{2n+1} = n/(2(2n+1))$ for all $n \geq 1$. Compare these with the Gragg-Warner truncation error bounds in Theorem 24.

(12) Let $\mathbf{K}(a_n/1)$ have all elements $a_n \in E = \{w \in \mathbf{C};\ |w-3-i| \leq 0.4\}$. Find a $C \in \mathbf{C}$ and an $R < |1 + C|$ such that E is a subset of the cartesian oval in Theorem 25. Use this to prove that $\mathbf{K}(a_n/1)$ converges to a finite value and to find truncation error bounds for suitably chosen approximants of $\mathbf{K}(a_n/1)$. (Hint: See Remark 4 to Theorem 25.)

(13) Use Theorem 28 to estimate the speed of convergence of $S_n(w)$ for the continued fraction $\mathbf{K}(a_n i/1)$ in Problem 11 when

$$w = (\sqrt{1 + i} - 1)/2, \qquad \Re(\sqrt{1 + i}) > 0.$$

(14) Let a_n be as in Problem 11. Does $\mathbf{K}(-a_n/1)$ converge?

(15) Let $\mathbf{K}(a_n/1)$ be given by

$$a_n = x(1 + x) + r^n \quad \text{where } 0 < |x| < |1 + x| \text{ and } 0 < |r| < 1.$$

Choose approximants $S_n(w_n^{(m)})$ for $\mathbf{K}(a_n/1)$ according to the scheme in Subsection 5.5 such that

$$\lim_{n \to \infty} \frac{f - S_n(w_n^{(m)})}{f - S_n(w_n^{(m-1)})} = 0 \quad \text{for } m = 1, 2, 3, \ldots.$$

Compare with the Bauer-Muir transformation in Example 11 of Chapter II.

(16) Let $\mathbf{K}(a_n/1)$ have real elements a_n such that $(-1)^n a_n > 0$ and

$$|a_{2n-1}| < 1 + a_{2n}, \quad |a_{2n+1}| < 1 + a_{2n} \quad \text{for all } n.$$

Prove that $\{S_{4n+p}(0)\}_{n=1}^{\infty}$ converges for $p = 1, 2, 3$ and 4.

(17) Suggest expressions for w_n such that the approximants $S_n(w_n)$ of

$$\mathbf{K} \frac{a_n}{1} = \frac{1^2}{1+} \frac{5 \cdot 2^2}{1} \frac{3^2}{+1+} \frac{5 \cdot 4^2}{1} \frac{5^2}{+1+} \frac{5 \cdot 6^2}{1} \frac{7^2}{+1+} \cdots$$

(hopefully) converge faster to the value of $\mathbf{K}(a_n/1)$ than $S_n(0)$. Compute the first 6 approximants of $S_n(0)$ and $S_n(w_n)$ and use the oscillating character to determine an error bound for $S_6(0)$ and $S_6(w_6)$.

Remarks

1. The idea of using value regions for continued fractions to derive convergence criteria was presented by Paydon and Wall in 1942, [PaWa42]. They had $V = \{w \in \mathbf{C}; |w - 1| \leq 1\}$ and $t_n(w) = 1/(1 + a_{n+1}w)$, and studied continued fractions

$$\frac{1}{1+} \frac{a_1}{1+} \frac{a_2}{1+} \frac{a_3}{1+} \cdots$$

   with the property that $t_n(V) \subseteq V$ for all n.

   This fruitful idea was further exploited by Wall and Wetzel, [WaWe44] in two papers on positive definite continued fractions.

   W. J. Thron realized the potential of this idea, and in a long series of important papers, some of which in collaboration with others, he refined it and used it to derive several useful convergence criteria. We can mention his work on the parabola theorem [Thron58] and twin convergence regions [Thron59]. A nice survey is given in [Thron74]. See also his book with W. B. Jones [JoTh80] for further references.

   In this classical work, value regions or value sets always contained the classical approximants $S_n(0)$ of the continued fractions. It was not until 1982 that one realized that value sets for other approximants $S_n(w_n)$ could be used in the same way, [Jaco82], [Jaco83], [Jaco86].

2. The demand for truncation error estimates became more prominent in the 1960s with the growing use of computers. W. J. Thron [Thron58] realized that value sets could also be used for the purpose of deriving such estimates. This started a series of useful publications in this area, such as [HePf66] and [JoTh76]. For further references we refer to [JoTh76, p. 298].

3. The first parabola theorem was published by Scott and Wall in 1940, [ScWa40]. It was proved by exploiting what they called the fundamental inequalities. The result was generalized almost immediately [PaWa42] and [LeTh42]. The most general parabola theorem is due to Jones and Thron, [JoTh68].

4. Limit periodic continued fractions, included those where $a_n \to \infty$ or $b_n \to \infty$, have been extensively studied, the reason being that so many useful continued fraction expansions have this form. One important issue here is the question of how to choose approximants $\{S_n(w_n)\}$ in order that $S_n(w_n)$ shall converge as fast as possible to the value of the continued fraction. The first known result in this direction dates back to Sylvester in 1869 [Sylv69]. More recently Wynn [Wynn59], Gill [Gill75], Masson [Mass83], [Mass85], Thron and Waadeland [ThWa80] and Jacobsen, Jones and Waadeland [JaJW87] contributed to this area.

The idea of using asymptotic side conditions to improve the speed of convergence even further, was published by Jacobsen and Waadeland [JaWa88], [JaWa90] and improved by Levrie [Levr89].

In the paper [Waad66], the idea of deriving analytic continuation of the value $f(z)$ of a continued fraction $\mathbf{K}(a_n(z)/b_n(z))$ by careful choice of approximants $S_n(w_n(z))$ was introduced. Independently, Masson came up with the same method, [Mass83]. A thorough presentation of the method can be found in [ThWa80], [ThWa81].

References

[Gill73]   J. Gill, *Infinite Compositions of Möbius Transformations*, Trans. Amer. Math. Soc. **176** (1973), 479–487.

[Gill75]   J. Gill, *The Use of Attractive Fixed Points in Accelerating the Convergence of Limit–Periodic Continued Fractions*, Proc. Amer. Math. Soc. **47** (1975), 119–126.

[Gill80]   J. Gill, *Convergence Acceleration for Continued Fractions* $\mathbf{K}(a_n/1)$ *with* $\lim a_n = 0$, "Analytic Theory of Continued Fractions", (W.B.Jones, W.J.Thron, H.Waadeland, eds), Lecture Notes in Mathematics **932**, Springer–Verlag, Berlin (1980), 67–70.

[GrWa83]   W. B. Gragg and D. D. Warner, *Two Constructive Results in Continued Fractions*, SIAM J. Numer. Anal. **20** (1983), 1187–1197.

[HePf66]   P. Henrici and P. Pfluger, *Truncation Error Estimates for Stieltjes Fractions*, Numer. Math. **9** (1966), 120–138.

[HiTh65]   K. L. Hillam and W. J. Thron, *A General Convergence Criterion for Continued Fractions* $\mathbf{K}(a_n/b_n)$, Proc. Amer. Math. Soc. **16** (1965), 1256–1262.

[Hille62]   E. Hille, "Analytic Function Theory", Vol **2**, Ginn, Boston (1962).

[Jaco82]   L. Jacobsen, *Some Periodic Sequences of Circular Convergence Regions*, "Analytic Theory of Continued Fractions",

Lecture Notes in Mathematics **932** (W. B. Jones, W. J. Thron and H. Waadeland eds.), Springer-Verlag, Berlin (1982), 87–98.

[Jaco83] L. Jacobsen, *Convergence Acceleration and Analytic Continuation by Means of Modification of Continued Fractions*, Det Kgl. Norske Vid. Selsk. Skr. No **1** (1983), 19–33.

[Jaco86] L. Jacobsen, *General Convergence of Continued Fractions*, Trans. Amer. Math. Soc. **294**(2) (1986), 477–485.

[JaJW87] L. Jacobsen, W. B. Jones and H. Waadeland, *Convergence Acceleration for Continued Fractions* $\mathbf{K}(a_n/1)$ *where* $a_n \to \infty$, "Rational Approximation and Its Applications in Mathematics and Physics", Lecture Notes in Mathematics **1237** (J. Gilewicz, M. Pindor and W. Siemaszko eds.) Springer-Verlag, Berlin (1987), 177–187.

[JaMa90] L. Jacobsen and D. R. Masson, *On the Convergence of Limit Periodic Continued Fractions* $\mathbf{K}(a_n/1)$, *where* $a_n \to -1/4$. Part III., Constr. Approx. **6** (1990), p.363–374.

[JaTh86] L. Jacobsen and W. J. Thron, *Oval Convergence Regions and Circular Limit Regions for Continued Fractions* $\mathbf{K}(a_n/1)$, "Analytic Theory of Continued Fractions" II, Lecture Notes in Mathematics **1199** (W. J. Thron ed.), Springer-Verlag, Berlin (1986), 90–126.

[JaWa86] L. Jacobsen and H. Waadeland, *Even and Odd Parts of Limit Periodic Continued Fractions*, J. Comp. Appl. Math. **15** (1986), 225–233.

[JaWa88] L. Jacobsen and H. Waadeland, *Convergence Acceleration of Limit Periodic Continued Fractions under Asymptotic Side Conditions*, Numer. Math. **53** (1988), 285–298.

[JaWa90] L. Jacobsen and H. Waadeland, *An Asymptotic Property for Tails of Limit Periodic Continued Fractions*, Rocky Mountain J. of Math. **20**(1) (1990), 151–163.

[JoTh68] W. B. Jones and W. J. Thron, *Convergence of Continued Fractions*, Canad. J. of Math. **20** (1968), 1037–1055.

[JoTh70]   W. B. Jones and W. J. Thron, *Twin-Convergence Regions for Continued Fractions* $\mathbf{K}(a_n/1)$, Trans. Amer. Math. Soc. **150** (1970), 93–119.

[JoTh76]   W. B. Jones and W. J. Thron, *Truncation Error Analysis by Means of Approximant Systems and Inclusion Regions*, Numer. Math. **26** (1976), 117–154.

[JoTh80]   W. B. Jones and W. J. Thron, "Continued Fractions: Analytic Theory and Applications", Encyclopedia of Mathematics and its Applications, **11**, Addison-Wesley Publishing Company, Reading, Mass. (1980). Now distributed by Cambridge University Press, New York.

[Lane45]   R. E. Lane, *The Convergence and Values of Periodic Continued Fractions*, Bull. Amer. Math. Soc. **51** (1945), 246–250.

[LaWa49]   R. E. Lane and H. S. Wall, *Continued Fractions with Absolutely Convergent Even and Odd Parts*, Trans. Amer. Math. Soc. **67** (1949), 368–380.

[LeTh42]   W. Leighton and W. J. Thron, *Continued Fractions with Complex Elements*, Duke. Math. J. **9** (1942), 763–772.

[Levr89]   P. Levrie, *Improving a Method for Computing Non-dominant Solutions of Certain Second-Order Recurrence Relations of Poincaré-Type*, Numer. Math. **56** (1989), 501–512.

[Lore]   L. Lorentzen, *Analytic Continuation of Functions Represented by Continued Fractions, Revisited.* To be published in Rocky Mountain J. of Math.

[Lore]   L. Lorentzen, *Bestness of the Parabola Theorem for Continued Fractions.* To be published.

[LoRu]   L. Lorentzen and St. Ruscheweyh, *Simple Convergence Sets for Continued Fractions* $\mathbf{K}(a_n/1)$. To be published.

[Mass83]   D. Masson, *The Rotating Harmonic Oscillator Eigenvalue Problem. 1. Continued Fractions and Analytic Continuation*, J. Math. Phys. **24** (8) (1983), 2074–2088.

[Mass85]   D. Masson, *Convergence and Analytic Continuation for a Class of Regular C-fractions*, Canad. Math. Bull. **28**(4) (1985), 411–421.

[PaWa42]   J. F. Paydon and H. S. Wall, *The Continued Fraction as a Sequence of Linear Transformations*, Duke. Math. J. **9** (1942), 360–372.

[ScWa40]   W. T. Scott and H. S. Wall, *A Convergence Theorem for Continued Fractions*, Trans. Amer. Math. Soc. **47** (1940), 155–172.

[Seid46]   L. Seidel, *Untersuchungen über die Konvergenz und Divergenz der Kettenbrüche*, Habilschrift München (1846).

[Stern48]   M. A. Stern, *Über die Kennzeichen der Konvergenz eines Kettenbruchs*, J. Reine Angew. Math. **37** (1848).

[Stern60]   M. A. Stern, "Lehrbuch der Algebraischen Analysis", Leipzig (1860).

[Stie94]   T. J. Stieltjes, *Recherches sur les fractions continues*, Ann. Fac. Sci. Toulouse **8** (1894), J, 1–122; **9** (1894), A, 1–47; Oeuvres **2**, 402–566. Also published in Memoires Présentés par divers savants à l'Académie de sciences de l'Institut National de France **33** (1892), 1–196.

[Stolz86]   O. Stolz, "Vorlesungen über allgemeine Arithmetic", Teubner, Leipzig (1886).

[Sylv69]   J. J. Sylvester, *Note on a New Continued Fraction Applicable to the Quadrature of the Circle*, Philos. Mag. Ser. 4 (1869), 373–375.

[Thie79]   T. N. Thiele, *Bemærkninger om periodiske Kjædebrøkers Konvergens*, Tidsskrift for Mathematik (4)**3** (1879).

[Thron58]   W. J. Thron, *On Parabolic Convergence Regions for Continued Fractions*, Math. Zeitschr. **69** (1958), 173–182.

[Thron59]   W. J. Thron, *Zwillingskonvergenzgebiete für Kettenbrüche* $1 + \mathbf{K}(a_n/1)$, *deren eines die Kreisscheibe* $|a_{2n-1}| \leq \rho^2$ *ist*, Math. Zeitschr. **70** (1959), 310–344.

[Thron74]   W. J. Thron, *A Survey of Recent Convergence Results for Continued Fractions*, Rocky Mountain J. Math. 4(2) (1974), 273–282.

[ThWa80]   W. J. Thron and H. Waadeland, *Accelerating Convergence of Limit Periodic Continued Fractions* $\mathbf{K}(a_n/1)$, Numer. Math. **34** (1980), 155–170.

[ThWa80]   W. J. Thron and H. Waadeland, *Analytic Continuation of Functions Defined by Means of Continued Fractions*, Math. Scand. **47** (1980), 72–90.

[ThWa81]   W. J. Thron and H. Waadeland, *Convergence Questions for Limit Periodic Continued Fractions*, Rocky Mountain J. Math.**11** (1981), 641–657.

[Waad66]   H. Waadeland, *A Convergence Property of Certain T-fraction Expansions*, Det Kgl. Norske Vid. Selsk. Skr. **9** (1966), 1–22.

[WaWe44]   H. S. Wall and M. Wetzel, *Contributions to the Analytic Theory of J-Fractions*, Trans. Amer. Math. Soc. **55** (1944), 373–397.

[WaWe44]   H. S. Wall and M. Wetzel, *Quadratic Forms and Convergence Regions for Continued Fractions*, Duke. Math. J. **11** (1944), 89–102.

[Wynn59]   P. Wynn, *Converging Factors for Continued Fractions*, Numer. Math. **1** (1959), 272–320.

Chapter IV

Continued fractions and three-term recurrence relations

About this chapter

The fact that the canonical numerators $\{A_n\}$ and denominators $\{B_n\}$ of the continued fraction $\mathbf{K}(a_n/b_n)$ satisfy the equalities

$$A_n = b_n A_{n-1} + a_n A_{n-2}, \quad B_n = b_n B_{n-1} + a_n B_{n-2} \quad \text{for } n = 1, 2, 3, \dots;$$

i.e. that $\{A_n\}$ and $\{B_n\}$ are solutions of the three-term recurrence relation

$$X_n = b_n X_{n-1} + a_n X_{n-2} \quad \text{for } n = 1, 2, 3, \dots$$

with initial values $A_{-1} = 1$, $A_0 = 0$, $B_{-1} = 0$ and $B_0 = 1$, is very useful. The solution space of this recurrence relation has a very nice structure: it is a linear space. And this fact can be used in the convergence theory for continued fractions with "surprisingly" good results, as we shall see. But we can also make use of this connection in the opposite direction, as the basis for a continued fraction method to compute certain solutions of three-term recurrence relations. So useful is this connection that one has asked the question of whether there are some continued fraction-like structures which corresponds to "longer" recurrence relations. This will be touched upon at the end of this chapter.

Readers familiar with hypergeometric functions and/or orthogonal polynomials will probably guess that this link between continued fractions and three-term recurrence relations provides a link between continued fractions and hypergeometric functions and between continued fractions and orthogonal polynomials. This is indeed so, but will be treated in later chapters.

1 Three-term recurrence relations

1.1 The structure of the solution space

Let us take a closer look at the three-term recurrence relation

$$X_n = b_n X_{n-1} + a_n X_{n-2} \quad \text{for } n = 1, 2, 3, \dots, \qquad (1.1.1)$$

where all a_n and b_n are complex numbers and all $a_n \neq 0$. A sequence $\{X_n\}_{n=-1}^{\infty}$ of complex numbers is called a *solution* of (1.1.1) if its elements satisfy this equality for all $n \in \mathbf{N}$.

Example 1 The sequence $\{X_n\}_{n=-1}^{\infty} = \left\{ \left(\frac{1-\sqrt{5}}{2} \right)^{n+1} \right\}_{n=-1}^{\infty}$ is a solution of the three-term recurrence relation

$$X_n = X_{n-1} + X_{n-2} \quad \text{for } n = 1, 2, 3, \dots .$$

We shall see this by checking that its elements X_n satisfy the relation. We have, for $n \geq 1$

$$
\begin{aligned}
X_{n-1} + X_{n-2} &= \left(\frac{1-\sqrt{5}}{2} \right)^{n} + \left(\frac{1-\sqrt{5}}{2} \right)^{n-1} \\
&= \left(\frac{1-\sqrt{5}}{2} + 1 \right) \left(\frac{1-\sqrt{5}}{2} \right)^{n-1} \\
&= \frac{3-\sqrt{5}}{2} \left(\frac{1-\sqrt{5}}{2} \right)^{n-1}
\end{aligned}
$$

and

$$
\begin{aligned}
X_n &= \left(\frac{1-\sqrt{5}}{2} \right)^{n+1} = \frac{1 - 2\sqrt{5} + 5}{4} \left(\frac{1-\sqrt{5}}{2} \right)^{n-1} \\
&= \frac{3-\sqrt{5}}{2} \left(\frac{1-\sqrt{5}}{2} \right)^{n-1},
\end{aligned}
$$

which proves the assertion. By the same method we also find that $\{Y_n\}_{n=-1}^{\infty} = \{\left(\frac{1+\sqrt{5}}{2}\right)^{n+1}\}_{n=-1}^{\infty}$ is a solution of the same recurrence relation. A third solution is

$$\{F_n\}_{n=-1}^{\infty} = (1, 1, 2, 3, 5, 8, 13, 21, 34, 55, 89, \ldots).$$

This solution is obtained by starting with $F_{-1} = F_0 = 1$ and using the recurrence relation recursively: each element F_n is the sum of the two previous ones. We recognize $\{F_n\}$ to be the sequence of Fibonacci numbers. (See Problem 1 in Chapter I.)

$\diamondsuit$

The set of all solutions $\{X_n\}_{n=-1}^{\infty}$ of (1.1.1) is called the *solution space* of (1.1.1). It has some nice properties:

1) $\{0\}_{n=-1}^{\infty}$ is a solution and thus belongs to the solution space.

2) If $\{X_n\}$ is a solution of (1.1.1), then so is $a\{X_n\} = \{aX_n\}$ for every fixed complex number a.

3) If $\{X_n\}$ and $\{Y_n\}$ are two solutions of (1.1.1), then $\{X_n\}+\{Y_n\} = \{X_n + Y_n\}$ is also a solution.

This means that the solution space is a *linear space* (vector space) with 0-element $\{0\}_{n=-1}^{\infty}$:

Theorem 1 *The solution space for the three-term recurrence relation*

$$X_n = b_n X_{n-1} + a_n X_{n-2}, \quad a_n, b_n \in \mathbf{C}, \quad a_n \neq 0 \quad for\ n = 1, 2, 3, \ldots$$

is a linear space of dimension 2. The canonical numerators $\{A_n\}$ and denominators $\{B_n\}$ of $\mathbf{K}(a_n/b_n)$ form a basis for the solution space.

Proof : We have seen that the solution space is a linear space. We also know that a solution $\{X_n\}$ is uniquely determined if X_{-1} and X_0 are given. In particular this means that $\{A_n\}$ and $\{B_n\}$ are uniquely determined, $\{A_n\} = (1, 0, \ldots)$ and $\{B_n\} = (0, 1, \ldots)$. An arbitrary solution $\{X_n\}$ can therefore be written as the linear combination $\{X_n\} = X_{-1} \cdot \{A_n\} + X_0 \cdot \{B_n\}$. Hence, the dimension of the solution space is ≤ 2. It remains to prove that $\{A_n\}$ and $\{B_n\}$ are linearly independent. Let $c_1\{A_n\} + c_2\{B_n\} = \{0\}$ for two complex constants c_1 and c_2. Then, in particular, $c_1 A_{-1} + c_2 B_{-1} = 0$ and $c_1 A_0 + c_2 B_0 = 0$; i.e. $c_1 = c_2 = 0$.

∎

Example 2 The two solutions

$$\{X_n\} = \left\{ \left(\frac{1 - \sqrt{5}}{2} \right)^{n+1} \right\}, \qquad \{Y_n\} = \left\{ \left(\frac{1 + \sqrt{5}}{2} \right)^{n+1} \right\}$$

from Example 1 are linearly independent, and thus form a basis for the solution space of

$$X_n = X_{n-1} + X_{n-2} \quad \text{for } n = 1, 2, 3, \ldots . \tag{1.1.2}$$

The Fibonacci sequence $\{F_n\}$ can therefore be written $\{F_n\} = c_1\{X_n\} + c_2\{Y_n\}$ for some complex constants c_1 and c_2. To determine these constants we check the equality for $n = -1$ and $n = 0$:

$$F_{-1} = c_1 X_{-1} + c_2 Y_{-1}, \quad \text{that is } 1 = c_1 + c_2$$
$$F_0 = c_1 X_0 + c_2 Y_0, \quad \text{that is } 1 = c_1 \frac{1 - \sqrt{5}}{2} + c_2 \frac{1 + \sqrt{5}}{2}.$$

Solving these equations for c_1 and c_2 gives $c_1 = -(1 - \sqrt{5})/2\sqrt{5}$ and $c_2 = (1 + \sqrt{5})/2\sqrt{5}$. We thus have the following closed form for the Fibonacci numbers:

$$F_n = \frac{1}{\sqrt{5}} \left(\frac{1 + \sqrt{5}}{2} \right)^{n+2} - \frac{1}{\sqrt{5}} \left(\frac{1 - \sqrt{5}}{2} \right)^{n+2}.$$

This formula goes by the name of Binet's formula.

◇

Example 3 The canonical numerators $\{A_n\}$ and denominators $\{B_n\}$ of the continued fraction

$$\mathbf{K}(1/1) = \frac{1}{1+} \frac{1}{1+} \frac{1}{1+} \cdots$$

are solutions of the recurrence relation (1.1.2). By Theorem 1 they actually form a basis for the solution space of this recurrence relation. However, so do also the two solutions $\{X_n\}$ and $\{Y_n\}$ from Example 2 since they are linearly independent. Therefore there exist complex constants α_1, α_2, β_1 and β_2 such that

$$A_n = \alpha_1 \left(\frac{1 - \sqrt{5}}{2} \right)^{n+1} + \alpha_2 \left(\frac{1 + \sqrt{5}}{2} \right)^{n+1} ,$$

$$B_n = \beta_1 \left(\frac{1 - \sqrt{5}}{2} \right)^{n+1} + \beta_2 \left(\frac{1 + \sqrt{5}}{2} \right)^{n+1}$$

for all n. Checking these equalities for $n = -1$ and $n = 0$ gives that $\alpha_1 = (1 + \sqrt{5})/2\sqrt{5}$, $\alpha_2 = -(1 - \sqrt{5})/2\sqrt{5}$, $\beta_1 = -1/\sqrt{5}$ and $\beta_2 = 1/\sqrt{5}$. The approximants of $\mathbf{K}(1/1)$ can therefore be written in closed form:

$$\frac{A_n}{B_n} = \frac{\alpha_1 X_n + \alpha_2 Y_n}{\beta_1 X_n + \beta_2 Y_n} = \frac{\dfrac{1 + \sqrt{5}}{2\sqrt{5}} \left(\dfrac{1 - \sqrt{5}}{2} \right)^{n+1} - \dfrac{1 - \sqrt{5}}{2\sqrt{5}} \left(\dfrac{1 + \sqrt{5}}{2} \right)^{n+1}}{-\dfrac{1}{\sqrt{5}} \left(\dfrac{1 - \sqrt{5}}{2} \right)^{n+1} + \dfrac{1}{\sqrt{5}} \left(\dfrac{1 + \sqrt{5}}{2} \right)^{n+1}} .$$

(This expression may of course be considerably simplified.) Since $((1 - \sqrt{5})/2)^{n+1} \to 0$ as $n \to \infty$ and $((1 + \sqrt{5})/2)^{n+1} \to \infty$ as $n \to \infty$, it follows that A_n/B_n converges to $(\sqrt{5} - 1)/2$ which is consistent with Theorem 6C in Chapter III.

                                                                      $\diamond$

1.2   *Approximants for periodic continued fractions in closed form*

The idea of Example 3 can be extended to more general periodic continued fractions. For simplicity we limit ourselves to the 1-periodic ones:

Theorem 2 *Let the linear fractional transformation $s(z) = a/(b + z)$ have two distinct fixed points x and y. Then the 1-periodic continued fraction $\mathbf{K}(a/b)$ has approximants*

$$f_n = \frac{A_n}{B_n} = -xy \frac{(-y)^n - (-x)^n}{(-y)^{n+1} - (-x)^{n+1}} \quad \text{for } n = 0, 1, 2, \dots . \quad (1.2.1)$$

Proof : Since $x = a/(b+x)$ and $y = a/(b+y)$, it follows that $a = -xy$, $b = -(x+y)$ and thus that $\{(-x)^{n+1}\}_{n=-1}^{\infty}$ and $\{(-y)^{n+1}\}_{n=-1}^{\infty}$ are two linearly independent solutions of the three-term recurrence relation

$$X_n = bX_{n-1} + aX_{n-2} \quad \text{for } n = 1, 2, 3, \ldots .$$

Hence the canonical numerators $\{A_n\}$ and denominators $\{B_n\}$ of $\mathbf{K}(a/b)$ can be written

$$\begin{aligned} A_n &= \alpha_1(-x)^{n+1} + \alpha_2(-y)^{n+1}, \\ B_n &= \beta_1(-x)^{n+1} + \beta_2(-y)^{n+1}, \quad \text{for } n = -1, 0, 1, \ldots . \end{aligned}$$

In particular

$$\begin{aligned} A_{-1} &= 1 = \alpha_1 + \alpha_2, & B_{-1} &= 0 = \beta_1 + \beta_2, \\ A_0 &= 0 = -\alpha_1 x - \alpha_2 y, & B_0 &= 1 = -\beta_1 x - \beta_2 y, \end{aligned}$$

so that $\alpha_1 = -y/(x - y)$, $\alpha_2 = x/(x - y)$, $\beta_1 = -1/(x - y)$ and $\beta_2 = 1/(x - y)$. Hence (1.2.1) follows. ∎

Remark: Note that the closed expression (1.2.1) for f_n can be written

$$f_n = x \frac{1 - (x/y)^n}{1 - (x/y)^{n+1}} \quad \text{for } n = 0, 1, 2, \ldots . \tag{1.2.2}$$

Hence, if $|x| < |y|$ then $\lim f_n = x$, just as proved in Theorem 6C in Chapter III. Note also that in Problem 5 in Chapter I you were asked to prove Theorem 2 (by a different method, induction).

Example 4 The periodic continued fraction $\mathbf{K}(3/2)$ has approximants

$$f_n = \frac{A_n}{B_n} = 3 \frac{3^n - (-1)^n}{3^{n+1} - (-1)^{n+1}} = \frac{1 - (-1/3)^n}{1 - (-1/3)^{n+1}},$$

since the linear fractional transformation $s(z) = 3/(2 + z)$ has the two fixed points $x = 1$ and $y = -3$. See also Example 1 in Chapter I.

◇

1.3   Linear independence of two solutions

How can we easily see that two solutions of a three-term recurrence relation are linearly independent? Intuitively one would say that $\{X_n\}$ and $\{Y_n\} \neq \{0\}$ are linearly dependent if and only if $X_{N-1} = \alpha Y_{N-1}$ and $X_N = \alpha Y_N$ for some $\alpha \in \mathbf{C}$ and $N \in \mathbf{N}_0$. This is indeed the case. The following theorem provides this result together with some other useful characterizations of linear independence:

Theorem 3 *Let $\{X_n\}$ and $\{Y_n\}$ be solutions of the three-term recurrence relation*

$$X_n = b_n X_{n-1} + a_n X_{n-2}, \qquad a_n \neq 0 \quad for \ n = 1, 2, 3, \ldots . \qquad (1.3.1)$$

Then the following statements are equivalent.

(A) $\{X_n\}$ *and* $\{Y_n\}$ *are linearly independent.*

(B) *There exists an $N \in \mathbf{N}_0$ such that $X_N Y_{N-1} - Y_N X_{N-1} \neq 0$.*

(C) $X_0 Y_{-1} - Y_0 X_{-1} \neq 0$.

(D) $X_n Y_{n-1} - Y_n X_{n-1} \neq 0$ *for all $n \in \mathbf{N}_0$.*

(E) *If $\{U_n\}, \{V_n\}$ is a basis for the solution space of (1.3.1) and $\{X_n\} = \alpha_1\{U_n\} + \alpha_2\{V_n\}$, $\{Y_n\} = \beta_1\{U_n\} + \beta_2\{V_n\}$, then $\alpha_1\beta_2 - \alpha_2\beta_1 \neq 0$.*

To prove this theorem we use the following lemma which is a generalization of the determinant formula, formula (1.2.10) in Chapter I. It follows by induction on n, using the recurrence relation:

Lemma 4 *Let $\{X_n\}$ and $\{Y_n\}$ be solutions of (1.3.1). Then*

$$X_n Y_{n-1} - Y_n X_{n-1} = (X_0 Y_{-1} - Y_0 X_{-1}) \prod_{k=1}^{n} (-a_k) \quad for \ n = 1, 2, 3, \ldots .$$

Proof of Theorem 3: The equivalence of (B), (C) and (D) follows directly from Lemma 4.

(**A**) $\iff$ (**C**): Let $c_1\{X_n\} + c_2\{Y_n\} = \{0\}$. Then

$$c_1 X_{-1} + c_2 Y_{-1} = 0 \quad \text{and} \quad c_1 X_0 + c_2 Y_0 = 0 \,.$$

This is a system of linear equations in c_1 and c_2. It has a unique solution if and only if its determinant $X_0 Y_{-1} - Y_0 X_{-1} \neq 0$. This unique solution is $c_1 = c_2 = 0$.

(**C**) $\iff$ (**E**): The equivalence follows from the identity

$$
\begin{aligned}
X_0 Y_{-1} - Y_0 X_{-1} &= (\alpha_1 U_0 + \alpha_2 V_0)(\beta_1 U_{-1} + \beta_2 V_{-1}) \\
&\quad - (\beta_1 U_0 + \beta_2 V_0)(\alpha_1 U_{-1} + \alpha_2 V_{-1}) \\
&= (\alpha_1 \beta_2 - \alpha_2 \beta_1)(U_0 V_{-1} - V_0 U_{-1}),
\end{aligned}
$$

where $U_0 V_{-1} - V_0 U_{-1} \neq 0$ by the equivalence (A) $\iff$ (C). ■

1.4 The adjoint recurrence relation

The *adjoint* of the three-term recurrence relation

$$X_n = b_n X_{n-1} + a_n X_{n-2} \quad \text{where } a_n \neq 0, \quad \text{for } n = 1, 2, 3, \ldots \quad (1.4.1)$$

is by definition the recurrence relation

$$P_n = b_n P_{n+1} + a_{n+1} P_{n+2} \quad \text{for } n = 0, 1, 2, \ldots, \quad\quad (1.4.2)$$

where b_0 is some (arbitrary) complex constant. (It is not essential that n runs from $n = 0$ in (1.4.2), but it is convenient for our purpose.) Solutions of (1.4.2) have the form $\{P_n\}_{n=0}^{\infty}$. This recurrence relation is more natural for the hypergeometric functions:

Example 5 The confluent hypergeometric function

$$
\begin{aligned}
\Psi(c; z) &= 1 + \frac{1}{c}\frac{z}{1!} + \frac{1}{c(c+1)}\frac{z^2}{2!} + \frac{1}{c(c+1)(c+2)}\frac{z^3}{3!} \\
&\quad + \cdots + \frac{1}{c(c+1)\cdots(c+n-1)}\frac{z^n}{n!} + \cdots,
\end{aligned}
$$

where c is a fixed complex number $\neq 0, -1, -2, \ldots$, is an entire function. For such functions we have that

$$\Psi(c; z) = \Psi(c + 1; z) + \frac{z}{c(c + 1)} \Psi(c + 2; z).$$

This can be seen by comparing the coefficient of z^k in the power series on both sides of the equality sign, for $k = 0, 1, 2, \ldots$. Setting $P_n(z) = \Psi(c + n; z)$ for $n = 0, 1, 2, \ldots$, we see that

$$P_n(z) = P_{n+1}(z) + \frac{z}{(c + n)(c + n + 1)} P_{n+2}(z) \quad \text{for } n = 0, 1, 2, \ldots.$$

$$(1.4.3)$$

That is, $\{\Psi(c + n; z)\}_{n=0}^{\infty}$ is a solution of this recurrence relation for every $z \in \mathbf{C}$.

$\diamond$

There is a strong connection between solutions of (1.4.1) and its adjoint (1.4.2):

Theorem 5

(A) $\{P_n\}_{n=0}^{\infty}$ is a solution of (1.4.2) where all $a_n \neq 0$ if and only if $P_0 = b_0 P_1 + a_1 P_2$ and $\{P_{n+2} \prod_{j=1}^{n+1}(-a_j)\}_{n=-1}^{\infty}$ is a solution of (1.4.1).

(B) $\{X_n\}_{n=-1}^{\infty}$ is a solution of (1.4.1) if and only if $\{P_n\}_{n=0}^{\infty}$ given by $P_n = X_{n-2} / \prod_{j=1}^{n-1}(-a_j)$ for all $n \in \mathbf{N}$, $P_0 = b_0 P_1 + a_1 P_2$ is a solution of (1.4.2).

(The empty product $\prod_{j=1}^{0} \cdots = 1$ by definition.)

Proof :
(A): Let $\{P_n\}$ be a solution of (1.4.2). Then

$$P_{n+2} = -\frac{b_n}{a_{n+1}} P_{n+1} + \frac{1}{a_{n+1}} P_n = b_n \frac{P_{n+1}}{-a_{n+1}} + a_n \frac{P_n}{(-a_n)(-a_{n+1})}$$

so that

$$P_{n+2} \prod_{j=1}^{n+1} (-a_j) = b_n P_{n+1} \prod_{j=1}^{n} (-a_j) + a_n P_n \prod_{j=1}^{n-1} (-a_j).$$

That is, $\{P_{n+2} \prod_{j=1}^{n+1} (-a_j)\}$ is a solution of $(1.4.1)$. The if-part follows from the same relations.

(B): This is a simple corollary of part (A) since

$$X_n = P_{n+2} \prod_{j=1}^{n+1} (-a_j) \quad \text{if and only if} \quad P_{n+2} = X_n / \prod_{j=1}^{n+1} (-a_j).$$

■

Example 6 In Example 5 we saw that $P_n(z) = \Psi(c+n; z)$ is a solution of the three-term recurrence relation $(1.4.3)$ for all $z \in \mathbf{C}$. Hence

$$X_n(z) = \Psi(c+n+2; z) \cdot \frac{\Gamma(c)\Gamma(c+1)(-z)^{n+1}}{\Gamma(c+n+1)\Gamma(c+n+2)} \quad \text{for } n = -1, 0, 1, \ldots$$

is a solution of

$$X_n(z) = X_{n-1}(z) + \frac{z}{(c+n-1)(c+n)} X_{n-2}(z) \quad \text{for } n = 1, 2, 3, \ldots .$$

(Remember that $\Gamma(x+1) = x\Gamma(x)$ for the gamma function $\Gamma(x)$. By using the Pochhammer symbol $(a)_n = a(a+1)\cdots(a+n-1) = \Gamma(a+n)/\Gamma(a)$, the expression for $X_n(z)$ can be somewhat simplified.) This illustrates that it sometimes is easier to look for solutions of the adjoint equation.

◇

1.5   Recurrence relations in a field **F**

Until now our recurrence relations have had coefficients a_n, b_n from **C**, and their solutions $\{X_n\}$ have been sequences of complex numbers. In some cases, for instance in the Examples 5 and 6, this is too special. Let us look closer at Example 5: A better approach is to regard (1.4.3) as a recurrence relation in the field **M** of functions $f(z)$ meromorphic at $z = 0$. That is, its coefficients $a_n z = z/((c + n)(c + n + 1)) \in$ **M** and $b_n(z) \equiv 1 \in$ **M**. Its solutions $\{P_n(z)\}$ also consists of elements from **M** in this case; for instance the confluent hypergeometric functions $P_n(z) = \Psi(c + n; z)$.

More rewarding is to regard (1.4.3) as a recurrence relation in the field **L** of formal power series $\sum_{n=n_0}^{\infty} c_n z^n$ with $n_0 \in$ **Z**. That is, we regard $a_n(z)$ and $b_n(z)$ as elements from **L** ("very short" power series) and get solutions $\{P_n(z)\} \subseteq$ **L**, as for instance the confluent hypergeometric series $P_n(z) = \Psi(c + n; z)$.

To cover this case, we let **F** denote a field which is either **C** or **L**. We further let $\|\cdot\|_{\mathbf{F}}$ be a norm in **F**, as for instance $|\cdot|$ in **C** (the usual absolute value; i.e. the euclidean norm). We shall return to the norm in **L** later. For the time being we just think of $(\mathbf{F}, \|\cdot\|_{\mathbf{F}})$ as $(\mathbf{C}, |\cdot|)$.

Inspired by the notation $\hat{\mathbf{C}} = \mathbf{C} \cup \{\infty\}$, we write $\hat{\mathbf{L}} = \mathbf{L} \cup \{l_\infty\}$ where l_∞ is the equivalence class of formal power series $\sum_{n=-\infty}^{\infty} c_n z^n$ where $c_n \neq 0$ for arbitrary small indices n. (The reason for this choice will become clear later.) That is, for short, $\hat{\mathbf{F}} = \mathbf{F} \cup \{\infty\}$.

Remark: Theorems 1, 2, 3 and 5 still hold when the three-term recurrence relations are in **F**. We could of course have considered even more general fields **F**, but we shall not need that in this exposition.

2 Convergence of continued fractions

2.1 Pincherle's theorem

The clue to Pincherle's theorem [Pinc94] can be found in the proof of Theorem 2 combined with the subsequent remark. The fact that the canonical approximants of $\mathbf{K}(a_n/b_n)$ can be written

$$f_n = \frac{A_n}{B_n} = \frac{\alpha_1 X_n + \alpha_2 Y_n}{\beta_1 X_n + \beta_2 Y_n} = \frac{\alpha_1 X_n/Y_n + \alpha_2}{\beta_1 X_n/Y_n + \beta_2}, \qquad (2.1.1)$$

where $\{X_n\}$ and $\{Y_n\}$ are solutions of

$$X_n = b_n X_{n-1} + a_n X_{n-2} \quad \text{where } a_n \neq 0 \quad \text{for } n = 1, 2, 3, \ldots, \qquad (2.1.2)$$

leads to convergence of $\{f_n\}$ if $X_n/Y_n \to 0$. And not only that, in such a case we have that $\lim f_n = \alpha_2/\beta_2$. This is essentially the idea we shall pursue in this section.

Let $\{X_n\}$ and $\{Y_n\}$ be two linearly independent solutions of (2.1.2). Then X_n and Y_n can not be zero for the same index n, by Theorem 3. Hence X_n/Y_n is well defined in $\hat{\mathbf{F}}$ for all n.

Definition *We say that $\{X_n\}$ is a minimal (or subdominant) solution of the linear three-term recurrence relation (2.1.2) if $\{X_n\}$ is non-trivial (i.e. $\neq \{0\}$), and there exists a solution $\{Y_n\}$ of (2.1.2) such that $\lim_{n \to \infty} X_n/Y_n = 0$. The solution $\{Y_n\}$ is then said to be dominant.*

Let $\{X_n\}$ be a minimal and $\{Y_n\}$ be a dominant solution of (2.1.2) Then all solutions $c\{X_n\} = \{cX_n\}$, where $c \in \mathbf{F}\backslash\{0\}$, are also minimal. All other non-trivial solutions $\{Z_n\}$ are dominant since they can be written

$$\{Z_n\} = c_1\{X_n\} + c_2\{Y_n\} \quad \text{where } c_1, c_2 \in \mathbf{F}, \quad c_2 \neq 0.$$

From this we also see that $Z_n/Y_n \to c_2 \neq 0$. The converse is also true; i.e. if $Z_n/Y_n \to c_2 \in \mathbf{F}\backslash\{0\}$, then $\{Z_n\}$ and $\{Y_n\}$ are both dominant:

Theorem 6 *Let $\{Y_n\}$ and $\{Z_n\}$ be two linearly independent solutions of (2.1.2) such that $\lim Y_n/Z_n = R$ exists in $\hat{\mathbf{F}}$. Then (2.1.2) has a minimal solution.*

Proof :   If $R = \infty$ then $\{Z_n\}$ is minimal. If $R \neq \infty$ then $X_n = Y_n - RZ_n$ is a non-trivial solution of (2.1.2) with $X_n/Z_n \to 0$; i.e. $\{X_n\}$ is a minimal solution.   ∎

Theorem 7 (Pincherle's theorem) *Let* $\mathbf{K}(a_n/b_n)$ *be a continued fraction with elements* a_n *and* b_n *from* $\mathbf{F}$ *and all* $a_n \neq 0$. *Then:*

(A) $\mathbf{K}(a_n/b_n)$ *converges in* $\hat{\mathbf{F}}$ *if and only if the corresponding linear three-term recurrence relation (2.1.2) has a minimal solution.*

(B) *If (2.1.2) has a minimal solution* $\{X_n\}$, *then* $\mathbf{K}(a_n/b_n)$ *converges to* $-X_0/X_{-1} \in \hat{\mathbf{F}}$.

(C) *If (2.1.2) has a minimal solution* $\{X_n\}$, *and* $\{Y_n\}$ *is a dominant solution, then the approximants* f_n *of* $\mathbf{K}(a_n/b_n)$ *satisfy*

$$
\begin{aligned}
(f_n + X_0/X_{-1}) &\sim C_1(X_n/Y_n) \quad \text{as } n \to \infty \quad \text{if } X_{-1} \neq 0, \\
f_n &\sim C_2(Y_n/X_n) \quad \text{as } n \to \infty \quad \text{if } X_{-1} = 0
\end{aligned}
\qquad (2.1.3)
$$

for some constants C_1 *and* C_2 *from* $\mathbf{F}\backslash\{0\}$.

Remark: The notation in (2.1.3) is to be understood as follows:

$$
t_n \sim Cy_n \quad \text{as } n \to \infty \iff \lim_{n \to \infty} t_n/y_n = C
$$

for a $C \in \mathbf{F}\backslash\{0\}$. We say that t_n is asymptotically equal to Cy_n. If $\mathbf{F} = \mathbf{C}$ and $\|\cdot\|_{\mathbf{F}} = |\cdot|$, the usual absolute value norm, then (2.1.3) expresses the speed of convergence of the continued fraction $\mathbf{K}(a_n/b_n)$.

Proof of Theorem 7:   Let $f_n = A_n/B_n$ be the nth canonical approximant of $\mathbf{K}(a_n/b_n)$.

(A): Assume first that (2.1.2) has a minimal solution $\{X_n\}$. Let $\{Y_n\}$ denote a dominant solution of (2.1.2). Then $\{X_n\}$ and $\{Y_n\}$ form a basis for the solution space of (2.1.2), and there exist elements α_1, α_2, β_1 and β_2 from $\mathbf{F}$ such that

$$\begin{array}{ll} A_n = \alpha_1 X_n + \alpha_2 Y_n, \\ B_n = \beta_1 X_n + \beta_2 Y_n \end{array} \quad \text{for } n = -1, 0, 1, \ldots. \qquad (2.1.4)$$

Hence f_n can be written on the form (2.1.1). If $\beta_2 \neq 0$ it follows that $f_n \to \alpha_2/\beta_2$. If $\beta_2 = 0$ we necessarily have that $\beta_1 \neq 0$ and $\alpha_2 \neq 0$ since $\{A_n\}$ and $\{B_n\}$ are linearly independent. Hence $f_n \to \infty$ if $\beta_2 = 0$. That is, the continued fraction converges to $\alpha_2/\beta_2 \in \hat{\mathbf{F}}$.

To prove the "only if" part we first assume that $\mathbf{K}(a_n/b_n)$ converges to a value $f \in \hat{\mathbf{F}}$. That is, $\lim_n \to \infty A_n/B_n = f$. Since $\{A_n\}$ and $\{B_n\}$ are two linearly independent solutions of (2.1.2), the existence of a minimal solution then follows from Theorem 6.

(**B**): In view of the observations above, it suffices to prove that $\alpha_2/\beta_2 = -X_0/X_{-1}$: Setting $n = -1$ and $n = 0$ in (2.1.4) gives the equations

$$\begin{array}{llll} A_{-1} &= 1 = \alpha_1 X_{-1} + \alpha_2 Y_{-1}, & B_{-1} &= 0 = \beta_1 X_{-1} + \beta_2 Y_{-1}, \\ A_0 &= 0 = \alpha_1 X_0 + \alpha_2 Y_0, & B_0 &= 1 = \beta_1 X_0 + \beta_2 Y_0, \end{array}$$

which has the solution $\alpha_1 = Y_0/\Delta$, $\alpha_2 = -X_0/\Delta$, $\beta_1 = -Y_{-1}/\Delta$, $\beta_2 = X_{-1}/\Delta$, where $\Delta = Y_0 X_{-1} - Y_{-1} X_0 \neq 0$, since $\{X_n\}$ and $\{Y_n\}$ are linearly independent.

(**C**): Let $X_{-1} \neq 0$. Then

$$\begin{aligned} f - f_n &= \frac{\alpha_2}{\beta_2} - \frac{\alpha_1 X_n + \alpha_2 Y_n}{\beta_1 X_n + \beta_2 Y_n} = \frac{(\alpha_2 \beta_1 - \alpha_1 \beta_2) X_n}{\beta_2(\beta_1 X_n + \beta_2 Y_n)} \\ &= \frac{\alpha_2 \beta_1 - \alpha_1 \beta_2}{\beta_2(\beta_2 + \beta_1 X_n/Y_n)} \frac{X_n}{Y_n}, \end{aligned}$$

where $X_n/Y_n \to 0$ and $\beta_2 \neq 0$. For the case $X_{-1} = 0$, the approximants satisfy

$$f_n = \frac{\alpha_1 X_n + \alpha_2 Y_n}{\beta_1 X_n + \beta_2 Y_n} = \frac{Y_0 X_n - X_0 Y_n}{-Y_{-1} X_n} = \frac{Y_0 - X_0 Y_n/X_n}{-Y_{-1}},$$

where $X_0 \neq 0$ and $Y_{-1} \neq 0$ by virtue of Theorem 3. ∎

The value $-X_0/X_{-1}$ is unique, since every minimal solution of (2.1.2) is proportional to $\{X_n\}$. If both a minimal solution $\{X_n\}$ and a dominant solution $\{Y_n\}$ are explicitly known, then the approximants $f_n = A_n/B_n$ and their limit $f = -X_0/X_{-1}$ are also known, and the truncation error estimate in part (C) is of no interest. What is often the case, however, is that $\{Y_n\}$ is only known in the sense that we have an estimate for the speed by which X_n/Y_n approaches 0. Or, alternatively, that the expressions for Y_n are so complicated that we prefer to use such estimates.

Example 7 In Ramanujan's second notebook we find the formula

$$\frac{x+a+1}{x+1} = \frac{x+a}{x-1} + \frac{x+2a}{x+a-1} + \frac{x+3a}{x+2a-1} + \cdots$$

for $a \in \mathbf{C}\backslash\{0\}$ and $x \in \mathbf{C}\backslash\{-a, -2a, -3a, \ldots\}$. The meaning of this formula is that the continued fraction on the right side converges to the value on the left side. Unfortunately Ramanujan very rarely indicated how his formulas could be proved! So how could he have found and proved his result?

Well, a clue is that $\{X_n\}$ given by

$$X_n = (-1)^n(x + na + a + 1) \quad \text{for } n = -1, 0, 1, 2, \ldots$$

is a solution of the three-term recurrence relation

$$X_n = (x + na - a - 1)X_{n-1} + (x + na)X_{n-2} \quad \text{for } n = 1, 2, 3, \ldots.$$

If $\{X_n\}$ is a *minimal* solution, then Ramanujan's formula follows by Pincherle's theorem. Is $\{X_n\}$ minimal? The answer is yes. One can prove (by Birkhoff's method which is described in [Wimp84]) that there is a solution $\{Y_n\}$ such that the limit

$$\lim_{n \to \infty} Y_n/(n+1)! \, a^n n^{(x/a)-2}$$

exists in $\mathbf{C}\backslash\{0\}$. That is, $Y_n \sim C(n-1)! \, a^n n^{x/a}$ and $\{Y_n\}$ is dominant.

The speed of convergence of the continued fraction is of the order $\left((n-2)! \, |a|^{n-1} n^{\Re(x/a)}\right)^{-1}$.

◇

Pincherle's theorem can also be stated for $b_0 + \mathbf{K}(a_n/b_n)$ and the adjoint recurrence relation

$$P_n = b_n P_{n+1} + a_n P_{n+2} \quad \text{for } n = 0, 1, 2, \ldots : \tag{2.1.5}$$

Corollary 8 *Let $b_0 + \mathbf{K}(a_n/b_n)$ be a continued fraction with elements a_n and b_n from $\mathbf{F}$ and all $a_n \neq 0$. Then:*

(A) $b_0 + \mathbf{K}(a_n/b_n)$ *converges in $\hat{\mathbf{F}}$ if and only if the three-term recurrence relation (2.1.5) has a minimal solution.*

(B) *If $\{P_n\}$ is a minimal solution of (2.1.5), then $b_0 + \mathbf{K}(a_n/b_n)$ converges to P_0/P_1.*

(C) *If $\{P_n\}$ is a minimal solution of (2.1.5) and $\{Q_n\}$ is a dominant solution, then*

$$\begin{aligned}
(f_n - P_0/P_1) &\sim C_1(P_{n+2}/Q_{n+2}) &&\text{as } n \to \infty &&\text{if } P_1 \neq 0, \\
f_n &\sim C_2(Q_{n+2}/P_{n+2}) &&\text{as } n \to \infty &&\text{if } P_1 = 0
\end{aligned}$$

for some constants C_1 and C_2 from $\mathbf{F}\backslash\{0\}$.

Proof :
(A): By Pincherle's theorem we know that $b_0 + \mathbf{K}(a_n/b_n)$ converges if and only if (2.1.2) has a minimal solution. From Theorem 5A it follows that $\{P_n\}$ is a solution of (2.1.5) if and only if $P_0 = b_0 P_1 + a_1 P_2$ and $\{X_n\}_{n=-1}^{\infty}$ given by

$$X_n = P_{n+2} \prod_{j=1}^{n+1} (-a_j) \quad \text{for } n = -1, 0, 1, \ldots \tag{2.1.6}$$

is a solution of (2.1.2). It follows from (2.1.6) that $\{P_n\}$ is a minimal solution of (2.1.5) if and only if $\{X_n\}$ is a minimal solution of (2.1.2).

(B): Let $\{P_n\}$ be a minimal solution of (2.1.5). By the arguments above and Pincherle's theorem we then know that $b_0 + \mathbf{K}(a_n/b_n)$ converges to

$$b_0 - \frac{X_0}{X_{-1}} = b_0 - \left(\frac{-a_1 P_2}{P_1}\right) = \frac{b_0 P_1 + a_1 P_2}{P_1} = \frac{P_0}{P_1}.$$

(C): This follows from Theorem 7C and the connection (2.1.6) between solutions of the two recurrence relations.   ■

2.2   Auric's theorem

Application of Pincherle's theorem requires knowledge of *two* solutions of the corresponding recurrence relation. What can we do if we only can find one? And how can we decide whether this is a minimal solution or not?

Theorem 9 *Let* $\{X_n\}$ *be a solution of the three-term recurrence relation*

$$X_n = b_n X_{n-1} + a_n X_{n-2} \quad \text{where } a_n, b_n \in \mathbf{F},\ a_n \neq 0 \quad \text{for } n = 1, 2, 3, \dots$$
(2.2.1)

with $X_n \neq 0$ *for all* n. *Then* $\{X_n\}$ *is a minimal solution of (2.2.1) if and only if*

$$\sum_{n=0}^{\infty} \frac{\prod_{m=1}^{n}(-a_m)}{X_{n-1}X_n} = \infty .$$
(2.2.2)

Proof : Let $\{X_n\}$ be a minimal solution. Then it follows by Pincherle's theorem, Theorem 7, that $\mathbf{K}(a_n/b_n)$ converges to $f = -X_0/X_{-1} \neq \infty$ since $X_{-1} \neq 0$. Let $f_n = A_n/B_n$ be the approximants of $\mathbf{K}(a_n/b_n)$ in canonical form. Since $\{A_n\}$ and $\{B_n\}$ are linearly independent solutions of (2.2.1), it follows that $\{B_n\}$ is minimal if and only if $A_n/B_n \to \infty$. We have $f_n = A_n/B_n \to f \neq \infty$. Hence $\{B_n\}$ is dominant and $B_n/X_n \to \infty$. Using Lemma 4 we find that

$$\frac{B_k}{X_k} = \sum_{n=0}^{k}\left(\frac{B_n}{X_n} - \frac{B_{n-1}}{X_{n-1}}\right) = \sum_{n=0}^{k}\frac{B_n X_{n-1} - B_{n-1}X_n}{X_{n-1}X_n}$$
(2.2.3)

$$= \sum_{n=0}^{k}(B_0 X_{-1} - B_{-1}X_0)\frac{\prod_{m=1}^{n}(-a_m)}{X_{n-1}X_n} = X_{-1}\sum_{n=0}^{k}\frac{\prod_{m=1}^{n}(-a_m)}{X_{n-1}X_n} ,$$

where we have used that $B_{-1} = 0$ and $B_0 = 1$. This proves (2.2.2).

Assume next that (2.2.2) holds. From (2.2.3) it follows then that $B_k/X_k \to \infty$. But this means that $\{X_n\}$ is a minimal solution. ∎

This result leads directly to the following useful theorem [Auric07]:

Theorem 10 (Auric's theorem) *Let* $\mathbf{K}(a_n/b_n)$ *be a continued fraction with elements* $a_n, b_n \in \mathbf{F}$ *and all* $a_n \neq 0$, *and let* $\{X_n\}$ *be a solution of the corresponding recurrence relation (2.2.1) such that all* $X_n \neq 0$ *and*

$$\sum_{n=0}^{\infty} \frac{\prod_{m=1}^{n}(-a_m)}{X_{n-1}X_n} = \infty. \tag{2.2.4}$$

Then $\mathbf{K}(a_n/b_n)$ *converges to the finite value* $-X_0/X_{-1}$, *and*

$$(f_k + \frac{X_0}{X_{-1}}) \sim C \left(\sum_{n=0}^{k} \frac{\prod_{m=1}^{n}(-a_m)}{X_{n-1}X_n} \right)^{-1} \quad as \ n \to \infty \tag{2.2.5}$$

for some constant $C \in \mathbf{F} \backslash \{0\}$.

Proof : That $\mathbf{K}(a_n/b_n)$ converges to $-X_0/X_{-1}$ is a direct consequence of Theorem 9 and Pincherle's theorem, Theorem 7. From the proof of Theorem 9 it follows that $\{B_n\}$ is a dominant solution of the recurrence relation. Hence, the order of the speed of convergence follows by Theorem 7C and (2.2.3). ∎

Example 8 The continued fraction

$$\mathbf{K}(a_n/1) = \frac{18}{1} + \frac{40}{1} + \cdots + \frac{4n^2 + 10n + 4}{1} + \cdots$$

converges by virtue of the parabola theorem, Theorem 20 in Chapter III. However, since $\{X_n\}$ given by

$$X_n = \prod_{k=0}^{n}(-2k - 3) \quad \text{for } n = -1, 0, 1 \ldots$$

is a solution of the corresponding three-term recurrence relation

$$X_n = X_{n-1} + (4n^2 + 10n + 4)X_{n-2} \quad \text{for } n = 1, 2, 3, \ldots,$$

we can also use Auric's theorem to study $\mathbf{K}(a_n/1)$. We have

$$
\sum_{n=0}^{\infty} \frac{\prod_{m=1}^{n}(-a_m)}{X_n X_{n-1}} = -\sum_{n=0}^{\infty} \frac{\prod_{m=1}^{n}(-4m^2 - 10m - 4)}{(2n+3)\prod_{m=0}^{n-1}(2m+3)^2}
$$

$$
= -\sum_{n=0}^{\infty} \frac{1}{2n+3} \prod_{m=1}^{n} \frac{-(2m+4)(2m+1)}{(2m+1)^2}
$$

$$
= -\sum_{n=0}^{\infty} \frac{(-1)^n}{2n+3} \prod_{m=1}^{n} \frac{2m+4}{2m+1}
$$

$$
= -\sum_{n=0}^{\infty} \frac{(-1)^n}{3} \prod_{m=1}^{n} \frac{2m+4}{2m+3}.
$$

This series diverges to $\infty \in \hat{\mathbf{C}}$ since its partial sums S_k satisfy

$$
S_{2k-1} = -\frac{1}{3} \sum_{n=0}^{k-1} \left\{ \prod_{m=1}^{2n} \frac{2m+4}{2m+3} - \prod_{m=1}^{2n+1} \frac{2m+4}{2m+3} \right\}
$$

$$
= \frac{1}{3} \sum_{n=0}^{k-1} \frac{1}{4n+5} \prod_{m=1}^{2n} \frac{2m+4}{2m+3} \to +\infty
$$

and

$$
S_{2k} = -\frac{1}{3} + \frac{1}{3} \sum_{n=1}^{k} \left\{ \prod_{m=1}^{2n-1} \frac{2m+4}{2m+3} - \prod_{m=1}^{2n} \frac{2m+4}{2m+3} \right\}
$$

$$
= -\frac{1}{3} - \frac{1}{3} \sum_{n=1}^{k} \frac{1}{4n+3} \prod_{m=1}^{2n-1} \frac{2m+4}{2m+3} \to -\infty.
$$

Hence $\mathbf{K}(a_n/1)$ converges to $-X_0/X_{-1} = 3$.

$\diamondsuit$

Corresponding results for the adjoint recurrence relation

$$
P_n = b_n P_{n+1} + a_{n+1} P_{n+2} \quad \text{where } a_n, b_n \in \mathbf{C}, \, a_n \neq 0 \quad \text{for } n = 1, 2, 3, \ldots
$$
$$
(2.2.6)
$$

are obtained by use of Theorem 5A. The analogue to Auric's theorem, Theorem 10 is

Corollary 11 *Let $b_0 + \mathbf{K}(a_n/b_n)$ be a continued fraction with elements $a_n, b_n \in \mathbf{F}$ and all $a_n \neq 0$, and let $\{P_n\}$ be a solution of the corresponding recurrence relation (2.2.6) such that all $P_n \neq 0$ and*

$$\sum_{n=1}^{\infty} \left(P_n P_{n+1} \prod_{m=1}^{n} (-a_m) \right)^{-1} = \infty . \tag{2.2.7}$$

Then $b_0 + \mathbf{K}(a_n/b_n)$ converges to the finite value P_0/P_1 and

$$f_k - \frac{P_0}{P_1} \sim C \left(\sum_{n=1}^{k+1} P_n P_{n+1} \prod_{m=1}^{n} (-a_m) \right)^{-1} \quad as \ n \to \infty \tag{2.2.8}$$

for some constant $C \in \mathbf{F} \backslash \{0\}$.

Proof : Let $X_n = P_{n+2} \prod_{j=1}^{n+1} (-a_j)$ for $n = -1, 0, 1, \dots$. Then $\{X_n\}$ is a solution of (2.2.1) by Theorem 5A. Moreover

$$\sum_{n=0}^{k} \frac{\prod_{m=1}^{n}(-a_m)}{X_{n-1} X_n} = \sum_{n=0}^{k} \frac{1}{P_{n+1} P_{n+2} \prod_{m=1}^{n+1}(-a_m)} \to \infty \quad as \ k \to \infty .$$
$$\tag{2.2.9}$$

Hence $\mathbf{K}(a_n/b_n)$ converges to $-X_0/X_{-1}$ by Theorem 10 and $b_0 + \mathbf{K}(a_n/b_n)$ converges to

$$b_0 - \frac{X_0}{X_{-1}} = b_0 + \frac{a_1 P_2}{P_1} = \frac{b_0 P_1 + a_1 P_2}{P_1} = \frac{P_0}{P_1} .$$

The estimate (2.2.8) follows from combining (2.2.5) and (2.2.9). ∎

3 Tail sequences once more

3.1 *Connection to recurrence relations*

Let $\mathbf{K}(a_n/b_n)$ be a continued fraction with elements a_n and b_n from $(\mathbf{F}, \|\cdot\|)$ with $a_n \neq 0$ for all n. Recall that $\{t_n\}_{n=0}^{\infty}$ is a *tail sequence* for $\mathbf{K}(a_n/b_n)$ if

$$t_{n-1} = a_n/(b_n + t_n) \quad \text{for } n = 1, 2, 3, \dots \tag{3.1.1}$$

with the usual interpretation if $t_{n-1} = 0$ or ∞. It is called a *right tail sequence* if $\mathbf{K}(a_n/b_n)$ converges to t_0.

Let $\{X_n\}$ be a non-trivial solution of the corresponding three-term recurrence relation

$$X_n = b_n X_{n-1} + a_n X_{n-2} \quad \text{for } n = 1, 2, 3, \ldots . \tag{3.1.2}$$

If $X_{n-1} \neq 0$, then we can divide this relation by X_{n-1}. Rearranging its terms gives us

$$-\frac{X_{n-1}}{X_{n-2}} = \frac{a_n}{b_n - X_n/X_{n-1}} \quad \text{for } n = 1, 2, 3, \ldots . \tag{3.1.3}$$

This equation is also valid (with the usual interpretation) if $X_{n-2} = 0$. What happens if $X_{n-1} = 0$? First we note that since $\{X_n\}$ is non-trivial and all $a_n \neq 0$, there are no two consecutive X_n which are both 0. Hence X_n/X_{n-1} is always well defined in $\hat{\mathbf{F}}$ for all n. Furthermore, the left side will be 0 and the right side $a_n/\infty = 0$. So, (3.1.3) holds without exceptions when $\{X_n\}$ is non-trivial and all $a_n \neq 0$. Comparison of (3.1.1) and (3.1.3) shows us that $\{-X_n/X_{n-1}\}_{n=0}^{\infty}$ is a tail sequence for $b_0 + \mathbf{K}(a_n/b_n)$. It is a right tail sequence if $\mathbf{K}(a_n/b_n)$ converges to $-X_0/X_{-1}$; i.e. if $\{X_n\}$ is a minimal solution of (3.1.2).

We say that $\{T_n\}_{n=0}^{\infty}$ is a sequence of *Perron-tails* for $b_0 + \mathbf{K}(a_n/b_n)$ if

$$T_{n-1} = b_{n-1} + \frac{a_n}{T_n} \quad \text{for } n = 1, 2, 3, \ldots . \tag{3.1.4}$$

Also here we allow $T_n = \infty$. We say that $\{T_n\}$ is a sequence of *right Perron-tails* if $b_0 + \mathbf{K}(a_n/b_n)$ converges to T_0.

As earlier, let us assume that all $a_n \neq 0$. Then $\{T_n\}$ is connected with the three-term recurrence relation

$$P_n = b_n P_{n+1} + a_{n+1} P_{n+2} \quad \text{for } n = 0, 1, 2, \ldots \tag{3.1.5}$$

in the following way. Let $\{P_n\}$ be a non-trivial solution of (3.1.5). Then, as before, P_n/P_{n+1} is well defined in $\hat{\mathbf{F}}$ for all n, and rearranging (3.1.5) shows that $\{P_n/P_{n+1}\}_{n=0}^{\infty}$ is a sequence of Perron-tails for $b_0 + \mathbf{K}(a_n/b_n)$. It is a sequence of right Perron-tails if $\{P_n\}$ is a minimal solution of (3.1.5) so that $b_0 + \mathbf{K}(a_n/b_n)$ converges to P_0/P_1.

3.2   *Minimal solutions and value sets*

So far we have seen two methods to determine whether a solution $\{X_n\}$ of the recurrence relation

$$X_n = b_n X_{n-1} + a_n X_{n-2}, \quad a_n, b_n \in \mathbf{C}, \quad a_n \neq 0 \quad \text{for } n \in \mathbf{N} \quad (3.2.1)$$

is minimal or not. Either by comparing $\{X_n\}$ to another solution $\{Y_n\}$ to see if $X_n/Y_n \to 0$, or by using Theorem 9. Here comes a third one for the special case when $(\mathbf{F}, \|\cdot\|) = (\mathbf{C}, |\cdot|)$. It is often easy to apply if we already know that $\mathbf{K}(a_n/b_n)$ converges generally, and if we know a sequence $\{V_n\}$ of value sets for $\mathbf{K}(a_n/b_n)$ with $\bar{V}_0 \neq \hat{\mathbf{C}}$. ($\bar{A}$ denotes the closure of a set A in $\hat{\mathbf{C}}$.) So assume that this is so.

We plan to use Theorem 9 from Chapter III, which essentially says that if V_n is not "to small" for large n, and $w_n \in V_n$ has a positive distance to the boundary ∂V_n of V_n, uniformly with respect to n, then

$$\lim_{n \to \infty} S_n(w_n) \to f^{(0)} \quad \text{and} \quad f^{(n)} \in \bar{V}_n \quad \text{for all } n, \quad (3.2.2)$$

where $f^{(n)}$ is the value of the nth tail of $\mathbf{K}(a_n/b_n)$. This leads to the following strategy: Form the tail sequence $t_n = -X_n/X_{n-1}$. If $t_n \notin \bar{V}_n$ for arbitrary large n, then $\{X_n\}$ is dominant. If $t_n \in V_n$ for all n and has a positive distance to the boundary ∂V_n (measured in the chordal metric), uniformly with respect to n, then $\{X_n\}$ is minimal since $S_n(t_n) = t_0 \to f^{(0)}$. Let us see how this works in an example:

Example 9 We want to prove that the solution

$$X_n = \prod_{j=0}^{n} \left(\frac{1}{4} - \frac{8}{j+1} \right)$$

of the recurrence relation

$$X_n = \left(2 - \frac{8}{n+1} \right) X_{n-1} + \left(-\frac{7}{16} + \frac{14}{n} \right) X_{n-2} \quad \text{for } n = 1, 2, 3, \dots$$

is a minimal solution. The corresponding continued fraction $\mathbf{K}(a_n/b_n)$ satisfies

$$|b_n| = \left| 2 - \frac{8}{n+1} \right| \geq |a_n| + 1 = \left| -\frac{7}{16} + \frac{14}{n} \right| + 1$$

from some n on. Hence we know from the Slészyński-Pringsheim theorem (Theorem 1 in Chapter I) that $\mathbf{K}(a_n/b_n)$ converges and that the unit disk $V = \{w \in \mathbf{C}; |w| < 1\}$ is a value set for some tail of $\mathbf{K}(a_n/b_n)$. We form the tail sequence

$$t_n = -X_n/X_{n-1} = -\frac{1}{4} + \frac{8}{n+1}.$$

Since $t_n \in V$, bounded away from the boundary ∂V from some n on, it follows that $\{X_n\}$ is minimal.

$\diamondsuit$

In some cases we know in addition that $\{S_n(\bar{V}_n)\}$ converges to a one-point set (the limit point case). Then $\{X_n\}$ is minimal if $t_n \in \bar{V}_n$ from some n on (and thus for all n). Take for instance the parabola theorem, Theorem 20 in Chapter III:

If all a_n are contained in the parabolic region P_α for a fixed angle α and $\sum(n|a_n|^{-1}) = \infty$, then $S_n(\bar{V}_\alpha)$ converges to a one-point set, where V_α is the halfplane which is the value set for P_α. In other words, *if $\{X_n\}_{n=-1}^{\infty}$ is a solution of*

$$X_n = X_{n-1} + a_n X_{n-2} \quad \text{for } n = 1, 2, 3, \dots$$

where all $a_n \in P_\alpha$ and $\sum(n|a_n|)^{-1} = \infty$, and $t_n = -X_n/X_{n-1} \in \bar{V}_\alpha$ for all n, then $\{X_n\}$ is minimal.

Example 10 $\{X_n\}_{n=-1}^{\infty}$ where $X_{2n-1} = -1$, $X_{2n} = n + 1$ is a solution of

$$X_n = X_{n-1} + a_n X_{n-2} \quad \text{for } n = 1, 2, 3, \dots$$

where $a_{2n} = 1 + 2/n$ and $a_{2n+1} = n + 2$ for all n. Since $a_n \in P_\alpha$ for all n, $\sum(n|a_n|)^{-1} = \infty$ and $t_n = -X_n/X_{n-1} > 0$ (and thus $\in V_0$), it follows that $\{X_n\}$ is minimal.

$\diamondsuit$

3.3   Tails and convergence

The connection between tails and solutions of the corresponding recurrence relations makes it possible to state Pincherle's and Auric's theorems in terms of tails. The first result is based on Pincherle's theorem, Theorem 7:

Theorem 12 *Let $\{t_n\}$ and $\{u_n\}$ be two tail sequences with finite elements for the continued fraction $\mathbf{K}(a_n/b_n)$, where $t_0 \neq u_0$. Then:*

(A) $\mathbf{K}(a_n/b_n)$ *converges if and only if the limit*

$$\lim_{n \to \infty} R_n \quad where \ R_n = \prod_{k=0}^{n} t_k/u_k$$

exists in $\hat{\mathbf{F}}$.

(B) *If* $\lim R_n = R$, *then* $\mathbf{K}(a_n/b_n)$ *converges to*

$$f = \frac{t_0 - Ru_0}{1 - R}$$

with the usual interpretations if $R = 1$ or $R = \infty$, and its speed of convergence is given by

$$\begin{aligned} f - f_n &\sim C_1(R_n - R) & as \ n \to \infty & \quad if \ R \neq \infty, 1, \\ f - f_n &\sim C_2/R_n & as \ n \to \infty & \quad if \ R = \infty, \\ f_n &\sim C_3/(R_n - 1) & as \ n \to \infty & \quad if \ R = 1 \quad and \ thus \ f = \infty \end{aligned}$$

for some constants C_1, C_2 and C_3 from $\mathbf{F} \setminus \{0\}$.

Proof : We first observe that since all $t_n \neq \infty$ and $u_n \neq \infty$, we also have all $t_n \neq 0$ and $u_n \neq 0$ since $t_n = a_{n+1}/(b_{n+1} + t_{n+1})$ and $u_n = a_{n+1}/(b_{n+1} + u_{n+1})$. Hence, R_n is well defined. Next we know that there exist solutions $\{X_n\}$ and $\{Y_n\}$ of the recurrence relation (3.1.2) such that $t_n = -X_n/X_{n-1}$ and $u_n = -Y_n/Y_{n-1}$ for all n. Clearly, all X_n and Y_n are non-zero. We have

$$\frac{X_n}{Y_n} = \frac{X_{-1}}{Y_{-1}} \prod_{j=0}^{n} \frac{-X_j/X_{j-1}}{-Y_j/Y_{j-1}} = \frac{X_{-1}}{Y_{-1}} \prod_{j=0}^{n} \frac{t_j}{u_j} = \frac{X_{-1}}{Y_{-1}} R_n. \tag{3.3.1}$$

Moreover, $\{X_n\}$ and $\{Y_n\}$ are linearly independent since $t_0 \neq u_0$.

(A): This part follows directly from (3.3.1), Theorem 6 and Pincherle's theorem, Theorem 7.

(B): If $R = \infty$, then $\{Y_n\}$ is minimal and $\mathbf{K}(a_n/b_n)$ converges to $f = -Y_0/Y_{-1} = u_0$ by Theorem 7B. Otherwise, $\{X_n - Y_n R X_{-1}/Y_{-1}\}$ is a minimal solution, and $\mathbf{K}(a_n/b_n)$ converges to

$$f = -\frac{X_0 - Y_0 R X_{-1}/Y_{-1}}{X_{-1} - Y_{-1} R X_{-1}/Y_{-1}} = \frac{t_0 - R u_0}{1 - R}.$$

Further it follows from Theorem 7C that as $n \to \infty$

$$f_n - f \;\sim\; C_1'(Y_n/X_n) = \tilde{C}_1/R_n$$
$$\text{if } \{Y_n\} \text{ is minimal; i.e. } R = \infty,$$

$$f_n - f \;\sim\; C_2'(X_n/Y_n) = \tilde{C}_2 R_n$$
$$\text{if } \{X_n\} \text{ is minimal; i.e. } R = 0,$$

$$f_n - f \;\sim\; C_3'\left(\frac{X_n - Y_n R X_{-1}/Y_{-1}}{X_n}\right) = C_3'\left(1 - \frac{R}{R_n}\right)$$
$$\sim\; \tilde{C}_3(R_n - R) \quad \text{if } R \neq 0, \infty, 1$$

and

$$f_n \;\sim\; C_4'\left(\frac{X_n}{X_n - Y_n X_{-1}/Y_{-1}}\right) = C_4'\left(\frac{R_n}{R_n - 1}\right) \sim \frac{\tilde{C}_4}{R_n - 1}$$
$$\text{if } R = 1 \quad \text{and thus } f = \infty.$$

∎

Remarks:

1. A similar result is also valid for Perron-tails $\{T_n\}$ and $\{U_n\}$ of $b_0 + \mathbf{K}(a_n/b_n)$, since $T_n \neq \infty$ can be written $T_n = b_n + t_n$ and $U_n \neq \infty$ can be written $U_n = b_n + u_n$ where $\{t_n\}$ and $\{u_n\}$ are as in Theorem 12. Hence the value f of $b_0 + \mathbf{K}(a_n/b_n)$ is given by

$$f = \frac{T_0 - R U_0}{1 - R}$$

and

$$R_n = \prod_{k=0}^{n} \frac{t_k}{u_k} = \prod_{k=0}^{n} \frac{\dfrac{a_{k+1}}{b_{k+1} + t_{k+1}}}{\dfrac{a_{k+1}}{b_{k+1} + u_{k+1}}} = \prod_{k=1}^{n+1} \frac{U_k}{T_k}.$$

2. In Problem 12 you are asked to prove that

$$f - f_n = (t_0 - u_0) \cdot \frac{R - R_n}{(1 - R)(1 - R_n)} \quad \text{if } R \neq 1,$$

and

$$f_n = \frac{t_0 - R_n u_0}{1 - R_n},$$

where $R = \lim R_n$ and R_n is as in Theorem 12.

Example 11 Let $a < b$ and $d > 0$ be arbitrary real constants, such that $a + nd \neq 0, b + nd \neq 0$ for all nonnegative integers n. We shall prove that the continued fraction

$$a + b - \frac{ab}{a+b+d-} \frac{(a+d)(b+d)}{a+b+3d} - \frac{(a+2d)(b+2d)}{a+b+5d} - \cdots$$

converges to b. We find that $\{T_n\}$ and $\{U_n\}$ given by

$$T_0 = b, \quad U_0 = a, \quad T_n = b + nd - d, \quad U_n = a + nd - d \quad \text{for } n = 1, 2, 3, \ldots$$

are Perron-tails for this continued fraction. Since

$$R_n = \prod_{k=1}^{n+1} \frac{U_k}{T_k} = \prod_{k=1}^{n+1} \left(1 - \frac{b-a}{b+kd-d}\right) \to R = 0 \quad \text{as } n \to \infty,$$

the continued fraction converges to

$$f = \frac{T_0 - R U_0}{1 - R} = T_0 = b.$$

$\diamond$

If we only know one sequence of tails for $b_0 + \mathbf{K}(a_n/b_n)$, we are in a situation where Auric's theorem can be used. For tails it takes the form:

Theorem 13 *Let $\{t_n\}$ with all $t_n \neq \infty$, be a tail sequence for the continued fraction $\mathbf{K}(a_n/b_n)$ such that all $a_n \neq 0$. Then $\mathbf{K}(a_n/b_n)$ converges if and only if the limit*

$$\lim_{n \to \infty} R_n, \quad \text{where } R_n = \sum_{k=0}^{n} \prod_{j=1}^{k} \kappa_j \quad \text{and } \kappa_j = \frac{b_j + t_j}{-t_j}$$

exists in $\hat{\mathbf{F}}$. If $\lim R_n = R \in \hat{\mathbf{F}}$, then $\mathbf{K}(a_n/b_n)$ converges to

$$f = t_0(1 - 1/R)$$

with the usual interpretation if $R = 0$ or $R = \infty$, and

$$
\begin{aligned}
f - f_n &= t_0(1/R_n - 1/R) &&\text{if } R \neq 0\,,\\
f_n &= t_0(1 - 1/R_n) &&\text{if } R = 0 \quad \text{and thus } f = \infty\,.
\end{aligned}
$$

This was proved in [Waad84]. To prove Theorem 13 we shall use the formulas

$$B_n + B_{n-1} t_n = \prod_{j=1}^{n}(b_j + t_j) \quad \text{for } n = 0, 1, 2, \ldots. \tag{3.3.2}$$

$$A_n - B_n t_0 = \prod_{j=0}^{n}(-t_j) \quad \text{for } n = -1, 0, 1, 2, \ldots. \tag{3.3.3}$$

$$B_n = \sum_{k=0}^{n}\left(\prod_{j=1}^{k}(b_j + t_j) \cdot \prod_{j=k+1}^{n}(-t_j)\right)$$
$$\text{for } n = -1, 0, 1, 2, \ldots. \tag{3.3.4}$$

for the canonical numerators A_n and denominators B_n of $\mathbf{K}(a_n/b_n)$ when $\{t_n\}$ is a tail sequence with finite elements. These formulas can be proved by straight forward induction on n. (See also Problem 3 in Chapter II.) Similar expressions for $A_n + A_{n-1} t_n$ and for A_n are easy to derive since $A_n = a_1 B_{n-1}^{(1)}$ where $\{B_k^{(1)}\}$ are the canonical denominators of

$$\mathbf{K}(a_{n+1}/b_{n+1}) = \frac{a_2}{b_2 +} \frac{a_3}{b_3 +} \frac{a_4}{b_4 +} \cdots.$$

Proof of Theorem 13: By using $(3.3.3)$–$(3.3.4)$ we find that

$$\frac{A_n}{B_n} - t_0 = \frac{A_n - B_n t_0}{B_n} = \frac{\prod_{j=0}^n (-t_j)}{\sum_{k=0}^n \left(\prod_{j=1}^k (b_j + t_j) \prod_{j=k+1}^n (-t_j) \right)} = \frac{-t_0}{R_n}.$$

This proves the assertions. ∎

Remark: The corresponding result for Perron-tails $\{T_n\}$ follows immediately by using the connection $T_n = b_n + t_n$. In particular we then have

$$\kappa_j = -T_j / (T_j - b_j).$$

Example 12 The continued fraction

$$x - a_1 + \mathop{\mathbf{K}}_{n=1}^{\infty} \frac{a_n x}{x - a_{n+1}} = x - a_1 + \frac{a_1 x}{x - a_2 +} \frac{a_2 x}{x - a_3 +} \cdots \qquad \text{for } x \neq 0, \ a_n \neq 0,$$

is one of the many continued fractions studied by Ramanujan. We find that $t_n = a_{n+1}$ is a sequence of tails. Hence, the continued fraction converges if and only if

$$R_n = \sum_{k=0}^n \prod_{j=1}^k \kappa_j \to R; \quad \kappa_j = \frac{x - a_{j+1} + a_{j+1}}{-a_{j+1}} = -\frac{x}{a_{j+1}},$$

that is, if and only if

$$R_n = \sum_{k=0}^n \left((-x)^k \bigg/ \prod_{j=2}^{k+1} a_j \right) \to R$$

for some $R \in \hat{\mathbf{F}}$. For instance, if $\{a_j\}$ is bounded, real and alternating in sign, then $R = \infty$ and the continued fraction converges to $x - a_1 + t_0 = x$ for all $x > \limsup |a_j|$. ◇

4   An application to linear recurrence relations

4.1   Forward stability of recurrence relations

Let us consider the three-term recurrence relation

$$X_n = b_n X_{n-1} + a_n X_{n-2}\,; \quad a_n, b_n \in \mathbf{C}, \quad a_n \neq 0 \quad \text{for } n = 1, 2, 3, \dots,$$

which we assume has a minimal solution $\{X_n\}$. Assume further that the two first elements of this solution, X_{-1} and X_0, are known, and that we want to compute X_n for $n \geq 1$. A simple method seems to be to use the recurrence relation directly, and compute $X_1, X_2, \dots$ recursively.

However, this method does not work in practice. The computation is unstable; i.e. roundoff errors "blow up" when we deal with minimal solutions. This can be seen by the following argument. Assume that all the computations we are doing are totally accurate, and that X_0 is given with its exact value. But in X_{-1} we have a small roundoff error, such that we begin with the values $\hat{X}_{-1} = X_{-1} + \epsilon$ and $\hat{X}_0 = X_0$. Then we are getting a sequence $\{\hat{X}_n\}$ which is no longer minimal, since it is not proportional to $\{X_n\}$. If it is not minimal (and not trivial), it has to be dominant; i.e. $X_n / \hat{X}_n \to 0$). The relative error for our values $\hat{X}_n$ (after our exact computations) will therefore blow up:

$$\left| \frac{\hat{X}_n - X_n}{X_n} \right| \to \infty.$$

We need another method to compute $\{X_n\}$. For dominant solutions this forward computation is in general stable. Under the same assumptions as above, only with $\{X_n\}$ dominant instead of minimal, we have the relative error

$$\left| \frac{\hat{X}_n - X_n}{X_n} \right| = \left| \frac{X_n - \alpha X_n - \beta \tilde{X}_n}{X_n} \right| \to |1 - \alpha|,$$

where $\{\tilde{X}_n\}$ denotes a minimal solution and $\hat{X}_n = \alpha X_n + \beta \tilde{X}_n$.

Example 13   The three-term recurrence relation

$$X_n = X_{n-1} + X_{n-2} \quad \text{for } n = 1, 2, 3, \dots \tag{4.1.1}$$

has the minimal solution $\{((1 - \sqrt{5})/2)^{n+1}\}_{n=-1}^{\infty}$. (See Example 1.) The following table shows the actual value of $X_n = ((1 - \sqrt{5})/2)^{n+1}$, and what we get when we use the recurrence relation to compute $\{X_n\}$. The computation is done with 4 decimals precision.

n	$X_n = ((1 - \sqrt{5})/2)^{n+1}$	$X_n = X_{n-1} + X_{n-2}$
-1	1.0000	
0	-0.6180	
1	0.3820	0.3820
2	-0.2361	-0.2360
3	0.1459	0.1460
4	-0.0902	-0.0900
5	0.0557	0.0560
6	-0.0344	-0.0340
7	0.0213	0.0220
8	-0.0132	-0.0120
9	0.0081	0.0100
10	-0.0050	-0.0020
11	0.0031	0.0080
12	-0.0019	0.0060
13	0.0012	0.0140
14	-0.0007	0.0200
15	0.0005	0.0340
16	-0.0003	0.0540
17	0.0002	0.0880
18	-0.0001	0.1420
19	0.0001	0.2300
20	-0.0000	0.3720

The value of X_n keeps on decreasing in absolute value, whereas the value obtained by use of the recurrence relation, will stay positive and keep on increasing, "faster and faster", as n increases from $n = 12$ on.

$\diamond$

What one therefore can do is to compute two linearly independent dominant solutions $\{Y_n\}$ and $\{Z_n\}$ and then find the minimal solution as a linear combination of these. This requires that we know, say, X_{-1} and X_0.

4.2   A method for computing minimal solutions

We want to compute the first terms of a minimal solution $\{X_n\}$ of the three-term recurrence relation.

$$X_n = b_n X_{n-1} + a_n X_{n-2}\,; \quad a_n, b_n \in \mathbf{C}, \quad a_n \neq 0, \quad \text{for } n = 1, 2, 3, \ldots.$$
$$(4.2.1)$$

(We assume that such a solution exists.) Strictly speaking, $\{X_n\}$ is uniquely determined if we choose one of its elements X_n, say X_{-1}. But how can we find this solution? Pincherle's theorem, Theorem 7, tells us that the continued fraction $\mathbf{K}(a_n/b_n)$ connected with (4.2.1) converges to $-X_0/X_{-1}$. Similarly, using Pincherle's theorem on the Nth tail

$$\underset{n=1}{\overset{\infty}{\mathbf{K}}}\left(a_{N+n}/b_{N+n}\right) = \frac{a_{N+1}}{b_{N+1}} + \frac{a_{N+2}}{b_{N+2}} + \cdots \qquad (4.2.2)$$

of $\mathbf{K}(a_n/b_n)$ shows that this tail converges to $-X_N/X_{N-1}$. We can therefore compute X_n (approximately) by the following method:

1) Compute approximants $f_k^{(N)}$ or $S_k^{(N)}(w_{N+k})$ of (4.2.2) for sufficiently large k, for $N = 0, 1, 2, \ldots, n$.

2) Set $-X_N/X_{N-1} \approx f_k^{(N)}$ (or $S_k^{(N)}(w_{N+k})$).

3) Compute X_n from the relation

$$X_n = X_{-1} \prod_{N=0}^{n} X_N/X_{N-1}$$

where X_{-1} is given.

This method was suggested by Gautschi, [Gaut67].

Example 14 To see how this turns out in practice, let us return to the problem in Example 13, where the minimal solution $\{X_n\}$ is actually known in advance. The continued fraction corresponding to (4.1.1) is

$$\mathbf{K}(1/1) = \frac{1}{1} + \frac{1}{1} + \frac{1}{1} + \cdots.$$

Using its 10th approximant f_{10} to approximate its value, we get

$$f = -X_0/X_{-1} \approx f_{10} = \frac{55}{89} \approx 0.61798.$$

Indeed, we also have

$$f^{(N)} = -X_N/X_{N-1} \approx f_{10}^{(N)} = f_{10} = \frac{55}{89} \approx 0.61798.$$

Choosing $X_{-1} = 1$ we thus get that

$$X_n = X_{-1} \prod_{N=0}^{n} X_N/X_{N-1} \approx (-f_{10})^{n+1} = \left(-\frac{55}{89}\right)^{n+1} \approx (-0.61798)^{n+1},$$

which is very close to the exact value $((1-\sqrt{5})/2)^{n+1}) \approx (-0.61803)^{n+1}$. Of course, using the exact value $f = f^{(N)} = (1-\sqrt{5})/2$ for the value of $\mathbf{K}(1/1)$ leads to the exact values for X_n.

$\diamond$

Minimal solutions $\{P_n\}$ of the adjoint recurrence relation

$$P_n = b_n P_{n+1} + a_{n+1} P_{n+2} \quad \text{for } n = 0, 1, 2, \ldots,$$

can be computed similarly, since then

$$\frac{P_N}{P_{N+1}} = b_N + \frac{a_{N+1}}{b_{N+1}+} \frac{a_{N+2}}{b_{N+2}+\cdots} \quad \text{for } N = 0, 1, 2, \ldots.$$

Example 15 We want to compute the integral

$$I(a, b; x) = \int_0^\infty \frac{e^{-u} u^{a-1}}{(1+xu)^b} du \quad \text{where } b \in \mathbf{N}_0 \quad \text{and } a > 0, x > 0.$$

For $b = 0$ it reduces to the gamma function

$$I(a, 0; x) = \int_0^\infty e^{-u} u^{a-1} du = \Gamma(a).$$

To get a value if $b > 0$ we observe that

$$I(a, b; x) = \int_0^\infty \frac{e^{-u} u^{a-1}(1+xu)}{(1+xu)^{b+1}} du = I(a, b+1; x) + xI(a+1, b+1; x).$$

Further, by integration by part, we get (when we first integrate u^{a-1}) that

$$I(a,b+1;x) = \left[\frac{e^{-u}u^a/a}{(1+xu)^{b+1}}\right]_0^\infty$$

$$-\frac{1}{a}\int_0^\infty \frac{-e^{-u}u^a(1+xu)^{b+1} - e^{-u}u^a(b+1)x(1+xu)^b}{(1+xu)^{2b+2}}\,du$$

$$= \frac{1}{a}I(a+1,b+1;x) + \frac{b+1}{a}xI(a+1,b+2;x).$$

Hence, $\{P_n(x)\}$ given by

$$P_{2n}(x) = I(a+n,b+n;x), \qquad P_{2n+1}(x) = I(a+n,b+n+1;x),$$

satisfies the linear three-term recurrence relation

$$P_n(x) = b_n P_{n+1}(x) + a_{n+1} x P_{n+2}(x) \quad \text{for } n = 0,1,2,\dots,$$

where

$$b_{2n} = 1, \qquad b_{2n+1} = \frac{1}{a+n}, \qquad a_{2n} = \frac{b+n}{a+n-1}, \qquad a_{2n+1} = 1.$$

If $a \ge b > 0$, one idea could therefore be to compute $I(a,b;x)$ by means of this recurrence relation, beginning with $I(a-b,0;x) = \Gamma(a-b)$. There are however two problems connected with this procedure:

1) We do not know the value of $I(a-b,1;x)$.

2) $\{P_n(x)\}$ is a minimal solution of the recurrence relation, so the calculation is unstable. (This follows from Theorem 13.)

The continued fraction technique takes care of both these problems. We get

$$\frac{I(a+N,b+N;x)}{I(a+N,b+N+1;x)}$$

$$= 1 + \cfrac{x}{a+N} + \cfrac{\dfrac{b+N+1}{a+N}x}{1} + \cfrac{x}{a+N+1} + \cfrac{\dfrac{b+N+2}{a+N+1}x}{1} + \dots$$

$$\approx 1 + \frac{(a+N)x}{1} + \frac{(b+N+1)x}{1} + \frac{(a+N+1)x}{1} + \frac{(b+N+2)x}{1} + \dots,$$

and

$$\frac{I(a+N, b+N+1; x)}{I(a+N+1, b+N+1; x)} = \frac{1}{a+N} + \frac{\dfrac{b+N+1}{a+N}x}{1} + \cfrac{x}{1+\dfrac{1}{a+N+1}} + \cdots$$

$$\approx \frac{1}{a+N}\left\{1 + \frac{(b+N+1)x}{1} + \frac{(a+N+1)x}{1} + \frac{(b+N+2)x}{1} + \cdots\right\}.$$

So let us for instance find an approximate value for

$$I(5, 3; 1) = \int_0^\infty \frac{e^{-u}u^4}{(1+u)^3}du.$$

We find that

$$I(5, 3; 1)$$

$$= I(3, 0; 1)\left\{\frac{I(3, 0; 1)}{I(3, 1; 1)} \cdot \frac{I(3, 1; 1)}{I(4, 1; 1)} \cdot \frac{I(4, 1; 1)}{I(4, 2; 1)} \cdot \frac{I(4, 2; 1)}{I(5, 2; 1)} \cdot \frac{I(5, 2; 1)}{I(5, 3; 1)}\right\}^{-1},$$

where $I(3, 0; 1) = \Gamma(3) = 2! = 2$,

$$\frac{I(3+N, N; 1)}{I(3+N, N+1; 1)}$$
$$= 1 + \frac{3+N}{1} + \frac{N+1}{1} + \frac{4+N}{1} + \frac{N+2}{1} + \frac{5+N}{1} + \frac{N+3}{1} + \cdots$$

for $N = 0, 1, 2$ and

$$\frac{I(3+N, N+1; 1)}{I(4+N, N+1; 1)}$$
$$= \frac{1}{3+N}\left\{1 + \frac{N+1}{1} + \frac{4+N}{1} + \frac{N+2}{1} + \frac{5+N}{1} + \frac{N+3}{1} + \cdots\right\}$$

for $N = 0, 1$. Using the 20th approximants of these continued fractions to approximate their values gives

$$I(5, 3; 1) \approx 2\{3.3534 \cdot 0.42483 \cdot 3.6410 \cdot 0.37841 \cdot 3.8888\}^{-1} \approx 0.26202.$$

However, here we could have saved some work by using several different continued fractions instead of several tails of the same continued fraction. We have

$$I(5, 3; 1) = \frac{I(5, 3; 1)}{I(5, 2; 1)} \cdot \frac{I(5, 2; 1)}{I(5, 1; 1)} \cdot \frac{I(5, 1; 1)}{I(5, 0; 1)} \cdot \Gamma(5),$$

where

$$\frac{I(5, N; 1)}{I(5, N+1; 1)} = 1 + \frac{5}{1+} \frac{N+1}{1} \frac{6}{+1+} \frac{N+2}{1} \frac{7}{+1+} \frac{N+3}{1} +\cdots.$$

Using the 20th approximants for $N = 0, 1$ and 2 gives us

$$I(5, 3; 1) \approx \{3.8888 \cdot 4.5126 \cdot 5.2212\}^{-1} \cdot 4! \approx 0.26194.$$

---◇

5   Some generalizations of continued fractions

5.1   Introduction

We have seen that a continued fraction $\mathbf{K}(a_n/b_n)$ is closely related to the three-term recurrence relation

$$X_n = b_n X_{n-1} + a_n X_{n-2} \quad \text{for } n = 1, 2, 3, \ldots. \tag{5.1.1}$$

We have:

(i) $\{t_n\}$ is a tail sequence for $\mathbf{K}(a_n/b_n)$ if and only if there exists a nontrivial solution $\{X_n\}$ of (5.1.1) such that $t_n = -X_n/X_{n-1}$ for all n.

(ii) The approximants of $\mathbf{K}(a_n/b_n)$ can be written on the form $f_n = A_n/B_n$, where $\{A_n\}$ and $\{B_n\}$ are solutions of (5.1.1) with initial values

$$\begin{bmatrix} A_{-1} & A_0 \\ B_{-1} & B_0 \end{bmatrix} = \begin{bmatrix} 1 & 0 \\ 0 & 1 \end{bmatrix}.$$

Can we get something similar if we have a "longer" recurrence relation, say four-term? Or more generally, an $(N+1)$-term linear recurrence relation

$$\sum_{k=0}^{N} a_n^{(N-k+1)} X_{n-k} = 0, \quad \text{where } a_n^{(N+1)} a_n^{(1)} \neq 0, \quad \text{for } n = 1, 2, 3, \ldots.$$

$$\tag{5.1.2}$$

For $N > 2$ we can not combine (i) and (ii). We have to settle for one or the other.

5.2   G-continued fractions

Let us first introduce a generalized form of continued fractions which is connected with (5.1.2) in the sense of (i). The idea is to extend Gautschi's continued fraction method for computing minimal solutions to this longer recurrence relation. Let all $a_n^{(N+1)} = 1$. For the case $N = 2$ with $a_n^{(1)} = -a_n$, $a_n^{(2)} = -b_n$, the recurrence relation takes the familiar form

$$X_n = b_n X_{n-1} + a_n X_{n-2} \quad \text{where } a_n \neq 0, \quad \text{for } n = 1, 2, 3, \dots .$$

For a non-trivial solution $\{X_n\}$ we thus have that

$$-\frac{X_{n-1}}{X_{n-2}} = \frac{a_n}{b_n - X_n/X_{n-1}} = s_n(-X_n/X_{n-1}) \quad \text{for } n = 1, 2, 3, \dots,$$

where $s_n(w)$ is the linear fractional transformation $s_n(w) = a_n/(b_n + w)$. So, $\{-X_n/X_{n-1}\}$ is a tail sequence for the continued fraction $\mathbf{K}(a_n/b_n)$. In analogy with this, we have for $N > 0$,

$$X_n + a_n^{(N)} X_{n-1} + a_n^{(N-1)} X_{n-2} + \dots + a_n^{(1)} X_{n-N} = 0 \qquad (5.2.1)$$

where $a_n^{(1)} \neq 0$. Let $\{X_n\}$ be a solution with all $X_n \neq 0$. Dividing (5.2.1) by X_{n-N+1} leads to

$$-\frac{a_n^{(1)}}{X_{n-N+1}/X_{n-N}} = a_n^{(2)} + a_n^{(3)} \frac{X_{n-N+2}}{X_{n-N+1}} + a_n^{(4)} \frac{X_{n-N+2}}{X_{n-N+1}} \cdot \frac{X_{n-N+3}}{X_{n-N+2}}$$
$$+ \dots + \prod_{k=1}^{N-1} \frac{X_{n-N+k+1}}{X_{n-N+k}},$$

such that

$$\frac{X_{n-N+1}}{X_{n-N}} = \frac{-a_n^{(1)}}{\sum_{k=1}^{N} \left(a_n^{(k+1)} \prod_{j=1}^{k-1} \frac{X_{n-N+j+1}}{X_{n-N+j}} \right)} \qquad (5.2.2)$$

where $a_n^{(N+1)} = 1$.

Let us introduce the transformations s_n and S_n from $\mathbf{C}^{N-1}$ into $\hat{\mathbf{C}}$ given by

$$s_n(w_1,\ldots,w_{N-1}) = \frac{-a_n^{(1)}}{\sum_{k=1}^N \left(a_n^{(k+1)} \prod_{j=1}^{k-1} w_j\right)} \quad \text{for } n = 1,2,3,\ldots,$$

$$S_1(w_1,\ldots,w_{N-1}) = s_1(w_1,\ldots,w_{N-1})$$

and

$$S_n(w_1,\ldots,w_{N-1}) = S_{n-1}(s_n(w_1,\ldots,w_{N-1}),w_1,\ldots,w_{N-2})$$

for $n = 2,3,4,\ldots$. Then (5.2.2) can be written

$$\frac{X_{n-N+1}}{X_{n-N}} = s_n\left(\frac{X_{n-N+2}}{X_{n-N+1}},\frac{X_{n-N+3}}{X_{n-N+2}},\ldots,\frac{X_n}{X_{n-1}}\right)$$

and

$$\begin{aligned}
\frac{X_{-N+2}}{X_{-N+1}} &= S_1\left(\frac{X_{-N+3}}{X_{-N+2}},\ldots,\frac{X_1}{X_0}\right)\\
&= S_1\left(s_2\left(\frac{X_{-N+4}}{X_{-N+3}},\ldots,\frac{X_2}{X_1}\right),\frac{X_{-N+4}}{X_{-N+3}},\ldots,\frac{X_1}{X_0}\right)\\
&= S_2\left(\frac{X_{-N+4}}{X_{-N+3}},\ldots,\frac{X_2}{X_1}\right)\\
&= \ldots = S_n\left(\frac{X_{-N+n+2}}{X_{-N+n+1}},\frac{X_{-N+n+3}}{X_{-N+n+2}},\ldots,\frac{X_n}{X_{n-1}}\right).
\end{aligned}$$

Following the idea of Levrie and Piessens [LePi87] we define the *G-continued fraction* (G in honor of W. Gautschi) $\mathbf{K}_G(-a_n^{(1)}/a_n^{(2)};\ldots;a_n^{(N)})$ by its approximations $f_n = S_n(0,0,\ldots,0)$. $\{X_{-N+n+2}/X_{-N+n+1}\}_{n=0}^\infty$ acts as a tail sequence for $\mathbf{K}_G(-a_n^{(1)}/a_n^{(2)};\ldots;a_n^{(N)})$. Note that for $N = 2$ we have $\mathbf{K}_G(-a_n^{(1)}/a_n^{(2)})$ defined by

$$s_n(w) = \frac{-a_n^{(1)}}{a_n^{(2)} + w} \quad \text{for } n = 1,2,3,\ldots$$

and

$$S_n(w) = \frac{-a_1^{(1)}}{a_1^{(2)}} + \frac{-a_2^{(1)}}{a_2^{(2)}} + \cdots + \frac{-a_n^{(1)}}{a_n^{(2)} + w},$$

whereas the classical continued fraction $\mathbf{K}(-a_n^{(1)}/-a_n^{(2)})$ has approximations $\tilde{S}_n(w)$ given by

$$\tilde{s}_n(w) = \frac{-a_n^{(1)}}{-a_n^{(2)} + w}$$

and

$$\tilde{S}_n(w) = \frac{-a_1^{(1)}}{-a_1^{(2)} +} \frac{-a_2^{(1)}}{-a_2^{(2)} +} + \cdots + \frac{-a_n^{(1)}}{-a_n^{(2)} + w}$$

$$\approx -\frac{-a_1^{(1)}}{a_1^{(2)} +} \frac{-a_2^{(1)}}{a_2^{(2)} +} + \cdots + \frac{-a_n^{(1)}}{a_n^{(2)} - w}$$

That is, $S_n(w) = -\tilde{S}_n(-w)$ and $\{X_n/X_{n-1}\}_{n=0}^{\infty}$ is a tail sequence for $\mathbf{K}_G(-a_n^{(1)}/a_n^{(2)})$ if and only if $\{-X_n/X_{n-1}\}_{n=0}^{\infty}$ is a tail sequence for $\mathbf{K}(-a_n^{(1)}/-a_n^{(2)})$. This change in sign is also reflected in the theorem below. We say that the G-continued fraction *converges* and has the *value* f, if $f = \lim f_n$ exists in $\hat{\mathbf{C}}$. For such continued fractions we have the following theorem of "Pincherle"-type due to Zahar [Zahar68]:

Theorem 14 *The G-continued fraction* $\mathbf{K}(-a_n^{(1)}/a_n^{(2)}; \ldots; a_n^{(N)})$ *converges if and only if the solution space of (5.2.1) has a basis* $\{X_n^{(1)}\}, \ldots, \{X_n^{(N)}\}$ *such that*

$$\lim_{n \to 0} \frac{\Delta_n^{(i)}}{\Delta_n} = 0 \quad \text{for } i = 1, 2, \ldots, N - 1, \qquad (5.2.3)$$

where

$$\Delta_n = \begin{vmatrix} X_n^{(1)} & \cdots & X_n^{(N-1)} \\ \vdots & & \vdots \\ X_{n-N+2}^{(1)} & \cdots & X_{n-N+2}^{(N-1)} \end{vmatrix},$$

and $\Delta_n^{(i)}$ *is the determinant we obtain by replacing column number i in* Δ_n, *by the column* $\left(X_n^{(N)}, \ldots, X_{n-N+2}^{(N)}\right)^t$.

If (5.2.3) holds, then the G-continued fraction has the value

$$f = X_{-N+2}^{(N)}/X_{-N+1}^{(N)}.$$

We shall not give the proof of this theorem, but rather indicate an application. The forward computation of a solution $\{X_n^{(N)}\}$ of (5.2.1), which satisfies (5.2.3), is unstable. Hence, we prefer to compute approximants $f_k^{(m)}$ of the G-continued fractions

$$\mathop{\mathbf{K}}_{n=1}^{\infty}\left(-a_{n+m}^{(1)}/a_{n+m}^{(2)};\ldots;a_{n+m}^{(N)}\right) \quad \text{for } m = 0, 1, 2, \ldots. \tag{5.2.4}$$

Since (5.2.4) converges to $X_{m+2-N}^{(N)}/X_{m+1-N}^{(N)}$, we can then use

$$X_{m+2-N}^{(N)} = X_{-N+1}^{(N)} \cdot \prod_{j=0}^{m} \frac{X_{j+2-N}^{(N)}}{X_{j+1-N}^{(N)}} \approx X_{-N+1}^{(N)} \prod_{j=0}^{m} f_k^{(j)}$$

for some suitable $k \in \mathbf{N}$. The computation of $f_k^{(m)}$ for given m and k can be done recursively by the formulas

$$f_1^{(m+k-1)} = S_1^{(m+k-1)}(0,\ldots,0) = s_{m+k}(0,\ldots,0) = -\frac{a_{k+m}^{(1)}}{a_{k+m}^{(2)}},$$

$$f_n^{(m+k-n)} = s_{m+k-n+1}(f_{n-N+1}^{(m+k-n+N-1)}, f_{n-N+2}^{(m+k-n+N-2)},$$
$$\ldots, f_{n-1}^{(m+k-n+1)}) \quad \text{for } n = 2, 3, \ldots, k,$$

where $f_j^{(n+k-j)} = 0$ for $j \geq 0$. (For $N = 2$ this corresponds to using the backwards recurrence algorithm to compute $\mathbf{K}(-a_n^{(1)}/-a_n^{(2)})$.) To improve these approximations one can also use modified approximants $S_n^{(m)}(w_{m+n}^{(1)}, w_{m+n}^{(2)}, \ldots, w_{m+n}^{(N-1)})$ instead of $f_n^{(m)} = S_n^{(m)}(0, 0, \ldots, 0)$.

5.3 Generalized (or vector valued) continued fractions

A *generalized continued fraction of dimension* $N - 1$, is given by its approximants

$$f_n = \begin{bmatrix} f_n^{(1)} \\ \vdots \\ f_n^{(N-1)} \end{bmatrix} = \begin{bmatrix} A_n^{(1)}/B_n \\ \vdots \\ A_n^{(N-1)}/B_n \end{bmatrix},$$

where $\{A_n^{(i)}\}$ and $\{B_n\}$ are solutions of the linear $(N+1)$-term recurrence relation

$$X_n = b_n X_{n-1} + a_n^{(N-1)} X_{n-2} + \cdots + a_n^{(1)} X_{n-N} \quad \text{for } n = 1, 2, 3, \ldots$$

where all $a_n^{(1)} \neq 0$, with initial values

$$\begin{bmatrix} A_{-N+1}^{(1)} & \cdots & A_0^{(1)} \\ \vdots & & \vdots \\ A_{-N+1}^{(N-1)} & \cdots & A_0^{(N-1)} \\ B_{-N+1} & \cdots & B_0 \end{bmatrix} = \begin{bmatrix} 1 & \cdots & 0 & 0 \\ \vdots & \ddots & \vdots & \vdots \\ 0 & \cdots & 1 & 0 \\ 0 & \cdots & 0 & 1 \end{bmatrix}.$$

Hence, this type of continued fraction is connected with this longer recurrence relation, in the sense of (ii) in Subsection *5.1*. Its approximants are $(N-1)$-dimensional vectors. We see that a generalized continued fraction belonging to a given $(N+1)$-term recurrence relation may fail to exist. (We may get $A_n^{(i)} = B_n = 0$ for some natural numbers n and $1 \leq i \leq N-1$.) This can be counteracted by using modified approximants

$$S_n(w) = \left(S_n^{(1)}(w_1, \ldots, w_{N-1}), \ldots, S_n^{(N-1)}(w_1, \ldots, w_{N-1}) \right)$$

where

$$S_n^{(i)}(w_1, \ldots, w_{N-1}) = \frac{A_n^{(i)} + A_{n-1}^{(i)} w_{N-1} + \ldots + A_{n-N+1}^{(i)} w_1}{B_n + B_{n-1} w_{N-1} + \ldots + B_{n-N+1} w_1}.$$

Such generalized continued fractions are used for simultaneous approximation of $(N-1)$ functions, when we require that the approximants are rational functions with common denominators. They are written $\mathbf{K}(a_n^{(1)}, \ldots, a_n^{(N-1)}; b_n)$.

Problems

(1) Show that the Tchebycheff polynomials

$$T_n(x) = \sum_{k=0}^{[n/2]} \binom{n}{2k} x^{n-2k}(x^2 - 1)^k \quad \text{for } n = 0, 1, 2, \ldots$$

of the first kind satisfy the three-term recurrence relation

$$T_n(x) = 2xT_{n-1}(x) - T_{n-2}(x) \quad \text{for } n = 2, 3, 4, \ldots.$$

(Here $[p]$ denotes the largest integer $\leq p$. Hint: Use that $T_n(x) = \cos n\theta$ where $x = \cos \theta$.)

(2) Show that $\{\Gamma(z+n)\}_{n=0}^{\infty}$ is a solution of the three-term recurrence relation

$$P_n(z) = \frac{2z + 2n + 1}{(z + n)^2} P_{n+1}(z) - \frac{1}{(z + n)^2} P_{n+2}(z) \text{ for } n = 0, 1, 2, \ldots,$$

where $z \in \mathbf{C}\backslash\{-1, -2, -3, \ldots\}$.

(3) Show that the integrals

$$I_n(x) = \int_0^{\infty} e^{-xt} \tanh^n t \, dt$$

for $x > 0$ and $n \in \mathbf{N}$, satisfy the recurrence relation

$$I_{n+1}(x) = -\frac{x}{n} I_n(x) + I_{n-1}(x) \quad \text{for } n \geq 1.$$

(Hint: Use integration by part.) Prove further that they satisfy

$$I_{n+2}(x) = \frac{2n^2 + x^2}{n^2 + n} I_n(x) - \frac{n^2 - n}{n^2 + n} I_{n-2}(x) \quad \text{for } n \geq 2.$$

(4) The incomplete gamma function is given by

$$G(a, x) = \frac{1}{\Gamma(a)} \int_0^x e^{-t} t^{a-1} dt, \quad \text{where } a > 1, x > 0.$$

Prove that

$$X_n(x) = G(a + n, x) \quad \text{for } n = 0, 1, 2, \ldots$$

is a solution of the three-term recurrence relation

$$X_{n+1}(x) = \left(1 + \frac{x}{a + n}\right) X_n(x) - \frac{x}{a + n} X_{n-1}(x) \text{ for } n = 1, 2, 3, \ldots.$$

(5) The function

$$I(\alpha, \beta; x) = \frac{1}{\Gamma(\beta)} \int_0^\infty \frac{e^{-u} u^{\beta-1}}{(1+xu)^\alpha} du, \quad \beta > 1, \quad \alpha \in \mathbf{R},$$

is well defined for $x > 0$. Show that $\{P_n(x)\}_{n=0}^\infty$, where

$$P_{2n}(x) = I(\alpha + n, \beta + n; x), \qquad P_{2n+1} = I(\alpha + n, \beta + n + 1; x),$$

is a solution of the recurrence relation

$$P_n(x) = P_{n+1}(x) + a_{n+1} x P_{n+2}(x) \quad \text{for } n = 0, 1, 2, \ldots,$$

where

$$a_{2n} = \beta + n, \qquad a_{2n+1} = \alpha + n.$$

(6) Let

$$G(z) = 1 + \sum_{k=1}^\infty \frac{q^{k^2} z^k}{(q)_k} \quad \text{for } |q| < 1, \ |z| < 1,$$

where $(q)_k = \prod_{j=1}^k (1 - q^j)$. Show that $\{G(zq^n)\}_{n=0}^\infty$ is a solution of the three-term recurrence relation

$$P_n(z) = P_{n+1}(z) + q^{n+1} z P_{n+2}(z) \quad \text{for } n = 0, 1, 2, \ldots.$$

(7) Show that

$$x \int_0^\infty e^{-xt} \tanh t \, dt = \frac{1}{x+} \frac{1 \cdot 2}{x+} \frac{2 \cdot 3}{x+} \frac{3 \cdot 4}{x+} \cdots$$

for $x > 0$. (Hint: Use the result from Problem 3.)

(8) Show that the ratio $G(a+1, x)/G(a, x)$, where $G(a, x)$ denotes the incomplete gamma function

$$G(a, x) = \frac{1}{\Gamma(a)} \int_0^x e^{-t} t^{a-1} dt \quad \text{for } a > 1, \quad x > 0,$$

has a T-fraction expansion of the form

$$\frac{G(a+1, x)}{G(a, x)} = - \mathop{\mathbf{K}}_{n=1}^\infty \left(\frac{F_n x}{1 - F_n x} \right), \quad \text{where } F_n = \frac{-1}{n+a}.$$

(Hint: Use the result from Problem 4.)

(9) Show that

$$z = 2z + 1 - \frac{(z+1)^2}{2z+3} - \frac{(z+2)^2}{2z+5} - \frac{(z+3)^2}{2z+7} - \cdots; \qquad -z \notin \mathbf{N_0},$$

by use of the result from Problem 2.

(10) Assume that the continued fraction

$$\frac{a_1 + h}{1} + \frac{a_1}{b} + \frac{a_2 + h}{1} + \frac{a_2}{b} + \cdots$$

converges. Prove that then

$$h + \frac{a_1}{1} + \frac{a_1 + h}{b} + \frac{a_2}{1} + \frac{a_2 + h}{b} + \cdots$$

converges to the same value if it converges.

(11) Show that the recurrence relation

$$X_n = 3X_{n-1} - 2X_{n-2} \quad \text{for } n = 0, 1, 2, \dots$$

has the general solution

$$X_n = C_1 + C_2 \cdot 2^n$$

where C_1 and C_2 are arbitrary complex constants.

What can therefore be said about the convergence/divergence of the 1-periodic continued fraction $\mathbf{K}(-2/3)$?

(12) Let $\{t_n\}$ and $\{u_n\}$ be two different tail sequences for the continued fraction $\mathbf{K}(a_n/b_n)$ with all $a_n \neq 0$, $t_n \neq \infty$ and $u_n \neq \infty$. Let further the limit

$$R = \lim_{n \to \infty} R_n \quad \text{where } R_n = \prod_{k=0}^{n} t_k / u_k$$

exist in $\hat{\mathbf{C}}$, so that $\mathbf{K}(a_n/b_n)$ converges to

$$f = \frac{t_0 - R u_0}{1 - R}$$

by Theorem 12. Prove that

$$f_n = \frac{t_0 - R_n u_0}{1 - R_n}$$

and

$$f - f_n = (t_0 - u_0) \cdot \frac{R - R_n}{(1 - R)(1 - R_n)} \quad \text{if } R \neq 1.$$

(13) Let $\{t_n\}$ be a tail sequence for the continued fraction $\mathbf{K}(a_n/b_n)$ with all $a_n \neq 0$ and $t_n \neq \infty$. Let further $u_0 \in \mathbf{C}$ be arbitrarily chosen. Prove that then $\{u_n\}$, where

$$u_n = S_n^{-1}(u_0) = t_n \frac{Q_1^{(n)} t_0 - Q_0^{(n)} u_0}{Q_1^{(n-1)} t_0 - Q_0^{(n-1)} u_0}$$

$$\text{and} \quad Q_r^{(m)} = \sum_{k=r}^{m} \prod_{j=1}^{k} \frac{b_j + t_j}{-t_j},$$

is also a tail sequence for $\mathbf{K}(a_n/b_n)$.

(14) We want to solve the differential equation

$$(1 - t^2)y'' - 2(1 - t)y' + \frac{1}{4}y = 0, \quad t > 1,$$

by Frobenius' method. That is, we try to find a formal power series solution $\sum c_k t^k$. Substituting $y(t) = \sum c_k t^k$ into the equation gives

$$(1 - t^2) \sum_{k=0}^{\infty} c_{k+2}(k + 2)(k + 1)t^k$$

$$- 2(1 - t) \sum_{k=0}^{\infty} c_{k+1}(k + 1)t^k + \frac{1}{4} \sum_{k=0}^{\infty} c_k t^k = 0.$$

(a) Find the recurrence relation for $\{c_k\}$ by matching the coefficients for t^k for every $k \in \mathbf{N}_0$.

(b) Show that this recurrence relation has a minimal solution.

(c) Explain why the minimal solution is of particular interest for us, in this situation.

(d) How do you propose to find $c_0, c_1, \ldots, c_N$ if $\{c_k\}$ is a minimal solution?

(15) In each of the following cases, try to find a tail sequence for the given continued fraction, and use this to find its value.

(a) $\displaystyle \mathop{K}_{n=1}^{\infty} \frac{(x+n)(x+n+2)}{1}$.

(b) $\displaystyle \mathop{K}_{n=1}^{\infty} \frac{(x+n)^2}{1}$.

(c) $\displaystyle \mathop{K}_{n=1}^{\infty} \frac{n^2 - x^2}{2x+1}$.

(d) $\displaystyle \mathop{K}_{n=1}^{\infty} \frac{1}{b_n}$ where $b_n = \dfrac{(n+a)^2 + (n+a-1)}{n+a+1}$, and a is a complex constant $\neq -1, -2, -3, \ldots$.

Remarks

1. Linear recurrence relations are closely related to linear difference equations. Written on the form

$$A(n)\Delta^2 Y_n + B(n)\Delta Y_n + C(n)Y_n = 0,$$

where $\Delta Y_n = Y_{n+1} - Y_n$ and $\Delta^2 Y_n = \Delta(\Delta Y_n) = Y_{n+2} - 2Y_{n+1} + Y_n$, it is often called a linear, homogenous difference equation of order 2.

Linear recurrence relations may also be given on a matrix form

$$\mathbf{Y}_n = \mathbf{A}(n)\mathbf{Y}_{n-1},$$

where $\mathbf{A}(n)$ is an $(N \times N)$- matrix and $\mathbf{Y}_n = (Y_n^{(1)}, Y_n^{(2)}, \ldots, Y_n^{(N)})^t$. The three-term recurrence relation

$$Y_n = b_n Y_{n-1} + a_n Y_{n-2} \quad \text{for } n = 1, 2, 3, \ldots$$

may for instance be written

$$Y_n^{(1)} = Y_{n-1}^{(2)}, \qquad Y_n^{(2)} = b_n Y_{n-1}^{(2)} + a_n Y_{n-1}^{(1)},$$

that is

$$\begin{bmatrix} Y_n^{(1)} \\ Y_n^{(2)} \end{bmatrix} = \begin{bmatrix} 0 & 1 \\ a_n & b_n \end{bmatrix} \begin{bmatrix} Y_{n-1}^{(1)} \\ Y_{n-1}^{(2)} \end{bmatrix}.$$

For more information on recurrence relations we refer to [Batc27], [Mill68], [Wimp84].

2. Pincherle's theorem was proved already in 1894, [Pinc94]. In their book, [JoTh80], Jones and Thron presented a generalized version. This version is the basis for our presentation here.

3. The Indian mathematican Ramanujan left an overwhelming heritage of deep, interesting and useful formulas at his far too early death in 1920. Among these were more than 50 results on continued fractions, of complex and astonishing character, [Rama57]. Unfortunately, most of these results were left without proofs. Hence, mathematicians are still working to prove them. We refer in particular to Berndt, [ABBW85], [ABJL92] and [BeLW85].

4. Generalized continued fractions were introduced by de Bruin, [Bruin74], [Bruin78]. They are based on the Jacobi-Perron algorithm, [Perr07]. Independently of this, Graves-Morris introduced the vector valued continued fractions, [Grav83], [Grav84] based on a work of Wynn, [Wynn63]. These structures turn out to be the same.

References

[ABBW85]   C. Adiga, B. C. Berndt, S. Bhargava and G. N. Watson, *Chapter 16 of Ramanujan's Second Notebook: "Theta-functions and q-series"*, Memoirs Amer. Math. Soc., Vol. **53**, No 315 (1985), 1–85.

[ABJL92]   G. E. Andrews, B. C. Berndt, L. Jacobsen and R. L. Lamphere, *The Continued Fractions Found in the Unorganized Portions of Ramanujan's Notebooks* (to appear in Mem. Amer. Math. Soc. 1992).

[Auric07]   A. Auric, *Recherches sur les fractions continues algébriques*, J. Math. Pures et App. (6) **3** (1907), 105–206.

[Batc27]   P. M. Batchelder, "An Introduction to Linear Difference Equations", Cambridge, Mass. (1927), Dover Publications, Inc., New York (1927).

[Bern89]   B. C. Berndt, "Ramanujan's Notebook. Part II", Springer-Verlag (1989).

[BeLW85]   B. C. Berndt, R. L. Lamphere and B. M. Wilson, *Chapter 12 of Ramanujan's Second Notebook: "Continued Fractions"*, Rocky Mountain J. Math., Vol. **15**, No 2 (1985), 235–310.

[Bruin74]   M. G. de Bruin,, "Generalized C-fractions and a Multidimensional Padé Table", Dissertation, Universiteit van Amsterdam (1974).

[Bruin78]   M. G. de Bruin, *Convergence of Generalized C-fractions*, J. of Approx. Theory **24** (1978), 177-207.

[Cruy79a]   P. van der Cruyssen, *Linear Difference Equations and Generalized Continued Fractions*, Computing **22** (1979), 269–278.

[Cruy79b]   P. van der Cruyssen, "Computing the Minimal Solution of a Certain Matrix-Vector Recursion", Report no **79-34**, Universiteit van Antwerpen (1979).

[Gauss13]   C. F. Gauss, "Disquisitiones generales circa seriem infinitam $1 + \frac{\alpha\beta}{1\cdot\gamma}x + \frac{\alpha(\alpha+1)\beta(\beta+1)}{1\cdot2\cdot\gamma\cdot(\gamma+1)}xx + \frac{\alpha(\alpha+1)(\alpha+2)\beta(\beta+1)(\beta+2)}{1\cdot2\cdot3\cdot\gamma\cdot(\gamma+1)(\gamma+2)}x^3$ etc", Commentationes Societatis Regiae Scientiarium Gottingensis Recentiores", Vol. **2** (1813), 1–46, Werke, Vol. **3** Göttingen (1876), 134–138.

[Gaut67]   W. Gautschi, *Computational Aspects of Tree-Term Recurrence Relations*, SIAM Review **9** (1967), 24–82.

[Grav83]   P. Graves-Morris, *Vector Valued Rational Interpolants I*, Numer. Math. **42** (1983), 331–348.

[Grav84]   P. Graves-Morris, *Vector Valued Rational Interpolants II*, IMA J. Num. Analy. **4** (1984), 209–224.

[JoTh80]   W. B. Jones and W. J. Thron, "Continued Fractions: Analytic Theory and Applications", Encyclopedia of Mathematics and its Applications **11**, Addison-Wesley Publishing Co., Reading, Mass. (1980). Now distributed by Cambridge University Press, New York.

[Levr87]   P. Levrie, "Het numeriek oplossen van lineaire recursiebetrekkingen: Een veralgemening van de kettingbreukmethode van Gautschi", Dissertation, Katholieke Universiteit Leuven, Faculteit Wetenschappen (1987).

[LePi87]   P. Levrie and R. Piessens, "Convergence Acceleration for Miller's Algorithm", Report TW88, Department of Computer Science, K. U. Leuven (February 1987).

[Mill68]   K. S. Miller, "Linear Difference Equations", Benjamin, New York (1968).

[Perr07]   O. Perron, *Über die Konvergenz der Jacobi-Ketten-algorithmen mit komplexen Elementen*, Sitzungsber. der Bayer. Akad. Wiss., Math. Naturwiss. Klasse **37** (1907) 401–481.

[Pinc94]   S. Pincherle, *Delle Funzioni ipergeometriche e di varie questioni ad esse attinenti*, Giorn. Mat. Battaglini **32** (1894), 209–291, Opere Selecte, Vol **1**, 273–357.

[Rama57]   S. Ramanujan, "Notebooks", Vol. **2**, Tata Institute of Fundamental Research, Bombay (1957).

[Waad84]   H. Waadeland, *Tales About Tails*, Proc. Amer. Math. Soc. **90** (1984), 57–64.

[Wimp84]   J. Wimp, "Computation with Recurrence Relations", Pitman Advanced Publishing Program, Pitman Publishing Inc., Boston, London, Melbourne (1984).

[Wynn63]   P. Wynn, *Continued Fractions whose Coefficients Obey a Non-commutative Law of Multiplication*, Arch. Rat. Mech. Anal. **12** (1963), 273–312.

[Zahar68]   R. V. M. Zahar, *Computational Algorithms for Linear Difference Equations*, Thesis, Purdue University (1968).

Chapter V

Correspondence of continued fractions

About this chapter

In Chapter I correspondence was a link between formal power series $L(z)$ and continued fractions $\mathbf{K}(a_n(z)/b_n(z))$ with polynomial elements $a_n(z)$ and $b_n(z)$. Judging from the examples there, it may seem as if $\mathbf{K}(a_n(z)/b_n(z)) \sim L(z)$ (i.e. correspondence) implies that $\mathbf{K}(a_n(z)/b_n(z))$ and $L(z)$ converge to the same function, or, at least that if $\mathbf{K}(a_n(z)/b_n(z))$ converges to a function $f(z)$, then $L(z)$ is a power series expansion of $f(z)$. It is important to know that this is not always so. In this chapter we shall look closer at what kind of conditions that enters the picture. Results of this type can be used to sum divergent series, as in Chapter I, and to find the value of a continued fraction by identifying the function to which it corresponds.

1   The normed field $(\mathbf{L}, \|\cdot\|)$

1.1   *Introducing the normed field*

Let $\mathbf{K}(a_n(z)/b_n(z))$ be a continued fraction with polynomial elements $a_n(z)$ and $b_n(z)$. We are interested in the following question: Does $\mathbf{K}(a_n(z)/b_n(z))$ correspond to a power series

$$L(z) = \sum_{n=m}^{\infty} c_n z^n \quad \text{where } m \in \mathbf{Z}, \ c_n \in \mathbf{C}, \ c_m \neq 0 \,? \tag{1.1.1}$$

That is, do the classical approximants

$$f_k(z) = \frac{a_1(z)}{b_1(z)} \frac{a_2(z)}{+ \, b_2(z) \, +} \cdots \frac{a_k(z)}{+ \, b_k(z)} = \frac{A_k(z)}{B_k(z)} \tag{1.1.2}$$

have power series expansions which coincide with $L(z)$ as far out as we want for k large enough?

Let us introduce some notation. Let $\mathbf{L}$ denote the field of all power series (1.1.1) with zero element $l_0 = \sum 0z^n$. (We do not require that these series converge at any point z, we are for the moment regarding them as mathematical objects in their own right. They are what we call *formal* power series.) For a function $f(z)$, meromorphic at $z = 0$, we let $\mathcal{L}(f)$ denote the Laurent series expansion of f in a neighborhood of $z = 0$. That is, $\mathcal{L}(f) \in \mathbf{L}$. Finally, the degree of the first non–zero term of an $L \in \mathbf{L}$ shall be denoted by $\lambda(L)$, that is

$$\lambda(L) = \begin{cases} m & \text{for } L(z) = \sum\limits_{n=m}^{\infty} c_n z^n \text{ with } c_m \neq 0, \\ \infty & \text{for } L(z) = l_0(z) = \sum 0z^n. \end{cases} \tag{1.1.3}$$

Then $\mathbf{K}(a_n(z)/b_n(z))$ corresponds at $z = 0$ to $L(z)$ iff

$$\nu_n := \lambda(L - \mathcal{L}(f_n)) \to \infty. \tag{1.1.4}$$

The number ν_n is called the *order of correspondence* of $f_n(z)$ to $L(z)$.

If $B_k(z) \not\equiv 0$, then the classical approximant $f_k(z)$ in (1.1.2) is a rational function, and $\mathcal{L}(f_k)$ is well defined. It is simpler though to regard a_n,

b_n, A_n and B_n as elements from $\mathbf{L}$ directly. Then the approximants

$$L_k(z) = \frac{a_1(z)}{b_1(z)} \frac{a_2(z)}{+ b_2(z)} + \cdots + \frac{a_k(z)}{b_k(z)} = \frac{A_k(z)}{B_k(z)} \qquad (1.1.5)$$

are again elements from $\mathbf{L}$ if $B_k \neq l_0$. Clearly, $L_k = \mathcal{L}(f_k)$. For convenience we shall use this way of presentation, that is, we shall use the same notation for a polynomial $p(z)$ and its Laurent series $\mathcal{L}(p(z)) = p(z)$.

It was the idea of Jones and Thron [JoTh80, p. 148] to regard (1.1.4) as convergence in $\mathbf{L}$. They did so by introducing the norm

$$\|L\| = 2^{-\lambda(L)} \quad \text{for } L \in \mathbf{L}. \qquad (1.1.6)$$

(Recall that $\|L\|$ is a norm in the field $\mathbf{L}$ by definition if

(i) $\|L\| \geq 0$,

(ii) $\|L\| = 0$ if and only if $L = l_0$,

(iii) $\|L_1 L_2\| = \|L_1\| \cdot \|L_2\|$, and

(iv) $\|L_1 + L_2\| \leq \|L_1\| + \|L_2\|$

for all L, L_1 and L_2 in $\mathbf{L}$. We see that our norm (1.1.6) meets the requirements.) Then (1.1.4) is equivalent to

$$\|L - L_n\| \to 0, \qquad (1.1.4')$$

where L_n is given by (1.1.5); i.e. $\mathbf{K}(a_n(z)/b_n(z))$ corresponds to L if and only if $\mathbf{K}(a_n(z)/b_n(z))$ converges to L in this norm.

1.2 *Correspondence at $z = \infty$*

Sometimes it is convenient to consider correspondence at other points than $z = 0$. Correspondence at a point $z = a$ is the same as convergence in the normed field $(\mathbf{L}(a), \|\cdot\|_a)$, where $\mathbf{L}(a)$ consists of Laurent series

$$\begin{cases} L(z) = \sum_{n=m}^{\infty} c_n(z-a)^n & \text{if } a \neq \infty \quad \text{and } c_m \neq 0, \\ L(z) = \sum_{n=m}^{\infty} c_n z^{-n} & \text{if } a = \infty \quad \text{and } c_m \neq 0, \end{cases} \qquad (1.2.1)$$

and $\|L\|_a = 2^{-m}$ as before. (A polynomial $p(z)$ can always be regarded as an element in $\mathbf{L}(a)$.) For correspondence at $z = \infty$ we have in particular:

Lemma 1 *The continued fraction* $\mathbf{K}(a_n(z)/b_n(z))$ *with polynomial elements* $a_n(z)$ *and* $b_n(z)$ *corresponds at* $z = \infty$ *to* $L \in \mathbf{L}(\infty)$ *if and only if* $\mathbf{K}(a_n(1/z)/b_n(1/z))$ *corresponds to* $L(1/z)$ *at* $z = 0$.

Example 1 The continued fraction

$$\frac{1}{1} \frac{1}{+z} + \frac{1}{1} \frac{2}{+z} + \frac{2}{1} \frac{3}{+z} + \frac{3}{1} \frac{4}{+z} + \cdots \qquad (1.2.2)$$

corresponds at $z = \infty$ to the power series

$$1 - \frac{1!}{z} + \frac{2!}{z^2} - \frac{3!}{z^3} + \frac{4!}{z^4} - \cdots . \qquad (1.2.3)$$

This result is due to Stieltjes [Stie18]. It can be proved in several ways. We shall here only observe that the classical approximants $f_n(z)$ of (1.2.2) can be written

$$f_1(z) = \frac{1}{1} = 1 ,$$

$$f_2(z) = \frac{z}{z+1} \sim 1 - \frac{1}{z} + \frac{1}{z^2} - \cdots ,$$

$$f_3(z) = \frac{z+1}{z+2} \sim 1 - \frac{1}{z} + \frac{2}{z^2} - \frac{4}{z^3} + \cdots ,$$

and so on. The order of correspondence turns out to be $\nu_n = n$.

Another matter is that the power series (1.2.3) diverges for all $z \in \mathbf{C}$. It is known to be an asymptotic expansion of the function

$$F(z) = \int_0^\infty \frac{e^{-t} dt}{1 + zt} .$$

The corresponding continued fraction (1.2.2) converges to $F(z)$ for all z in the cut plane $|\arg z| < \pi$. For more information on how continued fractions can be used to sum asymptotic series, we refer to Chapter VII.

$\diamondsuit$

Example 2 The T-fraction in Example 7 in Chapter I corresponds to one power series at $z = 0$ and to another power series at $z = \infty$. We shall see later (in Example 6) that this is always so for general T-fractions

$$\mathbf{K}\left(\frac{F_n z}{1 + G_n z}\right) = \frac{F_1 z}{1 + G_1 z} + \frac{F_2 z}{1 + G_2 z} + \cdots \tag{1.2.4}$$

if all $G_n \neq 0$. The following (modified) T-fraction

$$\frac{c}{c + (1-a)z} - \frac{1(c-a+1)z}{c+1+(2-a)z} - \frac{2(c-a+2)z}{c+2+(3-a)z} - \frac{3(c-a+3)z}{c+3+(4-a)z} - \cdots \tag{1.2.5}$$

where we assume that neither a, $-c + 1$ nor $a - c$ is a natural number, corresponds at $z = 0$ to the power series

$$L(z) = 1 + \frac{a}{c+1}z + \frac{a(a+1)}{(c+1)(c+2)}z^2 + \frac{a(a+1)(a+2)}{(c+1)(c+2)(c+3)}z^3 + \cdots. \tag{1.2.6}$$

That this is so will be evident in Theorem 4 in Chapter VI. Here we shall just check the first 3 approximants to see the pattern:

$$L_1(z) = \frac{c}{c + (1-a)z}$$

$$= 1 - \frac{1-a}{c}z + \left(\frac{1-a}{c}\right)^2 z^2 - \cdots,$$

$$L_2(z) = \frac{c(c+1) + c(2-a)z}{c(c+1) + 2c(1-a)z + (1-a)(2-a)z^2}$$

$$= 1 + \frac{a}{c+1}z - \frac{(1-a)(2c+2-a+ac)}{c(c+1)^2}z^2 + \cdots,$$

$$L_3(z) = \frac{c(c+1)(c+2) + c(3c+3-2ac-a)z + c(2-a)(3-a)z^2}{Q(z)}$$

$$= 1 + \frac{a}{c+1}z + \frac{a(a+1)}{(c+1)(c+2)}z^2 + \cdots, \quad \text{where}$$

$$Q(z) = c(c+1)(c+2) + 3c(c+1)(1-a)z$$
$$+ 3c(1-a)(2-a)z^2 + (1-a)(2-a)(3-a)z^3.$$

The order of correspondence is $\nu_n = n$. At $z = \infty$ (1.2.5) corresponds to the power series

$$\frac{c}{1-a}\frac{1}{z} + \frac{c(1-c)}{(1-a)(2-a)}\frac{1}{z^2} + \frac{c(1-c)(2-c)}{(1-a)(2-a)(3-a)}\frac{1}{z^3} + \cdots. \tag{1.2.7}$$

This is also something we will return to later. (Theorem 4 in Chapter VI.) At the moment we only check the first 3 approximants again. Let $\hat{L}_k(z)$ be the approximants regarded as elements in $\mathbf{L}(\infty)$. Then

$$\hat{L}_1(z) = \frac{c}{(1-a)z+c}$$

$$= \frac{c}{1-a}\frac{1}{z} - \left(\frac{c}{1-a}\right)^2\frac{1}{z^2} + \cdots,$$

$$\hat{L}_2(z) = \frac{c(2-a)z+c(c+1)}{(1-a)(2-a)z^2+2c(1-a)z+c(c+1)}$$

$$= \frac{c}{1-a}\frac{1}{z} + \frac{c(1-c)}{(1-a)(2-a)}\frac{1}{z^2} + \cdots,$$

$$\hat{L}_3(z) = \frac{c(2-a)(3-a)z^2+c(3c+3-2ac-a)z+c(c+1)(c+2)}{Q(z)}$$

$$= \frac{c}{1-a}\frac{1}{z} + \frac{c(1-c)}{(1-a)(2-a)}\frac{1}{z^2}$$

$$\quad + \frac{c(1-c)(2-c)}{(1-a)(2-a)(3-a)}\frac{1}{z^3} + \cdots, \quad \text{where}$$

$$Q(z) = (1-a)(2-a)(3-a)z^3+3c(1-a)(2-a)z^2$$

$$\quad + 3c(c+1)(1-a)z+c(c+1)(c+2).$$

The order of correspondence is $\nu_n = n+1$.

———◇

Remark: In this chapter we shall always let $a_n(z)$ and $b_n(z)$ be polynomials, as they are in most applications. However, it is straight forward to see that most of what we do also holds for the more general situation where a_n and b_n are functions of z which are meromorphic at $z = 0$ (or at some other fixed point $z = a \in \hat{\mathbf{C}}$).

1.3   *Properties of the normed field* $(\mathbf{L}, \|\cdot\|)$

In view of the previous section we are led to study convergence in $(\mathbf{L}, \|\cdot\|)$. Let us first note some properties of the functional λ acting on $\mathbf{L}$. For $L_1, L_2 \in \mathbf{L}$ we have

$$\lambda(L_1 L_2) = \lambda(L_1) + \lambda(L_2), \tag{1.3.1}$$

$$\lambda(L_1/L_2) = \lambda(L_1) - \lambda(L_2), \tag{1.3.2}$$

$$\lambda(L_1 + L_2) \text{ is } \begin{cases} = \lambda(L_1) & \text{if } \lambda(L_1) < \lambda(L_2), \\ \geq \lambda(L_1) & \text{if } \lambda(L_1) = \lambda(L_2). \end{cases} \tag{1.3.3}$$

This can be verified by inspection.

The normed field $(\mathbf{L}, \|\cdot\|)$ is known to be complete. Hence a sequence $\{L_n\}$ of elements from $\mathbf{L}$ converges to an element $L \in \mathbf{L}$ if and only if $\{L_n\}$ is a *Cauchy sequence*; that is, if and only if to every $\epsilon > 0$ there exists an $N \in \mathbf{N}$ such that

$$\|L_{n+m} - L_n\| < \epsilon \quad \text{for all } m \text{ and } n \in \mathbf{N} \quad \text{with } n \geq N. \tag{1.3.4}$$

In our field, this condition can be simplified:

Lemma 2 $\{L_n\}$ *is a Cauchy sequence in* $(\mathbf{L}, \|\cdot\|)$ *if and only if* $\|L_{n+1} - L_n\| \to 0$.

Proof : The "only if"-part follows immediately from $(1.3.4)$ with $m = 1$. To prove the "if"-part we assume that $\|L_{n+1} - L_n\| \to 0$. That is, to every $\epsilon > 0$ there exists an $N \in \mathbf{N}$ such that $\|L_{n+1} - L_n\| < \epsilon$ for all $n \geq N$. From $(1.3.3)$ it follows that

$$\lambda(L_{n+m} - L_n) = \lambda \left(\sum_{j=1}^{m} (L_{n+j} - L_{n+j-1}) \right) \geq \min_{1 \leq j \leq m} \lambda(L_{n+j} - L_{n+j-1})$$

$$\tag{1.3.5}$$

for $m \in \mathbf{N}$. That is

$$\|L_{n+m} - L_n\| \leq \max_{1 \leq j \leq m} \|L_{n+j} - L_{n+j-1}\| < \epsilon \quad \text{for } n \geq N. \tag{1.3.6}$$

This proves that $\{L_n\}$ is a Cauchy sequence.   ∎

If $\lambda(L_{n+1} - L_n) \to \infty$ strictly monotonely, then $(1.3.5)$ can be written

$$\lambda(L_{n+m} - L_n) = \lambda(L_{n+1} - L_n) \quad \text{for all } n \text{ and } m. \tag{1.3.7}$$

In particular this means:

Lemma 3 *If* $\| L_{n+1} - L_n \| \to 0$ *strictly monotonely, then* $\nu_n = \lambda(L - L_n) = \lambda(L_{n+1} - L_n)$ *for all* n*, where* L *is the limit of* $\{L_n\}$ *in* $(\mathbf{L}, \| \cdot \|)$*.*

2   Classification of continued fractions

2.1   *Criteria for correspondence*

We return to our original question: Which continued fractions $\mathbf{K}(a_n(z)/b_n(z))$ correspond to some $L \in \mathbf{L}$? We can not expect that every one does since we can not expect that every sequence $\{L_k(z)\}$ from $\mathbf{L}$ converges in $(\mathbf{L}, \| \cdot \|)$. The first example shows a simple continued fraction which does not correspond to any $L \in \mathbf{L}$:

Example 3 The Thiele interpolating continued fraction is given by

$$b_0 + \frac{z - z_0}{b_1} + \frac{z - z_1}{b_2} + \frac{z - z_2}{b_3} + \cdots ; \quad b_n \in \mathbf{C},$$

where all $z_n \in \mathbf{C}$ are given, distinct points, [Thie09]. We shall assume that all $z_k \neq 0$. Its approximants can be written

$$
\begin{aligned}
f_0(z) &= b_0, \\
f_1(z) &= \left(b_0 - \frac{z_0}{b_1}\right) + \frac{1}{b_1} z, \\
f_2(z) &= \frac{(b_0 b_1 b_2 - b_2 z_0 - b_0 z_1) + (b_0 + b_2)z}{(b_1 b_2 - z_1) + z}, \\
f_3(z) &= \frac{(b_0 b_1 b_2 b_3 - b_2 b_3 z_0 - b_0 b_3 z_1 - b_0 b_1 z_2 + z_0 z_2)}{(b_1 b_2 b_3 - b_3 z_1 - b_1 z_2) + (b_1 + b_3)z} \\
&\quad + \frac{(b_0 b_1 + b_0 b_3 + b_2 b_3 - z_0 - z_2)z + z^2}{(b_1 b_2 b_3 - b_3 z_1 - b_1 z_2) + (b_1 + b_3)z},
\end{aligned}
$$

etc. so that

$$
\begin{aligned}
\mathcal{L}(f_0) &= b_0, \\
\mathcal{L}(f_1) &= \left(b_0 - \frac{z_0}{b_1}\right) + \frac{1}{b_1} z,
\end{aligned}
$$

$$\mathcal{L}(f_2) = (b_0 - \frac{b_2 z_0}{b_1 b_2 - z_1}) + (\frac{b_2}{b_1 b_2 - z_1} + \frac{b_2 z_0}{(b_1 b_2 - z_1)^2})z + \cdots,$$

$$\mathcal{L}(f_3) = (b_0 - \frac{b_2 b_3 z_0 - z_0 z_2}{b_1 b_2 b_3 - b_3 z_1 - b_1 z_2}) + \cdots \quad \text{etc.}$$

As we see, the constant term changes each time we increase the index, so there is no chance that this continued fraction corresponds to a power series.

For the sake of justice, let it be mentioned that the application of this continued fraction has nothing to do with this kind of correspondence. It interpolates a function $f(z)$ with given values $f(z_n) = f_n(z_n)$ at the interpolation points z_0, z_1, $z_2,\ldots$. The coefficients b_n are calculated from the equations $f(z_n) = f_n(z_n)$ for $n = 0, 1, 2,\ldots$.

$\diamond$

So, what does it take for $\mathbf{K}(a_n(z)/b_n(z))$ with polynomial elements to correspond to some $L \in \mathbf{L}$? Or in other words, what does it take for its approximants $L_k(z) = \mathcal{L}(f_k(z))$ to be a Cauchy sequence in $(\mathbf{L}, \|\cdot\|)$? According to Lemma 2 we ought to look at the difference

$$L_{n+1}(z) - L_n(z) = \frac{A_{n+1}(z)}{B_{n+1}(z)} - \frac{A_n(z)}{B_n(z)} = \frac{(-1)^n \prod_{k=1}^{n+1} a_k(z)}{B_n(z)B_{n+1}(z)}, \quad (2.1.1)$$

where the last step follows from the determinant formula (1.2.10) in Chapter I. Using $(1.3.1) - (1.3.2)$ this leads to

$$\lambda(L_{n+1} - L_n) = \sum_{k=1}^{n+1} \lambda(a_k) - \lambda(B_n) - \lambda(B_{n+1}),$$
$$\text{if } B_n \neq l_0, \quad B_{n+1} \neq l_0, \quad (2.1.2)$$

or, translated to the language of continued fractions:

Theorem 4 *Let $\mathbf{K}(a_n(z)/b_n(z))$ be a continued fraction with polynomial elements $a_n(z) \not\equiv 0$ and $b_n(z)$, and let $A_n(z)/B_n(z)$ be its approximants in canonical form. Then the following statements hold.*

(A) $\mathbf{K}(a_n(z)/b_n(z))$ *corresponds to some* $L \in \mathbf{L}$ *if and only if*

$$\sum_{k=1}^{n+1} \lambda(a_k) - \lambda(B_n) - \lambda(B_{n+1}) \to \infty \quad \text{as } n \to \infty. \qquad (2.1.3)$$

(B) *If* $\mathbf{K}(a_n(z)/b_n(z))$ *corresponds to* $L \in \mathbf{L}$, *then* L *is uniquely determined.*

(C) *If (2.1.3) tends strictly monotonely to* ∞, *then the order of correspondence of* $A_n(z)/B_n(z)$ *to* L *is given by*

$$\nu_n = \sum_{k=1}^{n+1} \lambda(a_k) - \lambda(B_n) - \lambda(B_{n+1}). \qquad (2.1.4)$$

Proof :   **(A):** Let $\mathbf{K}(a_n(z)/b_n(z))$ correspond to an $L \in \mathbf{L}$. Then $\{L_n\}$ converges to L in $(\mathbf{L}, \|\cdot\|)$, which implies that $B_n(z) \not\equiv 0$ from some n on. Hence (2.1.3) follows from (2.1.2) and Lemma 2. Conversely, if (2.1.3) holds, then, again, $B_n(z) \not\equiv 0$ from some n on, and the correspondence follows similarly.

(B): Since L is the limit of a convergent sequence in $(\mathbf{L}, \|\cdot\|)$, it is unique.

(C): This follows from Lemma 3 and (2.1.2).   ∎

Theorem 4 is essentially due to Jones and Thron [JoTh80, p. 151–153].

Example 4 How does this fit in with the observation in Example 3? For the Thiele interpolating continued fraction we have $\lambda(a_k) = \lambda(z - z_{k-1}) = 0$ unless $z_{k-1} = 0$, but that can happen for at most one index. Hence $\sum_{k=1}^{n+1} \lambda(a_k) \leq 1$. At the same time $B_n(z)$ are polynomials so that $\lambda(B_n) \geq 0$ for all n. So, there is no way (2.1.3) can be satisfied, not even if we allow one of the points z_k to be $= 0$.

                                                                        ◇

Example 5 In Chapter I we saw examples of regular C-fractions $1 + \mathbf{K}(a_n z/1)$ which correspond to power series. For these we have

$\sum_{k=1}^{n+1} \lambda(a_k z) = n+1$ whereas $B_n(z)$ are polynomials with $B_n(0) = 1$ so that $\lambda(B_n) = 0$ for all n. This means that regular C-fractions always correspond to power series. The order of correspondence is $\nu_n \geq n + 1$ by virtue of Theorem 4C.

$\diamondsuit$

Example 6 Let us look at the correspondence properties of a T-fraction (1.2.4). Also here $\sum_{k=1}^{n+1} \lambda(F_k z) = n+1$ and $B_n(0) = 1$ so that $\lambda(B_n) = 0$ and $\nu_n = n + 1$ if all $F_k \neq 0$. That is, T-fractions always correspond at $z = 0$ to a power series.

To study possible correspondence at $z = \infty$ we shall apply Lemma 1. We have

$$\frac{F_1/z}{1 + G_1/z +} \frac{F_2/z}{1 + G_2/z +} \frac{F_3/z}{1 + G_3/z +} \cdots \approx \frac{F_1}{z + G_1 +} \frac{F_2 z}{z + G_2 +} \frac{F_3 z}{z + G_3 +} \cdots$$

where $\lambda(F_1) = 0$, $\sum_{k=2}^{n+1} \lambda(F_k z) = n$ and $B_n(0) = G_1 G_2 \cdots G_n$. Hence, at least if all $G_n \neq 0$, the T-fraction corresponds to a power series at $z = \infty$. The order of correspondence is $\nu_n = n$ if all $F_k \neq 0$ and all $G_k \neq 0$.

$\diamondsuit$

2.2 Terminating continued fractions

There is always a question of how to deal with continued fractions $\mathbf{K}(a_n(z)/b_n(z))$ where $a_N(z) \equiv 0$ for some $N \in \mathbf{N}$. Different authors have chosen different ways to handle this. Following Henrici and Pfluger we defined a continued fraction in Chapter I, Subsection *1.2* by requiring that all $a_n \neq 0$. However, it is convenient to allow $a_n = 0$ at times. We shall say that if

$$a_N(z) \equiv 0, \quad a_n(z) \not\equiv 0 \quad \text{for } 1 \leq n < N, \tag{2.2.1}$$

then the continued fraction terminates after $N - 1$ terms. That is, we make no distinction between the two structures

$$\mathop{\mathbf{K}}_{n=1}^{\infty} (a_n(z)/b_n(z)) \quad \text{with } a_N(z) \equiv 0 \tag{2.2.2}$$

and

$$\overset{N-1}{\underset{n=1}{\mathbf{K}}}(a_n(z)/b_n(z)) = \cfrac{a_1(z)}{b_1(z)} + \cfrac{a_2(z)}{b_2(z)} + \cdots + \cfrac{a_{N-1}(z)}{b_{N-1}(z)} = f_{N-1}(z). \quad (2.2.3)$$

In fact, we also say that two continued fractions (2.2.2) are equal if they have equal elements up to and including the first partial numerator which is 0. From (2.2.3) we can see that a terminating continued fraction with polynomial elements always corresponds to a power series if $f_{N-1}(z) \not\equiv \infty$, namely $L_{N-1} = \mathcal{L}(f_{N-1})$.

2.3   Why classifications?

The correspondence properties of a continued fraction $c_0(z) + \mathbf{K}(a_n(z)/b_n(z))$ with polynomial elements are closely connected to the degree and form of these polynomials. For instance, comparing Examples 4 and 5 shows that the two continued fractions

$$\mathbf{K}((a_n + z)/1) \quad \text{and} \quad \mathbf{K}(a_n z/1) \quad \text{where } 0 \neq a_n \in \mathbf{C}$$

have very different properties. The first one does not in general correspond to any power series $L \in \mathbf{L}$ at $z = 0$, whereas the second one always corresponds to some $L \in \mathbf{L}$. This has led to the introduction of a long list of various "types" of continued fractions with different correspondence properties. We shall here look at only one of these, and see what kind of questions one can answer, and how.

2.4   C-fractions

A C-fraction is a continued fraction of the form

$$c_0 + \mathbf{K}(a_n z^{\alpha_n}/1) \quad \text{where } c_0 \in \mathbf{C}, \quad a_n \in \mathbf{C} \quad \text{and } \alpha_n \in \mathbf{N}. \quad (2.4.1)$$

(Also the form $c_0/(1 + \mathbf{K}(a_n z^{\alpha_n}/1))$ is called a C-fraction. We shall even say that $c_0 + \mathbf{K}(a_n(z)/b_n(z))$ is a C-fraction (modified) if it can be brought to one of these forms by an equivalence transformation and/or a change of variable $z = 1/\zeta$. In the following, though, we shall use the form (2.4.1).) A regular C-fraction, as described in Example 5, is a special kind of C-fraction with all $\alpha_n = 1$. The importance of C-fractions lies in their powerful correspondence properties combined with their simple form. Before stating these properties, it is convenient to define the subset

$$\mathbf{L}_0 = \{L \in \mathbf{L}; \lambda(L) \geq 0\} \tag{2.4.2}$$

of $\mathbf{L}$. That is, series from $\mathbf{L}_0$ are Taylor series in the sense that they do not contain terms with negative exponents. $\mathbf{L}_0$ is no longer a field. The following can be found in [LeSc39]:

Theorem 5

(A) *To every C-fraction (2.4.1), terminating or not, there corresponds a uniquely determined $L \in \mathbf{L}_0$. The order of correspondence of the nth approximant $f_n(z)$ is*

$$\nu_n = \sum_{k=1}^{n+1} \alpha_k \tag{2.4.3}$$

(for $n < N - 1$ if the C-fraction terminates with $a_N = 0$).

(B) *To every $L \in \mathbf{L}_0$ there corresponds a uniquely determined C-fraction, terminating or not.*

(C) *$L \in \mathbf{L}_0$ is the Taylor expansion at $z = 0$ of a rational function holomorphic at $z = 0$ if and only if its corresponding C-fraction terminates.*

Proof : (A): C-fractions (2.4.1) have canonical denominators $B_n(z)$ such that $B_{-1}(z) \equiv 0$, $B_0(z) \equiv 1$ and $B_n(z) = B_{n-1}(z) + a_n z^{\alpha_n} B_{n-2}(z)$ for $n = 1, 2, 3 \ldots$. This means that $B_n(0) = 1$ for all $n \geq 0$, so that $\lambda(B_n) = 0$ for all $n \geq 0$. In the same way $\lambda(A_n) \geq 0$ for all $n \geq 0$ and thus $\lambda(A_n/B_n) \geq 0$; i.e. $A_n/B_n \in \mathbf{L}_0$ for all $n \geq 0$. The correspondence follows therefore trivially if the C-fraction terminates. Assume next that all $a_n \neq 0$. Then (2.1.3) in Theorem 4 holds since

$$\sum_{k=1}^{n+1} \lambda(a_k z^{\alpha_k}) - \lambda(B_n) - \lambda(B_{n+1}) = \sum_{k=1}^{n+1} \alpha_k ,$$

where α_k are natural numbers ≥ 1. In fact, $\sum \alpha_k \to \infty$ strictly monotonely, and the result follows from Theorem 4.

(B): Let $L \in \mathbf{L}_0$ be given, say

$$L(z) = \sum_{k=0}^{\infty} c_k z^k .$$

If $c_k = 0$ for all $k \in \mathbf{N}$, then $L(z) = c_0$, which can be regarded as a terminating C-fraction. Otherwise, let n be the first positive index for which $c_n \neq 0$. Then $L(z)$ can be written

$$L(z) = c_0 + c_n z^n \left\{ 1 + \sum_{k=1}^{\infty} \frac{c_{k+n}}{c_n} z^k \right\} ,$$

and we choose $a_1 = c_n$ and $\alpha_1 = n$. The power series in the brackets can be inverted in $\mathbf{L}_0$, and we obtain

$$L(z) = c_0 + \frac{a_1 z^{\alpha_1}}{L^{(1)}(z)} \quad \text{where } L^{(1)}(z) = \left\{ 1 + \sum_{k=1}^{\infty} \frac{c_{k+n}}{c_n} z^k \right\}^{-1} \in \mathbf{L}_0 .$$

$$(2.4.4)$$

If all the coefficients of $L^{(1)}(z)$ are zero, apart from its constant term which is 1, then

$$L(z) \sim c_0 + \frac{a_1 z^{\alpha_1}}{1} ,$$

a terminating C-fraction, and we are finished. Otherwise we repeat the procedure with $L^{(1)}(z)$ to obtain

$$L^{(1)}(z) = 1 + \frac{a_2 z^{\alpha_2}}{L^{(2)}(z)} \qquad (2.4.5)$$

and thus

$$L(z) = c_0 + \frac{a_1 z^{\alpha_1}}{1} + \frac{a_2 z^{\alpha_2}}{L^{(2)}(z)} ,$$

and so on. Either the process stops, and we get a corresponding C-fraction which terminates, or it never stops, and we get an infinite C-fraction. That this C-fraction actually corresponds to $L(z)$ follows since $\lambda(L - L_n) = \lambda(L_{n+1} - L_n)$. It remains to prove that this C-fraction is unique. Assume that the C-fraction $d_0 + \mathbf{K}(b_n z^{\beta_n}/1)$ also corresponds to $L(z)$. Since

$$\mathbf{K}(a_n z^{\alpha_n}/1) \sim a_1 z^{\alpha_1} + \cdots \quad \text{and} \quad \mathbf{K}(b_n z^{\beta_n}/1) \sim b_1 z^{\beta_1} + \cdots$$

it follows directly that $d_0 = c_0$, $b_1 = a_1$ and $\beta_1 = \alpha_1$. Hence

$$L^{(1)}(z) \sim 1 + \mathop{\mathbf{K}}_{n=2}^{\infty} \left(a_n z^{\alpha_n}/1\right), \quad L^{(1)}(z) \sim 1 + \mathop{\mathbf{K}}_{n=2}^{\infty} \left(b_n z^{\beta_n}/1\right).$$

But then $b_2 = a_2$ and $\beta_2 = \alpha_2$ for the same reason as above. Continuation of this process shows that $b_n = a_n$ and $\beta_n = \alpha_n$ for all n (up to and including the first zero-term if one of the C-fractions terminates). That is, the C-fraction is unique.

(C): If the C-fraction terminates at $a_N = 0$, then it corresponds to $L_{N-1} = \mathcal{L}(A_{N-1}/B_{N-1})$ by definition. (See Subsection 2.2.) To prove the converse we assume that the non-terminating C-fraction $c_0 + \mathbf{K}(a_n z^{\alpha_n}/1)$ corresponds at $z = 0$ to $L(z) = P(z)/Q(z) = \sum_{k=0}^{\infty} c_k z^k$, where P and Q are polynomials. Without loss of generality we assume that $Q(0) = 1$. Then we can write

$$L(z) = \frac{P(z)}{Q(z)} = c_0 + \frac{a_1 z^{\alpha_1}}{L^{(1)}(z)} \tag{2.4.6}$$

where a_1 and α_1 are determined as in (2.4.4), and

$$L^{(1)}(z) = \frac{a_1 z^{\alpha_1}}{L(z) - c_0} = \frac{a_1 z^{\alpha_1} Q(z)}{P(z) - c_0 Q(z)} = \frac{Q(z)}{Q_1(z)}, \tag{2.4.7}$$

where $Q_1(z) = (P(z) - c_0 Q(z))/a_1 z^{\alpha_1}$ is a polynomial with $Q_1(0) = 1$. The degree $\deg(Q_1)$ of this polynomial must satisfy the inequality $\deg(Q_1) \leq \max\{\deg(P), \deg(Q)\} - \alpha_1$. Repeating this process with $L^{(1)}(z)$ we get from (2.4.5) that

$$L^{(2)}(z) = \frac{a_2 z^{\alpha_2}}{L^{(1)}(z) - 1} = \frac{a_2 z^{\alpha_2} Q_1(z)}{Q(z) - Q_1(z)} = \frac{Q_1(z)}{Q_2(z)}, \tag{2.4.8}$$

where $Q_2(z) = (Q(z) - Q_1(z))/a_2 z^{\alpha_2}$ is a polynomial with $Q_2(0) = 1$ and degree $\deg(Q_2) \leq \max\{\deg(Q), \deg(Q_1)\} - \alpha_2$, and so on. This can not go on for ever, since a polynomial $Q_n(z)$ always must have $\deg(Q_n) \geq 0$. Hence the continued fraction terminates.   ■

Part C of this theorem implies that a non-rational function, holomorphic at $z = 0$, always has a corresponding *non-terminating* C-fraction. This can be useful to know in advance.

Theoretically, the result in part C can also be used to determine whether a given power series $L(z)$ is the MacLaurin series of a rational function or not. However, this is often simpler to decide by other means.

The proof of part B is constructive in the sense that it actually describes a method to produce the C-fraction corresponding to a given power series. The algorithm is easy to program on a computer. It is described as Method 1 in Subsection *2.6*.

The proof of part C is also constructive. It describes a method to produce a C-fraction corresponding to a given rational function. In Subsection *2.6* we shall see how these methods can be generalized.

Another question is: if the C-fraction converges to a function $f(z)$ holomorphic in a neighborhood of $z = 0$, will then $L = \mathcal{L}(f)$? The answer is in fact yes, at least if the convergence is uniform, as we shall see later in Theorem 10.

2.5   When does $f(z)$ have a regular C-fraction expansion?

Let $f(z)$ be holomorphic at $z = 0$ and let $L = \mathcal{L}(f)$. Then we know from Theorem 5 that L has a corresponding C-fraction $c_0 + \mathbf{K}(a_n z^{\alpha_n}/1)$. When is this C-fraction regular (all $\alpha_n = 1$), and when does this regular C-fraction converge to $f(z)$ in a neighborhood of $z = 0$?

One way to find out is to actually find the corresponding C-fraction and look at it. Is it regular? Does it converge? This can be used if the C-fraction terminates or if we can find a formula for its elements a_n.

What we would like to have, though, are criteria which can be checked before we go to the trouble of actually finding the C-fraction. We shall list four different results to this effect, all of them without proof.

Theorem 6 *The C-fraction expansion of the power series* $L(z) = \sum_{k=0}^{\infty} c_k z^k$ *is regular and non-terminating if and only if* $H_k^{(1)} H_k^{(2)} \neq 0$ *for* $k = 1, 2, 3, \ldots$, *where*

$$
H_k^{(n)} = \begin{vmatrix} c_n & c_{n+1} & \cdots & c_{n+k-1} \\ c_{n+1} & c_{n+2} & \cdots & c_{n+k} \\ \vdots & \vdots & & \vdots \\ c_{n+k-1} & c_{n+k} & \cdots & c_{n+2k-2} \end{vmatrix}.
$$

(2.5.1)

If $H_k^{(1)} H_k^{(2)} \neq 0$ *for all* $k \in \mathbf{N}$, *then* $L(z) \sim c_0 + \mathbf{K}(a_n z/1)$ *where*

$$
a_{2k} = -\frac{H_{k-1}^{(1)} H_k^{(2)}}{H_k^{(1)} H_{k-1}^{(2)}}, \qquad a_{2k+1} = -\frac{H_{k+1}^{(1)} H_{k-1}^{(2)}}{H_k^{(1)} H_k^{(2)}}.
$$

(2.5.2)

This is really not much improvement over the scheme already presented. Fortunately we have some more user-friendly results. The two next ones concern *S-fractions* (Stieltjes fractions) which are regular C-fractions with $c_0 > 0$ and all $a_n > 0$. (See for instance Subsection *4.3* in Chapter III.)

Theorem 7 (Stieltjes, [Stie18]) *The power series* $L(z) = \sum_{k=0}^{\infty} c_k z^k$ *has a corresponding S-fraction if and only if there exists a distribution function* $\Psi : [0, \infty] \rightarrow \mathbf{R}$ *such that*

$$
c_k = \int_0^{\infty} (-t)^k d\Psi(t) \quad \text{for } k = 0, 1, 2, \ldots.
$$

(2.5.3)

By a *distribution function* we mean a real function which is bounded, nondecreasing with infinitely many points of increase. We shall return to this in Subsection *3.1* in Chapter VII. A nice and simple consequence of Theorem 7 is obtained if $L(z)$ has a positive radius of convergence

R. Then $L(z)$ converges locally uniformly to a function $f(z)$ in $|z| < R$, and

$$f(z) = \sum_{k=0}^{\infty} c_k z^k = \sum_{k=0}^{\infty} z^k \int_0^{\infty} (-t)^k d\Psi(t)$$

$$= \int_0^{\infty} \sum_{k=0}^{\infty} (-tz)^k d\Psi(t) = \int_0^{\infty} \frac{d\Psi(t)}{1+tz}$$

on compact subsets of $|z| < R$. Hence, $f(z)$ has a corresponding S-fraction if and only if it can be written in the form

$$f(z) = \int_0^{\infty} \frac{d\Psi(t)}{1+tz}, \tag{2.5.4}$$

where $\Psi(t)$ is a distribution function on $[0, \infty)$.

Theorem 8 (Carleman [Carl26]) *If $L(z) = \sum_{k=0}^{\infty} c_k z^k$ satisfies (2.5.3) and*

$$\sum_{k=0}^{\infty} |c_k|^{-1/(2k)} = \infty, \tag{2.5.5}$$

then its corresponding S-fraction converges to $f(z)$ given by (2.5.4) locally uniformly in the cut plane $|\arg(z)| < \pi$.

For the proof of this theorem we refer to Wall's book [Wall48, p. 330].

A different approach was made by Lubinsky:

Theorem 9 (Lubinsky [Lubi85, Theorem 2]) *Let $f(z)$ be an entire function with $\mathcal{L}(f) = \sum_{k=0}^{\infty} c_k z^k$ where all $c_k \neq 0$ and*

$$\left| \frac{c_{k-1} c_{k+1}}{c_k^2} \right| \leq \rho^2 \quad \text{for } k = 1, 2, 3, \dots \tag{2.5.6}$$

Here $\rho = 0.4559\dots$ is the positive root of the equation

$$2 \sum_{k=1}^{\infty} \rho^{k^2} = 1. \tag{2.5.7}$$

Then $f(z)$ has a regular C-fraction expansion which converges locally uniformly to $f(z)$ in $\mathbf{C}$.

2.6 Algorithms for producing corresponding continued fractions

Let

$$L(z) = \sum_{k=m}^{\infty} c_k z^k \qquad (2.6.1)$$

be a given power series from **L**. We shall look at some methods to produce a continued fraction $b_0 + \mathbf{K}(a_n(z)/b_n(z))$ with polynomial elements of given degree and/or form, which corresponds to L at $z = 0$.

Method 1 The idea from the proof of Theorem 5B can be used to find a corresponding C-fraction. In fact, this was done in Chapter I, Subsection *3.2*, where it was used to find the beginning of the regular C-fraction expansion of

$$\log(1 + z) = z - \frac{z^2}{2} + \frac{z^3}{3} - \frac{z^4}{4} + \frac{z^5}{5} - \cdots . \qquad (2.6.2)$$

The method was called *successive substitutions*. This method can of course also be used to find other types of continued fractions.

Example 7 We want to expand (2.6.2) into a continued fraction of the form

$$c_0 + c_1 z + \frac{a_1 z^2 + b_1 z^3}{1} + \frac{a_2 z^2 + b_2 z^3}{1} + \frac{a_3 z^2 + b_3 z^3}{1} + \cdots . \qquad (2.6.3)$$

We have no guarantee that this is possible, even though (2.6.3) always corresponds to some power series. In fact, we have by Theorem 4C that

$$\nu_n = \sum_{k=1}^{n+1} \lambda(a_k z^2 + b_k z^3) - \lambda(B_n) - \lambda(B_{n+1}) \geq 2(n+1)$$

as long as not both a_k and b_k are equal to zero. (Why?) Say we want the fourth approximant of (2.6.3). Then we need to use the $2(4+1) = 10$ first terms of the series in (2.6.2) which we denote by $L(z)$. (We also count the constant term $c_0 = 0$.) We have

$$L(z) = z + \frac{-\dfrac{z^2}{2} + \dfrac{z^3}{3}}{L^{(1)}(z)}$$

where

$$L^{(1)}(z) = \cfrac{-\dfrac{z^2}{2} + \dfrac{z^3}{3}}{-\dfrac{z^2}{2} + \dfrac{z^3}{3} - \dfrac{z^4}{4} + \dfrac{z^5}{5} - \dfrac{z^6}{6} + \dfrac{z^7}{7} - \dfrac{z^8}{8} + \dfrac{z^9}{9} - \cdots}$$

$$= 1 - \frac{z^2}{2} + \frac{z^3}{15} - \frac{7z^4}{180} + \frac{5z^5}{189} - \frac{2221z^6}{113400} + \frac{2603z^7}{170100} - \cdots$$

$$= 1 + \cfrac{-\dfrac{z^2}{2} + \dfrac{z^3}{15}}{L^{(2)}(z)},$$

$$L^{(2)}(z) = \cfrac{-\dfrac{z^2}{2} + \dfrac{z^3}{15}}{-\dfrac{z^2}{2} + \dfrac{z^3}{15} - \dfrac{7z^4}{180} + \dfrac{5z^5}{189} - \dfrac{2221z^6}{113400} + \dfrac{2603z^7}{170100} - \cdots}$$

$$= 1 - \frac{7z^2}{90} + \frac{67z^3}{1575} - \frac{1297z^4}{47250} + \frac{2306z^5}{118125} - \cdots$$

$$= 1 + \cfrac{-\dfrac{7z^2}{90} + \dfrac{67z^3}{1575}}{L^{(3)}(z)},$$

and

$$L^{(3)}(z) = \cfrac{-\dfrac{7z^2}{90} + \dfrac{67z^3}{1575}}{-\dfrac{7z^2}{90} + \dfrac{67z^3}{1575} - \dfrac{1297z^4}{47250} + \dfrac{2306z^5}{118125} - \cdots}$$

$$= 1 - \frac{1297z^2}{3675} + \frac{10438z^3}{180075} - \cdots .$$

This means that

$$\log(1+z) \sim z + \cfrac{-\dfrac{z^2}{2} + \dfrac{z^3}{3}}{1} + \cfrac{-\dfrac{z^2}{2} + \dfrac{z^3}{15}}{1} + \cfrac{-\dfrac{7z^2}{90} + \dfrac{67z^3}{1575}}{1} +$$

$$\cfrac{-\dfrac{1297z^2}{3675} + \dfrac{10438z^3}{180075}}{1} + \cdots . \qquad (2.6.4)$$

Let us look a little closer at this continued fraction for $z = 1$. We used the first 10 terms of the Taylor series for $\log(1 + z)$. Summing these 10 terms with $z = 1$ gives $1627/2520 \approx 0.645635$ which is not a good approximation to $\log 2 \approx 0.6931472$. Computing the approximant $f_4(1)$ of (2.6.4) gives $f_4(1) = 545953/787182 \approx 0.6935537$ which is much better.

$\diamond$

The disadvantage of this method is that we repeatedly have to invert power series. We shall now see how this can be avoided:

Method 2 In the proof of Theorem 5C, $L(z)$ was a rational function, $L(z) = P(z)/Q(z)$. This in turn led to

$$L(z) = \frac{P(z)}{Q(z)} = c_0(z) + \frac{a_1(z)}{L^{(1)}(z)} \tag{2.6.5}$$

where

$$L^{(1)}(z) = \frac{a_1(z)Q(z)}{P(z) - c_0(z)Q(z)} = \frac{Q(z)}{Q_1(z)}, \tag{2.6.6}$$

where $Q_1(z)$ is again a polynomial if $c_0(z)$ is a polynomial and $a_1(z)$ is a polynomial which divides the polynomial $P(z) - c_0(z)Q(z)$. And so on.

What happens if we start with an arbitrary power series $L(z) \in \mathbf{L}_0$? Let us define $P(z) = L(z)$ and $Q(z) \equiv 1$. Then (2.6.5)–(2.6.6) still holds, only $P(z)$ and $Q_1(z)$ are no longer necessarily polynomials. The coefficients of the new series $Q_1(z)$ in (2.6.6) can be found from the relation

$$Q_1(z) = (P(z) - c_0(z)Q(z))/a_1(z) \tag{2.6.7}$$

which becomes particularly simple to compute if $a_1(z)$ is just a constant or a contant times a power of z. Repeating this process gives the continued fraction (if we have chosen a form for $a_n(z)$, $b_n(z)$ which works).

Example 8 Let $L(z) = \sum c_k z^k$ be a given power series from $\mathbf{L}_0$. Then we know from Theorem 5B that $L(z)$ has a corresponding C-fraction. Let us say that we know (or believe or hope) that this C-fraction is regular. That is, we want to find a continued fraction of the form $c_0 + \mathbf{K}(a_n z/1)$

corresponding to $L(z)$. Following (2.6.5) with $P(z) = L(z)$ and $Q(z) \equiv 1$ we get

$$L(z) = \frac{P(z)}{Q(z)} = \frac{c_0 + c_1 z + c_2 z^2 + \cdots}{1} = c_0 + \frac{c_1 z}{c_1 Q(z)/Q_1(z)}$$

where

$$Q_1(z) = \frac{1}{z}(P(z) - c_0 Q(z)) = \sum_{k=0}^{\infty} c_{k+1} z^k =: \sum_{k=0}^{\infty} c_{1,k} z^k.$$

Repeating the process we find that

$$\frac{c_1 Q(z)}{Q_1(z)} = 1 + \frac{c_1 Q(z) - Q_1(z)}{Q_1(z)} = 1 + \frac{c_{2,0} z}{c_{2,0} Q_1(z)/Q_2(z)}$$

where

$$Q_2(z) = \frac{1}{z}(c_1 Q(z) - Q_1(z)) \quad \text{and } c_{2,0} = Q_2(0)$$

and

$$\frac{c_{2,0} Q_1(z)}{Q_2(z)} = c_{1,0} + \frac{c_{2,0} Q_1(z) - c_{1,0} Q_2(z)}{Q_2(z)} = c_{1,0} + \frac{c_{3,0} z}{c_{3,0} Q_2(z)/Q_3(z)}$$

where $Q_3(z) = \frac{1}{z}(c_{2,0} Q_1(z) - c_{1,0} Q_2(z))$ and $c_{3,0} = Q_3(0)$.

Writing $Q_n(z) = \sum_{k=0}^{\infty} c_{n,k} z^k$, the general step becomes

$$Q_n(z) = \frac{1}{z}(Q_{n-2}(z)c_{n-1,0} - Q_{n-1}(z)c_{n-2,0}),$$

that is,

$$c_{n,k} = c_{n-1,0}\, c_{n-2,k+1} - c_{n-2,0}\, c_{n-1,k+1} \tag{2.6.8}$$

for $k = 0, 1, 2, \ldots$, $n = 2, 3, 4, \ldots$. From the expansion of $Q_1(z)$ we see that

$$c_{1,k} = c_{k+1} \quad \text{for } k = 0, 1, 2, \ldots. \tag{2.6.9}$$

Writing $Q_0(z) = Q(z) \equiv 1$ and $Q_{-1}(z) = P(z)$ so that

$$c_{0,0} = 1, \quad c_{0,k} = 0 \quad \text{and } c_{-1,k} = c_k \quad \text{for } k = 1, 2, 3, \ldots, \tag{2.6.10}$$

we find that also (2.6.9) follows from (2.6.8) with $n = 1$.

This particular recurrence system (2.6.8)–(2.6.10) goes by the name of *Viscovatov's algorithm*, [Visc06], [CuWu86, p. 16]. If all $c_{n,0} \neq 0$ we get

$$L(z) \sim c_0 + \frac{c_{1,0}z}{c_{0,0}} + \frac{c_{2,0}z}{c_{1,0}} + \frac{c_{3,0}z}{c_{2,0}} + \cdots$$

which is easily converted to a regular C-fraction by an equivalence transformation. If $c_{n,0} = 0$ for an index n, then either we have a terminating regular C-fraction (all $c_{n,k} = 0$ for this particular n), or the C-fraction corresponding to $L(z)$ is not regular after all. Let us try this algorithm on the power series

$$L(z) = 1 - z - z^2 + 3z^3 - z^4 - 5z^5 + 7z^6 + 3z^7 - 17z^8 + 11z^9 + \cdots,$$

where the coefficients c_k are given by the $(k \times k)$-determinant

$$c_k = (-1)^k \begin{vmatrix} 1 & 1 & 0 & 0 & \cdots & 0 \\ 2 & 1 & 1 & 0 & \cdots & 0 \\ 0 & 2 & 1 & 1 & \cdots & 0 \\ \vdots & \ddots & \ddots & \ddots & \ddots & \vdots \\ \vdots & & \ddots & \ddots & \ddots & 1 \\ 0 & \cdots & & 0 & 2 & 1 \end{vmatrix}$$

$$= \Re\left\{ \left(1 + \frac{i}{\sqrt{7}}\right)\left(\frac{-1 + i\sqrt{7}}{2}\right)^k \right\}.$$

The recursion (2.6.8) and initial values (2.6.9)–(2.6.10) then gives

$c_{-1,k}$:	1	−1	−1	3	−1	−5	7
$c_{0,k}$:	1	0	0	0	0	0	0
$c_{1,k} = c_{k+1}$:	−1	−1	3	−1	−5	7	
$c_{2,k} = -c_{1,k+1}$:	1	−3	1	5	−7		
$c_{3,k} = c_{1,k+1} + c_{2,k+1}$:	−4	4	4	−12			
$c_{4,k} = -4c_{2,k+1} - c_{3,k+1}$:	8	−8	−8				
$c_{5,k} = 8c_{3,k+1} + 4c_{4,k+1}$:	0	0					

Here the algorithm breaks down. Hence the C-fraction either terminates after these terms or the next term has a higher degree in z. We can determine which of these two cases we actually have by computing the general expression for $c_{5,k}$ to see if they all vanish or not. We get:

$$
\begin{aligned}
c_{2,k} &= c_{1,0}c_{0,k+1} - c_{0,0}c_{1,k+1} = -c_{1,k+1} = -c_{k+2} \\
&= -\Re\left\{\left(1 + \frac{i}{\sqrt{7}}\right)\left(\frac{-1+i\sqrt{7}}{2}\right)^{k+2}\right\} \Rightarrow c_{2,0} = 1, \\
c_{3,k} &= c_{2,0}c_{1,k+1} - c_{1,0}c_{2,k+1} = c_{k+2} - c_{k+3} \\
&= \Re\left\{\left(2 - \frac{2i}{\sqrt{7}}\right)\left(\frac{-1+i\sqrt{7}}{2}\right)^{k+2}\right\} \Rightarrow c_{3,0} = -4, \\
c_{4,k} &= c_{3,0}c_{2,k+1} - c_{2,0}c_{3,k+1} = 4c_{k+3} - c_{3,k+1} \\
&= \Re\left\{\left(2 + \frac{6i}{\sqrt{7}}\right)\left(\frac{-1+i\sqrt{7}}{2}\right)^{k+3}\right\} \Rightarrow c_{4,0} = 8, \\
c_{5,k} &= c_{4,0}c_{3,k+1} - c_{3,0}c_{4,k+1} = 8c_{3,k+1} + 4c_{4,k+1} = 0.
\end{aligned}
$$

Hence the continued fraction terminates, and we get

$$
L(z) \sim 1 - \frac{z}{1+} \frac{z}{-1-} \frac{4z}{1+} \frac{8z}{-4} \approx 1 - \frac{z}{1-} \frac{z}{1+} \frac{4z}{1-} \frac{2z}{1}.
$$

$\diamond$

Looking a little closer at method 2, we find from (2.6.5)–(2.6.6) that what we essentially do is setting

$$
\frac{P(z)}{Q(z)} = c_0(z) + \frac{a_1(z)}{Q(z)/Q_1(z)}
$$

and in the general step

$$
\frac{Q_n(z)}{Q_{n+1}(z)} = b_{n+1}(z) + \frac{a_{n+2}(z)}{Q_{n+1}(z)/Q_{n+2}(z)},
$$

that is,

$$
Q_n(z) = b_{n+1}(z)Q_{n+1}(z) + a_{n+2}(z)Q_{n+2}(z).
$$

This means that the sequence $\{P(z), Q(z), Q_1(z), Q_2(z), \ldots\}$ is essentially a solution of the three-term recurrence relation.

There exists a wide variety of algorithms for producing various types of continued fractions corresponding to a given power series. (See the remarks at the end of this chapter.)

3 Pincherle's and Auric's theorems in $(\mathbf{L}, \|\cdot\|)$

3.1 Interpretation

We have already defined correspondence of a continued fraction $c_0(z) + \mathbf{K}(a_n(z)/b_n(z))$ to a power series $L \in \mathbf{L}$ as convergence of its approximants $A_n(z)/B_n(z) = \mathcal{L}(f_n(z))$ to $L(z)$ in the norm $\|\cdot\|$ in $\mathbf{L}$. In Chapter IV we tied convergence of a continued fraction to the existence of dominant/minimal solutions of the corresponding three-term recurrence relation

$$X_n(z) = b_n(z)X_{n-1}(z) + a_n(z)X_{n-2}(z) \quad \text{for } n = 1, 2, 3, \dots \quad (3.1.1)$$

With further applications in mind we permitted convergence in some normed field $(\mathbf{F}, \|\cdot\|)$. So, what we want to do now, is to apply these results to the field $(\mathbf{L}, \|\cdot\|)$. For convenience we still restrict the elements a_n and b_n to be polynomials, and we use the simplified notation a_n, b_n, A_n and B_n to denote both the polynomials and their Taylor expansion at $z = 0$.

We recall that $\{X_n\}$ is a minimal solution of (3.1.1) if not all $X_n = 0$ and if there exists another solution $\{Y_n\}$ such that $X_n/Y_n \to 0$. If $\{X_n\}$ is minimal, then every solution $\{Z_n\}$ of (3.1.1) which is linearly independent of $\{X_n\}$ is dominant; i.e. $X_n/Z_n \to 0$. In the field $(\mathbf{L}, \|\cdot\|)$ we thus have that $\{X_n\}$; $X_n \in \mathbf{L}$ is a minimal solution of (3.1.1) if not all $X_n(z) = l_0$ and if there exists another solution $Y_n(z)$ of (3.1.1) such that $\| X_n/Y_n \| \to 0$, i.e. $\lambda(X_n) - \lambda(Y_n) \to \infty$.

Theorem 10 (Pincherle's theorem, modified)

(A) *The continued fraction* $b_0 + \mathbf{K}(a_n(z)/b_n(z))$ *with polynomial elements* $a_n(z) \not\equiv 0$ *and* $b_n(z)$ *corresponds to some formal power se-*

ries $L(z) \in \mathbf{L}$ if and only if its canonical denominators $\{B_n\}$ form a dominant solution of (3.1.1) in $(\mathbf{L}, \|\cdot\|)$.

(B) If $\{X_n\}$ is a minimal solution of (3.1.1) in $(\mathbf{L}, \|\cdot\|)$ with $X_{-1}(z) \neq l_0$, then $b_0(z) + \mathbf{K}(a_n(z)/b_n(z))$ corresponds to the formal power series $L = b_0 - X_0/X_{-1} \in \mathbf{L}$.

To determine whether a given solution $\{X_n\}$ is minimal or not is often easier in $(\mathbf{L}, \|\cdot\|)$ than in $(\mathbf{C}, |\cdot|)$. A typical situation is for instance that $B_n(0) \neq 0$ for all n, whereas $\lambda(X_n) \to \infty$. Then we immediately know that $\{X_n\}$ is a minimal solution.

Example 9 In Example 5 in Chapter IV we saw that $P_n(z) = \Psi(c + n; z)$ is a solution of the three-term recurrence relation

$$P_n(z) = P_{n+1}(z) + \frac{z}{(c + n)(c + n + 1)} P_{n+2}(z) \quad \text{for } n = 0, 1, 2, \dots,$$

where c is a complex constant $\neq 0, -1, -2, \dots$. This can be regarded as a recurrence relation in $\mathbf{L}$, and we want to use this fact to prove that $1 + \mathbf{K}(a_n z/1)$, where $a_n = 1/(c + n - 1)(c + n)$, corresponds to $\Psi(c; z)/\Psi(c + 1; z)$, [Gauss13]. From Example 6 in Chapter IV it follows that

$$X_n(z) = \Psi(c + n + 2; z)\frac{\Gamma(c)\Gamma(c + 1)(-z)^{n+1}}{\Gamma(c + n + 1)\Gamma(c + n + 2)} \quad \text{for } n = -1, 0, 1, \dots$$

is a solution of the recurrence relation

$$X_n(z) = X_{n-1}(z) + a_n z X_{n-2}(z) \quad \text{for } n = 1, 2, 3, \dots$$

for $\{B_n(z)\}$. Since $\lambda(X_n) = n + 1$ whereas $\lambda(B_n) = 0$ for all n, it follows that $\lambda(X_n/B_n) \to \infty$; i.e. $\|X_n/B_n\| \to 0$. Hence, $\{B_n\}$ is dominant, $\{X_n\}$ is minimal, and by Theorem 10, $1 + \mathbf{K}(a_n z/1)$ corresponds to

$$1 - X_0/X_{-1} = 1 + \frac{z}{c(c + 1)} \frac{\Psi(c + 2; z)}{\Psi(c + 1; z)} = \frac{\Psi(c; z)}{\Psi(c + 1; z)}.$$

———————————————————————————————◇

Theorem 11 (Auric's theorem, modified) *Let* $\mathbf{K}(a_n(z)/b_n(z))$ *be a continued fraction whose elements* $a_n, b_n \in \mathbf{L}$ *are polynomials with all* $a_n \neq l_0$. *Further let* $\{X_n\}$, *where* $X_n \in \mathbf{L}$, *be a solution of the corresponding recurrence relation (3.1.1) such that* $X_n \neq l_0$ *for all* n. *Then* $\mathbf{K}(a_n(z)/b_n(z))$ *corresponds to* $-X_0/X_{-1}$ *if and only if*

$$\lim_{n \to \infty} \| R_n \| = \infty \quad where \ R_n = \sum_{j=0}^{n} \frac{\prod_{m=1}^{j}(-a_m)}{X_{j-1}X_j} \in \mathbf{L}. \qquad (3.1.2)$$

Condition (3.1.2) may not be so easy to check. (Sums are always worse to control than products and ratios. See Subsection *1.3*.) We shall derive a useful consequence of this result:

Theorem 12 *Let* $\mathbf{K}(a_n(z)/b_n(z))$ *be a continued fraction where* $a_n, b_n \in \mathbf{L}$ *are polynomials with all* $a_n \neq l_0$, *and let* $\{X_n(z)\}$, *where* $X_n \in \mathbf{L}$, *be a solution of the corresponding recurrence relation (3.1.1) such that* $X_n \neq l_0$ *for all* n. *If*

$$\lambda(b_{n-1}) + \lambda(b_n) < \lambda(a_n) \quad for \ n = 1, 2, 3, \ldots, \qquad (3.1.3)$$

and

$$\lambda(b_n) < \lambda(X_n/X_{n-1}) \quad for \ n = 1, 2, 3, \ldots, \qquad (3.1.4)$$

then $\mathbf{K}(a_n(z)/b_n(z))$ *corresponds (at* $z = 0$) *to* $-X_0(z)/X_{-1}(z)$.

Proof : According to Theorem 11 we need that $\lambda(R_n) \to -\infty$, where R_n is given in (3.1.2) by a sum of $n + 1$ terms $\tilde{R}_j$ such that

$$d_j = \lambda(\tilde{R}_j) = \sum_{m=1}^{j} \lambda(a_m) - \lambda(X_{j-1}) - \lambda(X_j) \quad for \ j = 0, 1, 2, \ldots.$$

In view of the rule (1.3.3) for calculations of $\lambda(L_1 \pm L_2)$ it thus suffices to prove that $d_{j+1} < d_j$ under our conditions. We have

$$X_n = b_n X_{n-1} + a_n X_{n-2},$$

and thus

$$\frac{X_n}{X_{n-1}} = b_n + \frac{a_n X_{n-2}}{X_{n-1}}.$$

By use of (3.1.4) and the rules (1.3.1)–(1.3.3) of calulation it follows that

$$\lambda\left(\frac{a_n X_{n-2}}{X_{n-1}}\right) = \lambda(b_n),$$

so that

$$\lambda(X_{n-2}/X_{n-1}) = \lambda(b_n) - \lambda(a_n).$$

This gives

$$
\begin{aligned}
d_j - d_{j+1} &= -\lambda(a_{j+1}) - \lambda(X_{j-1}) - \lambda(X_j) + \lambda(X_j) + \lambda(X_{j+1}) \\
&= -\lambda(a_{j+1}) - \lambda(X_{j-1}/X_j) - \lambda(X_j/X_{j+1}) \\
&= -\lambda(a_{j+1}) - \lambda(b_{j+1}) + \lambda(a_{j+1}) - \lambda(b_{j+2}) + \lambda(a_{j+2}) \\
&= \lambda(a_{j+2}) - \lambda(b_{j+1}) - \lambda(b_{j+2}) > 0
\end{aligned}
$$

by (3.1.3). Hence $d_{j+1} < d_j$.   ∎

This result is essentially proved by Jones and Thron [JoTh80, Thm. 5.2, p. 152] (in a slightly different way).

By means of Theorem 5 of Chapter IV this result can be "translated" to solutions of the recurrence relation

$$P_n = b_n P_{n+1} + a_{n+1} P_{n+2} \quad \text{for } n = 0, 1, 2, \ldots. \tag{3.1.5}$$

Corollary 13 *Let* $\mathbf{K}(a_n(z)/b_n(z))$ *be a continued fraction, where* $a_n, b_n \in \mathbf{L}$ *are polynomials with all* $a_n \neq l_0$, *and let* $\{P_n\}; P_n \in \mathbf{L}$, *be a solution of (3.1.5) with* $b_0 = 0$ *and* $P_n \neq l_0$ *for all* n. *If*

$$\lambda(b_{n-1}) + \lambda(b_n) < \lambda(a_n) \quad \text{for } n = 1, 2, 3, \ldots, \tag{3.1.6}$$

and

$$\lambda(P_n/P_{n+1}) + \lambda(b_{n-1}) < \lambda(a_n) \quad \text{for } n = 1, 2, 3, \ldots, \tag{3.1.7}$$

then $\mathbf{K}(a_n(z)/b_n(z))$ *corresponds to* P_0/P_1.

Example 10 Let us use this to prove that the regular C-fraction $1 + \mathbf{K}(a_n z/1)$ in Example 9 corresponds to $\Psi(c; z)/\Psi(c + 1; z)$. In that example $P_n(z) = \Psi(c + n; z)$ so that $P_n(0) = 1$ for all n, and thus $\lambda(P_n/P_{n+1}) = 0$. Further $b_n = 1$ so that $\lambda(b_n) = 0$, and $a_n(z) = a_n z$ so that $\lambda(a_n(z)) = 1$. This means that (3.1.6)–(3.1.7) are satisfied, and the conclusion follows.

$\diamond$

Example 11 Recall that the hypergeometric function $_2F_1(a, b; c; z)$ is given by

$$_2F_1(a, b; c; z) = 1 + \frac{ab}{c}\frac{z}{1!} + \frac{a(a + 1)b(b + 1)}{c(c + 1)}\frac{z^2}{2!} + \cdots \quad \text{for } |z| < 1,$$

where a, b and c are complex constants and $c \neq 0, -1, -2, \ldots$. We can also regard $_2F_1(a, b; c; z)$ as an element from $\mathbf{L}_0$, in which case it is often referred to as the hypergeometric series. The following formulas due to Gauss [Gauss13] can be verified by comparing the coefficients of the series on each side of the equality signs:

$$(c-a)_2F_1(a-1, b; c; z) = (c-a-b)_2F_1(a, b; c; z) + b(1-z)_2F_1(a, b+1; c; z)$$

and

$$\begin{aligned}
(c - b - 1)_2F_1(a, b; c; z) &= (c - a - b - 1)_2F_1(a, b + 1; c; z) \\
&\quad + a(1 - z)_2F_1(a + 1, b + 1; c; z).
\end{aligned}$$

This means that the sequence $\{P_n(z)\}_{n=0}^{\infty}$ of elements from $\mathbf{L}_0$ given by

$$P_{2n}(z) = {}_2F_1(a + n - 1, b + n; c; z), \quad P_{2n+1}(z) = {}_2F_1(a + n, b + n; c; z)$$

is a solution of the three-term recurrence relation

$$P_n(z) = b_n P_{n+1}(z) + a_{n+1}(1 - z)P_{n+2}(z) \quad \text{for } n = 0, 1, 2, 3 \ldots,$$

where

$$\begin{aligned}
a_{2n+1} &= \frac{b + n}{c - a - n}, & b_{2n} &= \frac{c - a - b - 2n}{c - a - n}, \\
a_{2n} &= \frac{a + n - 1}{c - b - n}, & b_{2n-1} &= \frac{c - a - b - 2n + 1}{c - b - n}
\end{aligned}$$

if a, b, $a - c$ and $b - c$ are $\neq 0, -1, -2, \ldots$. It is therefore tempting to believe that the continued fraction

$$b_0 + \mathbf{K}\, \frac{a_n(1 - z)}{b_n} \;=\; \frac{c - a - b}{c - a} + \frac{\frac{b}{c - a}(1 - z)}{\frac{c - a - b - 1}{c - b - 1}+}$$

$$\frac{\frac{a}{c - b - 1}(1 - z)}{\frac{c - a - b - 2}{c - a - 1}} + \frac{\frac{b + 1}{c - a - 1}(1 - z)}{\frac{c - a - b - 3}{c - b - 2}} + \cdots$$

$$\approx \;\frac{1}{c - a}\left\{ c - a - b + \frac{b(c - b - 1)(1 - z)}{c - a - b - 1} \;+ \right.$$

$$\left. \frac{a(c - a - 1)(1 - z)}{c - a - b - 2} \;+\; \frac{(b + 1)(c - b - 2)(1 - z)}{c - a - b - 3} \;+\cdots \right\}$$

corresponds to $P_0(z)/P_1(z)$ at $z = 0$. According to Auric's Theorem for the adjoint recurrence relation as stated in Corollary 11 in Chapter IV, this is equivalent to

$$\lim_{n \to \infty} \| \hat{R}_n \| = \infty \quad \text{where } \hat{R}_n = \sum_{j=2}^{n} P_j P_{j+1}(1 - z)^j \prod_{m=1}^{j} (-a_m) \in \mathbf{L}\,.$$

This is impossible since

$$\hat{R}_n(0) = \sum_{j=2}^{n} P_j(0) P_{j+1}(0) \prod_{m=1}^{j} (-a_m) = \sum_{j=2}^{n} \prod_{m=1}^{j} (-a_m) \neq 0$$

for infinitely many indices n, except possibly for very special values of the parameters a, b and c. Hence, we do not have correspondence at $z = 0$. Another matter is that the correspondence at $z = 1$ follows easily if $P_j(1) \neq 0$ for infinitely many indices.

—— ◇

3.2   *A link between correspondence and classical convergence*

We have two ways of assigning a function $f(z)$ to a continued fraction $\mathbf{K}(a_n(z)/b_n(z))$: 1) Convergence of the continued fraction to $f(z)$ in some domain D, and 2) Correspondence of the continued fraction to $\mathcal{L}(f)$ at $z = 0$ if f is meromorphic at $z = 0$. Convergence is usually what one wants in applications. Correspondence is often what one has or what one is able to establish. So, the important question is then: say $\mathbf{K}(a_n(z)/b_n(z))$ corresponds at $z = 0$ to a power series $L(z)$ which is the Laurent expansion of a function $f(z)$ in a neighborhood of $z = 0$. Will then $\mathbf{K}(a_n(z)/b_n(z))$ converge to $f(z)$?

One has the feeling that if the classical approximants are uniformly bounded in a domain D with $0 \in D$, then the answer is yes. This feeling is based on the experience that uniform boundedness + "something else" often lead to uniform convergence of a sequence of analytic functions. And indeed, so it is:

Theorem 14 (Jones and Thron [JoTh80]) *Let D be a deleted neighborhood of the origin. Further let $\mathbf{K}(a_n(z)/b_n(z))$ with polynomial elements $a_n(z) \not\equiv 0$ and $b_n(z)$ correspond to an $L \in \mathbf{L}$ and have holomorphic approximants in D. Then the following statements hold.*

(A) $\mathbf{K}(a_n(z)/b_n(z))$ converges locally uniformly in D if and only if its approximants are uniformly bounded on every compact subset of D.

(B) If $\mathbf{K}(a_n(z)/b_n(z))$ converges locally uniformly in D, then its value $f(z)$ is holomorphic in D, meromophic at $z = 0$ and $L = \mathcal{L}(f)$.

Proof : (**A**): Let first C be a compact subset of D and $\mathbf{K}(a_n(z)/b_n(z))$ converge uniformly in C. Then $\{f_n(z)\}$ converges to a holomorphic function in C, and $|f(z) - f_n(z)| \leq \mu_n$ for all $z \in C$ where $\{\mu_n\}$ is a sequence of positive numbers converging to 0. Hence $|f_n(z)| \leq |f(z)| + \mu_n \leq \max\{|f(z)|; z \in C\} + \max\{\mu_n; n \in \mathbf{N}\} =: M < \infty$ for all n. This proves the "only if" part.

To prove the "if" part, we let $0 < \delta < r_1 < r_2 < R$ be positive numbers such that the annulus $A = \{z \in \mathbf{C}; \delta \leq |z| \leq R\} \subseteq D$, and $M_1 = \max\{|f_n(z)|; n \in \mathbf{N}$ and $z \in A\}$. Since each $f_n(z)$ is holomorphic in D

and meromorphic at $z = 0$, it has a Laurent expansion $L_n(z) = \mathcal{L}(f_n) \in$
L which converges to $f_n(z)$ in A. By Cauchy's estimate we thus know
that

$$\mathcal{L}(f_n) = L_n(z) = \sum_{k=m(n)}^{\infty} c_{n,k} z^k, \quad \text{where } |c_{n,k}| \leq M_1/r_2^k. \quad (3.2.1)$$

Let $\nu \in \mathbf{N}$ be chosen arbitrarily. Since $\| L - L_n \| \to 0$ and thus $\lambda(L - L_n) \to \infty$, there exists an $n \in \mathbf{N}$ such that $\lambda(L_{n+k} - L_n) \geq \nu$ for all
$n \geq N$ and $k \in \mathbf{N}$. This means that

$$|f_{n+k}(z) - f_n(z)| \leq \sum_{j=\nu}^{\infty} |c_{n+k,j} - c_{n,j}||z|^j \leq \sum_{j=\nu}^{\infty} \frac{2M_1}{r_2^j} r_1^j = M_2 \left(\frac{r_1}{r_2}\right)^\nu$$

$$(3.2.2)$$

for all $z \in A_1 = \{z \in \mathbf{C}; \delta \leq z \leq r_1\}$ and all $n \geq N$, $k \in \mathbf{N}$. This proves
that $\{f_n(z)\}$ is a (uniform) Cauchy sequence on A_1 and thus converges
on A_1. The locally uniform convergence of $\{f_n(z)\}$ on D follows then
by Stieltjes-Vitali's theorem.

(B): Let $\mathbf{K}(a_n(z)/b_n(z))$ converge locally uniformly to $f(z)$ in D. Then
$f(z)$ is holomorphic in D. We need to prove that f is meromorphic at
$z = 0$ and that $L = \mathcal{L}(f)$. Let $L(z) = \sum_{k=m}^{\infty} c_k z^k$. Since $\lambda(L - L_n) = \nu_n \to \infty$, it follows from (3.2.1) that $m = m(n)$ from some n on and
that $|c_k| \leq M_1/r_2^k$ for all k. Now

$$|f(z) - L(z)| \leq |f(z) - f_n(z)| + |f_n(z) - L_n(z)| + |L_n(z) - L(z)| \quad (3.2.3)$$

where $|f(z) - f_n(z)| \to 0$ uniformly in A_1, $|f_n(z) - L_n(z)| \equiv 0$ in A_1,
and $|L_n(z) - L(z)| \leq M_2(r_1/r_2)^{\nu_n}$ in A_1 by the same argument as in
(3.2.2). Hence, the right hand side of (3.2.3) can be made arbitrarily
small, so $L(z)$ converges to $f(z)$ in A_1. Since the inner radius δ in A,
can be made arbitrarily small, it follows that $L = \mathcal{L}(f)$. In particular
then f is meromorphic at $z = 0$. ∎

As we can see, the theorem is a simple consequence of the normality of
the family $\{f_n(z)\}$ of approximants for $\mathbf{K}(a_n(z)/b_n(z))$, combined with
the correspondence of $\mathbf{K}(a_n(z)/b_n(z))$. This theorem provides a useful
method for proving convergence of a continued fraction with polynomial
elements. In fact, there are even cases of continued fractions with constant elements where it pays to introduce an auxiliary variable z, just
to apply this method:

Example 12 We want to find the value of the so-called Rogers-Rama-nujan continued fraction

$$1 + \frac{q}{1} + \frac{q^2}{1} + \frac{q^3}{1} + \cdots \quad ; q \in \mathbf{C}, \ 0 < |q| < 1. \tag{3.2.4}$$

(We already know that it converges, since it has the form $\mathbf{K}(c_n/1)$ where $c_n \to 0$. See for instance Example 1 in Chapter II.) Trying to find explicit solutions of the three-term recurrence relation

$$P_n = P_{n+1} + q^{n+1} P_{n+2} \quad \text{for } n = 0, 1, 2, \ldots,$$

turns out to be rather difficult. So also for

$$X_n = X_{n-1} + q^n X_{n-2} \quad \text{for } n = 1, 2, 3, \ldots.$$

Let us introduce a complex variable z, to get the regular C-fraction

$$1 + \frac{qz}{1} + \frac{q^2 z}{1} + \frac{q^3 z}{1} + \cdots. \tag{3.2.5}$$

Then we find that

$$P_n(z) = \sum_{k=0}^{\infty} \frac{q^{k(k+n)}}{(q)_k} z^k,$$

where

$$(q)_0 = 1 \quad \text{and} \quad (q)_k = (1 - q)(1 - q^2) \cdots (1 - q^k),$$

is a solution of

$$P_n(z) = P_{n+1}(z) + q^{n+1} z P_{n+2}(z) \quad \text{for } n = 0, 1, 2, \ldots,$$

and that (3.2.5) corresponds to $P_0(z)/P_1(z)$. (See Problem 6 in Chapter IV.) We are interested in the convergence for $z = 1$. Let $D = \{z \in \mathbf{C}; |z| < R\}$ for an $R > 1$. Then there exists an $N \in \mathbf{N}$ such that $|q^n z| \le 1/4$ for all $n \ge N$. That is, the continued fraction

$$1 + \frac{q^N z}{1} + \frac{q^{N+1} z}{1} + \frac{q^{N+2} z}{1} + \cdots \tag{3.2.6}$$

satisfies the conditions of Worpitzky's theorem for $z \in D$, and thus its approximants are uniformly bounded by $1 + 1/2$ in D. Hence (3.2.6) converges to $P_{N-1}(z)/P_N(z)$ in D by virtue of Theorem 14. This in turn implies that (3.2.5) converges to $P_0(z)/P_1(z)$ in D and thus, in particular,

$$1 + \frac{q}{1} + \frac{q^2}{1} + \frac{q^3}{1} + \cdots = \left(\sum_{k=0}^{\infty} \frac{q^{k^2}}{(q)_k}\right) \Big/ \left(\sum_{k=0}^{\infty} \frac{q^{k(k+1)}}{(q)_k}\right),$$

where the equality sign stands for convergence.

$\diamond$

3.3 Tails and correspondence

In the previous subsections, correspondence of a continued fraction $\mathbf{K}(a_n(z)/b_n(z))$ was tied to properties of solutions of the corresponding three-term recurrence relation

$$X_n(z) = b_n(z)X_{n-1}(z) + a_n(z)X_{n-2}(z) \quad \text{for } n = 1, 2, 3 \ldots . \quad (3.3.1)$$

Of course, each time we have such a non-trivial solution $\{X_n(z)\}$ we also have a tail sequence $\{-X_n(z)/X_{n-1}(z)\}$. (See Subsection *3.3* in Chapter IV.) Hence, Theorem 12 is actually a theorem based on properties of a tail sequence. Similar relationships exist between Perron-tails and solutions of the adjoint of (3.3.1), so that Corollary 13 is in reality based on properties of Perron-tails.

4 Branched continued fractions

4.1 A simple example

The idea of correspondence can be extended to functions of several variables. The application is still the same, namely to find rational approximations to such functions or to sum divergent series. Let us look at a very simple example of how this can be done:

Example 13 We consider the function

$$f(x, y) = \frac{e^y}{1 - x}$$

of the two complex variables x and y. The Taylor expansion of $f(x,y)$ around $(0,0)$ can be found in the following way:

$$
\begin{aligned}
L(x,y) &= \mathcal{L}(e^y) \cdot \mathcal{L}\left(\frac{1}{1-x}\right) \\
&= \left\{1 + \frac{y}{1!} + \frac{y^2}{2!} + \frac{y^3}{3!} + \cdots\right\}\{1 + x + x^2 + x^3 + \cdots\} \\
&= 1 + \{x + x^2 + x^3 + \cdots\} + \left\{y + \frac{y^2}{2!} + \frac{y^3}{3!} + \cdots\right\} \\
&\quad + xy\left\{\frac{1}{1!} + \frac{y}{2!} + \frac{y^2}{3!} + \cdots\right\}\{1 + x + x^2 + x^3 + \cdots\}.
\end{aligned}
$$

Now

$$
x + x^2 + x^3 + \cdots \sim \frac{x}{1-x} = \frac{x}{1-}\frac{x}{1} =: \mathbf{K}\frac{a_n^{(0)}x}{1}
$$

which can be regarded as a terminating regular C-fraction, and

$$
y + \frac{y^2}{2!} + \frac{y^3}{3!} + \cdots \quad \sim \quad \frac{y}{1-}\frac{y}{2+}\frac{y}{3-}\frac{y}{2+}\frac{y}{5-}\frac{y}{2+}\frac{y}{7}-\cdots
$$

$$
=: \quad \mathbf{K}\frac{c_n^{(0)}y}{d_n^{(0)}},
$$

(see Problem 1), so

$$
\begin{aligned}
L(x,y) &\sim \quad 1 + \mathbf{K}\frac{a_n^{(0)}x}{1} + \mathbf{K}\frac{c_n^{(0)}y}{d_n^{(0)}} \\
&\quad + \cfrac{xy}{\cfrac{1}{1 + \frac{y}{2!} + \frac{y^2}{3!} + \cdots} \cdot \cfrac{1}{1 + x + x^2 + \cdots}} \\
&= \quad 1 + \mathbf{K}\frac{a_n^{(0)}x}{1} + \mathbf{K}\frac{c_n^{(0)}y}{d_n^{(0)}} + \frac{xy}{L_1(x,y)},
\end{aligned}
$$

where

$$
\begin{aligned}
L_1(x,y) &= \frac{1}{1 + \frac{y}{2!} + \frac{y^2}{3!} + \cdots} \cdot \frac{1}{1 + x + x^2 + \cdots} \\
&= \left\{1 - \frac{y}{2} + \frac{y^2}{2\cdot 3!} - \frac{y^4}{3!\,5!} + \frac{y^6}{3!\,7!} - \cdots\right\}\{1 - x\}
\end{aligned}
$$

$$= 1 - x + \left\{ -\frac{y}{2} + \frac{y^2}{2 \cdot 3!} - \frac{y^4}{3! \, 5!} + \cdots \right\}$$

$$+ xy \left\{ \frac{1}{2} - \frac{y}{2 \cdot 3!} + \frac{y^3}{3! \, 5!} - \cdots \right\}.$$

Again $-x$ can be regarded as a regular C-fraction $\mathbf{K}(a_n^{(1)} x / 1)$ and

$$-\frac{y}{2!} + \frac{y^2}{2 \cdot 3!} - \frac{y^4}{3! \, 5!} + \cdots \sim -\frac{y}{2} \frac{y}{+3} \frac{y}{-2} \frac{y}{+5} \frac{y}{-2} \frac{y}{+7} - \cdots =: \mathbf{K} \frac{c_n^{(1)} y}{d_n^{(1)}}.$$

Finally,

$$\cfrac{1}{\dfrac{1}{2!} - \dfrac{y}{2 \cdot 3!} - \dfrac{y^3}{3! \, 5!} - \cdots} \sim \cfrac{1}{\dfrac{1}{2} + \dfrac{y}{3} - \dfrac{y}{2} + \dfrac{y}{5} - \dfrac{y}{2} + \dfrac{y}{7} - \cdots}$$

$$= 2 + \frac{y}{3} \frac{y}{-2} \frac{y}{+5} \frac{y}{-2} \frac{y}{+7} - \cdots =: 2 + \mathbf{K} \frac{c_n^{(2)} y}{d_n^{(2)}}.$$

Therefore

$$f(x, y) \sim 1 + \mathbf{K} \frac{a_n^{(0)} x}{1} + \mathbf{K} \frac{c_n^{(0)} y}{d_n^{(0)}}$$

$$+ \cfrac{xy}{1 + \mathbf{K} \dfrac{a_n^{(1)} x}{1} + \mathbf{K} \dfrac{c_n^{(1)} y}{d_n^{(1)}} + \cfrac{xy}{2 + \mathbf{K} \dfrac{c_n^{(2)} y}{d_n^{(2)}}}} \qquad (4.1.1)$$

─── ◇

Let us look closer at what we got in (4.1.1). The right side can be regarded as a terminating continued fraction

$$B_0 + \frac{xy}{B_1} + \frac{xy}{B_2}$$

where the partial denominators B_0, B_1 and B_2 are (sums of) continued fractions. (4.1.1) is an example of what we call a branched continued fraction. More specifically it is a special case of what is called a TDCF (two–dimensional continued fraction), sometimes called a regular two–dimensional C–fraction. The structure of these regular two–dimensional C–fractions is in general

$$B_0 + \frac{xy}{B_1 +} \frac{xy}{B_2 +} \frac{xy}{B_3 +} \cdots,$$

(4.1.2)

where

$$B_k = b_0^{(k)} + \mathop{\mathbf{K}}_{n=1}^{\infty} \frac{a_n^{(k)} x}{b_n^{(k)}} + \mathop{\mathbf{K}}_{n=1}^{\infty} \frac{c_n^{(k)} y}{d_n^{(k)}} \quad \text{for } k = 0, 1, 2, \ldots ;$$

(4.1.3)

i.e. sums of regular C–fractions in x and y separately.

4.2   Approximants

To form approximants of (4.1.2) (or any other branched continued fraction), we need to truncate all the continued fractions involved. This can be done in many different ways, and it is not always easy to know which ones will serve meaningful convergence or correspondence purposes. If we truncate after the same number of terms in each branch of (4.1.2), then its 2. approximant (of classical type) would be

$$
F_2 = b_0^{(0)} + \cfrac{a_1^{(0)} x}{b_1^{(0)} + \cfrac{a_2^{(0)} x}{b_2^{(0)}}} + \cfrac{c_1^{(0)} y}{d_1^{(0)} + \cfrac{c_2^{(0)} y}{d_2^{(0)}}}
$$
$$
+ \cfrac{xy}{b_0^{(1)} + \cfrac{a_1^{(1)} x}{b_1^{(1)} + \cfrac{a_2^{(1)} x}{b_2^{(1)}}} + \cfrac{c_1^{(1)} y}{d_1^{(1)} + \cfrac{c_2^{(1)} y}{d_2^{(1)}}}}
$$
$$
+ \cfrac{xy}{b_0^{(2)} + \cfrac{a_1^{(2)} x}{b_1^{(2)} + \cfrac{a_2^{(2)} x}{b_2^{(2)}}} + \cfrac{c_1^{(2)} y}{d_1^{(2)} + \cfrac{c_2^{(2)} y}{d_2^{(2)}}}} \; .
$$

(4.2.1)

This choice is, however, usually not so good. Since ordinary, convergent continued fractions have values which depend mostly on their early elements, a better choice might be to include more terms in the first branches than in the later ones. The 2. approximant of (4.1.2) could then be for instance

$$
f_2 = b_0^{(0)} + \cfrac{a_1^{(0)} x}{b_1^{(0)} + \cfrac{a_2^{(0)} x}{b_2^{(0)}}} + \cfrac{c_1^{(0)} y}{d_1^{(0)} + \cfrac{c_2^{(0)} y}{d_2^{(0)}}}
$$

$$
+ \cfrac{xy}{b_0^{(1)} + \cfrac{a_1^{(1)} x}{b_1^{(1)}} + \cfrac{c_1^{(1)} y}{d_1^{(1)}}} + b_0^{(2)} \, . \tag{4.2.2}
$$

This choice, where branch number k in the nth approximant uses $(n-k)$ terms in each continued fraction is often the best choice.

Also for branched continued fractions one may choose to replace tails by some modifying factors $w_{n,p}^{(k)}$ when forming approximants, both for the main fraction and for the branches.

Example 14 We consider again the regular, branched C-fraction expansion (4.1.1) of $e^y/(1 - x)$, and we form approximants of the form (4.2.2). We get

$$
f_1(x, y) = 1 + \frac{a_1^{(0)} x}{1} + \frac{c_1^{(0)} y}{d_1^{(0)}} + \frac{xy}{1} = 1 + x + y + xy
$$

$$
f_2(x, y) = 1 + \cfrac{a_1^{(0)} x}{1 + \cfrac{a_2^{(0)} x}{1}} + \cfrac{c_1^{(0)} y}{d_1^{(0)} + \cfrac{c_2^{(0)} y}{d_2^{(0)}}} + \cfrac{xy}{1 + \cfrac{a_1^{(1)} x}{1} + \cfrac{c_1^{(1)} y}{d_1^{(1)}} + \cfrac{xy}{2}}
$$

$$
= 1 + \cfrac{x}{1 - x} + \cfrac{y}{1 - \frac{y}{2}} + \cfrac{xy}{1 - x - \frac{y}{2} + \frac{xy}{2}} = \frac{2 + y}{(2 - y)(1 - x)}
$$

$$
f_3(x, y) = 1 + \cfrac{x}{1 - x} + \cfrac{y}{1 - \frac{y}{2 + \frac{y}{3}}} + \cfrac{xy}{1 - x - \frac{y}{2 + \frac{y}{3}} + \frac{xy}{2 + \frac{y}{3}}}
$$

$$
= \frac{6 + 4y + y^2}{(1 - x)(6 - 2y)}
$$

and so on.

◇

4.3 Another example

A power series $L(z) = \sum c_n z^n$ has a natural ordering of its terms $c_n z^n$. This fact was the basis for our concept of correspondence. This is no longer so if we move on to power series $L(x, y)$. Correspondence between $L(x, y)$ and a TDCF will depend on how we order the terms in $L(x, y)$. The TDCF in $(4.1.2)$ corresponds to a power series with the ordering

$$c_{0,0} + \left(\sum_{n=1}^{\infty} c_{n,0} x^n + \sum_{n=1}^{\infty} c_{0,n} y^n \right) + \left(\sum_{n=1}^{\infty} c_{n,1} x^n y + \sum_{n=1}^{\infty} c_{1,n} x y^n \right) + \cdots .$$

(Correspondence is here tied to the main continued fraction $(4.1.2)$ and not to its branches $(4.1.3)$.)

Example 15 We return to the function

$$f(x, y) = \frac{e^y}{1 - x}$$

from Example 13, but this time we arrange its Taylor expansion $L(x, y)$ differently:

$$
\begin{aligned}
L&(x, y) \\
&= \left(1 + \frac{y}{1!} + \frac{y^2}{2!} + \frac{y^3}{3!} + \cdots \right) \left(1 + x + x^2 + x^3 + \cdots \right) \\
&= \left(1 + \frac{y}{1!} + \frac{y^2}{2!} + \frac{y^3}{3!} + \cdots \right) + x \left(1 + \frac{y}{1!} + \frac{y^2}{2!} + \frac{y^3}{3!} + \cdots \right) \\
&\quad + x^2 \left(1 + \frac{y}{1!} + \frac{y^2}{2!} + \frac{y^3}{3!} + \cdots \right) + \cdots \\
&= 1 + \mathbf{K} \frac{c_n y}{d_n} + \cfrac{x}{\cfrac{1}{\left(1 + \frac{y}{1!} + \frac{y^2}{2!} + \cdots \right) + x \left(1 + \frac{y}{1!} + \frac{y^2}{2!} + \cdots \right) + \cdots}} \\
&= 1 + \mathbf{K} \frac{c_n y}{d_n} + \cfrac{x}{\cfrac{1}{1 + \frac{y}{1!} + \frac{y^2}{2!} + \cdots} - x \cfrac{1}{1 + \frac{y}{1!} + \frac{y^2}{2!} + \cdots}}
\end{aligned}
$$

$$= \quad 1 + \mathbf{K} \, \frac{c_n y}{d_n} + \cfrac{x}{1 + \mathbf{K} \, \frac{-c_n y}{d_n} - \cfrac{x}{1 + \mathbf{K} \, \frac{-c_n y}{d_n}}} \quad ,$$

where $1 + \mathbf{K}(c_n y / d_n)$ is the regular C–fraction expansion of e^y as given in Problem 1. The TDCF has now the structure

$$b_0^{(0)} + \mathbf{K} \, \frac{c_n^{(0)} y}{d_n^{(0)}} + \cfrac{x}{b_0^{(1)} + \mathbf{K} \, \frac{c_n^{(1)} y}{d_n^{(1)}} +} \, \cfrac{x}{b_0^{(0)} + \mathbf{K} \, \frac{c_n^{(2)} y}{d_n^{(2)}} + \cdots} \quad .$$

_____◇

Problems

(1) The exponential function e^z has the corresponding regular C-fraction

$$e^z \sim 1 + \frac{z}{1} - \frac{z}{2} + \frac{z}{3} - \frac{z}{2} + \frac{z}{5} - \frac{z}{2} + \frac{z}{7} - \cdots$$

at $z = 0$. Find the Taylor series expansion of the first four approximants and compare to the Taylor series of e^z at $z = 0$.

(2) The function $f(z) = \log((1 + z)/(1 - z))$ has the corresponding C-fraction

$$\log \frac{1+z}{1-z} \sim \frac{2z}{1} - \frac{1^2 z^2}{3} - \frac{2^2 z^2}{5} - \frac{3^2 z^2}{7} - \cdots \qquad \text{at } z = 0.$$

Show that the continued fraction

$$\frac{2}{z} - \frac{1^2}{3z} - \frac{2^2}{5z} - \frac{3^2}{7z} - \cdots$$

corresponds to the function $\log((z + 1)/(z - 1))$ at $z = \infty$.

(3) P-fractions introduced by Magnus [Magn62A], [Magn62B] are continued fractions of the form

$$b_0(z) + \frac{1}{b_1(z)} + \frac{1}{b_2(z)} + \frac{1}{b_3(z)} + \cdots$$

where each $b_n(z)$ is a polynomial in $1/z$ with $\deg(b_k) \geq 1$ for $k \geq 1$,

$$b_n(z) = \sum_{k=0}^{N_n} a_k^{(n)} / z^k.$$

Show that a P-fraction always corresponds at $z = 0$ to a power series $L(z)$, and determine its order of correspondence.

(4) Regular δ-fractions introduced by Lange [Lange82] are continued fractions of the form

$$b_0 - \delta_0 z + \frac{d_1 z}{1 - \delta_1 z} + \frac{d_2 z}{1 - \delta_2 z} + \frac{d_3 z}{1 - \delta_3 z} + \cdots$$

where δ_n is either 0 or 1 for every n and where b_0, d_n are complex constants with $d_{n+1} = 1$ for each n such that $\delta_n = 1$.

(a) Show that a regular δ-fraction always corresponds at $z = 0$ to a power series $L(z)$, and determine its order of correspondence.

(b) Show that to every $L(z) \in \mathbf{L}_0$ there exists a corresponding regular δ-fraction.

(c) Show that $L(z) \in \mathbf{L}_0$ is the Taylor series expansion at $z = 0$ for a *rational* function if and only if its corresponding regular δ-fraction terminates.

(5) Hermitian PC-fractions (Perron-Carathéodory-fractions) are continued fractions of the form

$$\delta_0 - \frac{2\delta_0}{1} \frac{1}{+\bar{\delta}_1 z +} \frac{(1 - |\delta_1|^2)z}{\delta_1} \frac{1}{+\bar{\delta}_2 z +} \frac{(1 - |\delta_2|^2)z}{\delta_2} + \cdots$$

[JoNT86], [Perr57]. Show that if all $\delta_n \neq 0$ then the hermitian PC-fraction corresponds to a power series $L(z)$ at $z = 0$ and to a power series $\tilde{L}(1/z)$ at $z = \infty$. What is the connection between $L(z)$ and $\tilde{L}(1/z)$?

(6) Use Viscovatov's algorithm to develop the first 5 terms of the C-fraction corresponding to

$$L(z) = 1 - z + 2z^2 - 2z^3 - 4z^4 + 22z^5 + \cdots .$$

(7) Find the first 5 terms of the δ-fraction (see Problem 4 for definition) corresponding at $z = 0$ to the power series

$$L(z) = 1 + z - 2z^2 + 4z^3 - 14z^4 + 58z^5 + \cdots .$$

(8) Show that if

$$f(z) \sim \frac{a_0}{1+} \frac{a_1 z}{1+} \frac{a_2 z}{1+} \frac{a_3 z}{1+\cdots} \qquad \text{at } z = 0$$

where all $a_n \neq 0$, then

$$f(z) \sim a_0 - \frac{a_1 a_0 z}{1} \frac{c_2 z}{+1+} \frac{c_3 z}{1+} \frac{c_4 z}{1+\cdots} \qquad \text{at } z = 0$$

where $\{c_n\}$ is given by $c_2 = a_1 + a_2$ and

$$a_{2n} a_{2n+1} = c_{2n} c_{2n+1}, \qquad a_{2n+1} + a_{2n+2} = c_{n+1} + c_{n+2}$$

for all $n \geq 1$, if all $c_n \neq 0$.

Hint: Compare the odd part of the first continued fraction to the even part of the second one.

(9) Let $1 + \mathbf{K}(a_n z/1)$ be a non-terminating regular C-fraction corresponding (at $z = 0$) to the formal power series $L(z) = 1 + \sum c_n z^n$.

(a) Prove that $L(z)L(-z) = 1$ if and only if

$$a_2 = -a_1/2 \quad \text{and} \quad a_{2n} = -a_{2n-1} \quad \text{for all } n \geq 2.$$

(b) Prove that $c_{2k+1} = 0$ for all $k \in \mathbf{N}$ if and only if

$$a_{2k+1} = -a_{2k} \quad \text{for all } k \in \mathbf{N}.$$

Hint: Use the result in Problem (8).

(10) Show that the non-terminating general T-fraction $\mathbf{K}(F_n z/(1 + G_n z))$ with all $F_n \neq 0$, $G_n \neq 0$, corresponds to $\tilde{L}(1/z) \equiv -1$ at $z = \infty$ if and only if $G_n = -F_n$ for all n.

(11) Let the non-terminating general T-fraction $\mathbf{K}(F_n z/(1+G_n z))$ correspond to
$$L(z) = c_1 z + c_2 z^2 + c_3 z^3 + \cdots$$

at $z = 0$ and to

$$\tilde{L}(1/z) = d_0 + d_1 z^{-1} + d_2 z^{-2} + d_3 z^{-3} + \cdots$$

at $z = \infty$. Prove that $\tilde{L}(z) = -zL(1/z)$, i.e.
$$\tilde{L}(z) = -c_1 - c_2 z^{-1} - c_3 z^{-2} - \cdots$$

if and only if all $G_n = -1$.

(12) Find the first terms of the branched C-fraction described in (4.1.2) corresponding to:

(a) $f(x,y) = e^{x+y}$
(b) $\ln(x \ln y)$

Remarks

1. C-fractions go all the way back to Worpitzky, Pringsheim, Śleszyński and others, who used them extensively. Leighton and Scott
   made a systematic study of them in their paper from 1939, [LeSc39].
   One of the advantages of C-fractions is that every $L \in \mathbf{L}_0$ has a
   corresponding C-fraction.

2. T-fractions were closely examined by W. J. Thron, [Thron48],
   [Thron77]. As mentioned in Chapter I, O. Perron therefore suggested the name Thronsche Kettenbrüche (or T-fractions) for these
   structures, [Perr57, p. 174]. Their surprising correspondence property, that they correspond to *two* power series, one at $z = 0$ and
   one at $z = \infty$, was established by J. H. McCabe and J. A. Murphy [McMu76], [McCa78]. They called their continued fractions
   M-fractions, but M-fractions are essentially T-fractions. See also
   [Waad64].

3. There exist several algorithms for finding the continued fraction
   (of given type) which corresponds to a given power series $L(z)$.
   For instance:

   Regular C-fractions: The qd-algorithm introduced by Rutishauser [Ruti54]. This algorithm is also described in [JoTh80, p.
   227]. Henrici [Henr63] has written a very interesting survey
   on applications of this algorithm. The stability is discussed
   in [Ruti63], [Henr74, Sect. 7.6].

   The classical approximants of regular C-fractions are Padé
   approximants. Hence, one can also apply algorithms which
   produce these approximants directly. This is in particular
   useful if we do not need the continued fraction itself.

   C-fractions: E. Frank [Frank46] suggested a method for simultaneous computation of the denominators $B_n(z)$ and the elements $a_n z^{\alpha_n}$ of the C-fraction corresponding to a given power
   series. The method is also described in [Perr57, p. 111].

   T-fractions: The FG-algorithm was introduced by Jones and
   Thron, [JoTh80]. In [McCa83] McCabe showed that this algorithm can be regarded as an extension of the qd-algorithm.

4. There are several special examples of Theorem 7. For a survey we refer to [JaWa89].

5. Special versions of Theorem 14 have been known for a long time. See for instance [LeSc39].

6. Branched continued fractions were introduced by V. Ya. Skorobogat'ko. We refer to his book in Russian [Skor83] on the subject. He also wrote an article in English for the conference proceedings of a French-Polish meeting in Łańcut in Poland [Skor87]. The typical feature of such continued fractions are that the partial denominators of the main continued fraction $B_0 + \mathbf{K}(a_n/B_n)$ are again (sums of) continued fractions. Branched continued fractions have in general no natural connection to three-term recurrence relations. For more information we refer to the extensive works of Kutchminskaya, Cuyt, Wuytack, Verdonk, Siemasko and Bodnar. See for instance [Kuch78], [Kuch80], [Siem80], [Bodn86], [KuSi87], [CuWu86], [CuVe88] and the references therein. Let it merely be mentioned that the type of TDCF in Subsection *4.1* was introduced independently by O'Donoghue, Kutchminskaya and Cuyt and Verdonk.

References

[Bodn86]   D. I. Bodnar, "Branched Continued Fractions", Kiev Naukova Dumka (1986). (In Russian.)

[Carl26]   T. Carleman, "Les Fonctions Quasi Analytiques", Paris (1926), 78–96.

[CuVe88]   A. Cuyt and B. Verdonk, *A Review of Branched Continued Fraction Theory for the Construction of Multivariate Rational Approximants*, Appl. Numer. Math. 4 (1988).

[CuWu86]   A. Cuyt and L. Wuytack, "Nonlinear Methods in Numerical Analysis", North-Holland, Amsterdam (1986).

[Frank46] E. Frank, *Corresponding Type Continued Fractions*, Am. J. of Math. **68** (1946), 89–108.

[Gauss13] C. F. Gauss, *Disquisitiones generales circa seriem infinitam* $1 + \frac{\alpha\beta}{1\cdot\gamma}x + \frac{\alpha(\alpha+1a)\beta(\beta+1)}{1\cdot2\cdot\gamma(\gamma+1)}xx + \frac{\alpha(\alpha+1)(\alpha+2)\beta(\beta+1)(\beta+2)}{1\cdot2\cdot3\cdot\gamma(\gamma+1)(\gamma+2)}x^3 +$ *etc.*, Commentationes Societatis Regiae Scientiarum Goettingensis Recentiones, Vol. **2** (1813); Werke, Band **3**, Königlichen Gesellschaft der Wissenschaften, Göttingen (1876), 123–162.

[Henr63]   P. Henrici, *Some Applications of the Quotient-Difference Algorithm*, Proc. Symp. Appl. Math. **15**, Amer. Math. Soc., Providence, R.I., (1963), 159–183.

[Henr74]   P. Henrici, "Applied and Computational Complex Analysis", Vol. **1**, Wiley, New York (1974).

[JaWa89]   L. Jacobsen and H. Waadeland, *When does f(z) have a Regular C-Fraction or a Normal Padé Table?*, Journ. Comp. and Appl. Math. **28** (1989), 199–206.

[JoTh80]   W. B. Jones and W. J. Thron, "Continued Fractions: Analytic Theory and Applications", Encyclopedia of Mathematics and its Applications, Vol. **11**, Addison-Wesley (1980). Now distributed by Cambridge University Press.

[JoNT86]   W. B. Jones, O. Njåstad and W. J. Thron, *Schur Fractions, Perron-Carathéodory Fractions and Szegö Polynomials, a Survey*,"Analytic Theory of Continued Fractions II, Proceedings, Pitlochry and Aviemore", 1985 (W. J. Thron, ed.) Lecture Notes in Math., No. **1199**, Springer-Verlag Berlin, Heidelberg (1986), 127–158.

[Kuch78]   K. I. Kuchminskaya, *Corresponding and Associated Branched Continued Fractions for Double Power Series*, Dokl. Akad. Nauk Ukr. SSR, Ser. A **7** (1978), 614–617. (In Russian.)

[Kuch80]   K. I. Kuchminskaya, *On Approximation of Functions by Continued and Branched Continued Fractions*, Mat. Met. Fiz. Meh. Polya **12** (1980), 3–10.

[KuSi87]   K. I. Kuchminskaya and W. Siemasko, *Rational Approximation and Interpolation of Functions by Branched Continued Fractions*, "Rational Approximation and its Applications in Mathematics and Physics, Proceedings, Łańcut 1985", (J. Gilewicz, M. Pindor, W. Siemaszko, eds.), Lecture Notes in Math., No. **1237**, Springer-Verlag Berlin, Heidelberg (1987), 24–40.

[Lange82] L. J. Lange, *δ-Fraction Expansions of Analytic Functions*, "Analytic Theory of Continued Fractions, Proceedings, Loen, Norway 1981", (W. B. Jones, W. J. Thron and H. Waadeland, eds.), Lecture Notes in Math., No. **932**, Springer-Verlag Berlin, Heidelberg (1982), 152–175.

[LeSc39]   W. Leighton and W. T. Scott, *A General Continued Fraction Expansion*, Bull. Amer. Math. Soc. **45** (1939), 596–605.

[Lubi85]   D. S. Lubinsky, *Padé Tables of Entire Functions of Very Slow and Smooth Growth*, Constr. Approx. **1** (1985), 349–358.

[Magn62A]   A. Magnus, *Certain Continued Fractions Associated with the Padé Table*, Math. Zeitschr. **78** (1962), 361–374.

[Magn62B]   A. Magnus, *Expansion of Power Series into P-Fractions*, Math. Zeitschr. **80** (1962), 209–216.

[McCa78]   J. H. McCabe, *A Further Correspondence Property of M-Fractions*, Math. of Comp. **32** (1978), 1303–1305.

[McCa83]   J. H. McCabe, *The Quotient-Difference Algorithm and the Padé Table: An Alternative Form and a General Continued Fraction*, Math. of Comp. **41** (1983), 183–197.

[McMu76]   J. H. McCabe and J. A. Murphy, *Continued Fractions which Correspond to Power Series Expansions at Two Points*, J. Inst. Maths. Applics. **17** (1976), 233–247.

[Perr57]   O. Perron, "Die Lehre von den Kettenbrüchen", Band II, B. G. Teubner, Stuttgart (1957).

[Ruti54]   H. Rutishauser, *Anwendungen des Quotienten-Differenzen-Algorithmus*, Z. Angew. Math. Phys. **5** (1954), 496–508.

[Ruti63]   H. Rutishauser, *Stabile Sonderfälle des Quotienten-Differenzen-Algorithmus*, Numer. Math. **5** (1963), 95–112.

[Skor83]   V. Ya. Skorobogat'ko, *Theory of Branched Continued Fractions and Their Applications in Computational Mathematics*, ed. Nauka, Moscow (1983). (In Russian.)

[Skor87]   V. Ya. Skorobogat'ko, *Branched Continued Fractions and Convergence Acceleration Problems*, "Rational Approximation and its Applications in Mathematics and Physics, Proceedings, Lańcut 1985", (J. Gilewicz, M. Pindor, W. Siemaszko, eds.) Lecture Notes in Math., No. **1237**, Springer-Verlag Berlin, Heidelberg (1987), 46–50.

[Siem80]   W. Siemasko, *Branched Continued Fractions for Double Power Series*, J. Comp. Appl. Math. **6** (1980), 121–125.

[Stie18]   T. J. Stieltjes, *Recherches sur le fractions continues*, Ann. Fac. Sci. Toulouse Sci. Math. et Sci. Phys. **8** (1894), 1–122; **9** (1895), 1–47. Oevres complétes, Tome **2**, P. Noordhoff, Groningen (1918), 402–566. Also published in Memoirs présent'es par divers savants á l'Académie de Sciences de l'Institut National de France, **33**, 1–196.

[Thie09]   T. N. Thiele, "Interpolationsrechnung" Teubner, Leipzig (1909).

[Thron48] W. J. Thron, *Some Properties of Continued Fraction* $1 + d_0 z + \mathbf{K}(z/(1 + d_n z))$, Bull. Amer. Math. Soc. **54** (1948), 206–218.

[Thron77] W. J. Thron, *Two–Point Padé Tables, T–Fractions and Sequences of Schur*, "Padé and Rational Approximation", (E. B. Saff and R. S. Varga, eds.), Academic Press, New York (1977), 215–226.

[Visc06]   B. Viscovatov, *De la méthode générale pour reduire toutes sortes de quantités en fractions continues*, Mém. Acad. Impériale Sci. St. Petersburg **1** (1803–1806), 226–247.

[Waad64]   H. Waadeland, *On T–Fractions of Functions Holomorphic and Bounded in a Circular Disk*, Det Kgl. Norske Vid. Selsk. Skr. **8**, Trondheim (1964), 1–19.

[Waad66]   H. Waadeland, *A Convergence Property of Certain T–Fraction Expansions*, Det Kgl. Norske Vid. Selsk. Skr. **9**, Trondheim (1966), 1–22.

[Wall48]   H. S. Wall, "Analytic Theory of Continued Fractions", Van Nostrand, New York (1948).

Chapter VI

Hypergeometric functions

About this chapter

Hypergeometric functions $_2F_1$ form an important class of special functions. They satisfy three term recurrence relations which lead to very nice continued fraction expansions. This was pointed out already by Gauss in 1812, [Gauss12]. He obtained a regular C-fraction expansion of the ratio $_2F_1(a,b;c;z)/_2F_1(a,b+1;c+1;z)$, the so-called Gauss fraction. It has very nice convergence properties compared to the hypergeometric series itself. Also other types of continued fraction expansions for ratios of hypergeometric functions have been developed. We shall present some of them here.

The basic hypergeometric functions (or q-hypergeometric functions) $_2\varphi_1$ also have natural connections to continued fractions. The regular C-fraction expansion of $_2\varphi_1(a,b;c;q;z)/_2\varphi_1(a,bq;cq;q;z)$ is the q-analogue of the Gauss fraction. It was developed by Heine in 1847 [Heine47], and we call it the Heine fraction.

As an illustration of the role the hypergeometric functions play in the continued fraction theory, we refer to the appendix. Most of the functions there are related to hypergeometric functions.

291

1   The hypergeometric functions $_2F_1$

1.1   Why and how

Let us look at the hypergeometric series

$$_2F_1(a,b;c;z) = 1 + \frac{ab}{c}\frac{z}{1!} + \frac{a(a+1)b(b+1)}{c(c+1)}\frac{z^2}{2!} + \cdots, \qquad (1.1.1)$$

where the parameters a, b and c are complex constants. For short we denote it by $F(a,b;c;z)$. For obvious reasons we assume that $c \notin \mathbf{Z} \setminus \mathbf{N}$. If $a \in \mathbf{Z} \setminus \mathbf{N}$ or $b \in \mathbf{Z} \setminus \mathbf{N}$ then $F(a,b;c;z)$ reduces to a polynomial. Otherwise the infinite series in (1.1.1) has radius of convergence $= 1$. This can be seen by the ratio test. It converges at $z = 1$ if $\Re(c-a-b) > 0$. (See for instance [AbSt64, p. 556].) The function to which it converges can be extended analytically to the cut plane

$$D = \{z \in \mathbf{C}; |\arg(1-z)| < \pi\}, \qquad (1.1.2)$$

that is, to the complement of the real interval $[1, \infty)$. It is known as the hypergeometric function, or more precisely, the principal branch of the hypergeometric function, and we use the same notation $F(a,b;c;z)$ for this function as for the series. Special examples of such functions are

$$F(1,1;2;z) = -z^{-1}\log(1-z), \qquad \text{(natural logarithm)}$$

$$F(\tfrac{1}{2},1;\tfrac{3}{2};z^2) = \frac{1}{2}z^{-1}\log\left(\frac{1+z}{1-z}\right),$$

$$F(\tfrac{1}{2},1;\tfrac{3}{2};-z^2) = z^{-1}\arctan z,$$

$$F(\tfrac{1}{2},\tfrac{1}{2};\tfrac{3}{2};z^2) = (1-z^2)^{1/2}F(1,1;\tfrac{3}{2};z^2) = z^{-1}\arcsin z,$$

$$F(\tfrac{1}{2},\tfrac{1}{2};\tfrac{3}{2};-z^2) = (1+z^2)^{1/2}F(1,1;\tfrac{3}{2};-z^2)$$

$$= z^{-1}\log\left[z + (1+z^2)^{1/2}\right],$$

$$F(a,b;b;z) = (1-z)^{-a} \qquad \text{for } b \notin \mathbf{Z}\setminus\mathbf{N}.$$

For special values of z we get for instance

$$F(a, b; c; 1) = \frac{\Gamma(c)\Gamma(c - a - b)}{\Gamma(c - a)\Gamma(c - b)} \quad \text{if} \quad \Re(c - a - b) > 0,$$

$$F(a, a + \tfrac{1}{2}; a + \tfrac{5}{6}; \tfrac{1}{9}) = \left(\frac{3}{4}\right)^a \sqrt{\pi} \frac{\Gamma(\tfrac{5}{6} + \tfrac{2}{3}a)}{\Gamma(\tfrac{1}{2} + \tfrac{a}{3})\Gamma(\tfrac{5}{6} + \tfrac{a}{3})}$$

$$\text{if} \quad \frac{5}{6} + \frac{2}{3}a \neq 0, -1, -2, \ldots,$$

$$F(a, 1 - a; c; \tfrac{1}{2}) = \frac{2^{1-c}\Gamma(c)\Gamma(\tfrac{1}{2})}{\Gamma(\tfrac{a+c}{2})\Gamma(\tfrac{c-a+1}{2})}. \tag{1.1.3}$$

For more examples we refer for instance to [AbSt64, p. 556–557], [Bern89], [Erdé53], [Bail64]. In Subsection *3.1* of Chapter I we claimed that

$$\frac{F(a, b; c; z)}{F(a, b + 1; c + 1; z)} \sim 1 + \frac{a_1 z}{1} + \frac{a_2 z}{1} + \frac{a_3 z}{1} + \cdots, \tag{1.1.4}$$

where

$$a_{2n+1} = -\frac{(a + n)(c - b + n)}{(c + 2n)(c + 2n + 1)},$$
$$a_{2n} = -\frac{(b + n)(c - a + n)}{(c + 2n - 1)(c + 2n)}. \tag{1.1.5}$$

We even indicated that $1 + \mathbf{K}(a_n z / 1)$ converges to the function on the left side of (1.1.4) in the cut plane D given by (1.1.2) (which of course is much larger than the convergence disk of radius 1 for the series). We shall justify this. Let us first assume that all $a_n \neq 0$.

Correspondence. By comparing the coefficients of z^n on both sides of the equality, we derived that

$$F(a, b; c; z) = F(a, b + 1; c + 1; z) - \frac{a(c - b)}{c(c + 1)} z F(a + 1, b + 1; c + 2; z). \tag{1.1.6}$$

Since $F(a, b; c; z) = F(b, a; c; z)$ we therefore also have

$$F(a, b + 1; c + 1; z) = F(a + 1, b + 1; c + 2; z)$$
$$- \frac{(b + 1)(c - a + 1)}{(c + 1)(c + 2)} z F(a + 1, b + 2; c + 3; z).$$

This means that $\{P_n(z)\}_{n=0}^{\infty}$, where

$$P_{2n}(z) = F(a+n, b+n; c+2n; z), \quad P_{2n+1} = F(a+n, b+n+1; c+2n+1; z)$$

is a solution of the three-term recurrence relation

$$P_n(z) = P_{n+1}(z) + a_{n+1} z P_{n+2}(z) \qquad \text{for} \quad n = 0, 1, 2, \ldots. \qquad (1.1.7)$$

The correspondence (1.1.4) follows therefore from Corollary 13 in Chapter V.

Convergence. We see from (1.1.5) that $a_n \to -1/4$ as $n \to \infty$. Hence $1 + \mathbf{K}(a_n z/1)$ is limit periodic of loxodromic type for $z \in D$, uniformly on compact subsets $C \subset D$ such that $\infty \notin f(C)$. (See Theorem 28 and Theorem 31 in Chapter III.) That $f(z) = P_0(z)/P_1(z)$ follows then by Theorem 14B in Chapter V.

The point $z = 1$. At this point $1 + \mathbf{K}(a_n z/1) = 1 + \mathbf{K}(a_n/1)$ is limit periodic with $a_n \to -1/4$. Therefore the continued fraction may converge or diverge, depending on how $\{a_n\}$ approaches $-1/4$. In Problem 10 you are asked to prove that $1 + \mathbf{K}(a_n/1)$ converges to $1 - a/c$ if $\Re(c - a - b) > 0$ or if $c = a + b$, and that $1 + \mathbf{K}(a_n/1)$ also converges if $\Re(c - a - b) < 0$. $1 + \mathbf{K}(a_n/1)$ diverges if $c - a - b = it$ with $t \in \mathbf{R} \setminus \{0\}$. For $\Re(c - a - b) > 0$ or $c = a + b$ the value agrees with (1.1.4) since

$$
\begin{aligned}
\frac{F(a, b; c; 1)}{F(a, b+1; c+1; 1)} &= \frac{\Gamma(c)\Gamma(c - a - b)}{\Gamma(c - a)\Gamma(c - b)} \Big/ \frac{\Gamma(c + 1)\Gamma(c - a - b)}{\Gamma(c + 1 - a)\Gamma(c - b)} \\
&= \frac{\Gamma(c)}{\Gamma(c + 1)} \frac{\Gamma(c - a + 1)}{\Gamma(c - a)} = \frac{c - a}{c}.
\end{aligned}
\qquad (1.1.8)
$$

The cut $z > 1$. We shall not go into details here. The fact of the matter is however that the continued fraction diverges for $z > 1$. (See [Lore].)

So far we have assumed that all the coefficients a_n of $\mathbf{K}(a_n z/1)$ are non-zero, such that the regular C-fraction is non-terminating. It remains to look at:

The terminating case. We have that $a_N = 0$ for some $N \in \mathbf{N}$ if either $a \in \mathbf{Z} \setminus \mathbf{N}$, or $b \in \mathbf{Z} \setminus \mathbf{N}_0$, or $c - b \in \mathbf{Z} \setminus \mathbf{N}$, or $c - a \in \mathbf{Z} \setminus \mathbf{N}_0$.

Case 1: $a \in \mathbf{Z} \setminus \mathbf{N}$. Let $a = -k$, $k \in \mathbf{N}_0$. Then $F(a+k, b+k; c+2k; z) = F(a+k, b+k+1; c+2k+1; z) = 1$ and $a_{2k+1} = 0$. By repeated use of (1.1.7) we find that

$$f(z) \;=\; \frac{P_0(z)}{P_1(z)} = 1 + \frac{a_1 z}{1} + \cdots \frac{a_{N-1} z}{1} \frac{a_N z}{+ P_N(z)/P_{N+1}(z)}. \quad (1.1.9)$$

Hence the choice $N = 2k$ in this relation gives

$$\frac{F(-k, b; c; z)}{F(-k, b+1; c+1; z)} \;=\; 1 + \frac{a_1 z}{1} + \frac{a_2 z}{1} + \cdots + \frac{a_{2k} z}{1} \quad (1.1.10)$$

for all $z \in \mathbf{C}$. On the other hand $a_{2k+1} = 0$ so the right hand side of (1.1.10) is equal to $1 + \mathbf{K}(a_n z/1)$. Hence we still have $f(z) = 1 + \mathbf{K}(a_n z/1)$. In a similar way we can prove that $1 + \mathbf{K}(a_n z/1)$ has the value as given by the left side of (1.1.4) if $b = -k$ for a $k \in \mathbf{N}$.

Case 2: $c - b \in \mathbf{Z} \setminus \mathbf{N}$. Let $c - b = -k$, $k \in \mathbf{N}_0$. We shall use the well known formula (see [Erdé53, p. 69])

$$F(a, b; c; z) = (1 - z)^{c-a-b} F(c - b, c - a; c; z).$$

Since $c - b = -k$, it follows that

$$F(a+k, b+k; c+2k; z) = (1-z)^{c-a-b} F(0, c-a+k; c+2k; z) = (1-z)^{c-a-b}$$

and similarly

$$F(a + k, b + k + 1; c + 2k + 1; z) = (1 - z)^{c-a-b}.$$

Further, $a_{2k+1} = 0$, so (1.1.10) still holds and its right side is equal to $1 + \mathbf{K}(a_n z/1)$. The argument for the case $c - a \in \mathbf{Z} \setminus \mathbf{N}_0$ is essentially the same.

Of course, if $z = 0$ then the continued fraction also terminates. Then both sides of (1.1.4) are equal to 1, and equality holds trivially. But this case is already covered by the previous arguments. ($z = 0 \in D$.)

We collect all these results to get:

Theorem 1 (Gauss fractions) *Let a, b and c be complex constants with $c \notin \mathbf{Z} \setminus \mathbf{N}$, and let $\{a_n\}$ be given by (1.1.5). Then:*

(A) $1 + \mathbf{K}(a_n z/1) \sim F(a, b; c; z)/F(a, b+1; c+1; z)$.

(B) $1 + \mathbf{K}(a_n z/1) = f(z) = F(a, b; c; z)/F(a, b+1; c+1; z)$ *in the cut plane* $D = \{z \in \mathbf{C}; |\arg(1-z)| < \pi\}$. *That is,* $1 + \mathbf{K}(a_n z/1)$ *converges to the well defined, meromorphic function* $f(z)$ *in* D. *The convergence is uniform on every compact subset of* $\{z \in D; f(z) \neq \infty\}$.

(C) $1 + \mathbf{K}(a_n/1) = f(1) = 1 - a/c$ *if* $\Re(c - a - b) > 0$ *or* $c = a + b$. *That is,* $1 + \mathbf{K}(a_n/1)$ *converges to* $f(1)$ *(given as in B) under these conditions. If* $\Re(c - a - b) < 0$ *then* $1 + \mathbf{K}(a_n/1)$ *converges to* $\lim_{z \to 1-} f(z)$.

(D) $1 + \mathbf{K}(a_n z/1) = f(z) = F(a, b; c; z)/F(a, b+1; c+1; z)$ *for all* $z \in \mathbf{C}$ *if the continued fraction terminates.*

(E) $1 + \mathbf{K}(a_n z/1)$ *diverges if all* $a_n \neq 0$ *and either* $z = 1$ *with* $c - a - b = it; t \in \mathbf{R} \setminus \{0\}$, *or* $z > 1$.

1.2   A special case

If $b = 0$, then $F(a, b; c; z) = F(a, 0; c; z) = 1$. This means that Theorem 1 can be used to obtain a continued fraction expansion of $F(a, 0; c; z)/F(a, 1; c+1; z) = 1/F(a, 1; c+1; z)$. Let us replace c by $c - 1$. Then we get

$$
\begin{aligned}
F(a, 1; c; z) &= \cfrac{1}{1-} \cfrac{\dfrac{a(c-1)}{(c-1)c}z}{1} - \cfrac{\dfrac{(c-a)1}{c(c+1)}z}{1} - \cfrac{\dfrac{(a+1)c}{(c+1)(c+2)}z}{1} -\cdots \\[2mm]
&\approx \cfrac{1}{1-} \cfrac{az}{c-} \cfrac{1(c-a)z}{c+1-} \cfrac{(a+1)cz}{c+2-} \\[2mm]
&\qquad \cfrac{2(c-a+1)z}{c+3-} \cfrac{(a+2)(c+1)z}{c+4-}\cdots \qquad (1.2.1)
\end{aligned}
$$

for $c \notin \mathbf{Z} \setminus \mathbf{N}$, $c \neq 1$, $z \in D = \{z \in \mathbf{C}; |\arg(1-z)| < \pi\}$.

Example 1 We apply Theorem 1 to the special examples mentioned in the beginning of Subsection *1.1*. We get

$$\log(1-z) \;=\; -zF(1,1;2;z) =$$

$$-\frac{z}{1-}\frac{1^2z}{2-}\frac{1^2z}{3-}\frac{2^2z}{4-}\frac{2^2z}{5-}\frac{3^2z}{6-}\frac{3^2z}{7}-\cdots$$

$$\text{for} \quad z \in D,$$

$$\log\left(\frac{1+z}{1-z}\right) \;=\; 2zF(\tfrac{1}{2},1;\tfrac{3}{2};z^2)$$

$$=\; \frac{2z}{1-}\frac{\frac{1}{2}z^2}{\frac{3}{2}-}\frac{1^2z^2}{\frac{5}{2}-}\frac{\left(\frac{3}{2}\right)^2z^2}{\frac{7}{2}-}\frac{2^2z^2}{\frac{9}{2}}-\cdots$$

$$\approx\; \frac{2z}{1-}\frac{1^2z^2}{3-}\frac{2^2z^2}{5-}\frac{3^2z^2}{7-}\frac{4^2z^2}{9}-\cdots, z^2 \in D,$$

$$\arctan z \;=\; zF(\tfrac{1}{2},1;\tfrac{3}{2};-z^2)$$

$$=\; \frac{z}{1+}\frac{1^2z^2}{3+}\frac{2^2z^2}{5+}\frac{3^2z^2}{7+}\frac{4^2z^2}{9}+\cdots$$

$$\text{for} \quad -z^2 \in D,$$

$$\frac{\arcsin z}{\left(1-z^2\right)^{1/2}} \;=\; zF(1,1;\tfrac{3}{2};z^2)$$

$$=\; \frac{z}{1-}\frac{z^2}{\frac{3}{2}-}\frac{1\cdot\frac{1}{2}z^2}{\frac{5}{2}-}\frac{2\cdot\frac{3}{2}z^2}{\frac{7}{2}-}\frac{2\cdot\frac{3}{2}z^2}{\frac{9}{2}-}\frac{3\cdot\frac{5}{2}z^2}{\frac{11}{2}}-\cdots$$

$$\approx\; \frac{z}{1-}\frac{1\cdot2z^2}{3-}\frac{1\cdot2z^2}{5-}\frac{3\cdot4z^2}{7-}$$

$$\frac{3\cdot4z^2}{9-}\frac{5\cdot6z^2}{11}-\cdots \qquad \text{for} \quad z^2 \in D,$$

$$\frac{\log\left[z+\left(1+z^2\right)^{1/2}\right]}{\left(1+z^2\right)^{1/2}} \;=\; zF(1,1;\tfrac{3}{2};-z^2)$$

$$=\; \frac{z}{1+}\frac{1\cdot2z^2}{3+}\frac{1\cdot2z^2}{5+}\frac{3\cdot4z^2}{7+}\frac{3\cdot4z^2}{9}+\cdots$$

$$\text{for} \quad -z^2 \in D.$$

◇

1.3   Choice of approximants

What kind of approximants should one choose for $1 + \mathbf{K}(a_n z/1)$ in Theorem 1? Since $a_n z \to -z/4$ we can use the idea from Subsection 5.5 in Chapter III and use

$$S_n(x(z)) = 1 + \frac{a_1 z}{1} + \frac{a_2 z}{1} + \cdots + \frac{a_n z}{1 + x(z)} \tag{1.3.1}$$

where

$$x(z) = \left(\sqrt{1-z} - 1\right)/2 \tag{1.3.2}$$

$$\text{with} \quad \Re(\sqrt{1-z}) > 0 \quad \text{for } z \in D.$$

(If one needs rational approximants, then $\sqrt{1-z}$ can be approximated by a constant, a polynomial or a rational function.) But we can do better. From (1.1.5) we find that

$$a_{2n+1} = -\frac{1}{4} + \frac{4n(b - a + \frac{1}{2}) + c^2 - 4a(c - b) + c}{4(c + 2n)(c + 2n + 1)}$$

and

$$a_{2n} = -\frac{1}{4} + \frac{4n(a - b - \frac{1}{2}) + c^2 - 4b(c - a) - c}{4(c + 2n - 1)(c + 2n)}.$$

Hence, writing

$$\delta_n^{(0)}(z) = a_n z - x(z)(1 + x(z)) = \left(a_n + \frac{1}{4}\right)z \tag{1.3.3}$$

we find that

$$\lim_{n \to \infty} \frac{\delta_{n+1}^{(0)}(z)}{\delta_n^{(0)}(z)} = \begin{cases} -1 & \text{if } a - b - \frac{1}{2} \neq 0, \\ 1 & \text{if } a - b - \frac{1}{2} = 0 \\ & \text{and } c^2 - c - 4bc + 4b^2 + 2b \neq 0, \end{cases} \tag{1.3.4}$$

so that

$$w_n^{(1)}(z) = \begin{cases} x(z) + \delta_{n+1}^{(0)}(z) & \text{if } a - b - \frac{1}{2} \neq 0, \\ x(z) + \dfrac{\delta_{n+1}^{(0)}(z)}{1 + 2x(z)} & \text{if } a - b - \frac{1}{2} = 0 \end{cases} \tag{1.3.5}$$

is an even better choice according to Theorem 33 in Chapter III. (If $a - b - \frac{1}{2} = 0$ and $c^2 - c - 4bc + 4b^2 + 2b = 0$, then all $\delta_n^{(0)} \equiv 0$ and $1 + \mathbf{K}(a_n z/1)$ is periodic with value $x(z)$.) Continuing this process we can write

$$\delta_n^{(1)}(z) = a_n z - w_{n-1}^{(1)}(z)\left(1 + w_n^{(1)}(z)\right) \tag{1.3.6}$$

to find $w_n^{(2)}$. We distinguish between two cases.

Case 1: $P = a - b - 1/2 = 0$.

Suppressing the variable z we have

$$
\begin{aligned}
\delta_n^{(1)} &= a_n z - \left(x + \frac{\delta_n^{(0)}}{1 + 2x}\right)\left(1 + x + \frac{\delta_{n+1}^{(0)}}{1 + 2x}\right) \\
&= \delta_n^{(0)} - \frac{x\delta_{n+1}^{(0)}}{1 + 2x} - \frac{\delta_n^{(0)}(1 + x)}{1 + 2x} - \frac{\delta_n^{(0)}\delta_{n+1}^{(0)}}{(1 + 2x)^2} \\
&= \frac{x}{1 + 2x}(\delta_n^{(0)} - \delta_{n+1}^{(0)}) - \frac{\delta_n^{(0)}\delta_{n+1}^{(0)}}{(1 + 2x)^2}, \tag{1.3.7}
\end{aligned}
$$

where $\delta_n^{(0)} = Qz/(c + n - 1)(c + n)$, $Q = (c - 2b)(c - 2b - 1)/4$. Hence $\delta_{n+1}^{(1)}/\delta_n^{(1)} \longrightarrow 1$. In fact, by induction one finds that $\delta_{n+1}^{(m)}/\delta_n^{(m)} \longrightarrow 1$ as $n \longrightarrow \infty$ for every $m \geq 0$, just as in Example 25 in Chapter III. So we choose

$$w_n^{(m+1)} = w_n^{(m)} + \delta_{n+1}^{(m)}/(1 + 2x). \tag{1.3.8}$$

Case 2: $P = a - b - 1/2 \neq 0$.

Now we get

$$
\begin{aligned}
\delta_n^{(1)} &= a_n z - (x + \delta_n^{(0)})(1 + x + \delta_{n+1}^{(0)}) \\
&= \delta_n^{(0)} - x\delta_{n+1}^{(0)} - \delta_n^{(0)}(1 + x) - \delta_n^{(0)}\delta_{n+1}^{(0)} \\
&= -x(\delta_n^{(0)} + \delta_{n+1}^{(0)}) - \delta_n^{(0)}\delta_{n+1}^{(0)}, \tag{1.3.9}
\end{aligned}
$$

where

$$\delta_{2n}^{(0)} = \frac{Pn + Q_0}{(c + 2n - 1)(c + 2n)}z, \qquad \delta_{2n+1}^{(0)} = \frac{-Pn + Q_1}{(c + 2n)(c + 2n + 1)}z$$

with $Q_0 = (c^2 - c)/4 - b(c - a)$ and $Q_1 = (c^2 + c)/4 - a(c - b)$. Hence

$$\delta_{2n}^{(1)} = -z\frac{[4(P + Q_0 + Q_1)x + P^2 z]n^2 + \quad \text{lower degree terms}}{(c + 2n - 1)(c + 2n)^2(c + 2n + 1)},$$

and

$$\delta_{2n+1}^{(1)} = -z\frac{[4(Q_0 + Q_1)x + P^2 z]n^2 + \quad \text{lower degree terms}}{(c + 2n)(c + 2n + 1)^2(c + 2n + 2)},$$

and thus

$$\lim_{n \to \infty} \frac{\delta_{2n}^{(1)}}{\delta_{2n-1}^{(1)}} = \frac{4(P + Q_0 + Q_1)x + P^2 z}{4(Q_0 + Q_1)x + P^2 z} = t_1, \qquad (1.3.10)$$

and

$$\lim_{n \to \infty} \frac{\delta_{2n+1}^{(1)}}{\delta_{2n}^{(1)}} = \frac{4(Q_0 + Q_1)x + P^2 z}{4(P + Q_0 + Q_1)x + P^2 z} = t_2 = \frac{1}{t_1}, \qquad (1.3.11)$$

just as in Example 26 in Chapter III. Hence, by Theorem 34 in Chapter III we get faster convergence to the right value if we choose the approximants $S_n(w_n^{(2)})$ where

$$w_n^{(2)} = w_n^{(1)} + \frac{\delta_{n+1}^{(1)}}{1 + x + xq_n}, \qquad (1.3.12)$$

and

$$q_{2n} = q_0 = \frac{(1 + x)t_1 - x}{1 + x - xt_1}, \quad q_{2n+1} = q_1 = \frac{1}{q_0}. \qquad (1.3.13)$$

Let us look at some examples.

Example 2 Let $a = 1/2$, $b = 3/2$ and $c = 5/2$ in Theorem 1, so that

$$\frac{F(\frac{1}{2}, \frac{3}{2}; \frac{5}{2}; z)}{F(\frac{1}{2}, \frac{5}{2}; \frac{7}{2}; z)} \sim 1 - \frac{\frac{1 \cdot 2}{5 \cdot 7}z}{1} - \frac{\frac{5 \cdot 6}{7 \cdot 9}z}{1} - \frac{\frac{3 \cdot 4}{9 \cdot 11}z}{1} - \frac{\frac{7 \cdot 8}{11 \cdot 13}z}{1} - \cdots. \qquad (1.3.14)$$

We have $P = a - b - 1/2 \neq 0$, so we can use

$$w_n^{(1)}(z) = x(z) + \delta_{n+1}^{(0)}(z) = x(z) + (a_{n+1} + 1/4)z.$$

Since by $(1.3.10) - (1.3.11)$

$$\lim_{n \to \infty} \frac{\delta_{2n+1}^{(1)}(z)}{\delta_{2n}^{(1)}(z)} = \frac{3z - 2x(z)}{3z - 10x(z)},$$

$$\lim_{n \to \infty} \frac{\delta_{2n+2}^{(1)}(z)}{\delta_{2n+1}^{(1)}(z)} = \frac{3z - 10x(z)}{3z - 2x(z)},$$

we choose

$$w_n^{(2)}(z) = w_n^{(1)}(z) + \frac{\delta_{n+1}^{(1)}(z)}{1 + x(z) + q_n(z)x(z)}$$

where

$$q_{2n}(z) = q_0(z) = \frac{3z - 10x(z) - 8x^2(z)}{3z - 2x(z) + 8x^2(z)},$$

$$q_{2n+1}(z) = q_1(z) = \frac{1}{q_0(z)}.$$

We stop here, although we could have continued the process. We shall instead study the effect numerically, for given values of z.

We first choose $z = -1$. Then $1 + \mathbf{K}(a_n z/1)$ is a continued fraction with positive elements, and we expect fast convergence. The first 8 approximants are given in Table 1. The value of the continued fraction is 1.0397662053001, correctly rounded to 14 digits.

n	$S_n(0)$	$S_n(x)$	$S_n(w_n^{(1)})$	$S_n(w_n^{(2)})$
1	1.057...	1.047...	1.03986...	1.039791...
2	1.0387...	1.0409...	1.03963...	1.03976603...
3	1.0401...	1.03988...	1.0397675...	1.03976642...
4	1.039736...	1.03978...	1.0397647...	1.0397662018...
5	1.039774...	1.0397685...	1.039766226...	1.0397662079...
6	1.0397653...	1.03976659...	1.039766182...	1.03976620523...

Table 1: $z = -1$. $f(-1) = 1.0397662053001$

Notice also the nice oscillation properties of $\{S_n(0)\}$, $\{S_n(w_n^{(1)})\}$ and $\{S_n(w_n^{(2)})\}$. In fact, Theorem 4 in Chapter III can be applied to determine when $S_n(w_n)$ oscillates regularly about its limit.

In Table 2 we show how fast the various types of approximants reach the value of the continued fraction, correctly rounded to the given number of digits for some values of z. The number N is the smallest index such that the approximants take this value for all indices $n \geq N$.

z	Value of the continued fraction	N for $S_n(0)$	N for $S_n(x)$	N for $S_n(w_n^{(1)})$	N for $S_n(w_n^{(2)})$
-1	1.039766	7	7	5	2
-2000	1.198201	202	161	141	93
$10 + 0.1i$	$1.2152 - 0.1424i$	2991	1149	291	37
$100 + 0.1i$	$1.21728 - 0.018383i$	$\gg 5000$	> 5000	1483	183

Table 2.

As expected, the convergence is slower when z is close to the cut $z \geq 1$ of D. But it is for such values of z that the gain by using $S_n(w_n^{(1)})$ or $S_n(w_n^{(2)})$ is most dramatic.

Another question is: How can we use the (approximate) value of (1.3.14) to find for instance $F(1/2, 5/2; 7/2; z)$? We have

$$F(\tfrac{1}{2}, \tfrac{5}{2}; \tfrac{7}{2}; z) = \frac{F(\tfrac{1}{2}, \tfrac{5}{2}; \tfrac{7}{2}; z)}{F(\tfrac{1}{2}, \tfrac{3}{2}; \tfrac{5}{2}; z)} \cdot \frac{F(\tfrac{1}{2}, \tfrac{3}{2}; \tfrac{5}{2}; z)}{F(-\tfrac{1}{2}, \tfrac{3}{2}; \tfrac{3}{2}; z)} F(-\tfrac{1}{2}, \tfrac{3}{2}; \tfrac{3}{2}; z)$$

where $F(-\tfrac{1}{2}, \tfrac{3}{2}; \tfrac{3}{2}; z)$ is known to be equal to $(1 - z)^{1/2}$ for $z \in D$ and where

$$\frac{F(-\tfrac{1}{2}, \tfrac{3}{2}; \tfrac{3}{2}; z)}{F(\tfrac{1}{2}, \tfrac{3}{2}; \tfrac{5}{2}; z)} = 1 - \frac{\frac{3 \cdot 4}{3 \cdot 5} z}{1} \frac{\frac{1 \cdot 2}{5 \cdot 7} z}{1} \frac{\frac{5 \cdot 6}{7 \cdot 9} z}{1} \frac{\frac{3 \cdot 4}{9 \cdot 11} z}{1} \frac{}{- \cdots} = 1 - \frac{\frac{3 \cdot 4}{3 \cdot 5} z}{f(z)}$$

by Theorem 1, and $f(z) = F(\tfrac{1}{2}, \tfrac{3}{2}; \tfrac{5}{2}; z)/F(\tfrac{1}{2}, \tfrac{5}{2}; \tfrac{7}{2}; z)$ is the value of (1.3.14). Hence

$$F(\tfrac{1}{2}, \tfrac{5}{2}; \tfrac{7}{2}; -1) \approx \frac{1}{1.039766} \cdot \frac{1}{1 + \frac{4/5}{1.039766}} \cdot \sqrt{2} \approx 0.768692.$$

◇

Example 3 According to Example 1 we have

$$\log\frac{1+z}{1-z} = 2zF(\tfrac{1}{2},1;\tfrac{3}{2};z^2) \sim \frac{2z}{1} - \frac{1^2z^2}{3} - \frac{2^2z^2}{5} - \frac{3^2z^2}{7} - \cdots$$

for $z^2 \in D = \{w \in \mathbf{C}; |\arg(1-w)| < \pi\}$. Or, equivalently, $\log((1+z)/(1-z)) \sim 2z/(1 + \mathbf{K}_{n=2}^{\infty}(a_n z/1))$ where

$$1 + \mathop{\mathbf{K}}_{n=2}^{\infty}\frac{a_n z}{1} \sim F(\frac{1}{2},0;\frac{1}{2};z^2)/F(\frac{1}{2},1;\frac{3}{2};z^2).$$

Therefore we are in Case 1 where $P = 0$, and by (1.3.8) we choose

$$w_n^{(1)}(z) = x(z) + \frac{\delta_{n+1}^{(0)}(z)}{1+2x(z)} \qquad \text{for} \quad n = 1,2,3,\ldots,$$

$$w_n^{(2)}(z) = w_n^{(1)}(z) + \frac{\delta_{n+1}^{(1)}(z)}{1+2x(z)},$$

and so on. For $z = 5i$ we have

$$\log\frac{1+z}{1-z} = \log\frac{1+5i}{1-5i} = \log\left(e^{i2\theta}\right) = i2\theta,$$

where $\theta = \arg(1+5i) \approx 1.37340077$. That is, the continued fraction converges to $2.7468015i$, correctly rounded to 8 digits. Its first approximants are given in Table 3.

n	$S_n(0)$	$S_n(x)$	$S_n(w_n^{(1)})$	$S_n(w_n^{(2)})$
1	$10.0\ldots i$	$3.27\ldots i$	$2.891\ldots i$	
2	$1.07\ldots i$	$2.679\ldots i$	$2.731\ldots i$	$2.7427\ldots i$
3	$4.79\ldots i$	$2.7658\ldots i$	$2.750\ldots i$	$2.7475\ldots i$
4	$1.85\ldots i$	$2.7399\ldots i$	$2.7457\ldots i$	$2.74660\ldots i$
5	$3.53\ldots i$	$2.7496\ldots i$	$2.74716\ldots i$	$2.746860\ldots i$
6	$2.308\ldots i$	$2.7454\ldots i$	$2.74665\ldots i$	$2.746780\ldots i$
7	$3.08\ldots i$	$2.7474\ldots i$	$2.746863\ldots i$	$2.7468095\ldots i$
8	$2.540\ldots i$	$2.74647\ldots i$	$2.746773\ldots i$	$2.7467982\ldots i$

Table 3: $z = 5i$. $f(z) = 2.7468015i$

A similar exposition as given in Table 2 is given for this continued fraction in Table 4.

$\diamondsuit$

z	Value of the continued fraction	N for $S_n(0)$	N for $S_n(x)$	N for $S_n(w_n^{(1)})$	N for $S_n(w_n^{(2)})$
$5i$	$2.7468\,i$	30	12	8	6
$1+i$	$0.804719 + 2.034444i$	15	9	7	6
$5+0.1i$	$0.4053 + 3.133i$	1448	69	21	10

Table 4.

1.4 Other continued fraction expansions

So far we have looked at regular C-fraction expansions of ratios $F(a, b; c; z)/F(a, b + 1; c + 1; z)$, usually called Gauss fractions. They are very useful. In this section we shall briefly mention two other classical expansions. The first one is due to Nörlund [Nörl24].

Theorem 2 (Nörlund fractions) *Let a, b and c be complex constants with $c \notin \mathbf{Z} \setminus \mathbf{N}$. Then:*

(A) The continued fraction

$$1 - \cfrac{a + b + 1}{c}z + \cfrac{\dfrac{(a+1)(b+1)}{c(c+1)}(z - z^2)}{1 - \dfrac{a+b+3}{c+1}z} + \cfrac{\dfrac{(a+2)(b+2)}{(c+1)(c+2)}(z - z^2)}{1 - \dfrac{a+b+5}{c+2}z} + \cdots$$

$$(1.4.1)$$

corresponds at $z = 0$ to the series

$$L(z) = \frac{F(a, b; c; z)}{F(a + 1, b + 1; c + 1; z)}. \qquad (1.4.2)$$

(B) The continued fraction (1.4.1) converges to the function in (1.4.2) if it terminates, or if $\Re(z) < 1/2$, or if $z = 1/2$ and $|\Im(a + b)| < \Re(2c - a - b - 1)$.

Proof : **(A):** $\{F(a+n, b+n; c+n; z)\}_{n=0}^{\infty}$ is a solution of the three-term recurrence relation

$$
\begin{aligned}
P_n(z) &= \left(1 - \frac{a+b+2n+1}{c+n}z\right)P_{n+1} \\
&+ \frac{(a+n+1)(b+n+1)}{(c+n)(c+n+1)}z(1-z)P_{n+2}(z) \\
&\quad \text{for} \quad n = 0, 1, 2, \ldots .
\end{aligned}
\tag{1.4.3}
$$

(This can be verified by comparing the coefficients of the power series involved.)

(B): An equivalence transformation brings (1.4.1) to the form $b_0(z) + \mathbf{K}(c_n(z)/1)$ where

$$
c_n(z) = \frac{(a+n)(b+n)(z-z^2)}{(c+n-(a+b+2n+1)z)(c+n-1-(a+b+2n-1)z)}
$$

for $n > 2$. Hence $\lim_{n \to \infty} c_n(z) = c^*(z) = (z - z^2)/(1 - 2z)^2$. Since $c^*(z)$ is real and negative $< -1/4$ if and only if $\Re(z) = 1/2$, $z \neq 1/2$, it follows that (1.4.1) converges to a meromorphic function for $\Re(z) \neq 1/2$. By Theorem 14B in Chapter V we find that this function is (1.4.2) in the domain $\Re(z) < 1/2$.

For $z = 1/2$ we find that $c_n(z)$ reduces to

$$
c_n(1/2) = \frac{(a+n)(b+n)}{(2c-a-b-1)^2} \quad \text{for} \quad n > 2.
$$

For $\Re(2c-a-b) > 1$ we let $\alpha = -\arg(2c-a-b-1)$ where $|\alpha| < \pi/2$. We plan to use the parabola theorem, Theorem 20 in Chapter III, to prove convergence of $b_0(1/2) + \mathbf{K}(c_n(1/2)/1)$. We have for $n > 2$ that $c_n(1/2) \in P_\alpha$ from some n on if and only if

$$
\left|\frac{(a+n)(b+n)}{(2c-a-b-1)^2}\right| - \Re\left(\frac{(a+n)(b+n)}{|2c-a-b-1|^2}\right) \leq \frac{1}{2}\cos^2\alpha,
$$

i. e.

$$
|n^2 + (a+b)n + ab| \leq n^2 + \Re((a+b)n + ab) + \frac{1}{2}|2c-a-b-1|^2\cos^2\alpha
$$

from som n on. Squaring this equation and collecting terms of the same degree in n gives

$$n^2 \left([\Im(a+b)]^2 - |2c - a - b - 1|^2 \cos^2 \alpha \right) + \quad \text{lower terms} \ \leq 0$$

which holds from some n on if $|\Im(a+b)| < |2c - a - b - 1| \cos \alpha = \Re(2c - a - b - 1)$.

The terminating case can be proved in the same way as in Theorem 1.

■

The Nörlund fraction (1.4.1) can be multiplied by c and simplified somewhat by means of an equivalence transformation to give

$$
\begin{aligned}
c - (a+b+1)z \ &+ \ \frac{(a+1)(b+1)(z-z^2)}{c+1-(a+b+3)z} + \\
&\frac{(a+2)(b+2)(z-z^2)}{c+2-(a+b+5)z} +\cdots \\
\sim cL(z) \ = \ &c\frac{F(a,b;c;z)}{F(a+1,b+1;c+1;z)}.
\end{aligned}
\tag{1.4.4}
$$

By substituting $z = 1 - u$ in (1.4.1) we can also clarify what happens if $\Re(z) > 1/2$ or if $z = 1/2$ with $\Re(2c - a - b) < 1$:

Corollary 3 *Let the continued fraction* (1.4.1) *be non-terminating. Then it converges to*

$$\frac{c - a - b - 1}{c} \cdot \frac{F(a,b;a+b+1-c;1-z)}{F(a+1,b+1;a+b+2-c;1-z)}$$

if either $\Re(z) > 1/2$ *or* $z = 1/2$ *with* $|\Im(a+b)| < -\Re(2c - a - b - 1)$.

Proof :   With $z = 1 - u$, the continued fraction (1.4.1) can be written

$$\frac{1}{c}\left\{ c - a - b - 1 + (a+b+1)u + \cfrac{(a+1)(b+1)(u-u^2)}{c+1-a-b-3+(a+b+3)u} \right.$$
$$\left. + \cfrac{(a+2)(b+2)(u-u^2)}{c+2-a-b-5+(a+b+5)u+\cdots} \right\}$$
$$= -\frac{1}{c}\left\{ c^* - (a+b+1)u - \cfrac{(a+1)(b+1)(u-u^2)}{-c^*-1+(a+b+3)u} \right.$$
$$\left. + \cfrac{(a+2)(b+2)(u-u^2)}{-c^*-2+(a+b+5)u+\cdots} \right\}$$

where $c^* = -c + a + b + 1$. An equivalence transformation brings this over to the form

$$-\frac{1}{c}\left\{ c^* - (a+b+1)u + \cfrac{(a+1)(b+1)(u-u^2)}{c^*+1-(a+b+3)u} \right.$$
$$\left. + \cfrac{(a+2)(b+2)(u-u^2)}{c^*+2-(a+b+5)u+\cdots} \right\}$$

which we know converges to

$$-\frac{1}{c}\cdot c^* \frac{F(a,b;c^*;u)}{F(a+1,b+1;c^*+1;u)}$$

for $\Re(u) < 1/2$ or for $u = 1/2$ with $\Im(a+b)| < \Re(2c^* - a - b - 1)$ by Theorem 2 and (1.4.4). Substituting $u = 1 - z$ and $c^* = -c + a + b + 1$ gives the result. ∎

Pfaff's transformation

$$F(a,b;c;z) = (1-z)^{-b} F(c-a,b;c;z/(z-1)) \qquad (1.4.5)$$

can be verified (formally) by comparing the coefficients of the power series $\in \mathbf{L}_0$ on both sides. If we apply this in Theorem 2 and use the substitution $z/(z-1) \to z$ in (1.4.1) − (1.4.2), we get a continued fraction which essentially is due to Euler, [Euler27], [Euler67]:

Theorem 4 (Euler fractions) *Let a, b and c be complex constants with $c \notin \mathbf{Z} \setminus \mathbf{N}$. Then:*

(A) The general T-fraction

$$c + (b - a + 1)z - \cfrac{(c - a + 1)(b + 1)z}{c + 1 + (b - a + 2)z} \cfrac{(c - a + 2)(b + 2)z}{-c + 2 + (b - a + 3)z} - \cdots$$

$$(1.4.6)$$

corresponds at $z = 0$ *to*

$$cF(a, b; c; z)/F(a, b + 1; c + 1; z) \qquad (1.4.7)$$

and converges to $(1.4.7)$ *if* $|z| < 1$, *or if* $z = -1$ *with* $|\Im(c - a + b)| < \Re(c + a - b - 1)$, *or if* $(1.4.6)$ *terminates.*

(B) If $b - a \neq -2, -3, -4, \ldots$, *then* $(1.4.6)$ *corresponds at* $z = \infty$ *to*

$$(b - a + 1)z \cdot \frac{F(b - c + 1, b; b - a + 1; 1/z)}{F(b - c + 1, b + 1; b - a + 2; 1/z)} \qquad (1.4.8)$$

and converges to $(1.4.9)$ *if* $|z| > 1$ *or if* $z = -1$ *with* $|\Im(c - a + b)| < -\Re(c + a - b - 1)$.

Remark: Strictly speaking it is $(1.4.6)$ divided by c which is a T-fraction in the usual sense, since this can be written on the form $1 + G_0 z + \mathbf{K}(F_n z/(1 + G_n z))$ where

$$F_n = -\frac{(c - a + n)(b + n)}{(c + n - 1)(c + n)}, \quad G_n = \frac{b - a + n + 1}{c + n}.$$

Proof (A): By Theorem 2 and (1.4.4) we know that

$$
c - (a+b+1)\xi + \cfrac{(a+1)(b+1)(1-\xi)\xi}{c+1-(a+b+3)\xi}
$$

$$
+ \cfrac{(a+2)(b+2)(1-\xi)\xi}{c+2-(a+b+5)\xi} + \cdots
$$

$$
\approx (1-\xi)\left\{ c\left(1+\frac{\xi}{1-\xi}\right) - (a+b+1)\frac{\xi}{1-\xi} \right.
$$

$$
+ \cfrac{(a+1)(b+1)\xi/(1-\xi)}{(c+1)\left(1+\dfrac{\xi}{1-\xi}\right) - (a+b+3)\dfrac{\xi}{1-\xi}}
$$

$$
\left. + \cfrac{(a+2)(b+2)\xi/(1-\xi)}{(c+2)\left(1+\dfrac{\xi}{1-\xi}\right) - (a+b+5)\dfrac{\xi}{1-\xi}} + \cdots \right\}
$$

$$
= (1-\xi)\left\{ c + (-c+a+b+1)z - \cfrac{(a+1)(b+1)z}{c+1+(a+b+3-c-1)z} \right.
$$

$$
\left. \cfrac{(a+2)(b+2)z}{-c+2+(a+b+5-c-2)z-\cdots} \right\}
$$

$$
= (1-\xi)\left\{ c + (-c+a+b+1)z - \cfrac{(a+1)(b+1)z}{c+1+(a+b-c+2)z} \right.
$$

$$
\left. \cfrac{(a+2)(b+2)z}{-c+2+(a+b-c+3)z-\cdots} \right\} \tag{1.4.9}
$$

where $z = -\xi/(1-\xi)$, converges to

$$
c\frac{F(a,b;c;\xi)}{F(a+1,b+1;c+1;\xi)} = c\frac{(1-\xi)^{-b}F(c-a,b;c;z)}{(1-\xi)^{-b-1}F(c-a,b+1;c+1;z)} \tag{1.4.10}
$$

for $\Re(\xi) < 1/2$ or $\xi = 1/2$ with $|\Im(a+b)| < \Re(2c-a-b-1)$. We can cancel the factor $(1-\xi)$. Replacing a by $c-a$ gives the result since $\Re(\xi) < 1/2 \iff |z| < 1$, and $\xi = 1/2 \iff z = -1$.

(B): By Corollary 3 we find that for $\Re(\xi) > 1/2$, i. e. for $|z| > 1$, or for $\xi = 1/2$ with $|\Im(a+b)| < -\Re(2c-a-b-1)$, the continued fraction

(1.4.9) converges to

$$-(a+b+1-c)\frac{F(a,b;a+b+1-c;1-\xi)}{F(a+1,b+1;a+b+2-c;1-\xi)}$$

$$= -(a+b+1-c)\frac{\xi^{-b}F\left(b+1-c,b;a+b+1-c;\frac{1-\xi}{-\xi}\right)}{\xi^{-b-1}F\left(b+1-c,b+1;a+b+2-c;\frac{1-\xi}{-\xi}\right)}.$$

Again we replace a by $c-a$ and $\xi/(1-\xi)$ by $-z$ to get the result. ∎

Example 4 Let $b=0$ in (1.4.1) and replace a by $a-1$. Then Theorem 2 leads to the expression

$$F(a,1;c+1;z) = \frac{c}{c-}\frac{a\cdot 1(z-z^2)}{az+c+1-}\frac{(a+1)\cdot 2(z-z^2)}{(a+2)z+c+2-(a+4)z}+\cdots$$
$$(1.4.11)$$

for $\Re(z)<1/2$. Similarly, with $b=0$ in (1.4.6) – (1.4.7) we find that

$$F(a,1;c+1;z) = \frac{c}{c+(1-a)z-}\frac{(c-a+1)\cdot 1z}{c+1+(2-a)z-}\frac{(c-a+2)\cdot 2z}{c+2+(3-a)z}-\cdots$$
$$(1.4.12)$$

for $|z|<1$. Hence, also counting (1.2.1), we have three different continued fraction expansions for $F(a,1;c+1;z)$. They converge in somewhat different domains, but let us compare them for the function

$$F(1,1;2;z) = -z^{-1}\log(1-z)$$

for some values of z where they all are valid. All three of them have the form $\mathbf{K}(a_n(z)/b_n(z))$ where the limits

$$\lim_{n\to\infty}\frac{a_n(z)}{b_{n-1}(z)b_n(z)} = a(z)$$

exist. Hence we use the approximants $S_n(b_n(z)x(z))$ where

$$x(z) = \left(\sqrt{1+4a(z)}-1\right)/2 \qquad \text{where} \quad \Re\sqrt{1+4a(z)}>0.$$

With N as used in Table 2 and 4, the results are given in Table 5.

z	$-z^{-1}\log(1-z)$	N for (1.2.1)	N for (1.4.11)	N for (1.4.12)
0.49	1.37417	7	139	13
$0.99i$	$0.788256 + 0.345024i$	8	23	299
-0.99	0.6950855	9	17	ca 700
$0.49 + 0.8i$	$0.941249 + 0.510734i$	10	278	83
-0.2	0.911608	5	6	7

Table 5.

Of course the convergence is slower the closer we are to the boundary of the convergence regions. Still the Gauss fraction seems to be doing very well.

$\diamond$

One can find several continued fraction expansions of similar nature for hypergeometric functions. For instance, in [AbSt64, p. 558]: one can find quite a number of three-term recurrence relations for such functions. In the cases where the hypergeometric functions are minimal solutions of these relations, minimal regarded as elements in $(\mathbf{F}, \|\cdot\|)$, (see Chapter V), we immediately have a corresponding continued fraction. The usefulness of this continued fraction is normally tied to its convergence properties.

2 Confluent hypergeometric functions

2.1 Notation

Let us introduce the Pochhammer symbols

$$(a)_0 = 1, \quad (a)_k = a(a+1)(a+2)\cdots(a+k-1) \qquad \text{for} \quad k \in \mathbf{N}. \quad (2.1.1)$$

Then the hypergeometric series in (1.1.1) can be written

$$F(a, b; c; z) = \sum_{k=0}^{\infty} \frac{(a)_k (b)_k}{(c)_k} \frac{z^k}{k!} \qquad \text{for} \quad c \notin \mathbf{Z} \setminus \mathbf{N}. \qquad (2.1.2)$$

A *generalized hypergeometric series* is defined for given numbers $p, q \in \mathbf{N}_0$:

$$_pF_q(a_1, \ldots, a_p; b_1, \ldots, b_q; z) = \sum_{k=0}^{\infty} \frac{(a_1)_k \cdots (a_p)_k}{(b_1)_k \cdots (b_q)_k} \frac{z^k}{k!} \qquad (2.1.3)$$

where $a_1, \ldots, a_p, b_1, \ldots, b_q$ are complex parameters with $b_1, \ldots, b_q \notin \mathbf{Z} \setminus \mathbf{N}$. The series in (2.1.2) is a $_2F_1$. The series $\psi(c; z)$ in Example 9 in Chapter V is a $_0F_1$. We shall look at some cases which can be derived from $_2F_1$.

2.2 $_1F_1(b; c; z)$

The series $_2F_1(a, b; c; z/a)$ converges locally uniformly for $|z/a| < 1$. Hence, we can let $a \to \infty$ termwise in this series. Since

$$\lim_{a \to \infty} \frac{(a)_n}{a^n} = \lim_{a \to \infty} \left(1 + \frac{1}{a}\right) \left(1 + \frac{2}{a}\right) \cdots \left(1 + \frac{n-1}{a}\right) = 1 \quad (2.2.1)$$

for all $n \in \mathbf{N}$, we find that

$$\lim_{a \to \infty} {_2F_1}(a, b; c; z/a) = \sum_{k=0}^{\infty} \frac{(b)_k}{(c)_k} \frac{z^k}{k!} = {_1F_1}(b; c; z) \qquad (2.2.2)$$

for $c \neq 0, -1, -2, \ldots$. If we let $a \to \infty$ term by term in the Gauss fraction $1 + \mathbf{K}((a_n z/a)/1)$ corresponding to

$$\frac{F(a, b; c; z/a)}{F(a, b+1; c+1; z/a)} = f(\frac{z}{a})$$

that is, a_n is given by (1.1.5), then it transforms into the continued fraction

$$1 + \mathbf{K} \frac{d_n z}{1} \quad \text{where} \quad d_{2n+1} = -\frac{c - b + n}{(c + 2n)(c + 2n + 1)},$$

$$d_{2n} = \frac{b + n}{(c + 2n - 1)(c + 2n)}. \qquad (2.2.3)$$

There are at least two ways in which we can prove:

Theorem 5 *Let b and c be complex constants with $c \notin \mathbf{Z} \setminus \mathbf{N}$. Then the continued fraction (2.2.3) corresponds (at $z = 0$) to the function*

$$g(z) := {}_1F_1(b; c; z) / {}_1F_1(b + 1; c + 1; z) \qquad (2.2.4)$$

and converges to $g(z)$ for all $z \in \mathbf{C}$. The convergence is uniform on compact subsets of $D_g = \{z \in \mathbf{C}; g(z) \neq \infty\}$.

Correspondence. We can either proceed as in the proof of Theorem 1 and start with the recurrence relation

$$P_n(z) = P_{n+1}(z) + d_{n+1} z P_{n+2}(z) \qquad \text{for} \quad n = 0, 1, 2, \dots,$$

or we can use the following theorem due to Perron, [Perr57, Satz 3.10, p. 112]:

Theorem 6 *Let*

$$1 + \frac{a_1 z^{\alpha_1}}{1} + \frac{a_2 z^{\alpha_2}}{1} + \frac{a_1 z^{\alpha_1}}{1} + \cdots \sim 1 + c_1 z + c_2 z^2 + c_3 z^3 + \cdots, \quad (2.2.5)$$

where the coefficients a_k and c_k are functions of a parameter p such that

$$\lim_{p \to p_0} a_k = a_k^* \quad \text{and} \quad \lim_{p \to p_0} c_k = c_k^* \qquad \text{for} \quad k = 1, 2, 3, \dots \quad (2.2.6)$$

where $p_0 \in \mathbf{C}$ and $a_k^* \neq 0$ if $a_k = a_k(p) \not\equiv 0$. Then

$$1 + \frac{a_1^* z^{\alpha_1}}{1} + \frac{a_2^* z^{\alpha_2}}{1} + \frac{a_3^* z^{\alpha_3}}{1} + \cdots \sim 1 + c_1^* z + c_2^* z^2 + c_3^* z^3 + \cdots. \quad (2.2.7)$$

It simply gives that the correspondence $1 + \mathbf{K}(d_n z / 1) \sim g(z)$ is inherited from the correspondence $1 + \mathbf{K}((d_n z / a) / 1) \sim f(z/a)$. Theorem 6 is a direct consequence of the following observation:

Lemma 7 *Let the C-fraction $1 + \mathbf{K}(a_n z / 1)$ correspond to the power series $L(z) \in \mathbf{L}_0$ and have approximants $L_n(z) \in \mathbf{L}_0$. Then*

$$L_{n+1}(z) - L_n(z) = d_n z^{\nu_n} + \text{higher order terms} \qquad (2.2.8)$$

and

$$L(z) - L_n(z) = d_n z^{\nu_n} + \text{higher order terms} \qquad (2.2.9)$$

where

$$d_n = (-1)^n \prod_{k=1}^{n+1} a_k \qquad \text{and} \qquad \nu_n = \sum_{k=1}^{n+1} \alpha_k . \qquad (2.2.10)$$

Proof of Lemma 7:   Let $L_n(z) = A_n(z)/B_n(z)$ (canonical form). Then we find, (as in Chapter V, formula (2.1.1)) that

$$
\begin{aligned}
L_{n+1}(z) - L_n(z) &= \frac{A_{n+1}(z)}{B_{n+1}(z)} - \frac{A_n(z)}{B_n(z)} \\
&= \frac{A_{n+1}(z)B_n(z) - A_n(z)B_{n+1}(z)}{B_{n+1}(z)B_n(z)} \\
&= \frac{(-1)^n \prod_{k=1}^{n+1}(a_k z^{\alpha_k})}{B_{n+1}(z)B_n(z)} \\
&= d_n z^{\nu_n} + \text{higher order terms.} \qquad (2.2.11)
\end{aligned}
$$

The expression for $L(z) - L_n(z)$ follows then by Theorem 4C in Chapter V.   ∎

Convergence. That $1 + \mathbf{K}(d_n z/1)$ in (2.2.3) converges to $g(z)$ in (2.2.4) can also be seen in two different ways. Either we observe that $d_n z \to 0$ locally uniformly with respect to z in $\mathbf{C}$. This means that $1 + \mathbf{K}(d_n z/1)$ converges to a meromorphic function in $\mathbf{C}$. If all $d_n \neq 0$, then this function must be $g(z)$ because of the correspondence.

The other way to see this, is that, on each compact set $C \subseteq \mathbf{C}$ with 0 as an interior point, $1 + \mathbf{K}((d_n z/a)/1)$ has a tail $1 + \mathbf{K}((d_{2N+n}z/a)/1)$ converging uniformly to $_2F_1(a + N, b + N; c + 2N; z/a)/_2F_1(a + N, b + N + 1; c + 2N + 1; z/a)$ in C by Theorem 1. (The index N depends on C, of course.) Moreover $z/a \to 0$ uniformly in $\mathbf{C}$. Hence the uniform convergence of $1 + \mathbf{K}(d_{2N+n}z/1)$ to $_1F_1(b + N; c + 2N; z)/_1F_1(b + N + 1; c + 2N + 1; z)$ follows. This proves that $1 + \mathbf{K}(d_n z/1)$ converges to $g(z)$.

The terminating case. $d_N = 0$ for some $N \in \mathbf{N}$ only if $a_N = 0$ in $1 + \mathbf{K}(a_n(z/a)/1)$. From Theorem 1D it follows therefore that $g(z) = 1 + \mathbf{K}(d_n z/1)$ for all $z \in \mathbf{C}$ if $d_N = 0$.

Now we have proved Theorem 5. The regular C-fraction converges to $g(z)$ for all $z \in \mathbf{C}$. On the other hand, the $_1F_1$-series in (2.2.2) also converges for all $z \in \mathbf{C}$, so we do not gain anything when it comes to domain of convergence. What we gain is speed of convergence and less chance of overflow. As approximants we choose $S_n(0)$ which normally converges reasonably well for these continued fractions. The squareroot modification as described in for instance Example 27 of Chapter III is not a good idea here because of the alternating character of $\{d_n\}$. But one might try $S_n(w_n)$ where w_n is the value of the 2-periodic continued fraction

$$w_n(z) = \frac{d_{n+1}z}{1} + \frac{d_{n+2}z}{1} + \frac{d_{n+1}z}{1} + \frac{d_{n+2}z}{1} + \cdots$$

if it converges.

Example 5 The confluent hypergeometric series $_1F_1(1; c+1; z)$ can be written $_1F_1(1; c+1; z)/_1F_1(0; c; z)$ since $_1F_1(0; c; z) = 1$. Application of Theorem 5 therefore leads to

$$_1F_1(1; c+1; z) =$$
$$\frac{1}{1-} \frac{z}{c+1+} \frac{1 \cdot z}{c+2-} \frac{(c+1)z}{c+3} \frac{2z}{+c+4-} \frac{(c+2)z}{c+5} \frac{3z}{+c+6-\cdots}.$$

Let $c = 2$. We shall compare the speed of convergence of the hypergeometric series $_1F_1(1; 3; z)$ and the corresponding C-fraction

$$\frac{1}{1-} \frac{z}{3+} \frac{1 \cdot z}{4} \frac{3z}{-5+} \frac{2z}{6} \frac{4z}{-7+} \frac{3z}{8} -\cdots$$

for some values of z. Of course we neither need the series nor the continued fraction to compute

$$_1F_1(1; 3; z) = 2 \left\{ \frac{1}{2!} + \frac{z}{3!} + \frac{z^2}{4!} + \cdots \right\} = \frac{2}{z^2} \left(e^z - 1 - z \right).$$

On the other hand, since we know the value of $_1F_1(1; 3; z)$ already, the convergence is easier to study. Table 6 gives the number of terms needed to reach the given accuracy.

$\diamond$

z	Value	Series	$S_n(0)$ for c.fr.
1	1.4365636569	$N = 12$	$N = 10$
10	440.3039	$N = 28$	$N = 24$
-10	0.180000908	$N = 38$	$N = 20$
$10 + 10i$	-119.9286+184.9278i	$N = 40$	$N = 29$
$10i$	0.03678143+0.2108804i	$N = 35$	$N = 23$
-1000	0.001998	overflow	$N = 116$

Table 6.

2.3 $_2F_0(a, b; z)$

If we replace z by cz in Theorem 1 and let $c \to \infty$, we obtain

Theorem 8 *Let a and b be complex constants. Then*

$$1 + \mathbf{K}\frac{p_n z}{1} \qquad where \quad p_{2n+1} = -(a + n), \, p_{2n} = -(b + n) \qquad (2.3.1)$$

corresponds (at $z = 0$) to

$$\frac{_2F_0(a, b; z)}{_2F_0(a, b + 1; z)} \qquad\qquad (2.3.2)$$

and converges in the cut plane $D = \{z \in \mathbf{C}; |\arg(-z)| < \pi\}$ to a function $h(z)$, meromorphic in D. The convergence is uniform on compact subsets of $D_h = \{z \in D; h(z) \neq \infty\}$.

In this case the hypergeometric series has radius of convergence zero. That is, $_2F_0(a, b; z)$ diverges for all $z \neq 0$ if $a, b \notin \mathbf{Z} \setminus \mathbf{N}$. However the continued fraction $1 + \mathbf{K}(p_n z/1)$ corresponds to (2.3.2) and converges to $h(z)$ in D. It turns out that the connection between $h(z)$ and the divergent series $_2F_0(a, b; z)$ is that $_2F_0(a, b; z)$ is an asymptotic series for $h(z)$ in D as $z \to \infty$. For more information on asymptotic series we refer to Chapter VII.

Since $p_n z \to \infty$, approximately as $(-nz)$ as $n \to \infty$, the squareroot modification is likely to work well for this continued fraction. That is, we may use approximants $S_n(w_n)$ where

$$w_n(z) = \frac{p_{n+1}z}{1} + \frac{p_{n+1}z}{1} + \frac{p_{n+1}z}{1} + \cdots = \frac{\sqrt{1 + 4p_{n+1}z} - 1}{2}, \qquad (2.3.3)$$

and expect these to converge faster to the value of (2.3.1) than the classical approximants $S_n(0)$.

2.4 $_0F_1(c; z)$

Still another confluent case arises if we replace z in Theorem 5 by z/b and let $b \to \infty$. We get:

Theorem 9 *Let c be a complex constant with $c \notin \mathbf{Z} \setminus \mathbf{N}$. Then*

$$1 + \mathbf{K}\frac{q_n z}{1} \qquad \text{where} \quad q_n = \frac{1}{(c + n - 1)(c + n)}, \qquad (2.4.1)$$

corresponds at $z = 0$ to

$$k(z) := \frac{_0F_1(c; z)}{_0F_1(c + 1; z)} \qquad (2.4.2)$$

and converges to $k(z)$ for all $z \in \mathbf{C}$. The convergence is uniform on compact subsets of $D_k = \{z \in \mathbf{C}; k(z) \neq \infty\}$.

This is exactly the continued fraction that we studied in Example 5 in Chapter IV and Example 9 in Chapter V. It is connected with the Bessel function $J_\nu(z)$ of the first kind of order ν. For $\nu \notin \mathbf{Z} \setminus \mathbf{N}_0$ we have

$$J_\nu(z) = \frac{(z/2)^\nu}{\Gamma(\nu + 1)} {_0F_1}(\nu + 1; -z^2/4),$$

so that

$$z\frac{J_\nu(z)}{J_{\nu+1}(z)}$$

$$= 2(\nu+1)\frac{{}_0F_1(\nu+1;-z^2/4)}{{}_0F_1(\nu+2;-z^2/4)}$$

$$= 2(\nu+1)\left\{1-\frac{z^2/4(\nu+1)(\nu+2)}{1}-\frac{z^2/4(\nu+2)(\nu+3)}{1}-\cdots\right\}$$

$$\approx 2(\nu+1)-\frac{z^2}{2(\nu+2)}-\frac{z^2}{2(\nu+3)}-\frac{z^2}{2(\nu+4)}-\cdots.$$

3   Basic hypergeometric functions

3.1   *Definition*

In 1847 Heine [Heine47] studied the series

$$1+\frac{(1-q^\alpha)(1-q^\beta)}{(1-q^\gamma)(1-q)}z+\frac{(1-q^\alpha)(1-q^{\alpha+1})(1-q^\beta)(1-q^{\beta+1})}{(1-q^\gamma)(1-q^{\gamma+1})(1-q)(1-q^2)}z^2+\cdots$$
$$(3.1.1)$$

where $|q| < 1$. By use of L'Hôpital's theorem we find that

$$\lim_{q\to 1}\frac{1-q^s}{1-q^t}=\lim_{q\to 1}\frac{-sq^{s-1}}{-tq^{t-1}}=\frac{s}{t}.$$

This means that (3.1.1) is transformed into ${}_2F_1(\alpha,\beta;\gamma;z)$ when $q\to 1$. (3.1.1) is called a *basic* hypergeometric series (or just a q-hypergeometric series). Many results for hypergeometric series have their counterpart for q-hypergeometric series.

Watson [Wats29] simplified the notation in (3.1.1) by writing $a=q^\alpha$, $b=q^\beta$ and $c=q^\gamma$. Inspired by the Pochhammer symbol (2.1.1) used in hypergeometric series, we define

$$(a;q)_0=1,\quad (a;q)_n=(1-a)(1-aq)(1-aq^2)\cdots(1-aq^{n-1})\quad\text{for }n\in\mathbf{N}.$$
$$(3.1.2)$$

Then (3.1.1) can be written

$$_2\varphi_1(a, b; c; q; z) = \sum_{n=0}^{\infty} \frac{(a; q)_n (b; q)_n}{(c; q)_n} \cdot \frac{z^n}{(q; q)_n}, \qquad (3.1.3)$$

which strongly suggests the connection to hypergeometric series. The parameters a, b, c and q are complex constants with $|q| < 1$. To avoid zeros in the denominators we assume that c is chosen such that $(c; q)_n \neq 0$ for all n; that is, $c \neq 1$, q^{-1}, q^{-2}, q^{-3}, $\ldots$. If the series in (3.1.3) is non-terminating, then it converges locally uniformly for $|z| < 1$. Hence, the function to which it converges is analytic in the unit disk $|z| < 1$. On the other hand, if $a \neq q$ and $b \neq q$ then

$$(1 - z) _2\varphi_1(a, b; c; q; z)$$

$$= 1 + \sum_{n=1}^{\infty} \frac{(a; q)_{n-1}(b; q)_{n-1}}{(c; q)_{n-1}(q; q)_{n-1}} \left(\frac{(1 - aq^{n-1})(1 - bq^{n-1})}{(1 - cq^{n-1})(1 - q^n)} - 1 \right) z^n$$

$$= 1 + \sum_{n=1}^{\infty} \frac{(a; q)_{n-1}(b; q)_{n-1}}{(c; q)_n(q; q)_n} \left(-aq^{n-1} - bq^{n-1} + abq^{2n-2} \right.$$

$$\left. + cq^{n-1} + q^n - cq^{2n-1} \right) z^n$$

$$= 1 + \frac{1}{(1 - a/q)(1 - b/q)} \sum_{n=1}^{\infty} \frac{(a/q; q)_n(b/q; q)_n}{(c; q)_n(q; q)_n} \cdot$$

$$\cdot \left\{ \frac{-a - b + c + q}{q}(qz)^n + \frac{ab - cq}{q^2}(q^2 z)^n \right\}$$

$$= 1 + \frac{-a - b + c + q}{q(1 - a/q)(1 - b/q)} \left\{ _2\varphi_1\left(\frac{a}{q}, \frac{b}{q}; c; q; qz \right) - 1 \right\}$$

$$+ \frac{ab - cq}{q^2(1 - a/q)(1 - b/q)} \left\{ _2\varphi_1\left(\frac{a}{q}, \frac{b}{q}; c; q; q^2 z \right) - 1 \right\}. \qquad (3.1.4)$$

If $a = q$ or $b = q$, similar expressions can be found. Dividing by $1 - z$ we thus find that the function, which we also denote by $_2\varphi_1(a, b; c; q; z)$, is analytic for $|z| < q^{-1}$ except for a simple pole at $z = 1$. Repeating the argument, we find that $_2\varphi_1(a, b; c; q; z)$ is analytic in the whole complex plane except at the points $z = 1$, $z = q^{-1}$, $z = q^{-2}$, $\ldots$ where it has simple poles. This function $_2\varphi_1(a, b; c; q; z)$ is called the basic hypergeometric function or the q-hypergeometric function.

Just as for hypergeometric functions, one may generalize to get

$$r\varphi_s(a_1, \ldots, a_r; b_1, \ldots, b_s; q; z) = \sum_{n=0}^{\infty} \frac{(a_1; q)_n \cdots (a_r; q)_n}{(b_1; q)_n \cdots (b_s; q)_n} \cdot \frac{z^n}{(q; q)_n}. \quad (3.1.5)$$

3.2 $_2\varphi_1(a, b; c; q; z)$

The q-analogue of the recurrence relation $(1.1.6)$ for $_2F_1$ is

$$_2\varphi_1(a, b; c; q; z) =$$
$$_2\varphi_1(a, bq; cq; q; z) + \frac{(1-a)(c-b)}{(1-c)(1-cq)} z \, _2\varphi_1(aq, bq; cq^2; q; z).$$
$$(3.2.1)$$

(Notice the simpe and straightforward transformation of $(1.1.6)$ into $(3.2.1)$! That $(3.2.1)$ really is true can for instance be checked by comparing the coefficients of z^n on each side of the equality.) Based on this relation, it is natural to study the continued fraction

$$1 + \cfrac{\frac{(1-a)(c-b)}{(1-c)(1-cq)} z}{1} + \cfrac{\frac{(1-bq)(cq-a)}{(1-cq)(1-cq^2)} z}{1} + \cfrac{\frac{(1-aq)(cq^2-bq)}{(1-cq^2)(1-cq^3)} z}{1} + \cdots$$
$$(3.2.2)$$

and its possible connection to the series (or function)

$$F(z) = \frac{_2\varphi_1(a, b; c; q; z)}{_2\varphi_1(a, bq; cq; q; z)}. \quad (3.2.3)$$

This was first done by Heine himself [Heine47], and $(3.2.2)$ goes by the name of Heine's continued fraction. By the same methods as used in Subsection *1.1* one finds:

Theorem 10 *Let a, b, c and q be complex constants with $|q| < 1$ and $c \neq q^{-n}$ for all $n \in \mathbf{N}_0$. Then the continued fraction in $(3.2.2)$ corresponds at $z = 0$ to $F(z)$ in $(3.2.3)$ and converges to $F(z)$ in the whole complex plane. The convergence is uniform on compact subsets of $\{z \in \mathbf{C}; F(z) \neq \infty\}$.*

The continued fraction in (3.2.2) has the form $1 + \mathbf{K}(a_n z/1)$ where

$$
\begin{aligned}
a_{2n+1} &= \frac{(1 - aq^n)(cq^n - b)q^n}{(1 - cq^{2n})(1 - cq^{2n+1})}, \\
a_{2n} &= \frac{(1 - bq^n)(cq^n - a)q^{n-1}}{(1 - cq^{2n-1})(1 - cq^{2n})}.
\end{aligned}
\tag{3.2.4}
$$

That is, $a_n \to 0$, and the classical approximants $\{S_n(0)\}$ represent a reasonable choice. For large $|z|$ one can also use the squareroot modification $S_n(w_n)$ where

$$
w_n(z) = \frac{a_{n+1}z}{1} + \frac{a_{n+1}z}{1} + \frac{a_{n+1}z}{1} + \cdots = \frac{\sqrt{1 + 4a_{n+1}z} - 1}{2}.
\tag{3.2.5}
$$

Some other continued fraction expansions for hypergeometric functions also have their counterparts for basic hypergeometric functions. The relation (1.4.3) has the analogue

$$
\begin{aligned}
{}_2\varphi_1(a, b; c; q; z) &= \left[1 - \frac{a + b - ab - abq}{1 - c}z\right] {}_2\varphi_1(aq, bq; cq; q; z) \\
&\quad + \frac{(1 - aq)(1 - bq)}{(1 - c)(1 - cq)}z(c - abqz){}_2\varphi_1(aq^2, bq^2; cq^2; q; z),
\end{aligned}
\tag{3.2.6}
$$

which leads to:

Theorem 11 *Let a, b, c and q be complex constants with $|q| < 1$ and $c \neq q^{-n}$ for all $n \in \mathbf{N}_0$. Then the continued fraction $b_0 + \mathbf{K}(a_n/b_n)$ where*

$$
\begin{aligned}
a_n(z) &= (1 - aq^n)(1 - bq^n)q^{n-1}z(c - zabq^n), \\
b_n(z) &= 1 - cq^n - (a + b - abq^n - abq^{n+1})q^n z
\end{aligned}
\tag{3.2.7}
$$

corresponds at $z = 0$ to the function

$$
F(z) = (1 - c){}_2\varphi_1(a, b; c; q; z)/{}_2\varphi_1(aq, bq; cq; q; z)
\tag{3.2.8}
$$

and converges to $F(z)$ for all $z \in \mathbf{C}$. The convergence is uniform on compact subsets of $\{z \in \mathbf{C}; F(z) \neq \infty\}$.

The proof is left as an exercise.

Here also the classical approximants represent a reasonably good choice since $a_n \to 0$. The continued fraction $b_0 + \mathbf{K}(a_n/b_n)$ is the q-analogue of Nörlunds continued fraction (1.4.1).

4 Continued fractions $b_0 + \mathbf{K}(a_n/b_n)$ where a_n and b_n are polynomials in n

4.1 Introduction

The continued fraction expansions of (ratios of) hypergeometric functions presented in the previous sections have all had the form $b_0(z) + \mathbf{K}(a_n(z)/b_n(z))$, where $a_{2n+p}(z)$ and $b_{2n+p}(z)$ have been polynomials in n for $p = 0, 1$. This leaves the impression that maybe every continued fraction $b_0 + \mathbf{K}(a_n/b_n)$ where a_n and b_n (or a_{2n+p}, b_{2n+p}; $p = 0, 1$) are polynomials in n, is related to hypergeometric functions? This question is easy to answer affirmatively in some special cases.

4.2 Some special cases

Let

$$a_n = \sum_{k=0}^r p_k n^k \quad \text{with} \quad p_r \neq 0 \,,$$

$$b_n = \sum_{k=0}^s q_k n^k \quad \text{with} \quad q_s \neq 0 \,. \tag{4.2.1}$$

Do we then know the value of the continued fraction $b_0 + \mathbf{K}(a_n/b_n)$ in terms of hypergeometric functions?

The case $r = s = 0$: Then all $a_n = p_0$ and $b_n = q_0$. This means that $b_0 + \mathbf{K}(a_n/b_n)$ is 1-periodic, and thus has the value

$$\frac{q_0}{2} \left(\sqrt{1 + 4p_0/q_0^2} + 1 \right) = \frac{q_0}{2} \left(1 + {}_1F_0(-1/2; -4p_0/q_0^2) \right); \quad \Re(\sqrt{}) > 0$$

if $|\arg(1 + 4p_0/q_0^2)| < \pi$.

The case $r = 0$, $s = 1$: From Theorem 9 it follows that

$$c + \cfrac{z}{c+1+} \cfrac{z}{c+2+} \cfrac{z}{c+3+} \cdots = ck(z) = c\frac{{}_0F_1(c; z)}{{}_0F_1(c+1; z)}$$

where the equality sign means convergence for all $z \in \mathbf{C}$ as long as $c \notin \mathbf{Z} \setminus \mathbf{N}$. Hence, if $q_0 \neq 0$ then

$$q_0 + \cfrac{p_0}{q_0 + q_1 +} \cfrac{p_0}{q_0 + 2q_1 +} \cfrac{p_0}{q_0 + 3q_1 +} \cdots$$

$$\approx q_1 \left\{ \frac{q_0}{q_1} + \cfrac{p_0/q_1^2}{\frac{q_0}{q_1} + 1 +} \cfrac{p_0/q_1^2}{\frac{q_0}{q_1} + 2 +} \cdots \right\}$$

$$= q_0 \frac{{}_0F_1(q_0/q_1; p_0/q_1^2)}{{}_0F_1(q_0/q_1 + 1; p_0/q_1^2)}.$$

If $q_0 = 0$ then

$$\frac{p_0}{q_1 +} \frac{p_0}{2q_1 +} \frac{p_0}{3q_1 +} \cdots \approx \frac{p_0/q_1}{1} + \frac{p_0/q_1^2}{2} + \frac{p_0/q_1^2}{3} + \cdots$$

$$= \cfrac{p_0/q_1}{1 + \cfrac{p_0/q_1^2}{2} + \cfrac{p_0/q_1^2}{3} + \cdots} = \cfrac{p_0/q_1}{1 \cdot \cfrac{{}_0F_1(1; p_0/q_1^2)}{{}_0F_1(2; p_0/q_1^2)}}$$

$$= \frac{p_0}{q_1} \frac{{}_0F_1(2; p_0/q_1^2)}{{}_0F_1(1; p_0/q_1^2)}.$$

The case $r = 1$, $s = 1$: If we replace z by z/a in the Euler fraction (1.4.6) and let $a \to \infty$, we get the continued fraction

$$c - z + \cfrac{(b+1)z}{c+1-z+} \cfrac{(b+2)z}{c+2-z+} \cfrac{(b+3)z}{c+3-z+} \cdots. \tag{4.2.2}$$

By the same argument as used in Subsection 2.2 one can prove that (4.2.2) converges to the function

$$\frac{c \, {}_1F_1(b; c; z)}{{}_1F_1(b+1; c+1; z)} \tag{4.2.3}$$

for all $z \in \mathbf{C}$. (This identity can be found in [Perr57, p. 279], and in formula (4.1.5) in the appendix in a slightly different form.) Our continued fraction is therefore

$$q_0 + \frac{p_0 + p_1}{q_0 + q_1} + \frac{p_0 + 2p_1}{q_0 + 2q_1} + \frac{p_0 + 3p_1}{q_0 + 3q_1} + \cdots$$

$$\approx q_1 \left\{ \frac{q_0}{q_1} + \frac{\dfrac{p_0}{q_1^2} + \dfrac{p_1}{q_1^2}}{\dfrac{q_0}{q_1} + 1} + \frac{\dfrac{p_0}{q_1^2} + 2\dfrac{p_1}{q_1^2}}{\dfrac{q_0}{q_1} + 2} + \frac{\dfrac{p_0}{q_1^2} + 3\dfrac{p_1}{q_1^2}}{\dfrac{q_0}{q_1} + 3} + \cdots \right\}$$

$$= \left(q_0 + \frac{p_1}{q_1} \right) \frac{{}_1F_1\left(\dfrac{p_0}{p_1} ; \dfrac{q_0}{q_1} + \dfrac{p_1}{q_1^2} ; \dfrac{p_1}{q_1^2} ; \right)}{{}_1F_1\left(\dfrac{p_0}{p_1} + 1 ; \dfrac{q_0}{q_1} + \dfrac{p_1}{q_1^2} + 1 ; \dfrac{p_1}{q_1^2} ; \right)} .$$

The case $r = 2$, $s = 1$: We want to find the value of

$$q_0 + \frac{p_0 + p_1 + p_2}{q_0 + q_1} + \frac{p_0 + 2p_1 + 2^2 p_2}{q_0 + 2q_1} + \frac{p_0 + 3p_1 + 3^2 p_2}{q_0 + 3q_1} + \cdots, \qquad (4.2.4)$$

and the general T-fraction (1.4.6) in Theorem 4 seems to be a nice continued fraction to compare with. It has parameters a, b, c and z which we will try to adjust to match (4.2.4). We evidently need $p_2 = -z$ and $q_1 = 1 + z$. But this can only be done if $q_1 + p_2 = 1$. To get enough freedom we introduce an extra unknown $\delta \neq 0$ and write (4.2.4) as

$$\frac{1}{\delta} \left\{ q_0 \delta + \frac{(p_0 + p_1 + p_2)\delta^2}{(q_0 + q_1)\delta} + \frac{(p_0 + 2p_1 + 2^2 p_2)\delta^2}{(q_0 + 2q_1)\delta} + \cdots \right\}. \qquad (4.2.5)$$

Then we need that $p_2\delta^2 = -z$, $p_1\delta^2 = -(c - a + b)z$, $p_0\delta^2 = -(c - a)bz$, $q_1\delta = 1 + z$ and $q_0\delta = c + (b - a + 1)z$. This is a system of 5 equations to determine our 5 unknowns a, b, c, z and δ. If we find one solution with $|z| \neq 1$, then Theorem 4 gives us the value of (4.2.4).

Example 6 To find the value of

$$\frac{1}{2} + \mathop{\mathbf{K}}_{n=1}^{\infty} \frac{-\frac{1}{2} + 2n^2}{\frac{1}{2} + n} \approx \frac{1}{\delta} \left\{ \frac{\delta}{2} + \mathop{\mathbf{K}}_{n=1}^{\infty} \frac{-\delta^2/2 + 2\delta^2 n^2}{\delta/2 + 2\delta n} \right\}$$

by comparing with (1.4.6) we get the system

$$2\delta^2 = -z, \quad 0 = -(c - a + b)z, \quad -\delta^2/2 = -(c - a)bz$$
$$\delta = 1 + z, \quad \delta/2 = c + (b - a + 1)z$$

of equations. This system has the solutions

(1) $\delta = \frac{1}{2}$, $z = -\frac{1}{2}$, $a = 1$, $b = \frac{1}{2}$, $c = \frac{1}{2}$,

(2) $\delta = \frac{1}{2}$, $z = -\frac{1}{2}$, $a = 0$, $b = -\frac{1}{2}$, $c = \frac{1}{2}$,

(3) $\delta = -1$, $z = -2$, $a = 0$, $b = -\frac{1}{2}$, $c = \frac{1}{2}$,

(4) $\delta = -1$, $z = -2$, $a = 1$, $b = \frac{1}{2}$, $c = \frac{1}{2}$.

The first solution leads to the value

$$2 \cdot \frac{1}{2} {}_2F_1(1, \tfrac{1}{2}; \tfrac{1}{2}; -\tfrac{1}{2})/{}_2F_1(1, \tfrac{3}{2}; \tfrac{3}{2}; -\tfrac{1}{2}) = 1$$

by Theorem 4A. So does also the second solution. If we choose solution (3), we apply Theorem 4B which gives the same value

$$(-\frac{1}{2})(-2){}_2F_1(1, -\tfrac{1}{2}; \tfrac{1}{2}; -\tfrac{1}{2})/{}_2F_1(0, \tfrac{1}{2}; \tfrac{3}{2}; -\tfrac{1}{2}) = 1,$$

as it should.

Problems

(1) Prove that $P_n(z) = {}_1F_1(a + n; c + n; z)$ where a and c are complex numbers $\neq 0, -1, -2, \ldots$, is a solution of the three-term recurrence relation

$$(c + n)P_n(z) = (c + n - z)P_{n+1}(z) + \frac{a + n + 1}{c + n + 1} z P_{n+2}(z)$$

$$\text{for} \quad n = 0, 1, 2, \ldots$$

in **L**. Determine the correspondence and convergence properties of the corresponding general T-fraction

$$c - z + \frac{(a + 1)z}{c + 1 - z +} \frac{(a + 2)z}{c + 2 - z +} \frac{(a + 3)z}{c + 3 - z +} \cdots.$$

(2) Compute the first two approximants of each type $S_n(0)$, $S_n(x)$, $S_n(w_n^{(1)})$ and $S_n(w_n^{(2)})$ for the C-fraction expansion of $\arctan z$ for $z = 1$ and compare with $\arctan 1$. (This C-fraction expansion can be found in Example 1.)

(3) Establish a formula for approximants of the type $S_n(w_n^{(2)})$ for the Nörlund fraction in Theorem 2.

(4) Establish a formula for approximants of the type $S_n(w_n^{(2)})$ for Euler's T-fraction in Theorem 4.

(5) Use Theorem 5 to determine the C-fraction expansion of e^z.

(6) Use Theorem 9 to determine the C-fraction expansion of $\tan z = \sin z / \cos z$.

(7) Prove Theorem 11.

(8) Express the value of

$$\frac{2}{1 +} \frac{2}{3 +} \frac{4}{5 +} \frac{6}{7 +} \frac{8}{9 +} \cdots$$

by means of hypergeometric functions.

(9) Express the value of

$$1 + \frac{(1^2 - t^2)z}{1} + \frac{(2^2 - t^2)z}{1} + \frac{(3^2 - t^2)z}{1} + \cdots \qquad \text{for} \quad |\arg z| < \pi.$$

by means of hypergeometric functions.

(10) Let $1 + \mathbf{K}(a_n/1)$ be given by (1.1.5); that is

$$a_{2n+1} = \frac{-(a+n)(c-b+n)}{(c+2n)(c+2n+1)}, \qquad a_{2n} = \frac{-(b+n)(c-a+n)}{(c+2n)(c+2n)},$$

and assume that neither $a, b, c - a, c - b$ nor c is a non-positive integer.

a) Prove that t_n, where

$$t_{2n} = -\frac{a+n}{c+2n}, \qquad t_{2n+1} = -\frac{b+n+1}{c+2n+1}$$

is a tail sequence for $1 + \mathbf{K}(a_n/1)$.

b) Prove that $1 + \mathbf{K}(a_n/1)$ converges to

$$1 + t_0 = 1 - a/c$$

if $\Re(c - a - b) > 0$ or if $c = a + b$.

c) Prove that $1 + \mathbf{K}(a_n/1)$ converges if $\Re(c - a - b) < 0$.

(Hint: Theorem 13 in Chapter IV may be of help.)

Remarks

1. Gauss' work on hypergeometric functions [Gauss12] is very useful. His contiguous relations lead to recurrence relations for hypergeometric functions which again lead to continued fraction expansions of ratios of such functions. The Gauss-fractions are the most well known and widely used of these expansions. But also the Euler fractions and the Nörlund fractions, among others, can be obtained from these relations. See for instance [AbSt64, p. 558].

   Another important source for continued fraction expansions of ratios of hypergeometric functions or functions strongly related to these is Ramanujan's work. This extraordinary mathematician had a strong liking for continued fractions. In [ABBW85] and [Bern89] some of his results are presented with comments and proofs.

   Apart from this, almost every book on continued fractions contains a section on hypergeometric functions. We mention in particular [Perr57] and [JoTh80].

2. Quite a number of formulas for hypergeometric functions have an analogue valid for basic hypergeometric functions. From a continued fraction point of view the most striking is the Heine fraction [Heine47] which is the q-analogue of the Gauss fraction. But this is not the only one. Again we refer to Ramanujan's notebooks as described in [ABBW85] for a wide selection of continued fraction expansions related to basic hypergeometric functions.

3. The art of finding the value of a given continued fraction $b_0 + \mathbf{K}(a_n/b_n)$ where a_n and b_n are polynomials (or rational functions) in n, is well described in [Perr57, p. 276→].

References

[AbSt64]   M. Abramowitz and I. A. Stegun, "Handbook of Mathe-
             matical Functions with Formulas, Graphs and Mathemati-
             cal Tables", National Bureau of Standards, Appl. Math.
             Ser. **55**, U.S. Govt. Printing Office, Washington, D.C.
             (1964).

[ABBW85]   C. Adiga, B. C. Berndt, S. Bhargava and G. N. Watson,
             "Chapter 16 of Ramanujan's Second Notebook: Theta-
             Functions and q-Series", Memoirs, Amer. Math. Soc., **315**,
             Providence (1985).

[Bail64]   W. N. Bailey, "Generalized Hypergeometric Series",
             Stechert-Hafner, New York (1964).

[Bern89]   B. C. Berndt, "Ramanujan's Notebooks", Part II, Springer-
             Verlag (1989).

[Erdé53]   A. Erdélyi et al, "Higher Transcendental Functions",
             Vol. **1**, McGraw-Hill, New York (1953).

[Euler27]   L. Euler, "Institutiones Calculi Integralis", Vol. **2**, 3. ed.,
             Impensis Academiae Imperialis Scientiarum, Petropoli
             (1827); Opera Omnia, Ser. 1, Vol. **12**, B. G. Teubner, Lip-
             siae (1914), 1–413.

[Euler67]   L. Euler, *De fractionibus continuis observationes*, Comm.
             Acad. Sci. Imp. St. Pétersbourg, 11(1767), 32–81 Opera
             Omnia, Ser. 1, Vol. **14**, B. G. Teubner, Lipsiae (1925),
             291–349.

[Gauss12]   C. F. Gauss, *Disquisitiones Generales circa Seriem Infini-*
tam $1 + \frac{\alpha\beta}{1\cdot\gamma}x + \frac{\alpha(\alpha+1)\beta(\beta+1)}{1\cdot2\cdot\gamma(\gamma+1)}xx + \frac{\alpha(\alpha+1)(\alpha+2)\beta(\beta+1)(\beta+2)}{1\cdot2\cdot3\cdot\gamma(\gamma+1)(\gamma+2)}x^3 +$
etc., Pars prior, Comm. soc. regiae sci. Gottingensis rec. **2**
(1812), 1–46; Werke, Band **3**, Königliche Gesellschaft der
Wissenschaften, Göttingen (1876), 123–162.

[Heine47]   E. Heine,   *Untersuchungen*   *über*   *die*   *Reihe*
$1 + \frac{(1-q^\alpha)(1-q^\beta)}{(1-q)(1-q^\gamma)}x + \frac{(1-q^\alpha)(1-q^{\alpha+1})(1-q^\beta)(1-q^{\beta+1})}{(1-q)(1-q^2)(1-q^\gamma)(1-q^{\gamma+1})}x^2 + \cdots,$ J.
reine angew. Math. **34** (1847), 285–328.

[JoTh80]   W. B. Jones and W. J. Thron, "Continued Fractions. Analytic Theory and Applications", Encyclopedia of Mathematics and its Applications **11**, Addison-Wesley, Reading, MA (1980). Now distributed by Cambridge University
Press, New York.

[Lore]   L. Lorentzen, *Divergence of Continued Fractions Related*
to Hypergeometric Series. To appear.

[Nörl24]   N. E. Nörlund, "Vorlesungen über Differenzenrechnung",
Springer-Verlag, OHG, Berlin (1924).

[Perr57]   O. Perron, "Die Lehre von den Kettenbrüchen", Band II,
B. G. Teubner, Stuttgart (1957).

[Wats29]   G. N. Watson, *A New Proof of the Rogers-Ramanujan Iden-*
tities, Journal London Math. Soc. **4** (1929), 4–9.

Chapter VII

Moments and orthogonality

About this chapter

The threefold, rather modest, intention of this chapter is reflected in its three sections: In the first one the connection between orthogonal polynomials and continued fractions is established, key words being Favard's theorem and Jacobi fractions. In the second section the denominators of the approximants of the Jacobi fraction are used to construct the classical Gaussian quadrature formula. The third section is different: For the classical Stieltjes moment problem necessary and sufficient conditions for existence of a solution may be expressed in terms of continued fractions. This is also true for uniqueness.

The chapter contains very few proofs. Examples are used to illustrate the theorems. Even the topics chosen are meant merely as examples of connections between orthogonality, moments and continued fractions.

1   Orthogonality and continued fractions

1.1   Three examples

Example 1 The Tchebycheff polynomials $U_n(x)$ of the seond kind are defined by

$$U_0(x) = 1, \quad U_1(x) = 2x,$$

and generally

$$U_n(x) = \frac{\sin(n+1)\theta}{\sin\theta}, \quad x = \cos\theta.$$

We easily find, e.g.,

$$
\begin{aligned}
U_2(x) &= 4x^2 - 1, \\
U_3(x) &= 8x^3 - 4x.
\end{aligned}
$$

We shall first establish two properties: a) They satisfy a certain three-term recurrence relation. b) They are orthogonal with respect to a certain weight function.

a) From

$$\sin((n+2)\theta) + \sin(n\theta) = 2\sin((n+1)\theta)\cos\theta$$

we find

$$U_{n+1}(x) = 2xU_n(x) - U_{n-1}(x),$$

valid for $n \geq 0$ if we define $U_{-1}(x) = 0$.

b) From

$$\int_0^\pi \sin(k\theta) \cdot \sin(p\theta)d\theta = 0$$

for integers k, p, $k \neq p$ we get

$$\int_0^\pi \frac{\sin(n+1)\theta}{\sin\theta} \cdot \frac{\sin(m+1)\theta}{\sin\theta} \cdot \sin\theta \cdot \sin\theta d\theta =$$

$$\int_{-1}^1 U_n(x)U_m(x)(1-x^2)^{1/2}dx = 0$$

for $m \neq n$, which means *orthogonality* on $[-1, +1]$ with respect to the *weight* function $w(x) = (1-x^2)^{1/2}$. For $m = n$ the integral $= \pi/2$.

A natural connection to continued fractions is through the recurrence relation, from which it follows that $U_n(x)$ is the canonical denominator of the nth approximant of the continued fraction

$$\frac{-1}{2x} + \frac{-1}{2x} + \frac{-1}{2x} + \cdots + \frac{-1}{2x} + \cdots .$$

The polynomial $U_n(x)$ is of degree n, and the coefficient of x^n is 2^n. The polynomials

$$\tilde{U}_n(x) = 2^{-n} U_n(x) \tag{M$_1$}$$

are monic, i.e. the coefficient of x^n is 1. We have

$$\tilde{U}_0(x) = 1 , \quad \tilde{U}_1(x) = x ,$$

and generally

$$\tilde{U}_{n+1}(x) = x \cdot \tilde{U}_n(x) - \frac{1}{4}\tilde{U}_{n-1}(x) \tag{R$_1$}$$

for $n \geq 0$ if $\tilde{U}_{-1}(x) = 0$. The monic polynomials $\tilde{U}_n(x)$ are the canonical denominators of the approximants of the equivalent continued fraction

$$\frac{-\dfrac{1}{2}}{x} + \frac{-\dfrac{1}{4}}{x} + \frac{-\dfrac{1}{4}}{x} + \cdots + \frac{-\dfrac{1}{4}}{x} + \cdots . \tag{J$_1$}$$

The reason for using the symbol J is that the continued fraction here is a special case of what is called a Jacobi continued fraction. We shall return to other special cases as well as to the general Jacobi fraction later.

We still have the orthogonality, in fact

$$\int_{-1}^{1} \tilde{U}_n(x) \cdot \tilde{U}_m(x)(1 - x^2)^{1/2} dx = \frac{\pi}{2} 2^{-(m+n)} \delta_{mn} . \tag{O$_1$}$$

(The symbol δ_{mn} is called the *Kronecker delta*, and has the value 0 for $m \neq n$ and 1 for $m = n$.) In conclusion we emphasize the following key points: The properties of being monic (M$_1$) and orthogonal (O$_1$), the recurrence relation (R$_1$) and the Jacobi continued fraction (J$_1$). These points will also be present in the other examples, and will play a crucial role in the general theory.

$\diamondsuit$

Example 2 The Legendre polynomials $P_n(x)$ are given by

$$P_0(x) = 1, \quad P_1(x) = x,$$

and generally

$$(n+1)P_{n+1}(x) = (2n+1)x P_n(x) - nP_{n-1}(x)$$

for $n \geq 0$, if we define $P_{-1}(x) = 0$. We easily find the first polynomials

$$
\begin{aligned}
P_2(x) &= \frac{3}{2}x^2 - \frac{1}{2}, \\
P_3(x) &= \frac{5}{2}x^3 - \frac{3}{2}x, \\
P_4(x) &= \frac{35}{8}x^4 - \frac{15}{4}x^2 + \frac{3}{8}.
\end{aligned}
$$

The Legendre polynomials are known to be orthogonal on the interval $[-1, +1]$ with respect to the weight function $w(x) = 1$, we have in fact

$$\int_{-1}^{1} P_m(x) \cdot P_n(x)dx = \frac{2}{2n+1}\delta_{mn}.$$

The recurrence relation can be written

$$P_{n+1}(x) = \frac{2n+1}{n+1}x \cdot P_n(x) - \frac{n}{n+1}P_{n-1}(x).$$

For a continued fraction $\mathbf{K}(a_n/b_n)$ with arbitrary $a_1 \neq 0$ and

$$b_{n+1} = \frac{2n+1}{n+1}x, \quad a_{n+1} = -\frac{n}{n+1},$$

the recurrence relation above is exactly the recurrence relation for the canonical numerators and denominators of the approximants. The sequence $\{P_n(x)\}_{n=1}^{\infty}$ is thus the sequence of canonical denominators for the approximants of the continued fraction

$$\cfrac{a_1}{x + \cfrac{-\frac{1}{2}}{\frac{3}{2}x + \cfrac{-\frac{2}{3}}{\frac{5}{3}x + \cfrac{-\frac{3}{4}}{\frac{7}{4}x + \cdots}}}}.$$

By an equivalence transformation this continued fraction can be changed to

$$\cfrac{a_1}{x} \; \genfrac{}{}{0pt}{}{-\dfrac{1^2}{1\cdot 3}}{+\; x} \; \genfrac{}{}{0pt}{}{-\dfrac{2^2}{3\cdot 5}}{+\; x} \; \genfrac{}{}{0pt}{}{-\dfrac{3^2}{5\cdot 7}}{+\; x} \; +\cdots, \tag{J_2}$$

which is again a J-fraction. The recurrence relation (for the denominators) is

$$\tilde{P}_{n+1}(x) = x\tilde{P}_n(x) - \frac{n^2}{4n^2 - 1}\tilde{P}_{n-1}(x), \tag{R_2}$$

and the initial values are

$$\tilde{P}_0(x) = 1, \qquad \tilde{P}_1(x) = x.$$

We easily find that

$$\begin{aligned}
\tilde{P}_2(x) &= x^2 - \frac{1}{3}, \\
\tilde{P}_3(x) &= x^3 - \frac{3}{5}x, \\
\tilde{P}_4(x) &= x^4 - \frac{6}{7}x^2 + \frac{3}{35},
\end{aligned}$$

and generally

$$\tilde{P}_n(x) = x^n + \text{lower powers of } x, \tag{M_2}$$

i.e. the coefficient of x^n in $\tilde{P}_n(x)$ is 1, $\tilde{P}_n(x)$ is a monic polynomial. The orthogonality is of course preserved:

$$\int_{-1}^{1} \tilde{P}_m(x)\tilde{P}_n(x)dx = 0 \quad \text{for } m \neq n. \tag{O_2}$$

The same four key points as in Example 1 are indicated in a similar way, by writing (J), (R), (M), (O).

                                         ◇

Example 3 Let G be the following function of two variables x (real) and w (complex), given by

$$G(x, w) = e^{-w}(1 + w)^x.$$

The Taylor expansion at $w = 0$ is

$$G(x, w) = \left(1 - \frac{w}{1!} + \frac{w^2}{2!} - \cdots\right)\left(1 + \binom{x}{1}w + \binom{x}{2}w^2 + \cdots\right)$$

$$= \sum_{n=0}^{\infty} \frac{C_n(x)}{n!} w^n,$$

valid at least in $|w| < 1$. Here

$$C_0(x) = 1, \quad C_1(x) = x - 1,$$

and generally

$$C_n(x) = (-1)^n \cdot \left[1 + \sum_{k=1}^{n}(-1)^k \binom{n}{k} x(x-1)\cdots(x-k+1)\right].$$

From this formula it is not difficult to prove that these polynomials are determined by the initial values $C_0(x) = 1$, $C_1(x) = x - 1$ and the recurrence relation

$$C_{n+1}(x) = (x - n - 1)C_n(x) - nC_{n-1}(x), \quad n \geq 1. \qquad (\text{R}_3)$$

$C_n(x)$ is a polynomial of degree n, and the coefficient of x^n is 1, i.e.,

$$C_n(x) \text{ is monic}. \qquad (\text{M}_3)$$

We find from the recurrence relation that $C_n(x)$ is the canonical denominator of the nth approximant of the continued fraction

$$\frac{a_1}{x - 1 +} \frac{-1}{x - 2 +} \frac{-2}{x - 3 +} \frac{-3}{x - 4 +} \cdots + \frac{-n}{x - n - 1 +} \cdots, \qquad (\text{J}_3)$$

which is also a J-fraction. The value of $a_1 \neq 0$ can be chosen arbitrarily. Also for the polynomials $C_n(x)$ we have orthogonality:

$$\sum_{k=0}^{\infty} C_n(k) \cdot C_m(k) \frac{1}{k!} = e \cdot n! \cdot \delta_{mn}. \qquad (\text{O}_3)$$

The proof is left as an exercise, see Problem 3 (with hints). This is orthogonality on **R** with respect to the discrete mass distribution with mass $1/k!$ at $x = k$. An alternative way of writing the orthogonality relation is as a Riemann-Stieltjes integral with respect to the function ψ, defined by:

$$\psi(x) = \begin{cases} 0 & \text{for } x \in (-\infty, 0) \\ 1 & \text{for } x \in [0, 1) \\ 1 + \frac{1}{1!} & \text{for } x \in [1, 2) \\ \vdots \\ 1 + \frac{1}{1!} + \frac{1}{2!} + \cdots + \frac{1}{k!} & \text{for } x \in [k, k+1) \\ \vdots \end{cases}$$

$$\int_{-\infty}^{\infty} C_m(x) \cdot C_n(x) d\psi(x) = e \cdot n! \cdot \delta_{mn}. \qquad (O_3')$$

The polynomials $C_n(x)$ are special cases of the *monic Charlier polynomials*. The function $G(x, w)$ is a *generating* function for the polynomials

$$\frac{C_n(x)}{n!},$$

meaning that the Taylor expansion at $w = 0$ is such that these polynomials are the coefficients of w^n in the expansion.

——————————————————————◇

Remark: The polynomials in the Examples 1 and 2 can also be produced by generating functions:

$$\frac{1}{1 - 2xw + w^2} = \sum_{n=0}^{\infty} U_n(x) \cdot w^n \qquad \text{(See Problem 1b.)}$$

$$\frac{1}{\sqrt{1 - 2xw + w^2}} = \sum_{n=0}^{\infty} P_n(x) \cdot w^n.$$

For Charlier polynomials as well as other orthogonal polynomials and their elementary theory we refer to the first chapter of Chihara's book [Chih78].

1.2   *Moment sequences and moment functionals*

In the three examples in Subsection *1.1* we dealt with different integrals
of polynomials,

$$\int_{-1}^{1} Q(x)d\tilde{\psi}(x) \;=\; \int_{-1}^{1} Q(x)\sqrt{1-x^2}dx \quad \text{in Example 1,}$$

$$\int_{-1}^{1} Q(x)d\psi^*(x) \;=\; \int_{-1}^{1} Q(x)dx \qquad\qquad \text{in Example 2,}$$

$$\int_{-\infty}^{\infty} Q(x)d\psi(x) \qquad\qquad\qquad\qquad \text{in Example 3.}$$

If

$$Q(x) = a_0 + a_1 x + \cdots + a_n x^n \,,$$

we may in Example 1 write

$$\int_{-1}^{1} Q(x)\sqrt{1-x^2}dx$$

$$= \; a_0 \int_{-1}^{1} 1 \cdot \sqrt{1-x^2}dx + \cdots + a_n \int_{-1}^{1} x^n \sqrt{1-x^2}dx$$

$$= \; a_0\mu_0 + a_1\mu_1 + \cdots + a_n\mu_n \,,$$

where $\mu_k = \int_{-1}^{1} x^k \sqrt{1-x^2}dx = \int_{-1}^{1} x^k d\tilde{\psi}(x)$ for $k = 0,1,2,\ldots$ and
similarly for the other examples. The integrals are in all cases examples
of linear functionals acting on the space of polynomials. We shall now
look at this more generally.

Definition   *Let $\{\mu_n\}_{n=0}^{\infty}$ be a sequence of complex numbers and L a
complex linear functional defined on the space of all polynomials by*

$$L[x^n] = \mu_n \,, \qquad n = 0,1,2,\ldots .$$

*Then L is called the moment functional determined by the moment se-
quence $\{\mu_n\}$. μ_n is called the moment of order n.*

The polynomials $\sum_{k=0}^{n} c_k x^k$ to be considered have complex coefficients, whereas the symbol x is regarded as being real. Since the functional is linear we have

$$L\left[\sum_{k=0}^{n} c_k x^k\right] = \sum_{k=0}^{n} c_k \mu_k,$$

and, since x is real (and $\bar{z}$ means the complex conjugate of z),

$$L\left[\overline{\sum_{k=0}^{n} c_k x^k}\right] = \sum_{k=0}^{n} \bar{c}_k \mu_k.$$

We shall now define the concept of *orthogonal polynomial sequence*. The concept of orthogonality in itself, and in this setting will come later, after having introduced an inner product.

Definition *An orthogonal polynomial sequence $\{Q_n(x)\}_{n=0}^{\infty}$ for L is defined by the requirements*

$$Q_n(x) \text{ has exact degree } n,$$
$$L[Q_n(x)Q_m(x)] = 0 \text{ for } m \neq n,$$
$$L[Q_n(x)^2] \neq 0.$$

Since $\{Q_n(x)\}_{n=0}^{N}$ span the space of polynomials of degree $\leq N$, it is a consequence of this definition that also

$$L[x^n Q_N(x)] = 0 \quad \text{for } n < N$$

for every N. In the examples in Subsection *1.1* the moment functionals and the moment sequences are as follows:

In Example 1 we have

$$L[U_m(x)U_n(x)] = \int_{-1}^{1} U_m(x)U_n(x)(1-x^2)^{1/2}dx,$$
$$\mu_n = \int_{-1}^{1} x^n(1-x^2)^{1/2}dx.$$

For odd n we have $\mu_n = 0$, since the integrand in this case is an odd function. For even n, $n = 2k$, $k = 0, 1, 2, \ldots$ we have $\mu_0 = \pi/2$ and

$$
\begin{aligned}
\mu_{2k} &= \int_0^\pi \cos^{2k} \theta \cdot \sin^2 \theta d\theta \\
&= \int_0^\pi \cos^{2k} \theta d\theta - \int_0^\pi \cos^{2k+2} \theta d\theta \\
&= \frac{\pi}{2(k+1)} \cdot \frac{1 \cdot 3 \cdots (2k-1)}{2 \cdot 4 \cdots 2k}. \quad \text{for } k \geq 1.
\end{aligned}
$$

The details are left as an exercise (Problem 4).

In Example 2 we have

$$
\mu_n = \int_{-1}^1 x^n dx = \begin{cases} 0 & \text{for odd } n, \\ \dfrac{2}{2k+1} & \text{for } n = 2k, \ k = 0, 1, 2, \ldots. \end{cases}
$$

In Example 3 we have

$$
\mu_n = \int_{-\infty}^\infty x^n d\psi(t) = \sum_{k=0}^\infty \frac{k^n}{k!},
$$

in particular

$$
\begin{aligned}
\mu_0 &= e, \\
\mu_1 &= \sum_{k=1}^\infty \frac{k}{k!} = \sum_{k=1}^\infty \frac{1}{(k-1)!} = e, \\
\mu_2 &= \sum_{k=1}^\infty \frac{k^2}{k!} = \sum_{k=1}^\infty \frac{k}{(k-1)!} = \sum_{k=1}^\infty \frac{k-1+1}{(k-1)!} \\
&= \sum_{k=2}^\infty \frac{1}{(k-2)!} + \sum_{k=1}^\infty \frac{1}{(k-1)!} = 2e.
\end{aligned}
$$

Next we find $\mu_3 = 5e$. The proof is left as an exercise (Problem 5.)

In the examples in Subsection *1.1* the orthogonal polynomial sequence was the starting point, or more precisely: We were in each case given a polynomial sequence, which turned out to be an orthogonal polynomial sequence if the moment functional was properly defined. In the present subsection, however, we shall go in the opposite direction: We shall start with a moment functional L, or equivalently, with a moment sequence $\{\mu\}_{n=0}^{\infty}$, and ask for necessary and sufficient conditions for existence of an orthogonal polynomial sequence for L. Let us look at "the start of an answer": The two first polynomials (assumed to be monic) must be of the form $P_0(x) = 1$, $P_1(x) = x + a_1$ for some constant a_1. The conditions $L[P_0(x)^2] \neq 0$, $L[P_0(x)P_1(x)] = 0$, $L[P_1(x)^2] \neq 0$ take the form

$$L[P_0(x)^2] = L[1] = \mu_0 \neq 0,$$

$$L[P_0(x)P_1(x)] = L[x + a_1] = \mu_1 + a_1\mu_0 = 0, \quad \text{i.e. } a_1 = \frac{-\mu_1}{\mu_0},$$

$$L[P_1(x)^2] = L[x^2 + 2a_1 x + a_1^2] = \mu_2 + 2a_1\mu_1 + a_1^2\mu_0$$

$$= \mu_2 - \frac{2\mu_1^2}{\mu_0} + \frac{\mu_1^2}{\mu_0} = \frac{\mu_2\mu_0 - \mu_1^2}{\mu_0} \neq 0.$$

The two first conditions are thus

$$\Delta_0 := \mu_0 \neq 0, \qquad \Delta_1 := \begin{vmatrix} \mu_0 & \mu_1 \\ \mu_1 & \mu_2 \end{vmatrix} \neq 0.$$

The general answer is given in terms of the determinants (Hankel determinants)

$$\Delta_n = \begin{vmatrix} \mu_0 & \mu_1 & \cdots & \mu_n \\ \mu_1 & \mu_2 & \cdots & \mu_{n+1} \\ \vdots & \vdots & & \vdots \\ \mu_n & \mu_{n+1} & \cdots & \mu_{2n} \end{vmatrix}, \qquad (1.2.1)$$

by the following theorem, here stated without proof (see e.g. [Chih78], p. 11):

Theorem 1 *Let L be a moment functional and $\{\mu_n\}$ its moment sequence. Necessary and sufficient for existence of an orthogonal polynomial sequence is that*

$$\Delta_n \neq 0 \quad \text{for } n = 0, 1, 2, \ldots. \qquad (1.2.2)$$

Remark: A moment functional for which (1.2.2) is true, is called *quasi-definite*.

In the examples we met in Subsection *1.1* the moment functional L was defined by an integral: In Example 1 and Example 2 we had a Riemann integral with a non-negative weight function ($\sqrt{1-x^2}$ in Example 1 and 1 in Example 2), in Example 3 we had a Stieltjes integral with respect to a function ψ. The most important orthogonal polynomials are such that the moment functional L is defined by a Riemann integral with a weight function or more generally, as a Stieltjes integral

$$L[x^n] = \int_{-\infty}^{\infty} x^n d\psi(x) , \qquad (1.2.3)$$

where ψ is a bounded, non-decreasing function with an infinite number of points of increase, called *distribution* function. It can be proved, that this is the case, if and only if L is such that $L[p(x)] > 0$ for all polynomials $p(x)$ which are ≥ 0 for all real x and not identically zero, or equivalently, if and only if

$$\text{all moments are real and all } \Delta_n > 0. \qquad (1.2.4)$$

Such moment functionals are called *positive-definite*. They have some important properties, for instance having a corresponding orthogonal polynomial sequence of real polynomials with only real, simple zeros.

For a positive-definite moment functional it is easily verified that

$$(p, q) = L[p(x)\overline{q(x)}] \qquad (1.2.5)$$

defines an inner product on the space of all polynomials in one real variable. What we so far have called orthogonality of a sequence, without any reference to an inner product, is in fact orthogonality with respect to the inner product (1.2.5) in the case of a positive-definite moment functional. (We have in fact for polynomials P_m and P_n in an orthogonal polynomial sequence, that $(P_m, P_n) = L[P_m(x)\overline{P_n(x)}] = L[P_m(x)P_n(x)] = 0$ for $m \neq n$. This follows immediately from the property $L[P_m(x)x^n] = 0$ for all $n \leq m-1$.) The whole theory of inner product spaces will be at hand, e.g.: starting from

$$p_0(x) = 1/\sqrt{\mu_0}$$

we can by the Gram-Schmidt-process recursively construct an orthogonal sequence $\{p_n(x)\}$ of polynomials in the usual way:

$$P_1(x) = x - ap_0(x), \qquad a = L[xp_0(x)],$$

$$p_1(x) = P_1(x) \Big/ \sqrt{L[P_1(x)^2]},$$

and generally

$$\begin{cases} P_{n+1}(x) = x^{n+1} - \sum_{k=0}^{n} a_k p_k(x), & a_k = L[x^{n+1} p_k(x)], \\ p_{n+1}(x) = P_{n+1}(x) \Big/ \sqrt{L[P_{n+1}(x)^2]}. \end{cases} \qquad (1.2.6)$$

Observe that we, through the standard Gram-Schmidt process, get orthogonal polynomials which are normalized by the requirement $L[p_n(x)^2] = 1$, i.e. *orthonormal* polynomials, rather than by the requirement that the coefficients of x^n be 1. They are (if desired) transformed to *monic* orthogonal polynomials by suitable divisions by factors independent of x. Furthermore, having constructed an orthonormal sequence $\{p_n(x)\}$ we can find a Fourier expansion of an arbitrary polynomial $Q(x)$ of degree n:

$$Q(x) = \sum_{k=0}^{n} c_k p_k(x), \qquad c_k = L[Q(x)p_k(x)]. \qquad (1.2.7)$$

Example 4 In Example 1 we have

$$\int_{-1}^{1} U_n(x) U_n(x) (1 - x^2)^{1/2} dx = \int_{0}^{\pi} \sin^2((n+1)\theta) d\theta = \frac{\pi}{2}.$$

Hence the sequence $\{u_k(x)\}_{k=0}^{\infty}$, where

$$u_k(x) = \sqrt{\frac{2}{\pi}} U_k(x)$$

is an orthonormal sequence. We expand x^2 in a Fouries series based upon $\{\mu_k\}$, and find

$$x^2 = c_0 u_0(x) + c_1 u_1(x) + c_2 u_2(x),$$

where

$$c_k = \int_{-1}^{1} x^2 u_k(x)(1 - x^2)^{1/2} dx = \sqrt{\frac{2}{\pi}} \int_{0}^{\pi} \frac{\sin((k+1)\theta)}{\sin \theta} \cos^2 \theta \sin^2 \theta d\theta,$$

hence

$$c_0 = \frac{1}{8} \cdot \sqrt{2\pi}, \quad c_1 = 0, \quad c_2 = \frac{1}{8} \cdot \sqrt{2\pi},$$

and

$$x^2 = \frac{1}{8} \cdot \sqrt{2\pi}(u_0(x) + u_2(x)) \quad \left(= \frac{1}{4}(1 + 4x^2 - 1) = x^2 \right).$$

_____◇

Example 5 In Example 2 it can be proved (for instance by using the generating function) that

$$\int_{-1}^{1} P_k(x)^2 dx = \frac{2}{2k+1}.$$

Hence the sequence $\{p_k(x)\}_{k=0}^{\infty}$, where

$$p_k(x) = \sqrt{\frac{2k+1}{2}} P_k(x),$$

is an orthonormal sequence. A Fourier expansion based on these polynomials is then given by

$$x^2 = d_0 p_0(x) + d_1 p_1(x) + d_2 p_1(x),$$

where

$$d_k = \int_{-1}^{1} x^2 p_k(x) dx,$$

hence

$$d_0 = \frac{\sqrt{2}}{3}, \quad d_1 = 0, \quad d_2 = \frac{2}{15}\sqrt{10},$$

and

$$x^2 = \frac{\sqrt{2}}{3} p_0(x) + \frac{2}{15}\sqrt{10} p_2(x)$$

$$\left(\quad = \frac{\sqrt{2}}{3}\frac{1}{\sqrt{2}} + \frac{2}{15}\sqrt{10}\sqrt{\frac{5}{2}}\left(\frac{3}{2}x^2 - \frac{1}{2}\right) = x^2 \right).$$

_____◇

1.3 Favard's theorem and Jacobi fractions

In the three examples we have studied in Subsection *1.1*, the monic orthogonal polynomials satisfied a recurrence relation of the form

$$Q_n(x) = (x - c_n)Q_{n-1}(x) - \lambda_n Q_{n-2}(x), \qquad (1.3.1)$$

valid for $n \geq 1$ if we define $Q_{-1}(x) = 0$. In Example 1 we had $c_n = 0$ and $\lambda_n = 1/4$ for all n, in Example 2 we had $c_n = 0$ and $\lambda_n = n^2/(4n^2 - 1)$, whereas in Example 3 we had $c_n = n$ and $\lambda_n = -n+1$ for $n \geq 2$. This is actually a general property for monic orthogonal polynomial sequences for a quasi-definite moment functional. If in particular the functional is positive-definite, then c_n is real and $\lambda_{n+1} > 0$ for $n \geq 1$ (as in the three examples). See [Chih78, Thm. 4.1].

An important property in the theory of orthogonal polynomials is that a converse type of result is also true. The theorem carries the name of Favard. According to Chihara [Chih78, p. 21] it was discovered at about the same time independently by J. Shohat and I. Natanson. Jones and Thron point out that it can be deduced form a result in Perron's book on continued fractions [Perr57].

Theorem 2 (Favard's theorem) *Let $\{c_n\}_{n=1}^{\infty}$ and $\{\lambda_n\}_{n=1}^{\infty}$ be arbitrary sequences of complex numbers, and let $\{Q_n(x)\}_{n=0}^{\infty}$ be defined by the recurrence formula*

$$Q_n(x) = (x - c_n)Q_{n-1}(x) - \lambda_n Q_{n-2}(x), \quad n = 1, 2, 3, \ldots \qquad (1.3.1)$$

with $Q_{-1}(x) = 0$, $Q_0(x) = 1$. Then there is a unique moment functional L such that $L[1] = \lambda_1$, $L[Q_m(x)Q_n(x)] = 0$ for $m \neq n$, $m, n = 0, 1, 2, \ldots$. This L is quasi-definite and $\{Q_n(x)\}$ is the corresponding monic sequence of orthogonal polynomials if and only if $\lambda_n \neq 0$, and L is positive-definite if and only if c_n is real and $\lambda_n > 0$ for $n \geq 1$.

For the proof we refer to [Chih78], p. 22.

These two results show the close connection between monic orthogonal polynomials and Jacobi fractions:

a) For an arbitrary quasi-definite moment functional there is a J-fraction

$$\frac{\lambda_1}{x - c_1 +} \frac{-\lambda_2}{x - c_2 +} \cdots + \frac{-\lambda_n}{x - c_n +} \cdots \qquad (1.3.2)$$

such that the sequence of denominators of the approximants is the corresponding monic orthogonal polynomial sequence. If in particular the functional is positive-definite, the parameters c_n are all real and $\lambda_{n+1} > 0$ for $n \geq 1$.

b) For an arbitrary J-fraction (1.3.2) with all $\lambda_n \neq 0$ there exists a unique quasi-definite moment functional L with $L[1] = \lambda_1$ such that the sequence $\{Q_n(x)\}$ of denominators of the approximants is the sequence of monic orthogonal polynomials for L. If in particular c_n is real and $\lambda_n > 0$ for all $n \geq 1$, then the moment functional is positive-definite.

c) It follows from Theorem 4 in Chapter V that the J-fraction (1.3.2) corresponds at $x = \infty$ to a formal Laurent series

$$\sum_{n=0}^{\infty} \frac{\tilde{u}_n}{x^{n+1}}. \qquad (1.3.3)$$

This correspondence represents in fact an important link between continued fractions on the one hand and orthogonality on the other. Let us first look at the start of the corresponding series in the Examples 1, 2, 3, with $\lambda_1 = \mu_0 = L[1] > 0$ in all cases.

In Example 1: $\lambda_1 = \mu_0 = \int_{-1}^{1}(1 - x^2)^{1/2}dx = \pi/2$. J-fraction:

$$\frac{\pi}{2} \quad \frac{-\frac{1}{4}}{x +} \quad \frac{-\frac{1}{4}}{x +} \quad \frac{-\frac{1}{4}}{x +} \quad \frac{-\frac{1}{4}}{x +} \cdots.$$

Corresponding series:

$$\frac{\frac{\pi}{2}}{x} + \frac{\frac{\pi}{8}}{x^3} + \frac{\frac{\pi}{16}}{x^5} + \frac{\frac{\pi}{128}}{x^7} + \frac{\frac{\pi}{256}}{x^9} + \cdots.$$

With weight function

$$w(x) = (1 - x^2)^{1/2}, \quad x \in [-1,1],$$

the first moments are:

$$\frac{\pi}{2}, 0, \frac{\pi}{8}, 0, \frac{\pi}{16}, 0, \frac{\pi}{128}, 0, \frac{\pi}{256}, 0, \ldots .$$

◇

In Example 2: $\lambda_1 = \mu_0 = \int_{-1}^{1} dx = 2$. J-fraction:

$$\frac{2}{x+} \quad \frac{-\dfrac{1^2}{1\cdot 3}}{x} \quad \frac{-\dfrac{2^2}{3\cdot 5}}{x} \quad \frac{-\dfrac{3^2}{5\cdot 7}}{x} +\cdots .$$

Corresponding series:

$$\frac{2}{x} + \frac{\dfrac{2}{3}}{x^3} + \frac{\dfrac{2}{5}}{x^5} + \frac{\dfrac{2}{7}}{x^7} + \cdots .$$

With weight function

$$w(x) = 1, \quad x \in [-1,1],$$

the first moments are:

$$2, 0, \frac{2}{3}, 0, \frac{2}{5}, 0, \frac{2}{7}, 0, \ldots .$$

◇

In Example 3: $\lambda_1 = \mu_0 = e$. J-fraction:

$$\frac{e}{x-1+} \frac{-1}{x-2+} \frac{-2}{x-3+} \frac{-3}{x-4+} \cdots .$$

Corresponding series:

$$\frac{e}{x} + \frac{e}{x^2} + \frac{2e}{x^3} + \frac{5e}{x^4} + \frac{15e}{x^5} + \cdots .$$

With distribution function

$$\psi(x), \quad x \in \mathbf{R},$$

where ψ is defined as in Example 3, the first moments are:

$$e, e, 2e, 5e, 15e, \ldots .$$

◇

We observe in the three examples that the first coefficients of the corresponding series coincide with the first moments. It can be proved that this goes on, such that the sequence of coefficients coincides with the sequence of moments. And more so: This holds generally: If $\{Q_n\}$ is a sequence of polynomials satisfying a 3-term recurrence relation of the form (1.3.1) with real c_n and positive λ_n for all $n \geq 1$, then the J-fraction (1.3.2) given by the recurrence relation (1.3.1) corresponds at $x = \infty$ to a Laurent series (1.3.3), where the coefficients $\tilde{\mu}_n$ are the moments with respect to the unique moment functional of Favard's theorem. In our case the functional can be represented as an integral with respect to a distribution function $\psi(x)$, which means that we will have

$$\tilde{\mu}_n = \int_{-\infty}^{\infty} x^n d\psi(x), \qquad (1.3.4)$$

in particular $\lambda_1 = \int_{-\infty}^{\infty} d\psi(x)$. (In the Examples 1 and 2 this is to be understood as follows: Extend the definition of $w(x)$ in both cases to the whole real line by puttting $w(x) = 0$ for all $x \notin [-1, +1]$. Next take $\psi(x)$ to be the absolutely continuous function

$$\psi(x) = \int_{-\infty}^{x} w(t)dt .)$$

We leave out the proof, and refer to the monographs [Wall48], [Chih78] and [JoTh80] for a more thorough treatment of the subject.

2   Gaussian quadrature

2.1   A quadrature formula

We shall derive a formula for numerical integration of $\int_{-\infty}^{\infty} f(x)w(x)dx$, or more generally $\int_{-\infty}^{\infty} f(x)d\psi(x)$, where $\psi(x)$ is a distribution function. If $f(x)$ is a polynomial we can write this as $L[f(x)]$, where L is the positive definite moment functional corresponding to ψ.

We shall first get aquainted with the *Lagrange interpolation polynomial,* which is a polynomial $L_n(x)$ of degree $(n-1)$ taking prescribed values $y_1, y_2, \ldots, y_n$ at given points $x_1, x_2, \ldots, x_n$:

Let $x_1, x_2, \ldots, x_n$ be n distinct numbers, and let $F(x)$ be given by

$$F(x) = \prod_{k=1}^{n} (x - x_k). \tag{2.1.1}$$

For any k we find that

$$l_{nk}(x) = \frac{F(x)}{(x - x_k)F'(x_k)} \tag{2.1.2}$$

is a polynomial of degree $(n-1)$ (when the removable singularity at $x = x_k$ is removed, whereby $l_{nk}(x_k) = 1$). Moreover $l_{nk}(x_j) = 0$ for all $j \neq k$. Then the polynomial

$$L_n(x) = \sum_{k=1}^{n} y_k l_{nk}(x) \tag{2.1.3}$$

takes the value y_j, for $x = x_j$. This formula is called the *Lagrange interpolation polynomial*. The points $x_1, x_2, \ldots, x_n$ are called the *nodes*. The significance of the Lagrange polynomial is that it interpolates a function f at the points $x_1, x_2, \ldots, x_n$ by taking $y_k = f(x_k)$. In such a case the name Lagrange interpolation formula is used for

$$f(x) = \sum_{k=1}^{n} f(x_k) \cdot l_{nk}(x), \tag{2.1.4}$$

although we only know that it holds for $x = x_k$, $k = 1, 2, \ldots, n$. (But it is often very useful as an approximate formula for other x-values.)

We shall now use the Lagrange interpolation polynomials to compute $L[f(x)]$, where f is a polynomial of a real variable x, and where L is a positive-definite moment functional. We know in this case that L can be represented by an integral, as earlier mentioned:

$$L[f(x)] = \int_a^b f(x) d\psi(x). \tag{2.1.5}$$

Observe that whereas the lefthand side is defined only for polynomials $f(x)$, the righthand side is defined for all f, integrable with respect to ψ on the interval in question ($[a, b]$ if a and b are finite).

We shall use nodes, determined by L itself: It is known, that the orthogonal polynomials $Q_n(x)$ all are real for real x, and that they have simple, real zeros, located in the interval. Let the zeros of $Q_n(x)$ be

$$x_{n1}, x_{n2}, \ldots, x_{nk}.$$

We shall use them as nodes. This particular choice will prove to be very profitable compared to other choices. For a given polynomial $f(x)$ we replace $f(x)$ by the Lagrange interpolation polynomial, and find

$$L\left[\sum_{k=1}^{n} f(x_{nk}) \cdot l_{nk}(x)\right] = \sum_{k=1}^{n} A_{nk} \cdot f(x_{nk}), \qquad (2.1.6)$$

where

$$A_{nk} = L[l_{nk}(x)] = \int_{a}^{b} l_{nk}(x) d\psi(x).$$

For a polynomial of degree $\leq n - 1$ the Lagrange polynomial equals the polynomial, since it is uniquely dermined by its values at n points. Hence, for such polynomials the righthand side of (2.1.6) gives the exact value. This would be true for *any* choice of nodes. But with our particular choice we get much more: The formula (2.1.6) actually holds for all polynomials up to and including *the degree* $2n - 1$. This can be seen as follows:

Let $f(x)$ be a polynomial of degree $\leq 2n - 1$, and let $L_n(x)$ be the Lagrange interpolation polynomial constructed as above. Then

$$f(x) - L_n(x)$$

has degree $\leq 2n - 1$ and vanishes at the nodes. Hence it is equal to

$$Q_n(x) \cdot R(x),$$

where $R(x)$ is a polynomial of degree $\leq n - 1$. The Fourier expansion of $R(x)$ is a linear combination of the polynomials $Q_k(x)$, $k = 0, 1, \ldots, n-1$. Since $L[Q_n(x)Q_k(x)] = 0$ we find

$$L[f(x) - L_n(x)] = 0.$$

In conclusion we have: With the notation introduced the formula

$$\int_{a}^{b} f(x) d\psi(x) = \sum_{k=1}^{n} A_{nk} f(x_{nk}) \qquad (2.1.7)$$

holds when f is a polynomial of degree $\leq 2n - 1$. This is the Gauss quadrature formula. It has turned out to be of great use in numerical analysis, as an approximate formula in cases when it is not exact. We shall not go further into this here. Note that both the nodes x_{nk} and the weights A_{nk} are independent of $f(x)$.

2.2   An example

We shall illustrate the formula (2.1.7) on a specific example. For the example, as well as generally, we notice that the normalization of the orthogonal polynomials is insignificant for the formula.

Example 6 For the first Legendre polynomials we have

$$P_0(x) = 1, \quad P_1(x) = x, \quad P_2(x) = \frac{3}{2}x^2 - \frac{1}{2}, \quad P_3(x) = \frac{5}{2}x^3 - \frac{3}{2}x.$$

We shall find the quadrature formula in two cases. Here the weight function is 1 and the interval is $[-1, +1]$.

$n = 2$. The equation $P_2(x) = 0$ leads to the two nodes

$$x_{21} = -\frac{1}{\sqrt{3}}, \qquad x_{22} = \frac{1}{\sqrt{3}},$$

and (with the notation used)

$$l_{21}(x) = \frac{\frac{3}{2}\left(x - \frac{1}{\sqrt{3}}\right)}{-\frac{3}{\sqrt{3}}} = \frac{1}{2} - \frac{\sqrt{3}}{2}x,$$

$$l_{22}(x) = \frac{\frac{3}{2}\left(x + \frac{1}{\sqrt{3}}\right)}{\frac{3}{\sqrt{3}}} = \frac{1}{2} + \frac{\sqrt{3}}{2}x,$$

$$A_{21} = \int_{-1}^{1}\left(\frac{1}{2} - \frac{\sqrt{3}}{2}x\right) dx = 1,$$

$$A_{22} = \int_{-1}^{1}\left(\frac{1}{2} + \frac{\sqrt{3}}{2}x\right) dx = 1,$$

and the formula takes in this case the form

$$\int_{-1}^{1} f(x)dx = f\left(-\frac{1}{\sqrt{3}}\right) + f\left(\frac{1}{\sqrt{3}}\right).$$

It holds for polynomials up to and including degree 3 (which is also easily verified directly).

$n = 3$. The equation $P_3(x) = 0$ leads to the three nodes

$$x_{31} = -\sqrt{\frac{3}{5}}, \qquad x_{32} = 0, \qquad x_{33} = \sqrt{\frac{3}{5}}.$$

We furthermore get

$$l_{31}(x) = \frac{\frac{5}{2}x\left(x - \sqrt{\frac{3}{5}}\right)}{3} = \frac{5}{6}x^2 - \frac{\sqrt{15}}{6}x,$$

$$l_{32}(x) = \frac{\frac{5}{2}x^2 - \frac{3}{2}}{-\frac{3}{2}} = 1 - \frac{5}{3}x^2,$$

$$l_{33}(x) = \frac{\frac{5}{2}x\left(x + \sqrt{\frac{3}{5}}\right)}{3} = \frac{5}{6}x^2 + \frac{\sqrt{15}}{6}x,$$

and thus

$$A_{31} = \int_{-1}^{1}\left(\frac{5}{6}x^2 - \frac{\sqrt{15}}{6}x\right)dx = \frac{5}{9},$$

$$A_{32} = \int_{-1}^{1}\left(1 - \frac{5}{3}x^2\right)dx = 2 - \frac{10}{9} = \frac{8}{9},$$

$$A_{33} = \int_{-1}^{1}\left(\frac{5}{6}x^2 + \frac{\sqrt{15}}{6}x\right)dx = \frac{5}{9}.$$

In this case the formula takes the form

$$\int_{-1}^{1} f(x)dx = \frac{5}{9}f\left(-\sqrt{\frac{3}{5}}\right) + \frac{8}{9}f(0) + \frac{5}{9}f\left(\sqrt{\frac{3}{5}}\right).$$

It holds for all polynomials up to and including degree 5 (also easily verified directly).

——◇

An illustration with Tchebycheff polynomials of the second kind is left as an exercise (Problem 7).

3   Moment problems

3.1   *The Stieltjes moment problem*

A *moment problem* is roughly as follows: When is a sequence of numbers the sequence of moments with respect to some distribution function? And when is (in case of existence) the function unique up to an additive constant, except at the points of discontinuity? We shall make the first question more precise in one special case, which is a classical problem, the *Stieltjes moment problem*. Following Jones and Thron [JoTh80, p. 331] we shall for a, b with $-\infty \le a \le b \le +\infty$ let $\Phi(a, b)$ denote the family of all real-valued, bounded, non-decreasing functions ψ with infinitely many points of increase on $[a, b]$ if $a, b \in \mathbf{R}$, else on $(-\infty, b], [a, \infty)$ or $(-\infty, \infty)$.

Stieltjes moment problem: Find conditions on a sequence $\{c_n\}_{n=0}^{\infty}$ of real numbers to ensure the existence of a $\psi \in \Phi(0, \infty)$, such that

$$c_n = \int_0^{\infty} (-t)^n d\psi(t), \quad n = 0, 1, 2, \ldots . \tag{3.1.1}$$

(The factor $(-1)^n$ is included for practical reasons.) Such a function is called a *solution* of the moment problem.

There are several reasons for being interested in this problem, and in moment problems generally. Let it here merely be mentioned (as will also be seen in Theorem 3 and subsequent examples) that solutions of moment problems can be used to "sum" certain divergent series or to find closed integral representations of certain continued fractions.

We shall use the concept of *asymptotic expansion*: We say that the series

$$\sum_{n=0}^{\infty} c_n z^{-n} \tag{3.1.2}$$

is an asymptotic expansion of $F(z)$ at $z = \infty$ with respect to the angular region $|\arg z| < \alpha$, $0 < \alpha < \pi$, if there exist sequences of positive numbers $\{\eta_n\}$ and $\{R_n\}$, such that for each $n \geq 0$

$$\left| F(z) - \sum_{k=0}^{n} c_k z^{-k} \right| \leq \eta_n |z|^{-(n+1)} \qquad (3.1.3)$$

for $|z| > R_n$ and $|\arg z| < \alpha$.

Remarks:

1. The asymptotic expansion (3.1.2) may very well *diverge*, in fact: in many of the important cases it does. The point is, that for any *fixed* n the section $\sum_{k=0}^{n} c_k z^{-k}$ is an approximation to $F(z)$ that improves with increasing $|z|$ in the sense of (3.1.3).

2. Asymptotic expansions (3.1.2) may also be defined with respect to other angular regions, or other regions stretching to ∞ for that matter.

For the Stieltjes moment problem the following holds:

Theorem 3 *Let $\psi \in \Phi(0, \infty)$ be a solution of the Stieltjes moment problem for a sequence $\{c_n\}_{n=0}^{\infty}$. Then the integral*

$$\int_0^{\infty} \frac{z \, d\psi(t)}{z + t} \qquad (3.1.4)$$

is a holomorphic function $F(z)$ in the cut plane $|\arg z| < \pi$, and the series

$$\sum_{n=0}^{\infty} c_n z^{-n} \qquad (3.1.2)$$

is the asymptotic expansion of $F(z)$ at $z = \infty$ with respect to the angular region $|\arg z| < \alpha, 0 < \alpha < \pi$.

Step in the proof: Crucial in the proof is the connection between (3.1.4) and (3.1.2):

$$\int_0^\infty \frac{z\,d\psi(t)}{z+t} = \int_0^\infty \left(\sum_{k=0}^n \frac{(-t)^k}{z^k} + \frac{(-t)^{n+1}}{z^n(z+t)} \right) d\psi(t)$$

$$= \sum_{k=0}^n c_k z^{-k} + \int_0^\infty \frac{(-t)^{n+1}}{z^n(z+t)} d\psi(t) \,;$$

a correspondence we recognize from Subsection *1.3.*   ■

We shall illustrate Theorem 3 by an example:

Example 7 (first time). For the sequence $\{c_n\}_{n=0}^\infty$, where

$$c_n = (-1)^n n! \,,$$

we see by direct verification that

$$\psi(t) = 1 - e^{-t} \quad \text{for } 0 \leq t < \infty$$

is a solution of the Stieltjes moment problem. By Theorem 3 the function

$$F(z) = \int_0^\infty \frac{z e^{-t}}{z+t} dt$$

is holomorphic in the cut plane $|\arg z| < \pi$. The series

$$1 - 1! \, z^{-1} + 2! \, z^{-2} - 3! \, z^{-3} + \cdots$$

is an asymptotic expansion of $F(z)$ at $z = \infty$ with respect to the angular region $|\arg z| < \alpha$, $0 < \alpha < \pi$. This is easily verified. We have therefore, by determining $\psi(t)$, succeeded in summing the divergent series $\sum c_n z^{-n}$.

―――――――――――――――――――――――――――――――――――――◇

Let it briefly be mentioned, that for *bounded* intervals the following result by Markov holds: If in (3.1.1) and (3.1.4) the interval of integration is changed to $[a, b]$, $-\infty < a < b < \infty$, then the J-fraction corresponding to (3.1.2) converges to (3.1.4) for all $z \in \mathbf{C}$ not on the segment $[-b, -a]$ on the real line [Mark95].

3.2 Connection to continued fractions

When it comes to the solution of a moment problem there are in fact three questions to handle: existence, uniqueness and the actual construction of a possible solution.

Let us first consider the question of existence. In Favard's theorem we found that a positive definite moment functional L is uniquely determined by the J-fraction

$$\frac{\lambda_1}{x - c_1 -}\ \frac{\lambda_2}{x - c_2 -}\ \frac{\lambda_3}{x - c_3 -}\ldots\ ; \lambda_n > 0, \quad c_n \in \mathbf{R}.\qquad (3.2.1)$$

Moreover, in the subsequent discussion we found that the series

$$\sum_{n=0}^{\infty} \frac{\mu_n}{x^{n+1}}\qquad (3.2.2)$$

to which this J-fraction corresponds, has coefficients μ_n which actually are the moments of L; i.e. $\mu_n = L[x^n]$. Hence, a given sequence $\{\mu_n\}$ consists of the moments of a positive definite moment functional L (i.e. ψ exists) if and only if the series (3.2.2) has a corresponding J-fraction (3.2.1) with all $\lambda_n > 0$ and $c_n \in \mathbf{R}$. But this is a $\psi \in \Phi(-\infty, \infty)$. In the Stieltjes moment problem we are looking for a $\psi \in \Phi(0, \infty)$.

We shall first see that this is equivalent to the existence of a $\psi \in \Phi(-\infty, \infty)$ with $\psi(-x) = -\psi(x)$. Let $\psi \in \Phi(-\infty, \infty)$ with $\psi(-x) = -\psi(x)$. Then $d\psi(-x) = d\psi(x)$, and the moments are

$$\mu_n = \int_{-\infty}^{\infty} x^n d\psi(x) = \begin{cases} 0 & \text{if } n \text{ is odd,} \\ 2\int_0^{\infty} x^n d\psi(x) & \text{if } n \text{ is even.} \end{cases}\qquad (3.2.3)$$

Hence $2d\psi(\sqrt{x})$ restricted to $x \geq 0$ gives a function $\psi_1 \in \Phi(0, \infty)$ with moments $\tilde{\mu}_k = \mu_{2k}$. Similarly, given $\psi_1 \in \Phi(0, \infty)$, an odd extension of $\frac{1}{2}\psi_1(x^2)$ gives a $\psi \in \Phi(-\infty, \infty)$ with moments (3.2.3).

One can prove that $\psi \in \Phi(-\infty, \infty)$ is an odd function if and only if all c_k in the corresponding J-fraction (3.2.1) are zero. Hence the Stieltjes moment problem for a sequence $\{c_n\}_{n=0}^{\infty}$ has a solution if and only if the series

$$\sum_{n=0}^{\infty} \frac{(-1)^n c_n}{x \cdot x^{2n}} = -\sum_{n=0}^{\infty} \frac{c_n}{x(-x^2)^{n+1}}$$

has a corresponding J-fraction

$$\frac{\lambda_1}{x} - \frac{\lambda_2}{x} - \frac{\lambda_3}{x} - \cdots \approx \frac{\lambda_1/x}{1} + \frac{\lambda_2}{(-x^2)} + \frac{\lambda_3}{1} + \frac{\lambda_4}{(-x^2)} + \cdots.$$

One can also prove that the solution is unique if and only if this particular J-fraction converges. Multiplying by x and substituting $z = -x^2$ now gives:

Theorem 4 *The Stieltjes moment problem for a sequence $\{c_n\}_{n=0}^{\infty}$ has a solution if and only if the series*

$$c_0 + c_1 z^{-1} + c_2 z^{-2} + \cdots \tag{3.2.4}$$

corresponds at $z = \infty$ to a continued fraction of the form

$$\frac{a_1}{1} + \frac{a_2}{z} + \frac{a_3}{1} + \frac{a_4}{z} + \cdots, \tag{3.2.5}$$

where $a_n > 0$ for $n = 1, 2, 3, \ldots$. The solution is unique if and only if (3.2.5) converges for $|\arg z| < \pi$.

The continued fraction (3.2.5) is traditionally called a *modified Stieltjes fraction*, or more generally, if all $a_n \in \mathbf{C} \setminus \{0\}$, a *modified regular C-fraction*. Note also that (3.2.5) converges for all z with $|\arg z| < \pi$ if and only if it converges for one such z (see Theorem 22 in Chapter III).

We shall illustrate Theorem 4 on the Stieltjes problem for the sequence in Example 7.

Example 7 (second time). We have (again) the sequence $\{(-1)^n \cdot n!\}_{n=0}^{\infty}$. To the series

$$1 - 1! \, z^{-1} + 2! \, z^{-2} - 3! \, z^{-3} + \cdots$$

corresponds the modified regular C-fraction

$$\frac{1}{1} \frac{1}{+z} \frac{1}{+1} \frac{2}{+z} \frac{2}{+1} \frac{3}{+z} \frac{3}{+1} \frac{n}{+\cdots z} \frac{n}{+1} \frac{}{+\cdots},$$

which is a modified Stieltjes fraction since the coefficients all are > 0.
From Theorem 4 it then follows that the Stieltjes moment problem in
this particular case has a solution. From Theorem 22 in Chapter III it
follows that the continued fraction converges for $|\arg z| < \pi$, and hence
the solution of the Stieltjes problem is unique. (We already know one
solution, namely

$$\psi(t) = 1 - e^{-t}, \quad 0 \le t < \infty,$$

and hence *the* solution is

$$\psi(t) = K - e^{-t},$$

where K is arbitrary.)

_____◇

Again bounded intervals represent a simpler situation. Markov has
proved, that if there is a solution $\psi \in \Phi(0, b)$, $0 < b < \infty$, then (3.2.5)
converges for all $z \in \mathbf{C}$ not on the segment $[-b, 0]$ of the real line (to
(3.1.4), where ∞ is replaced by b) [Mark95].

In Theorem 4 necessary and sufficient conditions for

a) existence   and   b) uniqueness

of a solution to the Sieltjes moment problem were presented. In both
cases the conditions were expressed in terms of conditions on a *continued
fraction expansion*. We shall not here go into the question of the actual
construction of a solution. We shall merely indicate briefly one way,
where also in fact continued fractions are used as a tool: We assume
that the continued fraction (3.2.5) is a modified Stieltjes fraction, in
which case we know that there exists a solution. Let $\{A_n(z)/B_n(z)\}$ be
the sequence of approximants. It can be proved, that the zeros of all B_n
are real, negative and simple. The partial fraction decomposition of the

approximants can be written as a Stieltjes integral. To illustrate this take $A_4(z)/B_4(z)$ in Example 7.

$$\frac{1}{1+}\frac{1}{z+}\frac{1}{1+}\frac{2}{z} = \frac{z(z+3)}{z^2+4z+2}$$

$$= \frac{\frac{2+\sqrt{2}}{4}z}{z+2-\sqrt{2}} + \frac{\frac{2-\sqrt{2}}{4}z}{z+2+\sqrt{2}} = \int_0^\infty \frac{z\,d\psi_4(t)}{z+t}.$$

Here

$$\psi_4(t) = \begin{cases} 0 & \text{for } 0 \le t < 2-\sqrt{2}, \\ \frac{2+\sqrt{2}}{4} & \text{for } 2-\sqrt{2} \le t < 2+\sqrt{2}, \\ \frac{2+\sqrt{2}}{4} + \frac{2-\sqrt{2}}{4} = 1 & \text{for } 2+\sqrt{2} \le t < \infty. \end{cases}$$

If the continued fraction converges, the expression

$$\frac{A_n(z)}{B_n(z)} = \int_0^\infty \frac{z\,d\psi_n(t)}{z+t}$$

converges to

$$F(z) = \int_0^\infty \frac{z\,d\psi(t)}{z+t}.$$

From $F(z)$ the distribution function can be determined by using Stieltjes inversion formula, see for instance [Chih78, p. 90],

$$\psi(t) - \psi(s) = -\frac{1}{\pi} \lim_{y \to 0+} \int_s^t \Im\{F(x+iy)\}dx.$$

If the continued fraction diverges, the solution is no longer unique. In such a case its even and odd parts converge, and by the above procedure one can get two different solutions ψ_1 and ψ_2 (and hence infinitely many, $\alpha\psi_1 + (1-\alpha)\psi_2, 0 \le \alpha \le 1$).

Remarks:

1. Closely related to the Stieltjes moment problem is the *Hamburger moment problem*, in which the interval is $(-\infty, \infty)$ instead of $(0, \infty)$. Observe that a solution of the Stieltjes problem automatically gives a solution of the Hamburger problem, by defining

it also for $t < 0$ by $\psi(t) = 0$. Furthermore, if we have (in either problem) a double sequence

$$\ldots, c_{-2}, c_{-1}, c_0, c_1, c_2, \ldots$$

instead of a simple one $\{c_n\}_{n=0}^{\infty}$, the moment problems are called *strong* problems. The strong Stieltjes problem and the strong Hamburger problem have both been studied recently, see for instance the survey article [JoTh82]. All problems mentioned above, as well as other moment problems, play an important role in the analytic theory of continued fractions, and may be successfully dealt with by using continued fractions.

2. Some times it is useful to have simple tests. A simple test for uniqueness of solution of the Stieltjes problem (if we *know* the existence) is the Carleman criterion

$$\sum_{n=1}^{\infty} c_n^{-1/(2n)} = \infty,$$

see also Theorem 8 in Chapter V.

Problems

(1) (a) Prove that the monic Tchebycheff polynomials of the second kind, $\tilde{U}_n(x)$, can be expressed by the $(n \times n)$-determinant

$$\tilde{U}_n(x) = \begin{vmatrix} x & 1 & 0 & \cdots & 0 \\ \frac{1}{4} & x & 1 & \cdots & 0 \\ 0 & \frac{1}{4} & x & \cdots & \cdot \\ \vdots & \vdots & \vdots & & 1 \\ 0 & 0 & \cdot & \frac{1}{4} & x \end{vmatrix}.$$

(b) Prove that the function

$$\frac{1}{1 - 2xw + w^2}$$

is a generating function for the Tchebycheff polynomials of the second kind $U_n(x)$.

(2) (a) Take for granted the expression

$$\frac{1}{(1 - 2xw + w^2)^{1/2}} = \sum_{n=0}^{\infty} P_n(x) w^n$$

for the Legendre polynomials $P_n(z)$. Then establish the recurrence relation for the Legendre polynomials. Hint: Prove first that

$$(1 - 2xw + w^2) \cdot \frac{\partial}{\partial w} \left[\frac{1}{(1 - 2xw + w^2)^{1/2}} \right]$$
$$- \frac{x - w}{(1 - 2xw + w^2)^{1/2}} = 0.$$

(b) Find the coefficient of x^n for the Legendre polynomials $P_n(z)$.

(c) Prove the following connection between Legendre polynomials and Tchebycheff polynomials of the second kind:

$$U_n(x) = \sum_{k=0}^{n} P_k(x) \cdot P_{n-k}(x).$$

(3) Prove the orthogonality of the Charlier polynomials (i.e. the relation O_3). Hint: Establish the two expansions

$$\sum_{k=0}^{\infty} \frac{G(k,w)G(k,z)}{k!} = e \cdot e^{zw} = \sum_{n=0}^{\infty} \frac{e(zw)^n}{n!}$$

and

$$\sum_{k=0}^{\infty} \frac{G(k,w)G(k,z)}{k!} = \sum_{m,n=0}^{\infty} \left(\sum_{k=0}^{\infty} \frac{C_m(k)}{m!} \cdot \frac{C_n(k)}{n!} \cdot \frac{1}{k!} \right) z^m w^n ,$$

where $G(x,w)$ is the generating function $e^{-w}(1+w)^x$. Then compare coefficients.

(4) Compute the moments for the functional in Example 1. (See Subsection *1.2.*)

(5) Compute the moments μ_2 and μ_3 for the functional in Example 3. (See Subsection *1.2.*) What can you say about μ_n generally?

(6) Find the Fourier expansion of x^3

   (a) in terms of Tchebycheff polynomials of the second kind, and

   (b) in terms of the Legendre polynomials.

   Find in both cases the expansions also in terms of the corresponding orthonormal polynomials.

(7) Find the Gauss quadrature formula for $n = 2$ and $n = 3$ for the integral

$$\int_{-1}^{1} f(x)\sqrt{1-x^2}\,dx .$$

(8) In the continued fraction in Example 7 write $A_n(z)/B_n(z)$, $n = 2,3$ in the form

$$\int_{0-}^{\infty} \frac{z\,d\psi_n(z)}{z+t} .$$

(9) Use Carleman's test to prove the uniqueness of the solution of the Stieltjes problem for the sequence in Example 7. (Hint: Use Stirling's formula.)

Remarks

1. For a deeper study of orthogonal polynomials, including their connections to continued fractions, we refer to the book [Chih78] by T. S. Chihara. It also contains the concept of chain sequences, which will be briefly touched upon in Chapter X on zero-free regions. See also Wall's book [Wall48]. Other useful expositions are for instance [Nevai79] and [Lubi87]. As an example of orthogonal rational functions we refer to [HeNj89].

2. For moment problems we refer to the book by N. I. Akhiezer [Akhi65] and the book by U. Grenander and G. Szegö [GrSz58]. But the topic is treated in a large number of books and papers. A useful survey article is [JoTh82].

3. For the *Hamburger* moment problem a result related to the one for the Stieltjes moment problem in Theorem 3 holds. Essential differences are that the integral

$$\int_{-\infty}^{\infty} \frac{z\,d\psi(t)}{z+t}$$

represents two different functions in different regions (half-planes), and the series

$$\sum_{n=0}^{\infty} c_n z^{-n}$$

is an asymptotic expansion at $z = \infty$ to the two functions in the two regions.

For the *strong Stieltjes* problem we have two series, being asymptotic expansions of

$$\int_{0}^{\infty} \frac{z\,d\psi(t)}{z+t}$$

at 0 and ∞. For details, see [JoTh80], Section 9.2.

4. Moment theory can be established on different real or complex sets, and several things have been done, which will not be discussed here. Moment theory on the unit circle is one special topic which has attracted attention recently. The topic is old, but much of the theory developed from the old roots is rather new. We refer to

the paper [JoNT89] and to the bibliography therein. Orthogonal polynomials on a *circular arc* are studied in [Gaut89] and [DeBr90].

References

[Akhi65]   N. I. Akhiezer, "The Classical Moment Problem and Some
            Related Questions in Analysis", Hafner, New York (1965).

[Chih78]   T. S. Chihara, "An Introduction to Orthogonal Polynomi-
            als", Mathematics and Its Applications Series, Gordon and
            Breach, New York (1978).

[DeBr90]   M. G. de Bruin, *Polynomials Orthogonal on a Circular Arc*,
            J. Comp. and Appl. Math. **31** (1990), 253–266.

[Gaut89]   W. Gautschi, *On Zeros of Polynomials Orthogonal on the
            Semicircle*, SIAM J. Math. Anal. **20** (1989), 738–743.

[Gragg74]   W. B. Gragg, *Matrix Interpretations and Applications of
            the Continued Fraction Algorithm*, Rocky Mountain J.
            Math. 4 (1974), 213–225.

[GrSz58]   U. Grenander and G. Szegö, "Toeplitz Forms and their Ap-
            plications", University of California Press, Berkeley (1958).

[HeNj89]   E. Hendriksen and O. Njåstad, *A Favard Theorem for Ra-
            tional Functions*, J. Math. Anal. Appl., **142**, 2 (1989), 508–
            520.

[JoTh80]   W. B. Jones and W. J. Thron, "Continued Fractions: An-
            alytic Theory and Applications", Encyclopedia of Mathe-
            matics and its Applications **11**, Addison-Wesley Publish-
            ing Company, Reading, Mass. (1980). Now distributed by
            Cambridge University Press, New York.

365

[JoNT89]   W. B. Jones, O. Njåstad and W. J. Thron, *Moment Theory, Orthogonal Polynomials, Quadrature, and Continued Fractions associated with the Unit Circle*, Bull. London Math. Soc. **21** (1989), 113–152.

[JoTh82]   W. B. Jones and W. J. Thron, *Survey of Continued Fraction Methods for Solving Moment Problems and Related Topics*, "Analytic Theory of Continued Fractions", Lecture Notes in Math. **932** (W. B. Jones, W. J. Thron and H. Waadeland eds.) Springer-Verlag, Berlin (1982), 4–37.

[Lubi87]   P. S. Lubinsky, *A Survey of General Orthogonal Polynomials with Weight Functions on Finite and Infinite Intervals*, Acta Appl. Math. **10** (1987), 237–296.

[Mark95]   A. Markov, *Deux démonstrations de la convergence de certaines fractions continues*, Acta Math. **19** (1895), 93–104.

[Nevai79]   P. G. Nevai, *Orthogonal Polynomials*, Mem. Amer. Math. Soc. **213**, Providence, R. I. (1979).

[Perr57]   O. Perron, "Die Lehre von den Kettenbrüchen", Band II, B. G. Teubner, Stuttgart (1957).

[Wall48]   H. S. Wall, "Analytic Theory of Continued Fractions", Van Nostrand, New York (1948).

Chapter VIII

Padé approximants

About this chapter

Any decent book on continued fractions should contain a section on Padé approximants (and vice versa). Anything else would mean renouncing one's nearest of kin. On the other hand, the topic of Padé approximants or more generally rational approximations is treated in numerous expositions, such as the monograph [BaGr81], to name but one example. The many conferences on the subject illustrate the rapid development of the field, as well as the increased interest in applications, for instance in physics. One example is the 1985-conference in Łańcut, Poland, [GiPS87]. The field and its applications are thus pretty well taken care of. This justifies, in our opinion, the low profile we have chosen in our book. We have restricted ourselves to an example-based introduction to the basic, classical elements of the theory. Next we have emphasized connections to certain continued fraction expansions whose approximants follow certain paths in the Padé table. The Padé table and continued fraction expansions are based upon the same principle, the principle of correspondence. This means that convergence (or divergence) results for continued fraction expansions may lead to convergence (divergence) results for paths in the table, and vice versa.

Padé approximants, as well as continued fraction expansions for a given function, can be derived by certain practical algorithms. We decided to leave these out here, and merely refer to them in the remarks.

The rapid and fruitful development of the theory of Padé approximants has also lead to interesting generalizations. In the second part of the chapter some of them are mentioned in a way which is to be regarded as a list of keywords with some comments.

1 Classical Padé approximants

1.1 A creative problem

We shall let three examples serve as introduction to the main topics of this chapter.

Example 1 Given the formal power series

$$L(z) = 1 + \frac{z}{1!} + \frac{z^2}{2!} + \cdots + \frac{z^n}{n!} + \cdots \qquad (1.1.1)$$

(which happens to coincide with the Taylor expansion of $\exp(z)$ at $z = 0$). We want to find the rational function $R_{1,1}(z)$ (numerator degree ≤ 1, denominator degree ≤ 1) whose Taylor expansion at $z = 0$ agrees with the given formal series as far out as possible. More precisely: If the expansion is

$$R_{1,1}(z) = a_0 + a_1 z + a_2 z^2 + \cdots,$$

we want $a_0 = 1$, $a_1 = 1/1!$, ..., $a_n = 1/n!$ for an n as large as possible.

Since $R_{1,1}(0) = a_0 = 1$, the constant terms in numerator and denominator must be equal and $\neq 0$ (if we ignore the case when they both are zero). Without loss of generality we may assume them to be $= 1$, in which case we have, for $R_{1,1}$ and its Taylor expansion at $z = 0$:

$$R_{1,1}(z) = \frac{1 + az}{1 + bz} = 1 + (a - b)z + b(b - a)z^2 + b^2(a - b)z^3 + \cdots.$$

From

$$a - b = 1$$
$$b(b - a) = \frac{1}{2}$$

we find agreement up to and including the z^2-term if and only if

$$a = \frac{1}{2}, \qquad b = -\frac{1}{2},$$

i.e.

$$R_{1,1}(z) = \frac{1 + \frac{1}{2}z}{1 - \frac{1}{2}z} = 1 + z + \frac{z^2}{2} + \frac{z^3}{4} + \cdots$$

is the unique solution to our problem. Since $1/4 \neq 1/3!$, the agreement terminates at the z^2-term. We write this $L(z) - R_{1,1}(z) = \mathcal{O}[z^3]$.

We could raise the problem more generally for rational functions $R_{m,n}(z)$ (numerator degree $\leq m$, denominator degree $\leq n$). Let us briefly look at the solution in the case $m = 2$, $n = 1$, in which case $R_{2,1}(z)$ and its Taylor expansion may be assumed to be of the form

$$R_{2,1}(z) = \frac{1 + a_1 z + a_2 z^2}{1 + b_1 z} = 1 - (b_1 - a_1)z + (a_2 - a_1 b_1 + b_1^2)z^2$$
$$- (b_1^3 - a_1 b_1^2 + a_2 b_1)z^3 + \cdots .$$

Agreement with (1.1.1) in the coefficients of z, z^2 and z^3 is obtained if and only if

$$-(b_1 - a_1) = 1 ,$$
$$a_2 - a_1 b_1 + b_1^2 = \frac{1}{2} ,$$
$$-(b_1^3 - a_1 b_1^2 + a_2 b_1) = \frac{1}{3!} .$$

Simple computation leads to the unique solution

$$a_1 = \frac{2}{3} , \qquad a_2 = \frac{1}{6} , \qquad b_1 = -\frac{1}{3} ,$$

i.e. to the rational function

$$R_{2,1}(z) = \frac{1 + \frac{2}{3}z + \frac{1}{6}z^2}{1 - \frac{1}{3}z} = 1 + z + \frac{z^2}{2} + \frac{z^3}{6} + \frac{z^4}{18} + \cdots .$$

The agreement with the given series terminates at the z^3-term $(L(z) - R_{2,1}(z) = \mathcal{O}[z^4])$, since $1/18 \neq 1/4!$. Since the solution was unique, we did not have any possibility to require agreement any further. (Observe that the denominator in $R_{2,1}$ differs from the denominator in $R_{1,1}$.)

For large values of m and n the computation becomes more complicated. We shall (try to) require $L(z) - R_{m,n}(z) = \mathcal{O}[z^{m+n+1}]$. One case where we can write down the solution right away, is the case $n = 0$, i.e. where the rational function is a polynomial. In this case the solution is

$$R_{m,0}(z) = 1 + \frac{z}{1!} + \cdots + \frac{z^m}{m!} .$$

The solutions may be arranged in a table with increasing m going down and increasing n going to the right. In the present example the start of the table (i.e. the top left corner, actually the only corner) is easily computed, and is shown below. Observe in particular the symmetry property

$$R_{m,n}(z) = [R_{n,m}(-z)]^{-1} ,$$

which is a consequence of the property $\exp(z) = [\exp(-z)]^{-1}$ for the exponential function.

$\diagdown{}^{n}_{m}$	0	1	2	3	4
0	1	$\dfrac{1}{1-z}$	$\dfrac{1}{1-z+\frac{z^2}{2}}$	$\dfrac{1}{1-z+\frac{z^2}{2}-\frac{z^3}{6}}$	.
1	$1+z$	$\dfrac{1+\frac{1}{2}z}{1-\frac{1}{2}z}$	$\dfrac{1+\frac{1}{3}z}{1-\frac{2}{3}z+\frac{1}{6}z^2}$	$\dfrac{1+\frac{1}{4}z}{1-\frac{3}{4}z+\frac{1}{4}z^2-\frac{1}{24}z^3}$	.
2	$1+z+\frac{z^2}{2}$	$\dfrac{1+\frac{2}{3}z+\frac{1}{6}z^2}{1-\frac{1}{3}z}$	$\dfrac{1+\frac{1}{2}z+\frac{1}{12}z^2}{1-\frac{1}{2}z+\frac{1}{12}z^2}$	.	.
3	$1+z+\frac{z^2}{2}+\frac{z^3}{6}$	$\dfrac{1+\frac{3}{4}z+\frac{1}{4}z^2+\frac{1}{24}z^3}{1-\frac{1}{4}z}$	.	.	.
4	.	.	.	.	.

$\diamondsuit$

We shall look at a related problem, illustrated on the same formal power series as the one in Example 1.

Example 2 Let $L(z)$ be the formal power series of Example 1. Find polynomials ($\neq$ zero-polynomials)

$$\begin{aligned} P_2(z) &= a_0 + a_1 z + a_2 z^2 , \\ Q_1(z) &= b_0 + b_1 z , \end{aligned}$$

such that as many as possible of the first consecutive terms of the formal series

$$Q_1(z) \cdot L(z) - P_2(z)$$

vanish. We find for the start of this formal power series

$$(b_0 + b_1 z) \cdot \left(1 + z + \frac{1}{2}z^2 + \frac{1}{6}z^3 + \cdots\right) - (a_0 + a_1 z + a_2 z^2) \equiv$$

$$(b_0 - a_0) + (b_0 + b_1 - a_1)z + \left(\frac{b_0}{2} + b_1 - a_2\right)z^2 + \left(\frac{b_0}{6} + \frac{b_1}{2}\right)z^3 + \cdots,$$

and a system of equations with the following start:

$$
\begin{aligned}
b_0 - a_0 &= 0 \\
b_0 + b_1 - a_1 &= 0 \\
\frac{b_0}{2} + b_1 - a_2 &= 0 \\
\frac{b_0}{6} + \frac{b_1}{2} &= 0
\end{aligned}
$$

This is necessary and sufficient for vanishing of the first four coefficients of $Q_1 \cdot L - P_2$ (starting with the constant term). Since the total number of coefficients in P_2 and Q_1 is 5, this is all we can require. Simple computation leads to the values

$$
\begin{aligned}
b_1 &= -\frac{b_0}{3}, \\
a_0 &= b_0, \\
a_1 &= \frac{2}{3}b_0, \\
a_2 &= \frac{b_0}{6},
\end{aligned}
$$

i.e.

$$Q_1(z) = b_0 - \frac{b_0}{3}z, \qquad P_2(z) = b_0 + \frac{2}{3}b_0 z + \frac{b_0}{6}z^2,$$

where $b_0 \neq 0$ (since we do not accept the zero polynomial). We have $Q_1(z) \cdot L(z) - P_2(z) = \mathcal{O}[z^4]$. For the rational function $P_2(z)/Q_1(z)$ we find, after having cancelled the factor b_0

$$\frac{P_2(z)}{Q_1(z)} = \frac{1 + \frac{2}{3}z + \frac{1}{6}z^2}{1 - \frac{1}{3}z},$$

which is the same as the function $R_{2,1}(z)$ in Example 1. Actually, for the formal series in Examples 1 and 2 the problem of finding $P_m(z)$ and $Q_n(z)$ such that as many as possible of the first consecutive coefficients of $Q_n \cdot L - P_m$ vanish, leads to the same rational functions $R_{m,n}(z) = P_m(z)/Q_n(z)$, and thus to the same table of rational functions.

◇

Remark: Since the numerator and denominator polynomials in $R_{m,n}$ both depend upon m and n, it might have been better to write $P_{m,n}(z)$ and $Q_{m,n}(z)$, rather than $P_m(z)$ and $Q_n(z)$. The latter notation is chosen in order to avoid too many subscripts.

We conclude this section by making an observation which illustrates the connection between continued fraction expansions and tables as the one in Example 1. We shall use the same formal power series.

Example 3 Let $L(z)$ be the formal power series from Examples 1 and 2. We know from Chapter V, Problem 1, that this series has a corresponding regular C-fraction of the form

$$1 + \cfrac{z}{1+} \; \cfrac{-\frac{1}{2}z}{1} \; \cfrac{\frac{1}{6}z}{+1+} \; \cfrac{-\frac{1}{6}z}{1} \; \cfrac{\frac{1}{10}z}{+1} + \cdots .$$

The first approximants are

$$f_0 = 1, \qquad f_1 = 1 + z, \qquad f_2 = 1 + \cfrac{z}{1+} \; \cfrac{-\frac{1}{2}z}{1} = \frac{1 + \frac{z}{2}}{1 - \frac{z}{2}},$$

$$f_3 = 1 + \cfrac{z}{1+} \; \cfrac{-\frac{1}{2}z}{1} \; \cfrac{\frac{1}{6}z}{+1} = \frac{1 + \frac{2}{3}z + \frac{1}{6}z^2}{1 - \frac{1}{3}z},$$

$$f_4 = 1 + \cfrac{z}{1+} \; \cfrac{-\frac{1}{2}z}{1} \; \cfrac{\frac{1}{6}z}{+1+} \; \cfrac{-\frac{1}{6}z}{1} = \frac{1 + \frac{1}{2}z + \frac{1}{12}z^2}{1 - \frac{1}{2}z + \frac{1}{12}z^2},$$

$$f_5 = 1 + \cfrac{z}{1+} \; \cfrac{-\frac{1}{2}z}{1} \; \cfrac{\frac{1}{6}z}{+1+} \; \cfrac{-\frac{1}{6}z}{1} \; \cfrac{\frac{1}{10}z}{+1} = \frac{1 + \frac{3}{5}z + \frac{3}{20}z^2 + \frac{1}{60}z^3}{1 - \frac{2}{5}z + \frac{1}{20}z^2}.$$

◇

We look at the table in Example 1 and observe that the approximants f_0, f_1, f_2, f_3, f_4 all are in the table, they actually form a diagonal staircase in the table. If we had extended the table to the case $m = 3$, $n = 2$, we would also have found f_5 there. Their location in the table is illustrated below:

f_0			
f_1	f_2		
	f_3	f_4	
		f_5	

A natural question is if this holds more generally, and in fact, the answer is YES, under certain conditions. It represents an important example of the connection between such tables and continued fractions.

1.2   Padé approximants

We shall here, as in Subsection *1.1*, let $P_m(z)$ and $Q_n(z)$ denote polynomials of degree at most m and n respectively, and with complex coefficients. We shall furthermore assume $Q_n(z)$ to be different from the zero polynomial. Moreover, we shall regard two rational functions

$$R_{m,n}(z) = \frac{P_m(z)}{Q_n(z)} \quad \text{and} \quad R_{\tilde{m},\tilde{n}}(z) = \frac{P_{\tilde{m}}(z)}{Q_{\tilde{n}}(z)}$$

as identical iff $P_m Q_{\tilde{n}} - Q_n P_{\tilde{m}} = 0$. The three examples in Subsection *1.1* are different examples of approximating a formal power series by means of rational functions. In the first one a formal power series

$$L(z) = c_0 + c_1 z + c_2 z^2 + \cdots \tag{1.2.1}$$

is approximated by rational functions $R_{m,n}(z)$ in the metric defined in Chapter V, Subsection *1.1*, i.e. the metric "turning correspondence into convergence". The problem is called the *Hermite rational interpolation problem*, and is as follows: To a given formal power series (1.2.1) and given non-negative numbers m, n find an $R_{m,n}(z)$, such that when $R_{m,n}(z)$ is replaced by its Taylor expansion at 0, then

$$L(z) - R_{m,n}(z) = \mathcal{O}\left[z^{m+n+1}\right], \tag{1.2.2}$$

meaning that the series on the left-hand side starts with a term of degree at least $m + n + 1$. In the second one we approximated by making $Q_n(z)L(z) - P_m(z)$ small in the metric of Chapter V, Subsection *1.1*, i.e. starting with a high degree term. This again means, more precisely, to replace (1.2.2) by

$$Q_n(z) \cdot L(z) - P_m(z) = \mathcal{O}\left[z^{m+n+1}\right]. \tag{1.2.3}$$

Whereas the Hermite interpolating problem does not always have a solution (we shall soon see an example), it can be proved that the second problem always has a solution.

The rational functons $R_{m,n}(z) = P_m(z)/Q_n(z)$, where P_m and Q_n satisfy (1.2.3) and $Q_n(z) \neq 0$, are the *Padé approximants* of L, and the two-dimensional array

$$
\begin{array}{cccccc}
R_{0,0} & R_{0,1} & R_{0,2} & \cdot & \cdot & \cdot \\
R_{1,0} & R_{1,1} & R_{1,2} & \cdot & \cdot & \cdot \\
R_{2,0} & R_{2,1} & R_{2,2} & \cdot & \cdot & \cdot \\
\cdot & \cdot & \cdot & \cdot & \cdot & \cdot \\
\cdot & \cdot & \cdot & \cdot & \cdot & \cdot
\end{array}
$$

is called the *Padé table* of L.

In the Examples 1 and 2 it seemed that the Hermite problem had a solution, at least for the (m, n) we computed, and it seemed to lead to the Padé approximants. It is readily seen, that the solution of the Hermite interpolation problem, if it exists, is the Padé approximant. Observe that if $Q_n(0) \neq 0$ it follows from (1.2.3) that

$$L(z) - \frac{P_m(z)}{Q_n(z)} = \mathcal{O}\left[z^{m+n+1}\right].$$

We have used the term approximation problem, meaning approximation in the "correspondence metric". On the other hand we have the classical, well established name Hermite interpolating problem. The significance of the word *interpolation* is that if the formal power series represents a function f, we ask for a rational function $R(z)$, for which

$$R^{(k)}(0) = f^{(k)}(0) \quad \text{for } 0 \le k \le m + n.$$

(Here $f^{(k)}(0)$ denotes the kth derivative of $f(z)$ at $z = 0$.) The next example shows a case where the Hermite problem (1.2.2) has no solution.

Example 4 Given the formal power series

$$L(z) = 1 - \frac{z^2}{2} + \frac{z^4}{24} - \cdots (-1)^n \frac{z^{2n}}{(2n)!} + \cdots$$

(which is of course the well known Taylor series expansion of $\cos z$ at 0). Take $m = n = 1$. A possible solution of the Hermite problem is of the form

$$R_{1,1}(z) = \frac{a_0 + a_1 z}{b_0 + b_1 z}.$$

If we ignore the case $b_0 = 0$, which implies that $a_0 = 0$, and thus that z can be cancelled, then we must have $a_0 = b_0$ in order to have correspondence in the first term, and with

$$a := \frac{a_1}{a_0}, \qquad b := \frac{b_1}{b_0}$$

we find

$$R_{1,1}(z) = \frac{1 + az}{1 + bz} = 1 + (a - b)z + (b^2 - ab)z^2 + \cdots.$$

For the desired correspondence we need simultaneously

$$a - b = 0, \qquad b^2 - ab = -\frac{1}{2},$$

which is impossible.

For the other problem (1.2.3), i.e. the one leading to the Padé approximants, we must solve the equation

$$(b_0 + b_1 z)\left(1 - \frac{z^2}{2} + \cdots\right) - (a_0 + a_1 z) = \mathcal{O}[z^3],$$

i.e. the coefficient equations

$$
\begin{aligned}
b_0 - a_0 &= 0 && \text{(Constant term)}, \\
b_1 - a_1 &= 0 && (z\text{-coefficient}), \\
b_0 \cdot \left(-\tfrac{1}{2}\right) &= 0 && (z^2\text{-coefficient}).
\end{aligned}
$$

This system has the solution

$$
a_0 = b_0 = 0 , \qquad a_1 = b_1 \neq 0 .
$$

Hence (with the earlier notation)

$$
P_1(z) = a_1 z , \qquad Q_1(z) = a_1 z ,
$$

and for the Padé approximant we get in this case

$$
R_{1,1} = 1 .
$$

We observe that in this case the rational function in the $(1,1)$-place in the Padé table may be expressed as the ratio of polynomials of degree 0 (i.e. < 1). Observe also that $L(z) - R_1(z) = \mathcal{O}[z^2]$; i.e. the interpolation is not good enough as compared to $(1.2.2)$.

$\diamond$

The next example is less trivial.

Example 5 We shall look at the same two interpolation problems for a formal series starting with

$$
1 + z - \frac{z^2}{2} + \frac{z^4}{24} - \cdots
$$

for $m = 3$, $n = 1$. A solution of the Hermite problem must have the form

$$
R_{3,1}(z) = \frac{a_0 + a_1 z + a_2 z^2 + a_3 z^3}{b_0 + b_1 z} ,
$$

where $b_0 \neq 0$ and $a_0 = b_0$. Without loss of generality take $a_0 = b_0 = 1$, then $R_{3,1}(z)$ and its Taylor expansion at $z = 0$ must be of the form

$$
\begin{aligned}
R_{3,1}(z) &= \frac{1 + a_1 z + a_2 z^2 + a_3 z^3}{1 + b_1 z} \\
&= 1 + (a_1 - b_1)z + (a_2 - a_1 b_1 + b_1^2)z^2 \\
&\quad + (a_3 - a_2 b_1 + a_1 b_1^2 - b_1^3)z^3 \\
&\quad + (-a_3 b_1 + a_2 b_1^2 - a_1 b_1^3 + b_1^4)z^4 + \cdots .
\end{aligned}
$$

We have the desired correspondence if and only if the following equations are simultaneously satisfied:

$$
\begin{aligned}
a_1 - b_1 &= 1, \\
a_2 - a_1 b_1 + b_1^2 &= -\frac{1}{2}, \\
a_3 - a_2 b_1 + a_1 b_1^2 - b_1^3 &= 0, \\
-a_3 b_1 + a_2 b_1^2 - a_1 b_1^3 + b_1^4 &= \frac{1}{24}.
\end{aligned}
$$

We find successively from the first three equations

$$
\begin{aligned}
a_1 &= b_1 + 1, \\
a_2 &= b_1 - \frac{1}{2}, \\
a_3 &= -\frac{b_1}{2}.
\end{aligned}
$$

When this is inserted into the last equation we find the contradictive statement $0 = 1/24$, showing that the Hermite problem has no solution.

For the problem (1.2.3), i.e.

$$
(b_0 + b_1 z)\left(1 + z - \frac{z^2}{2} + \frac{z^4}{24} - \cdots\right) - (a_0 + a_1 z + a_2 z^2 + a_3 z^3) = \mathcal{O}[z^5]
$$

the solution is given by

$$
\begin{aligned}
b_0 - a_0 &= 0, \\
b_1 + b_0 - a_1 &= 0, \\
b_1 - \frac{b_0}{2} - a_2 &= 0, \\
-\frac{b_1}{2} - a_3 &= 0, \\
\frac{b_0}{24} &= 0.
\end{aligned}
$$

We find $b_0 = a_0 = 0$, $a_1 = a_2 = b_1$, $a_3 = -b_1/2$. Without loss of generality we may take $b_1 = 1$, and find

$$
P_3(z) = z + z^2 - \frac{1}{2}z^3, \qquad Q_1(z) = z,
$$

and hence (after cancelling of the factor z)

$$R_{3,1} = \frac{1 + z - \frac{1}{2}z^2}{1}.$$

Observe that the rational function in the $(3,1)$-place in the Padé table is expressed as the ratio of polynomials of lower degrees than 3 and 1, actually 2 and 0.

$\diamond$

We have here chosen to use (1.2.3) as the basis for our definition of the Padé table, but also (1.2.2) is widely used. Both approaches have their advantages and disadvantages. The use of (1.2.2) is in a way more natural, since we are aiming at a rational approximation (in the correspondence metric) to a formal series. On the other hand, as illustrated in the Examples 4 and 5, the (1.2.2)-approximation does not always exist. Important is, however, that when it exists, it coincides with the Padé approximation in our definition. This implies, for instance, that the table in Example 1 is a Padé table. If the (1.2.2)-approximation fails to exist for a pair (m, n), then the (m, n) Padé approximant is equal to a (1.2.2) approximant of lower order. This will be evident in the next subsection.

1.3 Normal tables. Block structure.

In the Padé table in Example 1 we have seen that our entries $R_{m,n}(z)$ are rational functions where the degrees in numerator and denominator are exactly m and n respectively, and can not be reduced by cancellation, and the entries are all different. Such a Padé table is called a *normal Padé table*. It can be proved that the table in Example 1 is normal (the whole table, not only the part we have seen). In Example 4 the rational function in the $(1,1)$-place was 1, the same as in the $(0,0)$-place. It is easily seen that we also get 1 in the places $(1,0)$ and $(0,1)$. The upper left corner of the Padé table in this case is

	0	1	2	...
0	1	1	*	
1	1	1	*	
2	*	*	*	
⋮				

Observe the square block of equal elements.

In Example 5 we found in the $(3,1)$-place the function

$$\frac{1 + z - \frac{1}{2}z^2}{1}.$$

This is obviously also the function in the $(2,0)$-place (since it is a section of the given series) and even the $(3,0)$-place, since the z^3-term in the series is 0. Simple computation (Problem 3) shows that it is also the function in the $(2,1)$-place. In the Padé table for the series in Example 5 we have a square block of equal elements as shown below.

	0	1	2	3	...
0	*	*	*	*	
1	*	*	*	*	
2	$\dfrac{1 + z - \frac{1}{2}z^2}{1}$	$\dfrac{1 + z - \frac{1}{2}z^2}{1}$	*	*	
3	$\dfrac{1 + z - \frac{1}{2}z^2}{1}$	$\dfrac{1 + z - \frac{1}{2}z^2}{1}$	*	*	
⋮					

These observations reflect a general structural pattern of the Padé table: Equal entries appear only in square blocks. In a normal table each block consists of only one function. The "block theorem" is as follows:

Theorem 1 (The block theorem) *Let $R(z) = P(z)/Q(z)$, where P and Q are relatively prime polynomials of degree m and n respectively. Assume furthermore that $R(z)$ occurs in the Padé table of a formal power series L. If, for a non-negative integer r the formal power series*

$$QL - P \qquad (1.3.1)$$

starts with the term of degree $m + n + r + 1$, then the set of places where $R(z)$ occurs is a square block with $(r + 1)^2$ places and opposite corner places in (m, n) and $(m + r, n + r)$.

For a proof we refer to [Gragg72].

Remark: Observe that in a square block of size > 1 the elements not in the upper, leftmost corner have numerator or denominator degrees (or both) lower than the place (m, n) "should indicate". This is illustrated in the Examples 4 and 5. The computation (solution of linear equations) in these cases indicate why it happens, and a corresponding *general* discussion is essential in the proof of the theorem.

The theorem extends to $r = \infty$, in which case the block is unbounded down and to the right and $QL - P$ is the zero series. This can only happen if $L(z)$ is the Maclaurin series of the rational function P/Q.

The word *normal* is used for the *approximants* $R(z) = P(z)/Q(z)$, where P and Q are relatively prime and of degree m and n respectively, meaning that $QL - P$ starts with the term of degree $m + n + 1$. It is used for the *table*, meaning that all elements are normal, and for the formal *power series*, meaning that the table is normal. Criteria for normality of a power series $\sum c_n z^n$ may be expressed in terms of the Toeplitz determinants

$$c_{m,n} = \begin{vmatrix} c_m & c_{m-1} & \cdots & c_{m-n+1} \\ c_{m+1} & c_m & \cdots & \\ \vdots & \vdots & & \vdots \\ c_{m+n-1} & & \cdots & c_m \end{vmatrix}, \qquad (1.3.2)$$

where $c_k = 0$ for $k < 0$, $c_{m,0} = 1$, $m = 0, 1, 2, \ldots$, in e.g. the following theorem:

Theorem 2 *An (m, n) Padé approximant of a formal power series*

$$c_0 + c_1 z + c_2 z^2 + \cdots, \quad c_0 \neq 0, \tag{1.3.3}$$

is normal if and only if the determinants

$$c_{m,n}, \qquad c_{m,n-1}, \qquad c_{m+1,n}, \qquad c_{m+1,n+1} \tag{1.3.4}$$

are all $\neq 0$.

For a proof we refer to [Gragg72].

It follows from Theorem 2 that a formal power series and its Padé table are normal if and only if

$$c_{m,n} \neq 0 \quad \text{for all } m, n = 0, 1, 2, \ldots. \tag{1.3.5}$$

This shows in particular that a formal power series with gaps, i.e. where at least one $c_k = 0$, $k \geq 1$, is *not normal*, since $c_{k,1} = c_k$.

1.4 Connection to continued fraction expansions

In Example 3 we saw (at least for the first entries) that the corresponding regular C-fraction to the given formal power series was such that the successive approximants coincided with the rational functions $R_{0,0}$, $R_{1,0}$, $R_{1,1}$, $R_{2,1}$, $R_{2,2}$, ... in the Padé table for the formal series. This property actually holds generally, under the condition of normality.

Theorem 3 *Let*

$$1 + c_1 z + c_2 z^2 + c_3 z^3 + \cdots \tag{1.4.1}$$

be a formal power series with the property that the Padé approximants

$$R_{0,0}(z), \ R_{1,0}(z), \ R_{1,1}(z), \ R_{2,1}(z), \ R_{2,2}(z), \ R_{3,2}(z), \ldots$$

all are normal. Then (1.4.1) has a corresponding regular C-fraction

$$1 + \overset{\infty}{\underset{n=1}{\mathbf{K}}} \, \frac{a_n z}{1}, \tag{1.4.2}$$

whose approximants f_n satisfy

$$f_{2m} = R_{m,m}, f_{2m+1} = R_{m+1,m}, \quad m = 0, 1, 2, \ldots. \tag{1.4.3}$$

For a proof we refer to [JoTh80], p.191. See also Problem 4.

Even a converse result holds:

Theorem 4 *Let (1.4.2) be a given regular C-fraction, and (1.4.1) the corresponding formal power series. Then the successive approximants f_n of the C-fraction come as a staircase in the Padé table of (1.4.1) by satisfying (1.4.3).*

Idea of proof: The correspondence of C-fractions (not only regular C-fractions) to power series is described in Theorem 5, Chapter V. We also know that the successive approximants of a regular C-fraction is such that $f_n = A_n/B_n$, where the degrees are given by $\deg(A_{2m+1}) = m+1$, $\deg(B_{2m+1}) \leq m$, $\deg(A_{2m}) \leq m$, $\deg(B_{2m}) = m$. Furthermore, the correspondence is such that the Taylor expansion at 0 of A_n/B_n agrees with the formal series up to and including the term z^N, where $N = m+n$. For a detailed proof we refer to [JoTh80, p. 192].

Theorem 4 tells about *one* illustration of the connection between Padé tables and continued fraction expansions. There are several. Generally, if $\{R_{m_k,n_k}\}$ is any path in the Padé table with $R_{m_{k+1},n_{k+1}} \neq R_{m_k,n_k}$, then there is a corresponding continued fraction with approximants $\{R_{m_k,n_k}\}_{k=0}^{\infty}$. (See Corollary 8 in Chapter II.) We shall not go into that here, only mention very briefly one interesting example due to Arne Magnus [Magn62a], [Magn62b]. He introduced the P-fractions (principal part continued fractions), which in a way is related to the regular continued fractions. Whereas the regular continued fraction may be constructed by repeatedly taking the integer part of a number and the reciprocal of the fractional part, the P-fraction is constructed in a similar way by letting the principal part plus the constant term play the role of the integer part, and the Taylor part minus the constant term play the role of the fractional part:

$$c_{-k}z^{-k} + \cdots + c_{-1}z^{-1} + c_0 + \cfrac{1}{\cfrac{1}{c_1 z + c_2 z^2 + \cdots}}.$$

For the formal power series of Example 1 we find the following start of the P-fraction:

$$1 + \cfrac{1}{\left(\dfrac{z}{1!} + \dfrac{z^2}{2!} + \dfrac{z^3}{3!} + \cdots\right)^{-1}} = 1 + \cfrac{1}{\frac{1}{z} - \frac{1}{2}} + \cfrac{1}{\frac{12}{z}} + \cdots .$$

For the first approximants we find:

$$f_0 = 1, \qquad f_1 = 1 + \cfrac{1}{\frac{1}{z} - \frac{1}{2}} = \frac{1 + \frac{z}{2}}{1 - \frac{z}{2}},$$

$$f_2 = 1 + \cfrac{1}{\frac{1}{z} - \frac{1}{2}} + \cfrac{1}{\frac{12}{z}} = \frac{1 + \frac{z}{2} + \frac{z^2}{12}}{1 - \frac{z}{2} + \frac{z^2}{12}} .$$

We observe that these elements are the first three diagonal elements in the Padé table for the series, and it can be proved that the successive approximants of the P-fraction in this case are in turn the diagonal elements of the Padé table. This actually holds generally, in the following sense: For any formal power series

$$c_0 + c_1 z + c_2 z^2 + \cdots , \qquad c_0 \neq 0 ,$$

let $f_n(z)$ denote the nth approximant of the corresponding P-fraction. Then f_n is the nth element in the main diagonal of the corresponding Padé table if we only count distinct elements (i.e. one element from each square block the main diagonal passes through). This result, proved by Arne Magnus [Magn62a], tells that the P-fraction picks up exactly one element from each block intersecting the diagonal. For a *normal* table $f_n(z)$ is the element on the (n, n)-place. In [Magn62a] it is proved that the P-fraction also can create the side-diagonals: For any integer s, take the nth approximant $f_n^{(s)}(z)$ of the P-fraction corresponding to the formal power series

$$z^s L(z) = z^s (c_0 + c_1 z + c_2 z^2 + \cdots) .$$

Then $f_n^{(s)} z^{-s}$ is the nth distinct element in the $(m, m - s)$-diagonal for $s \leq 0$ and the $(m + s, m)$-diagonal for $s \geq 0$.

1.5 A convergence result

There are several results on convergence of Padé approximants, and several open questions. They concern different types of convergence. We have of course some obvious results in the metric defined in Chapter V, Subsection *1.1*, the "correspondence metric". Any path in the Padé table with $m + n \to \infty$ is such that the sequence of corresponding approximants converges to the series from which the Padé table is constructed in this particular metric. But this is only a restatement of the correspondence property. Usually one wants more, for instance pointwise or uniform convergence. We have of course already some results: Convergence results for continued fraction expansions may lead to convergence results along paths in the Padé table, for instance in the case of regular C-fractions. Some examples are promising, such as for instance the Padé approximants to the circumference of the ellipse, discussed in Chapter I, Subsection *3.5*, but there are also some nasty results on "nonconvergence", for instance that there exists an entire function f such that the diagonal sequence of Padé approximants is divergent everywhere in the complex plane except at the origin [Wall74]. We shall here restrict ourselves to one single convergence result, perhaps the most famous one for Padé approximants. It is due to de Montessus de Ballore [Mont02], and concerns vertical sequences in the Padé table, i.e. sequences $\{R_{m,n}\}_{m=0}^{\infty}$.

Theorem 5 (Montessus de Ballore) *Let $f(z)$ be holomorphic in the disk*

$$|z| \leq R \,,$$

except for n simple poles $p_1, p_2, \ldots, p_n$, where

$$0 < |p_1| \leq |p_2| \leq \cdots \leq |p_n| < R \,.$$

Take the formal power series to be the Taylor expansion of $f(z)$ at 0, and let $R_{m,n}(z)$ denote the (m,n)-Padé approximant. Then for all z in the disk $|z| \leq R$ minus the poles we have

$$\lim_{m \to \infty} R_{m,n}(z) = f(z) \,,$$

and the convergence is uniform on compact subsets of this set.

Proofs may be found in for instance [Gragg72] and [Perr57]. In the
case $n = 0$ this reduces to convergence of the Taylor series of a function,
holomorphic in the disk $|z| \leq R$. For a fixed $n > 0$ we get a generalization
to functions, holomorphic in $|z| \leq R$, except for n simple poles in $0 <
|z| < R$. All the rational functions $R_{m,n}$ with this fixed n have n poles,
tending to the poles of $f(z)$ when $n \to \infty$.

2   Generalizations and extensions

2.1   Two-point Padé table

The ordinary Padé table interpolates *one* formal power series at *one*
point, usually, $z = 0$ (Hermite interpolation), and is connected to continued fractions like regular C-fractions and P-fractions in the way described in Section 1. For different types of continued fractions *correspondence* to a formal power series is a property which is appreciated,
and more so if the correspondence is strong enough to make the approximants entries in the Padé table. This is — as we have seen — the case
for regular C-fractions, whose approximants form a staircase in the Padé
table, and P-fractions, whose entries form a diagonal in the table. The
T-fractions

$$\frac{z}{1 + d_1 z +} \frac{z}{1 + d_2 z +} \frac{z}{1 + d_3 z +} \cdots,$$

introduced in 1948 by Thron [Thron48], have a simple structure, and
much can be said about convergence. It also corresponds to a power
series. However, none of the approximants are in the Padé table (except
for the case $d_n = 0$ for all n). (This was said with some implicit regret.)
Much later it was discovered that in addition to correspondence to a
power series

$$c_0 + c_1 z + c_2 z^2 + c_3 z^3 + \cdots \qquad \text{(correspondence at 0)}, \qquad (2.1.1)$$

the general T-fraction

$$c_0 + \frac{F_1 z}{1 + G_1 z +} \frac{F_2 z}{1 + G_2 z +} \frac{F_3 z}{1 + G_3 z +} \cdots, \qquad F_n \neq 0, \qquad (2.1.2)$$

also corresponds to a power series

$$c_0^* + c_{-1}^* z^{-1} + c_{-2}^* z^{-2} + \cdots \qquad \text{(correspondence at } \infty) \qquad (2.1.3)$$

under the additional condition $G_n \neq 0$ for all n. Under certain determinant conditions on the coefficients we also know the converse, i.e. that to a pair of series (2.1.1) and (2.1.3) there corresponds a general T-fraction (2.1.2). The interpolation provided by the general T-fraction is shared between interpolation at 0 and at ∞ (actually roughly equally shared as far as degree of correspondence is concerned). *The two-point Padé table* (the points being 0 and ∞) is constructed from *a pair of series* (2.1.1) and (2.1.3) in a way related to what is done for ordinary Padé tables (one-point tables) of *one* formal series. And *in such a* table we find the approximants of the general T-fraction (2.1.2). This was first observed by McCabe and Murphy [McMu76] (for M-fractions, which are closely related to general T-fractions). Let L and L^* denote the series (2.1.1) and (2.1.3) respectively. Then the two-point Padé approximant $P_{m,n}/Q_{m,n}$ of (L, L^*) is defined by simultaneous requirements on the orders of the first terms of the series

$$Q_{m,n}L - P_{m,n} \qquad \text{and} \qquad Q_{m,n}L^* - P_{m,n}.$$

It can be done in different ways. We want the correspondence to be close to be "equally shared" by the two interpolations, we require (for even $m + n + 1$)

$$Q_{m,n}L - P_{m,n} = \mathcal{O}\left[z^{\frac{m+n+1}{2}}\right], \qquad Q_{m,n}L^* - P_{m,n} = \mathcal{O}\left[z^{-\frac{m+n+1}{2}}\right],$$
$$(2.1.4)$$

(and a related condition for odd $m + n + 1$). We shall not go further into this. We refer to [Magn82] for a precise definition of the two-point Padé approximants and properties of the two-point table, as well as an example and references.

Another important connection between continued fractions and two-point Padé tables is given by the PC-fractions (Perron-Carathéodory fractions) introduced by Jones, Njåstad and Thron. They are continued fractions of the form

$$\beta_0 + \frac{\beta_1}{1} \frac{1}{+\beta_2 z} \frac{\alpha_3 z}{+\beta_3} \frac{1}{+\beta_4 z} \frac{\alpha_5 z}{+\beta_5} \frac{1}{+\beta_6 z +} \cdots \qquad (2.1.5)$$

where $\beta_1 \neq 0$ and $\alpha_{2n+1} = 1 - \beta_{2n}\beta_{2n+1} \neq 0$ for $n = 1, 2, 3, \ldots$. For these one has an even/odd correspondence as follows

$$L - \frac{P_{2n}}{Q_{2n}} = \mathcal{O}\left[z^{n+1}\right], \qquad L^* - \frac{P_{2n+1}}{Q_{2n+1}} = \mathcal{O}\left[z^{-(n+1)}\right]. \qquad (2.1.6)$$

For a description and also more correspondence properties of these and a proof of the connection to two-point tables we refer to [JoNT86]. Let it finally be mentioned, that the PC-fractions, or rather certain subfamilies are closely related to the trigonometric moment problem, Gaussian quadrature on the unit circle, and Szegö polynomials. The even and odd parts of PC-fractions are T- and M-fractions respectively if they exist. See also Remark 2 in Chapter V.

One may have other points of interpolation, and there may be more than two points. This leads to multiple point Padé approximants. The reference list in [JoTh80] provides a relevant bibliography on the subject. There are also bridges between continued fractions and multiple point Padé approximants. E. Hendriksen and O. Njåstad introduced in [HeNj89a] multipoint Padé fractions (to mention but one example). See also [HeNj89b] and the references therein.

One approach to multiple point Padé tables is through the formal Newton series, where the formal power series L is replaced by a formal Newton series

$$\tilde{L} = c_0 + \sum_{n=1}^{\infty} c_n \prod_{k=1}^{n} (z - \beta_k), \qquad (2.1.7)$$

where the points β_k, not necessarily distinct, are the interpolation points. Certain staircase sequences of normal *Newton-Padé*-approximants are the approximants of a Thiele continued fraction

$$a_0 + \frac{a_1(z - \beta_1)}{1} + \frac{a_2(z - \beta_2)}{1} + \cdots . \qquad (2.1.8)$$

Observe that if all interpolation points $\beta_k = 0$, we are back to the ordinary normal Padé table and the regular C-fractions. We refer to [CuWu87] and the references therein.

2.2 Padé type approximants

In the process of interpolating a function or a formal series by rational functions one sometimes want other conditions to be satisfied in addition to the correspondence. That, of course, has its price, the payment being a reduction in the degree of correspondence.

The Padé type approximants, introduced by Brezinski (see e.g. [Brez80]) represent such a case. The background for inventing and studying such a concept is that the poles of Padé approximants are essentially beyond control, since the Padé approximants are uniquely defined by the correspondence requirement. Sometimes, however, we want to choose some of the poles of the approximant and then determine the numerator and the denominator in such a way that the Taylor expansion at $z = 0$ matches the given formal series as far out as possible. We then get what is called Padé type approximants. There are two extreme cases: On one hand we can choose *all* the poles, on the other hand we choose *no* pole. In the latter case we are back to ordinary Padé approximants.

We shall not go further into this topic, only refer to [Brez80]. Let it also be mentioned that further generalizations have been made along the same line, by fixing not only poles, but also zeros of the approximants (pseudo-approximants).

2.3 Multivariate Padé approximants

There are different ways of obtaining the univariate Padé approximants, such as solving the system of equations for the coefficients, or to use continued fractions to produce staircases of diagonals, both mentioned earlier in the chapter. Other methods are using some kinds of recursive schemes to produce the table. These different approaches, being equivalent in the univariate case, have been generalized to the multivariate case. The equivalence between the different techniques, however, is no longer there in the multivariate case. Also in the multivariate case rational approximants and interpolants can be constructed by using continued fractions. In Chapter V, Subsections *4.1* and *4.2*, we have briefly touched upon branched continued fractions and different ways of defining approximants in order to obtain (a meaningful type of) correspondene.

The definition used in [CuWu87] preserves several properties of the univariate Padé approximants. The starting point is the bivariate function f with Taylor expansion

$$f(x,y) = \sum_{i,j=0}^{\infty} c_{i,j} x^i y^j \, . \tag{2.3.1}$$

From the determinant solution of the univariate Padé approximation problem (obtained from the system of equations by the Cramer rule) the multivariate Padé approximants are defined by analogy: The numerator and denominator polynomials $p(x,y)$ and $q(x,y)$ are given by determinants related to the ones for $P(x)$ and $Q(x)$ in the univariate case (approximant $P(x)/Q(x)$). We refer to the exposition [CuWu87], which also includes (among other things) multivariate versions of methods of computation, such as ε-algorithm and qd-algorithm (see remarks in Chapter V), and also examples.

Problems

(1) Use the method of Example 1 to compute $R_{2,3}(z)$ and $R_{3,2}(z)$ in the Padé table for the series (1.1.1).

(2) Use the method of Example 2 to compute $P_2(z)$ and $Q_3(z)$ of degrees ≤ 2 and 3 respectively, such that as many as possible of the first consecutive terms of

$$L(z)Q_3(z) - P_2(z)$$

vanish, when L is the power series (1.1.1). Compare $P_2(z)/Q_3(z)$ to $R_{2,3}(z)$ of Problem 1.

(3) Compute the function in the $(2, 1)$-place in Example 5.

(4) Given the formal power series (1.4.1). Assume that it has a corresponding regular C-fraction (1.4.2). Compute the approximant f_3, in terms of the coefficients of (1.4.1). Compute next the Padé approximant $R_{2,1}(z)$, and compare it to f_3.

(5) Determine the start of the P-fraction expansion of the series in Example 4 (Taylor-expansion of $\cos z$ at $z = 0$) up to and including the third term (i.e. the 0th, the 1st, the 2nd and the 3rd). Compute the approximants, and verify their positions in the Padé table.

Remarks

1. For practical computation of Padé approximants different algorithms are available. We refer to [CuWu87], Chapter II, Section 3 and to the references therein. Some key words deserve to be mentioned: The qd-algorithm, the method of Viscovatov, Gragg's algorithm, and the ε-algorithm.

2. The method of vector valued interpolation is introduced by [Wynn63] and further developed by Graves-Morris and others, see e.g. [Grav83]. See also Subsection *5.3* and Remark 4 in Chapter IV. For any proper vector v in a complex finite-dimensional linear space we define the *vector inverse*, called *Samelson inverse* by

$$v^{-1} = v^*/|v|^2 \, ,$$

where $*$ denotes complex conjugation. It is easily verified, that with this definition

$$v^{-1} \cdot v = 1 \qquad \text{and} \qquad (v^{-1})^{-1} = v \, .$$

As observed by Peter Wynn these inverses may be used to generalize the Thiele continued fraction to treat the case of vector valued interpolation.

References

[BaGr81]   G. Baker and P. Graves-Morris, "Padé Approximants: Basic Theory." Encyclopedia of Mathematics and its Applications, Vol. **13**, Addison-Wesley Publishing Co., Reading Mass. (1981).

[Brez80]   C. Brezinski, "Padé-Type Approximation and General Orthogonal Polynomials", International Series of Numerical Mathematics, Vol. **50**, Birkhäuser, Basel (1980).

[CuWu87]   A. Cuyt and L. Wuytack, "Nonlinear Methods in Numerical Analysis", North-Holland Mathematics Studies in Computational Mathematics **1**, Amsterdam (1987).

[Gile78]   J. Gilevicz, "Approximants de Padé", Lecture Notes in Mathematics **667**, Springer-Verlag, Berlin (1978).

[GiPS87]   J. Gilewicz, M. Pindor and W. Siemaszko, "Rational Approximation and its Applications in Mathematics and Physics", Proceedings, Łańcut 1985, Lecture Notes in Mathematics **1237**, Springer-Verlag, Berlin (1987).

[Gragg72]   W. B. Gragg, *The Padé Table and its Relation to Certain Algorithms in Numerical Analysis*, SIAM Review **14** (1972), 1–62.

[Grav83]   P. R. Graves-Morris, *Vector Valued Rational Interpolants I*, Num. Math. **42** (1983), 331–348.

393

[HeNj89a]   E. Hendriksen and O. Njåstad, *A Favard Theorem for Rational Functions*, Journal of Math. Anal. Appl., Vol. **142**, 2 (1989), 508–520.

[HeNj89b]   E. Hendriksen and O. Njåstad, *Positive Multipoint Padé Continued Fractions*, Proceedings of the Edinburgh Math. Soc. **32** (1989), 261–269.

[JoNT86]   W. B. Jones, O. Njåstad and W. J. Thron, *Continued Fractions Associated with Trigonometric and Other Strong Moment Problems*, Constructive Approx. **2** (1986), 197–211.

[JoTh80]   W. B. Jones and W. J. Thron, "Continued Fractions: Analytic Theory and Applications", Encyclopedia of Mathematics and its Applications Vol. **11**, Addison-Wesley Publishing Company, Reading, Mass. (1980). Now distributed by Cambridge University Press, New York.

[Magn62a]   A. Magnus, *Certain Continued Fractions Associated with the Padé Table*, Math. Zeitschr. **78** (1962), 361–374.

[Magn62b]   A. Magnus, *Expansions of Power Series into P-Fractions*, Math. Zeitschr. **80** (1962), 209–216.

[Magn82]   A. Magnus, *On the Structure of the Two-Point Padé Table*, "Analytic Theory of Continued Fractions" (W. B. Jones, W. J. Thron and H. Waadeland eds.), Lecture Notes in Mathematics No. **932**, Springer-Verlag, Berlin (1982), 176–193.

[McMu76]   J. H. McCabe and J. A. Murphy, *Continued Fractions which Correspond to Power Series Expansions at Two Points*, J. Inst. Maths. Applics. **17** (1976), 233–247.

[Mont02]   R. de Montessus de Ballore, *Sur les fractions continues algébriques*, Bull. Soc. Math. France **30** (1902), 28–36.

[Perr57]   O. Perron, "Die Lehre von den Kettenbrüchen", 3. Auflage, Band II, B. G. Teubner, Stuttgart (1957).

[Thron48]   W. J. Thron, *Some Properties of Continued Fraction* $1 + d_0 z + \mathbf{K}(z/(1 + d_n z))$, Bull. Amer. Math. Soc. **54** (1948), 206–218.

[Wall74]   H. Wallin, *The Convergence of Padé Approximants and the Size of the Power Series Coefficients*, Applicable Analysis 4 (1974), 235–251.

[Wall83]   H. Wallin, *Convergence of Multipoint Padé Approximants with a Fixed Number of Poles*, Det Kgl. Norske Vid. Selskabs Skrifter, No. 1 (1983), 151–158.

[Wynn63]   P. Wynn, *Continued Fractions whose Coefficients Obey a Non-Commutative Law of Multiplication*, Arch. Rat. Mech. Anal. **12** (1963), 273–312.

Chapter IX

Some applications in number theory

About this chapter

Books in Number Theory usually have a chapter, or at least some sections, on continued fractions, mostly restricted to *regular* continued fractions. This restriction is of course highly understandable, in view of the role these continued fractions have played (and play) in number theory. On the other hand, as an undesired side-effect, many people (meaning mathematicians) think of a continued fraction as "something" within number theory, and only there. They are highly surprised (hopefully *pleasantly* surprised) to see the many fields in mathematics and adjacent subjects where continued fractions are of use as a descriptive or problem-solving tool.

Here we have the opposite situation: A book on continued fractions, that contains a chapter on number theory, and a very restricted one, as far as topics are concerned. Nevertheless, we do not think we are running any noticeable risk of making people think that Number Theory, the "Queen of Mathematics", is a small subset of the Theory of Continued Fractions.

As in most books on Number Theory, this chapter uses exclusively regular continued fractions (although others would also have been of interest). We have chosen, in this chapter, which is placed in the "applied part" of the book, to present *two* applications in number theory: The simplest examples of the *very* classical field of diophantine equations, and the likewise old problem of factoring numbers, which however, in view of modern (but classically based) cryptography has caused renewed interest and has led to research, where the combination of classical mathematics and modern computer technology has demonstrated its power.

1 Some basics on regular continued fractions

1.1 The Euclidean algorithm

Let a and b be two positive integers. We want to find the greatest common divisor of a and b, here written

$$\gcd(a, b).$$

As will be seen in Section 3 we shall for practical reasons be very interested in finding this quantity, in particular for large numbers a, b. The algorithm used for this goes all the way back to Euclid's Elements, although in a slightly different form, and is usually called the *Euclidean algorithm*. It goes as follows: There is a unique non-negative integer q_0 and a unique integer $r_1, 0 \le r_1 \le b - 1$, such that

$$a = q_0 b + r_1. \tag{1.1.1}$$

If $r_1 = 0$, then b divides a, written $b|a$, and we have

$$\frac{a}{b} = q_0.$$

The process stops. In this case the greatest common divisor is b. If $r_1 \ne 0$, we let (b, r_1) replace (a, b) in the argument above:

$$b = q_1 r_1 + r_2, \qquad 0 \le r_2 \le r_1 - 1, \tag{1.1.2}$$

and, if $r_2 \ne 0$:

$$r_1 = q_2 r_2 + r_3, \qquad 0 \le r_3 \le r_2 - 1, \tag{1.1.3}$$

and so forth. Since $0 \le r_1 \le b - 1, 0 \le r_2 \le r_1 - 1$, and so on, we will, after a finite number of steps, reach an r_{k+1}, which is 0, whereas $r_i \ne 0$ for all $i \le k$, and we have

$$r_{k-1} = q_k r_k, \quad \text{where } q_k \ne 0. \tag{1.1.4}$$

Hence $r_k | r_{k-1}$, and from

$$r_{k-2} = q_{k-1} r_{k-1} + r_k$$

it follows that $r_k | r_{k-2}$. Step by step we reach a and b, and find that r_k is a common divisor of a and b. Conversely, if d is a common divisor of a and b, it must divide in turn $r_1, r_2, \ldots, r_k$. Hence the last non-zero residuum r_k is the greatest common divisor of a and b,

$$\gcd(a, b) = r_k . \qquad (1.1.5)$$

Example 1 Find $\gcd(2587, 1547)$.

$$
\begin{aligned}
2587 &= 1 \cdot 1547 + 1040 \\
1547 &= 1 \cdot 1040 + 507 \\
1040 &= 2 \cdot 507 + 26 \\
507 &= 19 \cdot 26 + 13 \\
26 &= 2 \cdot 13
\end{aligned}
$$

Hence: $\gcd(2587, 1547) = \underline{13}$.

───◇

Example 2 Find $\gcd(96577, 1155)$.

$$
\begin{aligned}
96577 &= 83 \cdot 1155 + 712 \\
1155 &= 1 \cdot 712 + 443 \\
712 &= 1 \cdot 443 + 269 \\
443 &= 1 \cdot 269 + 174 \\
269 &= 1 \cdot 174 + 95 \\
174 &= 1 \cdot 95 + 79 \\
95 &= 1 \cdot 79 + 16 \\
79 &= 4 \cdot 16 + 15 \\
16 &= 1 \cdot 15 + 1 \\
15 &= 15 \cdot 1
\end{aligned}
$$

Hence: $\gcd(96577, 1155) = \underline{1}$. We also say, that 1155 and 96577 are *coprime*.

───◇

The equalities of Example 2 can be written

$$\frac{96577}{1155} = 83 + \frac{712}{1155} = 83 + \frac{1}{\dfrac{1155}{712}}$$

$$\frac{1155}{712} = 1 + \frac{443}{712} = 1 + \frac{1}{\dfrac{712}{443}}$$

$$\vdots$$

$$\frac{79}{16} = 4 + \frac{15}{16} = 4 + \frac{1}{\dfrac{16}{15}}$$

$$\frac{16}{15} = 1 + \frac{1}{15}.$$

By repeated substitution of the fractions we find

$$\frac{96577}{1155} = 83 + \frac{1}{1+}\frac{1}{1+}\frac{1}{1+}\frac{1}{1+}\frac{1}{1+}\frac{1}{1+}\frac{1}{4+}\frac{1}{1+}\frac{1}{15}.$$

Quite similarly we find that Example 1 gives rise to the following terminating continued fraction

$$\frac{2587}{1547} = 1 + \frac{1}{1+}\frac{1}{2+}\frac{1}{19+}\frac{1}{2}.$$

Continued fractions where all partial numerators are 1 and all partial denominators are positive integers are called *regular continued fractions*. We permit a term in front (positive integer), and we permit it to terminate, in which case we call it a *terminating regular continued fraction*. We have seen, through the two examples, that the Euclidean algorithm gives rise to a terminating regular continued fraction

$$\frac{a}{b} = q_0 + \frac{1}{q_1+}\frac{1}{q_2+}\cdots+\frac{1}{q_k},$$

where q_0 is an integer ≥ 0 and q_i, $1 \leq i \leq k$ are positive integers. In expanding a fraction a/b in a terminating regular continued fraction as seen here, we often assume a and b to be coprime, $\gcd(a, b) = 1$, but it also works (of course) without this assumption.

1.2   Representation of positive numbers by regular continued fractions

An alternative way of obtaining a regular continued fraction expansion of a positive number x_0 is as follows, and please observe that this algorithm, contrary to the Euclidean algorithm, is not restricted to merely rational numbers x_0:

We let (as usual) $[a]$ mean the integer part of a, and start by writing x_0 in the following way:

$$x_0 = [x_0] + (x_0 - [x_0]) . \tag{1.2.1}$$

If x_0 is not an integer, we have $x_0 - [x_0] > 0$, and we define

$$x_1 = \frac{1}{x_0 - [x_0]} . \tag{1.2.2}$$

If x_1 is not an integer, define

$$x_2 = \frac{1}{x_1 - [x_1]} \tag{1.2.2'}$$

and so on. There are now two possibilities:

1) Either we hit, sooner or later, an integer x_n, or 2) no x_n will be an integer.

We shall comment on both possibilities:

1) Since

$$x_0 = [x_0] + \frac{1}{x_1}$$

$$x_1 = [x_1] + \frac{1}{x_2}$$

and finally

$$x_{n-1} = [x_{n-1}] + \frac{1}{x_n}$$

$$x_n = [x_n],$$

we have
$$x_0 = [x_0] + \cfrac{1}{[x_1]} + \cfrac{1}{[x_2]} + \cdots + \cfrac{1}{[x_n]}, \qquad (1.2.3)$$

in which case x_0 is a rational number a/b. On the other hand, if we start with a rational number $x_0 = a/b$, the steps indicated above all coincide with the steps of the Euclidean algorithm. It suffices to look at the first step: With $x_0 = a/b$, a, b positive integers we have from (1.1.1)

$$x_0 = \frac{a}{b} = q_0 + \frac{r_1}{b}.$$

Here q_0 must be the largest integer $\leq x_0$, i.e. $[x_0]$, and r_1/b the "fractional part" $x_0 - [x_0]$, since $0 \leq r_1 \leq b - 1$. For the later steps the argument is the same. This means that for rational numbers the last algorithm is the same as the Euclidean as far as regular continued fraction are concerned. We have also proved that x_0 is a rational number if and only if it has a terminating regular continued fraction expansion.

Is the expansion unique? Unfortunately the answer is No, as may be seen from the following example:

$$3 \qquad \text{and} \qquad 2 + \frac{1}{1}$$

are both regular continued fraction expansions of one and the same number 3. The left one is according to the two algorithms, since $[3] = 3$, the other one is not. The same goes for any positive integer $x_n \geq 2$, which may be written as

$$x_n \qquad \text{or} \qquad x_n - 1 + \frac{1}{1}.$$

But the expansion is unique if we require the last partial denominator to be > 1.

2) If we never hit an integer in the described algorithm we get a non-terminating continued fraction, which is a continued fraction in the proper sense, as defined in Chapter I, Subsection *1.2*. Since it is of the form $\mathbf{K}(1/b_n)$ where all b_n are positive and $\sum b_n = \infty$, it follows from Theorem 3 in Chapter III (the Seidel-Stern Theorem), that the continued fraction converges. See also Example 10 in

Chapter I. To see that it converges to the right value x_0, which in this case has to be an irrational number, we realize that

$$[x_n] < x_n < [x_n] + \frac{1}{[x_{n+1}]},$$

from which it follows that x_0 in value (for all n) lies (properly) between the two approximant values

$$[x_0] + \frac{1}{[x_1] + \cdots + [x_n]} \quad \text{and} \quad [x_0] + \frac{1}{[x_1] + \cdots + [x_{n+1}]}.$$

From this and the convergence, it follows that the convergence is to the right value. Uniqueness of the expansion follows as in the rational case. Let it also be mentioned, that since the regular continued fraction is a positive continued fraction where all $q_k \geq 1$, we have (from Theorem 2 in Chapter III) that the sequences of even order approximants and of odd order approximants both are monotone,

$$f_0 < f_2 < f_4 < \cdots < f_{2n} < \cdots < f_{2n+1} < f_5 < f_3 < f_1, \quad (1.2.4)$$

and that we, for the value x_0 have the truncation error estimate

$$|x_0 - f_n| < |f_n - f_{n+1}|. \tag{1.2.5}$$

We summarize the results of the discussions in two theorems [Perr54, Satz 2.2 and Satz 2.6, p. 25 and p. 33].

Theorem 1 *Every terminating regular continued fraction represents a unique positive rational number, and every positive rational number is uniquely represented by a terminating regular continued fraction where the last partial denominator is ≥ 2 (or where the last partial denominator is 1).*

Theorem 2 *Every non-terminating regular continued fraction represents (converges to) a positive irrational number, and to every positive irrational number there is a unique, non-terminating regular continued fraction converging to that number.*

To compute the approximants of a regular continued fraction, we can use the familiar recurrence relations, which in this case, since all $a_k = 1$, take the form

$$\begin{aligned} A_{k+1} &= b_{k+1} A_k + A_{k-1}, \\ B_{k+1} &= b_{k+1} B_k + B_{k-1}, \end{aligned}$$

and the initial conditions

$$\begin{aligned} A_{-1} &= 1, \quad A_0 = b_0, \\ B_{-1} &= 0, \quad B_0 = 1. \end{aligned}$$

We arrange the partial denominators and the approximants in a table as follows:

		b_0	b_1	b_2	$\ldots$
0	1	A_0	A_1	A_2	$\ldots$
1	0	B_0	B_1	B_2	

The lines indicate the relation

$$A_2 = b_2 A_1 + A_0.$$

We shall illustrate this on four examples.

Example 3 For the rational number $4199/1155$ we find the following regular continued fraction expansion:

$$\frac{4199}{1155} = 3 + \frac{1}{1} + \frac{1}{1} + \frac{1}{1} + \frac{1}{2} + \frac{1}{1} + \frac{1}{8} + \frac{1}{1} + \frac{1}{4} + \frac{1}{2}.$$

In this case the table looks as follows:

		3	1	1	1	2	1	8	1	4	2
0	1	3	4	7	11	29	40	349	389	1905	4199
1	0	1	1	2	3	8	11	96	107	524	1155

We shall use this example later.

$\diamond$

Example 4 The continued fraction

$$\frac{1}{1+}\frac{1}{1+}\frac{1}{1+}\cdots+\frac{1}{1+}\cdots$$

is known from Problem 3 in Chapter I, and we know that it converges to $(\sqrt{5}-1)/2$. The table now looks like

		0	1	1	1	1	1	1	1	1	1	$\cdots$
0	1	0	1	1	2	3	5	8	13	21	34	$\cdots$
1	0	1	1	2	3	5	8	13	21	34	55	

We know from Problem 1 in Chapter I that the numerators and denominators are the Fibonacci numbers F_n, and the sequence of approximants is $\{F_n/F_{n+1}\}$.

$\diamond$

Example 5 $x_0 = \sqrt{2}$. This is known from Chapter I, where the treatment was informal and heuristic. If we here use the integer part algorithm we find

$$x_0 = \sqrt{2} = 1 + (\sqrt{2}-1) = 1 + \frac{1}{\sqrt{2}+1},$$

$$x_1 = \sqrt{2}+1 = 2 + (\sqrt{2}-1) = 2 + \frac{1}{\sqrt{2}+1},$$

and as we see, all later x_k must be $\sqrt{2}+1$. We find the continued fraction

$$\sqrt{2} = 1 + \frac{1}{2+}\frac{1}{2+}\frac{1}{2+}\cdots+\frac{1}{2+}\cdots.$$

The table is as follows

		1	2	2	2	2	2	$\cdots$
0	1	1	3	7	17	41	99	$\cdots$
1	0	1	2	5	12	29	70	
		1	1.5	1.4	1.417	1.414	1.4143	$\cdots$

We have earlier observed how quickly the approximants approach $\sqrt{2}$.

—————————————————————————————————◇

Example 3 was a *terminating* continued fraction and represented a rational number. Examples 4 and 5 were periodic continued fractions, in both cases of period 1. For convergent periodic regular continued fractions we know from Theorem 6 in Chapter III that their value is of the form

$$S_N(x) = \frac{A_N + A_{N-1}x}{B_N + B_{N-1}x}$$

where x is the attractive fixed point of

$$T_k(w) = \frac{A_k^* + A_{k-1}^* w}{B_k^* + B_{k-1}^* w}.$$

That is, the value is of the form

$$\frac{C + \sqrt{D}}{E}$$

where D is a non-negative integer, C, E are integers, $E \neq 0$.

The next example is neither terminating nor periodic. It is the regular continued fraction expansion of π.

Example 6 With $x_0 = \pi = 3.1415926535\ldots$ we find the following (start of a) continued fraction:

$$\pi = 3 + \frac{1}{7+}\frac{1}{15+}\frac{1}{1+}\frac{1}{292+}\frac{1}{1+}\frac{1}{1+}\cdots.$$

The table with the approximants is as follows:

		3	7	15	1	292	1	·
0	1	3	22	333	355	103993	·	·
1	0	1	7	106	113	33102		
		3	3.143	3.1415...	3.1415929...	3.141592653...	·	·

Observe how good this is already for $n = 3$ and $n = 4$. That it is so good for $n = 3$ has to do with the very small tail, caused by the partial denominator 292.

—————————————————————————————————◇

1.3   Best approximation

Already in Chapter I, in Subsection *2.1*, we mentioned that a regular
continued fraction produces the *best* rational approximation to an irrational number. The main purpose of the present section is to *prove* this.
We first recall the definition of bestness (A, B, P, Q are integers):

Definition   *For fractions A/B, $\gcd(A, B) = 1$, $B > 0$ we use the term
best rational approximation to a real number ξ, if (and only if) every
other fraction P/Q, $Q > 0$, with $|\xi - P/Q| < |\xi - A/B|$, has a larger
denominator.*

The main theorem to be proved in the present subsection is the following
[Perr54, Sektion 15]:

Theorem 3 *The regular continued fraction approximants of order ≥ 1
for a positive number ξ are the best rational approximations to ξ.*

Remark: The lattice point illustration in Subsection *2.1* of Chapter
I illustrates the bestness. No lattice points (P, Q) are contained in the
polygon with corners in $(0, 0)$, $(1, 0)$, $(0, 1)$ and the points (A_n, B_n) corresponding to the nth regular continued fraction approximants of ξ (and
a point on the ray $y = \xi x$), as described in Example 2, Chapter I.

Our main tool in the proof of bestness is the following Lemma, which in
fact is a theorem of Lagrange, see for instance [Perr54, Satz 2.17]:

Lemma 4 *Let ξ be a positive number, and let A_n/B_n be the nth approximant ($n \geq 1$) of the regular continued fraction expansion of ξ in
canonical form. Let P and Q be positive integers such that $P/Q \neq
A_n/B_n$ with $0 < Q \leq B_n$. Then*

$$|Q\xi - P| \geq |B_{n-1}\xi - A_{n-1}| > |B_n\xi - A_n|. \qquad (1.3.1)$$

Proof of Lemma 4: This proof goes back to Legendre. Let M and N be such that

$$A_n M + A_{n-1} N = P \,,$$
$$B_n M + B_{n-1} N = Q \,. \tag{1.3.2}$$

Since the determinant of this 2×2-system of linear equations with unknowns M, N is

$$A_n B_{n-1} - B_n A_{n-1} = \pm 1 \,,$$

such numbers M, N exist uniquely, and they are integers. It is readily seen, that $N \neq 0$, since $N = 0$ would lead to $A_n/B_n = P/Q$, which is assumed not to be the case. Furthermore M is either $= 0$ or has opposite sign of N, else $Q > B_n$, contradicting the conditions.

With M and N as given above we study the identity

$$Q\xi - P = M(B_n\xi - A_n) + N(B_{n-1}\xi - A_{n-1}) \,. \tag{1.3.3}$$

Here the two expressions in parantheses have opposite signs by property (1.2.4), and M, N also have opposite signs (unless $M = 0$). Hence

$$|Q\xi - P| = |M(B_n\xi - A_n)| + |N(B_{n-1}\xi - A_{n-1})| \,,$$

and, since N is an integer $\neq 0$, we have

$$|Q\xi - P| \geq |B_{n-1}\xi - A_{n-1}| \,. \tag{1.3.4}$$

Since always $|B_{n-1}\xi - A_{n-1}| > |B_n\xi - A_n|$ (Problem 4), the lemma is proved. ∎

Proof Theorem 3: Let A_n/B_n be a continued fraction approximant for ξ (canonical form), and let $P/Q \neq A_n/B_n$ be closer to ξ or equally close, and $0 < Q \leq B_n$. That is,

$$|\xi - \frac{P}{Q}| \leq |\xi - \frac{A_n}{B_n}| \,.$$

Simultanous multiplication, left by Q and right by B_n, gives

$$|Q\xi - P| \leq |B_n\xi - A_n| \,.$$

This contradicts Lemma 4, and the theorem is thus proved. ∎

2 Some diophantine equations

2.1 Linear diophantine equations

Diophantine equations are algebraic equations in two or more unknowns, with integer coefficients, where one seeks integer solutions. They are named after Diophantos of Alexandria (around 250 A.D.). Linear equations in two unknowns are of the form

$$ax + by = c\,, \tag{2.1.1}$$

where a, b and c are integers, and where the problem is to find all integer solutions (x, y). In some cases one wants to find particular solution(s) satisfying certain conditions, for instance all positive solutions x, y, both less than some fixed N.

If $d \neq 1$ is a positive integer which is a common factor for a and b, then any combination $ax + by$ with integers x, y must be divisible by d. Hence, unless also d is a factor in c, the equation does not have any solution at all. On the other hand, if such a factor exists, it can be cancelled, and hence without loss of generality we assume a and b to be coprime, $\gcd(a, b) = 1$.

We shall at first find the "structure" of the set of solutions. Assume that we somehow have found a solution (x_0, y_0), i.e.

$$ax_0 + by_0 = c\,.$$

Then any other solution (x, y) must satisfy the equation

$$a(x - x_0) + b(y - y_0) = 0\,.$$

Since $\gcd(a, b) = 1$, the set of solutions is given by

$$x - x_0 = tb\,, \qquad y - y_0 = -ta\,,$$

where $t \in \mathbf{Z}$, i.e. the set of solutions is

$$\{(x_0 + tb, y_0 - ta); t \in \mathbf{Z}\}\,. \tag{2.1.2}$$

This means: If we have one solution, we have them all. Differently phrased: If we are able to find *one* solution, we have the *general solution*. We shall see how we can use regular continued fractions to find one solution.

In addition to the condition $\gcd(a, b) = 1$ we shall now assume a and b to be positive. (As we shall see later, this is no severe restriction.) We expand a/b into a regular continued fraction. Then the last approximant, A_n/B_n must be $= a/b$. Since $\gcd(a, b) = 1$ and $\gcd(A_n, B_n) = 1$, we have

$$A_n = a, \qquad B_n = b,$$

and hence, by the determinant formula (1.2.10) in Chapter I,

$$a B_{n-1} - b A_{n-1} = (-1)^{n-1}.$$

From this it immediately follows that

$$a \cdot (-1)^{n-1} B_{n-1} c + b \cdot (-1)^n A_{n-1} c = c.$$

Here we have our special solution

$$x_0 = (-1)^{n-1} B_{n-1} c, \qquad y_0 = (-1)^n A_{n-1} c, \tag{2.1.3}$$

and the following:

Theorem 5 *Let a and b be positive coprime integers, and let A_{n-1}/B_{n-1} be the second to last approximant in one of the two regular continued fraction expansions of a/b. Then the general solution of the diophantine equation*

$$ax + by = c$$

is

$$\begin{aligned} x &= (-1)^{n-1} B_{n-1} c + t \cdot b \\ y &= (-1)^n A_{n-1} c - t \cdot a \end{aligned}, \qquad t \in \mathbf{Z}. \tag{2.1.4}$$

We shall illustrate this by an example.

Example 7 We shall find all integer solutions of the diophantine equation

$$3x + 5y = 2.$$

The regular continued fraction expansion of 3/5 is

$$\frac{3}{5} = 0 + \frac{1}{1+}\frac{1}{1+}\frac{1}{2}.$$

The second to last approximant is $A_2/B_2 = 1/2$, and we have $A_2 = 1$, $B_2 = 2$. We then have the general solution $(n = 3)$

$$\begin{aligned} x &= 2 \cdot 2 + 5t = 4 + 5t \\ y &= -1 \cdot 2 - 3t = -2 - 3t \end{aligned} \quad , \quad t \in \mathbf{Z}.$$

$\diamond$

If a and b are not both positive it is just as simple. Since we disregard the case when a or b is 0, we may without loss of generality assume $a > 0$. If b is negative we get the same x, but opposite sign for y as compared to the case $a > 0, b > 0$. We have for instance that

$$3x - 5y = 2$$

has the general solution

$$\begin{aligned} x &= 4 + 5t \\ y &= 2 + 3t \end{aligned} \quad , \quad t \in \mathbf{Z}.$$

(Compare Example 7.) Rather than thinking in terms of formulas, such as the ones in Theorem 5 or some modification for other a, b-signs, we should keep the main ideas in mind:

1) To get a special solution by using the determinant formula (from the expansion of $|a|/|b|$ or $|b|/|a|$).

2) To get from one special solution to all of them.

We illustrate this by a final example.

Example 8 We shall find the general solution of the diophantine equation

$$4199x - 1155y = 3.$$

In Example 3 we found the regular continued fraction expansion of 4199/1155. The second to last approximant was found to be 1905/524. The determinant formula gives

$$4199 \cdot 524 - 1155 \cdot 1905 = 1 ,$$

and hence

$$4199 \cdot 1572 - 1155 \cdot 5715 = 3 .$$

The general solution is then

$$\begin{aligned} x &= 1572 + 1155t \\ y &= 5715 + 4199t \end{aligned} \quad , \qquad t \in \mathbf{Z} .$$

We find in particular that $t = -1$ gives the solution with smallest absolute value of x as well as y. This solution is

$$x' = 417 , \qquad y' = 1516 .$$

---◇

2.2   Pell's equation

The Pell equation is a diophantine equation of the form

$$x^2 - Dy^2 = 1 , \tag{2.2.1}$$

where D is a positive integer, and where we are looking for integer solutions (x, y) different from the trivial ones $(\pm 1, 0)$. If $D = C^2$ for an integer $C \neq 0$, then (2.2.1) can be written

$$(x - Cy)(x + Cy) = 1,$$

which has only the trivial solutions $(\pm 1, 0)$. Therefore we shall assume that D is not the square of an integer. We shall also be interested in diophantine equations

$$x^2 - Dy^2 = -1 . \tag{2.2.2}$$

Let us first look at a special case, which indicates how the regular continued fraction expansion of $\sqrt{D}$ enters into the process of solving the two equations.

Theorem 6 *Let D be of the form $m^2 + 1$, where m is a positive integer, and let A_n/B_n be the nth regular continued fraction approximant for $\sqrt{D}$, canonical form. Then the following holds for all $k = 0, 1, 2, 3, \ldots$:*

$$A_{2k}^2 - DB_{2k}^2 = -1, \qquad A_{2k+1}^2 - DB_{2k+1}^2 = 1. \qquad (2.2.3)$$

Proof : We find that the regular continued fraction is

$$\sqrt{D} = \sqrt{m^2 + 1} = m + \cfrac{1}{2m} \cfrac{1}{+2m} + \cdots + \cfrac{1}{2m} + \cdots.$$

Hence the sequence of right tails $f^{(n)}$ is $f^{(n)} = \sqrt{D} - m$ for all n. Therefore

$$\sqrt{D} = S_{n+1}(f^{(n+1)}) = \frac{A_{n+1} + A_n(\sqrt{D} - m)}{B_{n+1} + B_n(\sqrt{D} - m)}, \qquad (2.2.4)$$

and hence

$$
\begin{aligned}
& D - \frac{A_n^2}{B_n^2} \\
&= \left(\frac{A_{n+1} + A_n(\sqrt{D} - m)}{B_{n+1} + B_n(\sqrt{D} - m)} \right)^2 - \frac{A_n^2}{B_n^2} \\
&= \frac{A_{n+1}^2 B_n^2 - A_n^2 B_{n+1}^2 + 2A_n B_n(A_{n+1}B_n - A_n B_{n+1})(\sqrt{D} - m)}{B_n^2(B_{n+1} + B_n(\sqrt{D} - m))^2} \\
&= \frac{A_{n+1}B_n - A_n B_{n+1}}{B_n^2} \cdot \frac{A_{n+1}B_n + A_n B_{n+1} + 2A_n B_n(\sqrt{D} - m)}{(B_{n+1} + B_n(\sqrt{D} - m))^2} \\
&= \frac{(-1)^n}{B_n^2} \cdot \frac{(A_{n+1} + A_n(\sqrt{D} - m))B_n + (B_{n+1} + B_n(\sqrt{D} - m))A_n}{(B_{n+1} + B_n(\sqrt{D} - m))^2} \\
&= \frac{(-1)^n}{B_n^2} \cdot \frac{\sqrt{D}B_n + A_n}{B_{n+1} + B_n(\sqrt{D} - m)},
\end{aligned}
$$

where we used (2.2.4) to arrive at the last equality. From (2.2.4) it follows that $x = \sqrt{D}$ is a solution of the quadratic equation

$$x(B_{n+1} + B_n(x - m)) = A_{n+1} + A_n(x - m)$$

i.e.

$$B_n x^2 + (B_{n+1} - mB_n - A_n)x - A_{n+1} + A_n m = 0.$$

This means that $B_{n+1} - mB_n - A_n = 0$; i.e. $A_n = B_{n+1} - mB_n$. Inserted into the last expression for $D - A_n^2/B_n^2$ this gives

$$D - \frac{A_n^2}{B_n^2} = \frac{(-1)^n}{B_n^2} \cdot \frac{\sqrt{D}B_n + B_{n+1} - mB_n}{B_{n+1} + B_n(\sqrt{D} - m)} = \frac{(-1)^n}{B_n^2}$$

which proves (2.2.3). ∎

Remark: It can be proved, that the solutions given by Theorem 6 are the only non-trivial solutions.

Example 9 Take $D = 2$, that is: We study the equations

$$x^2 - 2y^2 = 1$$

and

$$x^2 - 2y^2 = -1 .$$

This example fits right into Theorem 6 with $m = 1$. We have in Example 5 the regular continued fraction expansion of $\sqrt{D} = \sqrt{2}$, as well as its first approximants in canonical form. From the table of approximants we find, by using Theorem 6:

Solutions (x, y) of

$$
\begin{aligned}
x^2 - 2y^2 &= 1 : & (3,2), (17,12), (99,70), \ldots \\
x^2 - 2y^2 &= -1 : & (1,1), (7,5), (41,29), \ldots
\end{aligned}
$$

Numerical verifications:

$$
\begin{aligned}
99^2 - 2 \cdot 70^2 &= 9801 - 2 \cdot 4900 = 1 , \\
41^2 - 2 \cdot 29^2 &= 1681 - 2 \cdot 841 = -1 .
\end{aligned}
$$

───◇

Example 10 $D = 50$ is also an example of Theorem 6. Here $m = 7$. The continued fraction expansion of $\sqrt{50}$ is

$$7 + \frac{1}{14+}\frac{1}{14+}\frac{1}{14+} \cdots$$

The first approximants are

$$\frac{7}{1} , \quad \frac{99}{14} , \quad \frac{1393}{197} , \quad \frac{19601}{2772} , \ldots$$

The first solutions are :

$$\text{For } x^2 - 50y^2 \;=\; 1 \;:\; (99, 14), (19601, 2772), \ldots$$
$$\text{For } x^2 - 50y^2 \;=\; -1 \;:\; (7, 1), (1393, 197), \ldots$$

We have, for instance,

$$19601^2 - 50 \cdot 2772^2 = 384199201 - 50 \cdot 7683984 = 1 ,$$

and

$$1393^2 - 50 \cdot 197^2 = 1940449 - 50 \cdot 38809 = -1 .$$

―――◇

We shall see that also for general D the solutions are found by using the regular continued fraction expansion of $\sqrt{D}$.

It is a well known fact that if D is a positive integer, not a perfect square, then $\sqrt{D}$ has a periodic regular continued fraction. (This is even true if D is a rational number such that $\sqrt{D}$ is non-rational, [Perr54, Satz 3.9, p. 79].) It turns out that the length of the period enters into the process of solving the equations and also into the solution itself. Theorem 6 deals with the very simplest case, with period length 1. We conclude this section by stating without proof a general theorem by Legendre [Perr54, Satz 3.18, p. 93], followed by two examples as illustration.

Theorem 7 *Let D be a positive integer, not the square of an integer. Further let k be the length of the primitive (shortest) period in the regular continued fraction expansion of $\sqrt{D}$ and let A_n/B_n be the approximants in canonical form. The Pell equation $x^2 - Dy^2 = 1$ is always solvable. The set of non-trivial solutions consists of all (x, y) with*

$$x = A_{nk-1} , \qquad y = B_{nk-1} ,$$

for $n = 1, 2, 3, \ldots$ for even k, $n = 2, 4, 6 \ldots$ for odd k. The equation $x^2 - Dy^2 = -1$ is only solvable for odd k, and the set of solutions consists of all (x, y) with

$$x = A_{nk-1} , \qquad y = B_{nk-1} , \qquad n = 1, 3, 5, \ldots .$$

(Observe again that Theorem 7 contains Theorem 6 as a special case ($k = 1$). Observe again that the Pell equation $x^2 - Dy^2 = 1$ always has the trivial solution $x = \pm 1, y = 0$.)

Example 11

$$x^2 - 51y^2 = \pm 1 \, .$$

$\sqrt{51}$ has the following 2-periodic regular continued fraction expansion (Problem 3b):

$$\sqrt{51} = 7 + \frac{1}{7+} \frac{1}{14+} \frac{1}{7+} \frac{1}{14+} \cdots .$$

The first approximants are listed in the table below :

		7	7	14	7	14	...
0	1	7	50	707	4999	70693	...
1	0	1	7	99	700	9899	

From Theorem 7 we know that the equation $x^2 - 51y^2 = -1$ has no solutions (the period length $k = 2$). The Pell equation $x^2 - 51y^2 = 1$ has the solutions $(x, y) = (A_{2m-1}, B_{2m-1})$ for $m = 1, 2, 3, \ldots$, the first ones being:

$$
\begin{aligned}
(x, y) = (50, 7) \quad &: \ 50^2 - 51 \cdot 7^2 \quad\ \ = 2500 - 2499 \quad\qquad = 1 \, , \\
(x, y) = (4999, 700) &: \ 4999^2 - 51 \cdot 700^2 = 24990001 - 24990000 = 1 \, .
\end{aligned}
$$

Let us, just for fun, se what happens to (A_{2m}, B_{2m}), i.e. we compute

$$A_{2m}^2 - DB_{2m}^2$$

for $D = 51$ for some small m-values :

$$
\begin{aligned}
m = 0 \ &: \ 7^2 - 51 \cdot 1^2 = -2 \, , \\
m = 1 \ &: \ 707^2 - 51 \cdot 99^2 = 499849 - 499851 = -2 \, , \\
m = 2 \ &: \ 70693^2 - 51 \cdot 9899^2 = 4997500249 - 4997500251 = -2 \, .
\end{aligned}
$$

(All of you will be able to *guess*, some of you may be able to *prove* what happens for larger m-values.)

◇

Example 12

$$x^2 - 53y^2 = \pm 1,$$

$\sqrt{53}$ has the following 5-periodic regular continued fraction expansion (Problem 3c):

$$\sqrt{53} = 7 + \cfrac{1}{3+}\cfrac{1}{1+}\cfrac{1}{1+}\cfrac{1}{3+}\underbrace{\cfrac{1}{14}}_{Period} + \cfrac{1}{3+}\cdots.$$

From Theorem 7 we know that

$$A_{10m-6}^2 - 53 \cdot B_{10m-6}^2 = -1$$

for $m = 1, 2, 3, \ldots$, and that

$$A_{10m-1}^2 - 53 \cdot B_{10m-1}^2 = 1$$

for $m = 1, 2, 3, \ldots$ The first approximants are listed in the table below:

		7	3	1	1	3	14	...
0	1	7	22	29	51	182	2599	
								...
1	0	1	3	4	7	25	357	

We have for instance

$$A_4^2 - 53 \cdot B_4^2 = 182^2 - 53 \cdot 25^2 = 33124 - 33125 = -1.$$

If we go on with the table we find $A_9/B_9 = 66249/9100$, and we have

$$A_9^2 - 53 \cdot B_9^2 = 66249^2 - 53 \cdot 9100^2 = 4388930001 - 4388930000 = 1.$$

$\diamond$

3   Factoring integers

3.1   Introduction

The problem of factoring integers with a large number of digits (20, 30, 40 and more) has become gradually more important, also from a practical point of view, throughout the last 10–15 years. The reason for this is the rapid development and increased use of number theoretic cryptosystems for secret communication in business, for instance in banking. Such systems are beautiful, interesting and, in some cases, rather simple examples of modern applications of classical (old) mathematics. We shall not go into this here, but refer the interested reader to the books [Kobl87], [Schr86].

One important feature of the crypto-system (RSA-cryptography) we here have in mind, is that the way of *encrypting* a message (in form of a number, a string of digits) is publicly known, whereas the way of *decrypting* an encrypted message is known only to the receiver of the message. The way to break the code depends upon the possibility of writing a certain known number (usually a product of two publicly unknown primes) as a product of at least two proper factors. A proper factor is a factor that is different from 1 and the number itself. The methods developed to factor large numbers are of course not aiming at a criminal type of application. They serve two purposes:

1) They represent examples of what can be accomplished by using mathematical algorithms adjusted to the potential and power of present computer technology.

2) They help draw the line for what is presently possible or not, a line which is vital for the use of cryptography.

With the technology present in the 1980's, the factoring of a number of 100 digits was estimated to take 74 years. A number of 200 digits would take $3.8 \cdot 10^9$ years. This means that the banks (or other users) were pretty safe in using as their crucial number for their crypto-system one that is a product of two prime numbers of approximately 50 digits each. (A "mere-luck" breaking of the code has probability of the order 10^{-50}.)

The time estimates do not exclude the possibility of finding one or more factors more quickly. We can e.g. see it immediately if a 200-digit is divisible by 2 and by 5, say, and if other small prime numbers are present, they are easily traced in a short time. But there still remains the factoring problem for a number with almost as many digits as the original one. Experience seems to indicate, that numbers with few, large factors are the worst.

A description of factoring methods can be found in for instance [Kobl87] and [Ries85]. In the present exposition we shall restrict ourselves to one single method, where a basic continued fraction property plays a crucial role.

3.2   Fermat factorization

In the rest of the chapter the problem to be discussed is how to factor a positive integer n. In most cases we will assume n to be an odd number, not a perfect square (i.e. not the square of an integer). These restrictions are not severe, they are only meant to rule out the two most trivial cases. If n is a perfect square, we already have a factorization, and we may proceed the factoring process on $\sqrt{n}$. Similarly with $n/2$, in case n had been even. If we do not test these things in advance, it will show up through the method. Another remark before we start: Once we have found a proper factor of n, we are through, because what we are aiming at here, is to describe a method for factoring a number n into two proper factors. If we need a further factorization, for instance down to the prime factors, we can use the method again and again on the factors we successively find (or a simpler method, when we come down to smaller numbers).

Assume now that n is an odd integer, not a perfect square. Fermat's method is to search for positive integers x, y, such that $x^2 - y^2 = n$, in which case we have

$$n = (x - y)(x + y).$$

Unless $x - y = 1$, this gives a proper factorization. Since $x > \sqrt{n}$ we start the search at $x = [\sqrt{n}] + 1$ and go on to $x = [\sqrt{n}] + 2$, $x = [\sqrt{n}] + 3, \ldots$ and every time ask the question: Is $x^2 - n$ a perfect square? When (if) we get a conclusive answer we are through: $x^2 - n = y^2$ gives $n = x^2 - y^2 = (x - y)(x + y)$. Even if $x - y = 1$ we are finished. If this is the very first x for which the answer is yes, then n is a prime. Hence: The search in Fermat's method will *always*, sooner or later, lead to a y with $x^2 - y^2 = n$. If it happens "as late as for" $x = (n + 1)/2$, then n is a prime. (Another story is that for numbers of more than 3–4 digits we would never go as far as that in the search. Already a prime number near 1000 would cost more than 450 steps in the process.)

Example 13

a) $n = 44377$, $\sqrt{n} = 210.658\ldots$

x	$x^2 - n$	Square?
211	144	YES (12^2)

$$n = 44377 = 211^2 - 12^2 = (211 + 12)(211 - 12) = \underline{223 \cdot 199}$$

b) $n = 1018579$, $\sqrt{n} = 1009.2457\ldots$

x	$x^2 - n$	Square?
1010	1521	YES (39^2)

$$n = 1018579 = 1010^2 - 39^2 = (1010 + 39)(1010 - 39) = \underline{1049 \cdot 971}$$

c) $n = 962001$, $\sqrt{n} = 980.816\ldots$

x	$x^2 - n$	Square?
981	360	No
982	2323	No
983	4288	No
984	6255	No
985	8224	No
$\vdots$	$\vdots$	$\vdots$
999	36000	No
1000	37999	No
1001	40000	YES (200^2)

$$n = 962001 = 1001^2 - 200^2 = (1001 + 200)(1001 - 200) = \underline{1201 \cdot 801}$$

$\diamondsuit$

In working out the last list, we have used the fact that the square of an integer never ends with 2, 3, 7 or 8.

The Fermat method obviously is useful only if n can be written as a product of two factors rather close to $\sqrt{n}$. If this is not so, the number of cases will be large.

A slight modification of Fermat's method is to take a small, positive k, and with $x = [\sqrt{kn}] + 1$, $x = [\sqrt{kn}] + 2, \ldots$ ask the question: Is $x^2 - kn$ a perfect square? If the answer is YES, we have

$$x^2 - kn = y^2 \tag{3.2.1}$$

for some positive integer, and hence

$$kn = (x + y)(x - y). \tag{3.2.2}$$

This is a factorization of kn. Since here k is known, (we have chosen it) we divide by k and get a factorization of n.

Example 14

a) $n = 2813$. Take $k = 3$. $\sqrt{kn} = 91.864\ldots$

x	$x^2 - 3n$	Square?
92	25	YES (5^2)

$$3n = 92^2 - 5^2 = 97 \cdot 87, \qquad n = \underline{97 \cdot 29}$$

b) Let us try the Fermat method unmodified on the same number: $n = 2813$. $\sqrt{n} = 53.0377\ldots$

x	$x^2 - n$	Square?
54	103	No
55	212	No
$\vdots$	$\vdots$	$\vdots$
62	1031	No
63	1156	YES (34^2)

$$n = (63 + 34)(63 - 34) = \underline{97 \cdot 29}$$

⎯⎯⎯⎯⎯⎯⎯⎯⎯⎯⎯⎯⎯⎯⎯⎯⎯⎯⎯⎯⎯⎯⎯⎯⎯⎯⎯⎯⎯⎯◇

Observe that the modified process in this example is much quicker than the ordinary Fermat method, the reason being that $n = a \cdot b$, where $3b$ is near a.

Since in most non-trivial cases we have no way of finding in advance a "good" k-value, like the one in Example 14, we need a further modification. Instead of looking for x, y such that $x^2 - y^2 = n$ or kn for a given k, we just look for x, y such that n is a factor in $x^2 - y^2$,

$$n | (x^2 - y^2), \qquad (3.2.3)$$

or written a different way:

$$x^2 \equiv y^2 \ (\text{mod } n). \qquad (3.2.3')$$

If we are able to find such x, y-values, we find a factor in n by determining $\gcd(x - y, n)$ or $\gcd(x + y, n)$ by using the Euclidean algorithm, which is a very simple and quick procedure. The rest of our discussion on factoring integers n will be to establish a procedure for finding such x, y-values.

3.3   Factor bases

In the search for x, y with $x^2 \equiv y^2 \ (\text{mod } n)$ we shall be helped by the concept of a *factor base*, which is a set

$$B = \{p_1, p_2, \ldots, p_h\}, \qquad (3.3.1)$$

where the numbers p_i are distinct primes, except that p_1 may be -1. For a given odd integer n and a given factor basis B an integer A shall be called a *B-number* iff the unique number a, given by

$$A^2 \equiv a \ (\text{mod } n), \qquad -\frac{n}{2} < a < \frac{n}{2} \qquad (3.3.2)$$

can be written as a product of factors from B.

A more precise name would perhaps have been B-number *with respect to* n, since n is a part of the definition. The way it will be used, though, is that we are keeping n fixed throughout the process (n is the number to be factored), whereas we may have to change from B to a $\tilde{B}$ or B^* etc. Accordingly, there will be B-numbers, $\tilde{B}$-numbers, B^*-numbers etc., all with respect to n.

We shall now illustrate the concepts of factor base and B-numbers through an example. In this example the base and the B-number seem to come out of the thin air. This example, however, will come back repeatedly, and in the end be the first one to illustrate the method we are aiming at. We will in congruences often omit $(\bmod\ n)$ when it is clear from the context.

Example 15 (First time.) Let $n = 6649$, and let

$$B = \{-1, 3, 5\}$$

be the factor base. Then, since

$$
\begin{aligned}
82^2 &\equiv 75 = 3 \cdot 5^2, \\
163^2 &\equiv -27 = (-1) \cdot 3^3, \\
1060^2 &\equiv -81 = (-1) \cdot 3^4,
\end{aligned}
$$

we see that 82, 163 and 1060 are B-numbers. Observe that if we switch over to the factor base $\tilde{B} = \{-1, 3\}$, the numbers 163 and 1060 are $\tilde{B}$-numbers, but not 82.

── ◇

Let h be the number of elements in the factor base B, and let $\mathcal{F}_h$ be the vector space of dimension h over the field of the two elements 0, 1 with operations $+ \pmod 2$ and $\cdot \pmod 2$. To each B-number A we associate a vector in $\mathcal{F}_h$, such that the ith coordinate counts the number of times $(\bmod\ 2)$ the base number p_i occurs as a factor in a (i.e. 0 if we have no p_i or p_i raised to an even number, 1 if we have a p_i raised to an odd number). We shall illustrate this:

Example 15 (Second time.)

$$n = 6649, \qquad B = \{-1, 3, 5\}.$$

From the factors of the a's we find that the B-numbers are associated with the following vectors in $\mathcal{F}_3$:

$$82 \longrightarrow \langle 0, 1, 0 \rangle,$$
$$163 \longrightarrow \langle 1, 1, 0 \rangle,$$
$$1060 \longrightarrow \langle 1, 0, 0 \rangle.$$

$\diamond$

We shall now see *how* these vectors are used. We still assume that we have an odd, positive number n, not a perfect square, and that we have a factor base B and some B-numbers A_i. Let a_i be the numbers given by

$$A_i^2 \equiv a_i, \qquad -\frac{n}{2} < a_i < \frac{n}{2}. \qquad (3.3.2')$$

i is supposed to belong to a certain index set I. Assume now, that the set of vectors associated with the set of A_i's is linearly dependent. Since we only can have 0 or 1 as coefficients in a linear combination, the linear dependence means that the sum of some of the vectors is $\langle 0, 0, 0, \ldots, 0 \rangle$ (we may as well assume *all*, since we may throw away the A_i's where the associated vector has the coefficient 0 in the linear combination in question). Since the sum of the vectors is the zero vector, the product of the a_i's must have all its factors $p_1, p_2, \ldots, p_h$ an even number of times, i.e. it must be a perfect square C^2. From

$$\prod_{i \in I} A_i^2 \equiv \prod_{i \in I} a_i \pmod{n} \qquad (3.3.3)$$

we find

$$A^2 \equiv C^2 \pmod{n}, \qquad (3.3.4)$$

where A is the product of the A_i's, reduced $\pmod{n}$ to a positive number $\leq n$. C is also reduced $\pmod{n}$ to $\pm$ a fixed positive number. But now we have found what we have been looking for, two squares such that n divides the difference:

$$n | (A - C)(A + C). \qquad (3.3.5)$$

Unless we have bad luck, meaning that $A \equiv \pm C \pmod{n}$ we find a factor of n by computing

$$\gcd(A + C, n) \qquad \text{or} \qquad \gcd(A - C, n) \,.$$

We illustrate this by going back to our example:

Example 15 (Third time.)

$$n = 6649 \,, \qquad B = \{-1, 3, 5\} \,.$$

B-numbers and their vectors:

$$
\begin{aligned}
82 &\longrightarrow \langle 0, 1, 0 \rangle \,, \\
163 &\longrightarrow \langle 1, 1, 0 \rangle \,, \\
1060 &\longrightarrow \langle 1, 0, 0 \rangle \,.
\end{aligned}
$$

The three vectors are linearly dependent, since their sum is $\langle 0, 0, 0 \rangle$. The product of the a_i's is (see Example 15, first time.)

$$
\begin{aligned}
75 \cdot (-27) \cdot (-81) &= 3 \cdot 5^2 \cdot (-1) \cdot 3^3 \cdot (-1) \cdot 3^4 \\
&= (-1)^2 \cdot 3^8 \cdot 5^2 \,,
\end{aligned}
$$

which is a square. We take

$$C \equiv (-1) \cdot 3^4 \cdot 5 = -405 \,.$$

We furthermore use

$$\prod_{i \in I} A_i = 82 \cdot 163 \cdot 1060 = 14167960 \equiv 5590 = A \,.$$

We now know, that

$$6649 \mid (5590 - 405)(5590 + 405) \,.$$

We seek the greatest common divisor of $5590 - 405 = 5185$ and 6649:

$$
\begin{aligned}
6649 &= 1 \cdot 5185 + 1464 \\
5185 &= 3 \cdot 1464 + 793 \\
1464 &= 1 \cdot 793 + 671 \\
793 &= 1 \cdot 671 + 122 \\
671 &= 5 \cdot 122 + 61 \\
122 &= 2 \cdot 61
\end{aligned}
$$

Hence the greatest common divisor is 61. We find

$$6649 = 61 \cdot 109 \,.$$

Since both factors are prime numbers, no further factoring is possible.

———————————————————————————————————————◇

What we have seen in Example 15 is of no use, unless we, to a given n, are able to pick a base B and B-numbers in a "good" way, i.e. such that we can find A, B with $A^2 \equiv C^2 \pmod{n}$ by using a reasonable amount of effort. Here we will be helped by the continued fractions. In the next section we shall present a lemma, which is the basis for the use of continued fractions in the problem of factoring numbers.

3.4   A lemma on continued fractions

Lemma 8 *Let $x > 1$ be a real number, and $\{A_i/B_i\}$ its sequence of regular continued fraction approximants in canonical form. Then*

$$|A_i^2 - x^2 B_i^2| < 2x \qquad \text{for all} \qquad i \in \mathbf{N} \,. \tag{3.4.1}$$

Proof :   From the proof of Theorem 3 in Chapter III we have, since the continued fraction is non-terminating,

$$\left| \frac{A_i}{B_i} - x \right| < \left| \frac{A_i}{B_i} - \frac{A_{i+1}}{B_{i+1}} \right| = \frac{1}{B_i B_{i+1}} \,, \tag{3.4.2}$$

from which it immediately follows that

$$\begin{aligned}
|A_i^2 - x^2 B_i^2| &= |A_i + xB_i||A_i - xB_i| = \left| \frac{A_i}{B_i} - x + 2x \right| \left| \frac{A_i}{B_i} - x \right| B_i^2 \\
&< \left(2x + \frac{1}{B_i B_{i+1}} \right) \frac{1}{B_i B_{i+1}} B_i^2 \\
&= 2x \frac{B_i}{B_{i+1}} + \frac{1}{B_{i+1}^2}
\end{aligned}$$

$$= 2x \left(\frac{B_i}{B_{i+1}} + \frac{1}{2x B_{i+1}^2} \right) \leq 2x \left(\frac{B_i}{B_{i+1}} + \frac{1}{B_{i+1}^2} \right)$$

$$\leq 2x \left(\frac{B_i + 1}{B_{i+1}} \right) \leq 2x,$$

since $B_{i+1} = b_{i+1} B_i + B_{i-1} \geq B_i + 1$ for $i \geq 1$. This proves the lemma.

∎

Our use of the lemma is for the case

$$x^2 = n,$$

where n is a positive integer, in which case we have

$$|A_i^2 - nB_i^2| < 2\sqrt{n}. \tag{3.4.3}$$

As before we assume that n is an odd number, not a perfect square. The significance of this result to us is that if we from such an A_i compute A_i^2 and reduce it $(\bmod\ n)$ to get an a_i with

$$-\frac{n}{2} < a_i < \frac{n}{2}, \tag{3.4.4}$$

we even get

$$-2\sqrt{n} < a_i < 2\sqrt{n}. \tag{3.4.5}$$

In connection with the choice of base one needs to factor the a_i's. And if n has say 40 digits, the a_i's we get by using the continued fraction approximant numerators will have at most roughly 20 digits rather than the expected 40. This makes a huge difference as far as computing the factors of a_i is concerned.

We are now ready to describe a continued fraction method for factoring large numbers.

3.5   The continued fraction factoring algorithm

Let n be the positive integer we want to factor. Without loss of generality we assume that n is an odd number, not a perfect square.

Step 1 Expand $\sqrt{n}$ into a regular continued fraction (meaning the *start* of the expansion, as far out as it turns out to be needed).

Step 2 Find the corresponding numerators A_i of the approximants (canonical form), if desired reduced (mod n).

Step 3 Compute A_i^2 and reduce (mod n) to a number a_i between $-n/2$ and $+n/2$, which, as we know from Lemma 8, is between $-2\sqrt{n}$ and $2\sqrt{n}$.

Except for the question about how far out one should go, these steps are straightforward and require no estimation or test. The next step is more of a trial-and-error step.

Step 4 By studying the factors of the computed a_i-values we try to choose a base B. We do this by picking the indices such that -1 (possibly) and primes occur more than once among the picked a_i's. Let B consist of these primes and possibly -1. Then all A_i with the corresponding indices are B-numbers. List the associated vectors in $\mathcal{F}_h$. If we find a linearly dependent set of vectors, take as A the corresponding product of the A_i's (mod n) and as C^2 the product of the corresponding a_i's (since we know that it is a square). Unless $A \equiv \pm C$ (mod n) we are through, since $\gcd(A + C, n)$ or $\gcd(A - C, n)$ is a proper factor in n, which is then easily found by the Euclidean algorithm.

Step 5 If we have not found a linearly dependent set of vectors in $\mathcal{F}_h$, or if we have reached the situation $A \equiv \pm C$ (mod n), we have to continue the expansion and run through steps 1, 2, 3, 4 (and in unfortunate cases 5) again.

We shall ilustrate this method on the number in Example 15.

Example 15 (Last time.)

$$n = 6649$$
$$\sqrt{6649} = b_0 + \mathbf{K}(1/b_n) = 81 + \frac{1}{1+}\frac{1}{1+}\frac{1}{5+}\frac{1}{1+}\cdots$$

i	0	1	2	3	4
b_i	81	1	1	5	1
A_i	81	82	163	897	1060
a_i	−88	75	−27	80	−81
		*	*		*

Observe the sizes of the a_i's, compared to

$$[\frac{n}{2}] = 3324 \quad \text{and} \quad [2\sqrt{n}] = 163.$$

Interesting i-values are indicated by *. They are interesting because the set of a_i-values $\{75, -27, -81\}$ is such that all factors occuring, i.e. $-1, 3, 5$, occur *more than once*, and hence it may be a good idea to try out

$$B = \{-1, 3, 5\}$$

as a factor base, in which case $A_1 = 82$, $A_2 = 163$, and $A_4 = 1060$ automatically are B-numbers. Now we are back to what we did in Example 15 (Third time). The rest of the argument is identical with what we did there, leading to the factorization

$$6649 = 61 \cdot 109.$$

_____◇

The only new thing in this version of Example 15 is that factor base and B-numbers are the result of a search based upon some simple principles.

A final remark in connection with "interesting i-values": Why could not other i-values have been chosen within the ones for which A_i is computed? Obviously $i = 0$ is out, since 11 is a factor in a_0 and nowhere else within the table. But $i = 3$ could have been chosen. We have 2 as a factor more than once, but 5 only once. Hence $i = 1$ has to be chosen, since also there we have 5 as a factor. But in order to match the factor 3 there we must choose $i = 2$ or $i = 4$. But if one of these is chosen, the other one must be chosen, in order to match the factor -1. These considerations suggest the factor base

$$\tilde{B} = \{-1, 2, 3, 5\}.$$

$\tilde{B}$-numbers and vectors are

$$82 \quad \longrightarrow \quad V_1 = \langle 0, 0, 1, 0 \rangle$$
$$163 \quad \longrightarrow \quad V_2 = \langle 1, 0, 1, 0 \rangle$$
$$897 \quad \longrightarrow \quad V_3 = \langle 0, 0, 0, 1 \rangle$$
$$1060 \quad \longrightarrow \quad V_4 = \langle 1, 0, 0, 0 \rangle.$$

We find that V_1, V_2, V_4 are linearly dependent, V_3 is not needed in the argument. The element $2 \in \tilde{B}$ is not needed to establish a working factor base. It is a good strategy to keep the factor base as simple as possible. Thus we are back to B.

We shall briefly present two more examples. They are worked out precisely as Example 15, and will be presented without comments.

Example 16 Factoring $n = 9073$ by the c.f. method:

$$\sqrt{9073} = 95 + \cfrac{1}{3 +} \cfrac{1}{1 +} \cfrac{1}{26 +} \cfrac{1}{2 +} \cdots$$

i	0	1	2	3	4
b_i	95	3	1	26	2
$A_i \pmod{n}$	95	286	381	1119	2619
$a_i \equiv A_i^2 \pmod{n}$	-48	139	-7	87	-27
	$*$				$*$

Interesting i-values are indicated by $*$. Suggested factor base: $B = \{-1, 2, 3\}$. B-numbers and vectors:

$$95 \quad \longrightarrow \quad \langle 1, 0, 1 \rangle$$
$$2619 \quad \longrightarrow \quad \langle 1, 0, 1 \rangle.$$

They are linearly dependent (Sum $= \langle 0, 0, 0 \rangle$).

$$C^2 = (-48) \cdot (-27) = (-1)^2 \cdot 2^4 \cdot 3^4$$
$$C = (-1) \cdot 2^2 \cdot 3^2 = \underline{-36}$$
$$A \equiv 95 \cdot 2619 \pmod{n} \qquad \underline{A = 3834}$$

We find $\gcd(A - C, n) = \gcd(3870, 9073) = \underline{43}$, for instance by the Euclidean algorithm. Conclusion: $\underline{9073 = 43 \cdot 211}$.

(Since 43 and 211 both are primes, no further factorization is possible.)

In the next example we show how the method at first fails (since $A \equiv \pm C$), and next how we can make it work again by increase of i and change of base.

Example 17 A case of bad luck...

$$n = 26069$$

$$\sqrt{26069} = 161 + \frac{1}{2} + \frac{1}{5} + \frac{1}{1} + \frac{1}{1} + \frac{1}{2} + \frac{1}{5} + \frac{1}{1} + \frac{1}{4} + \frac{1}{2}$$

$$+ \frac{1}{4} + \frac{1}{1} + \frac{1}{1} + \frac{1}{16} + \frac{1}{2} + \frac{1}{4} + \cdots$$

i	0	1	2	3	4	5	6
b_i	161	2	5	1	1	2	5
$A_i \pmod{n}$	161	323	1776	2099	3875	9849	982
a_i	-148	53	-173	140	-119	52	-229
				*		*	

7	8	9	10	11	12	13	·
1	4	2	4	1	1	16	·
10831	18237	21236	24974	20141	19046	12049	·
61	-133	65	-149	172	-19		·
	*	*			*		

Interesting i-values are indicated by a $*$. Suggested factor base:

$$B = \{-1, 2, 5, 7, 13, 19\}.$$

The B-numbers A_i, the corresponding $a_i \equiv A_i^2 \pmod{n}$ and the vectors are in the table below.

A_i	a_i	Vector
2099	$140 = 2^2 \cdot 5 \cdot 7$	$\langle 0, 0, 1, 1, 0, 0 \rangle$
9849	$52 = 2^2 \cdot 13$	$\langle 0, 0, 0, 0, 1, 0 \rangle$
18237	$-133 = (-1) \cdot 7 \cdot 19$	$\langle 1, 0, 0, 1, 0, 1 \rangle$
21236	$65 = 5 \cdot 13$	$\langle 0, 0, 1, 0, 1, 0 \rangle$
19046	$-19 = (-1) \cdot 19$	$\langle 1, 0, 0, 0, 0, 1 \rangle$

The vectors are linearly dependent (Sum $= \langle 0, 0, 0, 0, 0, 0 \rangle$).

$$
\begin{aligned}
A &\equiv \prod A_i \equiv \underline{8511} \\
C^2 &\equiv (-1)^2 \cdot 2^4 \cdot 5^2 \cdot 7^2 \cdot 13^2 \cdot 19^2 \\
C &\equiv (-1) \cdot 2^2 \cdot 5 \cdot 7 \cdot 13 \cdot 19 = -34580 \equiv -8511 . \quad \text{Too bad!}
\end{aligned}
$$

...followed by good luck:

Go on to $i = 13$. We find

$$A_{13} \equiv 12049 \pmod{n}$$

and

$$a_{13} \equiv A_{13}^2 \equiv 140 \pmod{n}.$$

Now the interesting i-values are 3 and 13. They suggest the factor base

$$\tilde{B} = \{2, 5, 7\}.$$

$\tilde{B}$-numbers A_i, corresponding a_i and vectors are

$$
\begin{array}{llll}
A_3 \equiv 2099 & a_3 = 140 & \langle 0, 1, 1 \rangle \\
A_{13} \equiv 12049 & a_{13} = 140 & \langle 0, 1, 1 \rangle
\end{array}
$$

The vectors are linearly dependent, sum $= \langle 0, 0, 0 \rangle$.

$$
\begin{aligned}
A &\equiv A_3 \cdot A_{13} \equiv 3921 \\
C^2 &\equiv 2^4 \cdot 5^2 \cdot 7^2, \qquad C = 2^2 \cdot 5 \cdot 7 = 140 \\
\gcd(A + C, n) &= \gcd(4061, 26069) = 131
\end{aligned}
$$

Conclusion: $\underline{26069 = 131 \cdot 199}$.

Since 131 and 199 both are primes, no further factorization is possible.

$\diamondsuit$

A final suggestion: Look at the a_i's, how small they are. They are always picked to be in each case the unique one for which

$$|a_i| \leq 13034 \; (= [\tfrac{n}{2}]),$$

but since we have regular continued fraction numerators as A_i's, we know they must satisfy

$$|a_i| \leq 322 \; (= [2\sqrt{n}]).$$

Problems

(1) Use the Euclidean algorithm to find:

    (a) gcd(119, 221),

    (b) gcd(3839, 1711),

    (c) gcd(49907, 22243).

(2) Find the regular continued fraction expansion and the approximants for the following numbers:

    (a) 47/99,

    (b) 3839/1711,

    (c) 15015/7429.

(3) Find the regular continued fraction expansion, including the period, and some of the first approximants for

    (a) $\sqrt{82}$,

    (b) $\sqrt{51}$,

    (c) $\sqrt{53}$.

(4) With notation and conditions as in Lemma 4 prove that

$$|B_{n-1}\xi - A_{n-1}| > |B_n\xi - A_n| \,.$$

(5) Find the general solution of the following linear diophantine equations:

    (a) $11x - 3y = 5$,

    (b) $99x - 47y = 3$,

    (c) $3839x - 1711y = 1$,

    (d) $3839x + 1711y = 1$.

In all cases find in addition the particular solution with the smallest positive x-value. In problem (c), *before* solving it, justify the existence of solutions from Problem 1b.

(6) Find explicit formulas for A_n and B_n for the continued fraction used in the proof of Theorem 6. Use these formulas to prove that

$$A_n^2 - DB_n^2 = (-1)^{n+1}.$$

(Hint: Put $r = \sqrt{m^2 + 1} - m$, and hence $r^{-1} = \sqrt{m^2 + 1} + m$.)

(7) Find two solutions of each of the following diophantine equations:

(a) $x^2 - 10y^2 = \pm 1$,

(b) $x^2 - 37y^2 = \pm 1$.

(Use Theorem 6.)

(8) Use Theorem 7 to decide which of the following diophantine equations have solutions and which ones do not. In cases of existence, find two solutions.

(a) $x^2 - 6y^2 = \pm 1$,

(b) $x^2 - 130y^2 = \pm 1$.

(9) (a) For $n = 2077$ take $B = \{-1, 2, 3, 13\}$. Prove that 45, 46 and 91 are B-numbers, and use this to factor 2077 as shown in Subsection *3.3* and Example 15.

(b) For $n = 6649$ take $B = \{-1, 2, 3, 5\}$. Prove that 75, 163, 1060 are B-numbers, and try to use this to factor 6649. (Compare with Example 15.) Why does it fail?

(10) Use the continued fraction factoring algorithm to factor

(a) 943,

(b) 2077.

(11) Pretend that you overlook the fact that 286 is an even number. Go ahead with the continued fraction factoring algorithm and see what happens.

Remarks

1) For historic information on the Pell equation we refer to Perron's book [Perr54, Sektion 27]. As remarked there, the name *Pell equation* is misleading. It was Euler, who incorrectly believed that a method, used by Wallis, was due to the contemporary mathematician Pell.

2) The continued fraction method for factoring integers, described in the present exposition, goes back to 1931, to D. H. Lehmer and R. E. Powers [LePo31], but was for a long time regarded as being of little practical value, because of its fallibility, and was not used. Towards the end of the 60's John Brillhart suggested that the advent of electronic computers might have changed the practical basis for the use of the method. He was in many ways supported by Lehmer and Donald Knuth. Together with Michael Morrison he worked out the method, programmed and tested it and attacked, by use of the IBM 360/91 at UCLA, the seventh Fermat number, $F_7 = 2^{128} + 1$, a 39 digit number. They succeeded in factoring it on September 13, 1970. This was the first successfull attempt to factor F_7, although it had been known since 1905 that it was not a prime. Brillhart and Morrison published a description of the method in 1975 [MoBr75]. They wrote in their paper that it took 90 minutes (over a period of 7 weeks) to factor F_7 and that this most likely could be pushed down to 50 minutes, without coming anywhere near the factoring of F_8, though. They emphasized strongly the importance of the "small" $A_i^2 \pmod{n}$.

The method has been followed up in different ways. On one hand running time analyses have been made [PoWa83], [Wund79], [Wund84], often heuristic. On the other hand, the method has been modified. One version is CFRAC. In the papers [PoWa83] and [Wund85] implementations of CFRAC on parallel machines are described.

3) Lenstra and Manasse in April 1989 completed the factorization of a 106 digit number. This was done by using 80 Firefly multiprocessor workstations in California, and by borrowing computers all over the world through electronic mail. They estimated that this would have taken a century on one processor operating at 1

million operations per second. These informations were given by Carl Pomerance in a talk at the American Mathematical Society Short Course in Cryptology and Computational Number Theory in Boulder, Colorado, August 6–7, 1989. In the same talk he also mentioned that Alford and Pomerance, by using 140 Zenith PC's have factored 95 digit numbers through nights and week-ends over a 5 week period, corresponding to 2 years on a processor as the one mentioned above.

The use of continued fractions for factoring numbers is likely to become history pretty soon. Other methods are about to take over, such as the quadratic sieve method and the elliptic curve method. But the Morrison-Brillhart method, being essentially the Lehmer-Powers continued fraction method with factor bases introduced to combine the congruences in an efficient way, certainly deserves its place in a chapter of applications of number theory as well as in the history of factoring numbers. Carl Pomerance, in his Boulder talk, said: "It can be safely said that the Morrison-Brillhart paper began the modern era of advances in factoring."

References

[Kobl87]   N. Koblitz, "A Course in Number Theory and Cryptography", Graduate Texts in Mathematics, Springer-Verlag, Berlin (1987).

[LePo31]   D. H. Lehner and R. E. Powers, *On Factoring Large Numbers*, Bull. Amer. Math. Soc., Vol. **37** (1931), 770–776.

[MoBr75]   M. A. Morrison and J. Brillhart, *A Method of Factoring and the Factorization of F_7*, Math. of Comp., Vol. **29** (1975), 183–205.

[Perr54]   O. Perron, "Die Lehre von den Kettenbrüchen", Band I, Dritte Aufl., B. G. Teubner, Stuttgart (1954).

[PoWa83]   C. Pomerance and S. S. Wagstaff, Jr., *Implementation of the Continued Fraction Integer Factoring Algorithm*, Proceedings of the 12th Winnipeg Conference on Numerical Methods and Computing, (1983).

[Ries85]   H. Riesel, "Prime Numbers and Computer Methods for Factorization", Birkhäuser, Boston (1985).

[Seid46]   L. Seidel, *Untersuchungen über die Konvergenz und Divergenz der Kettenbrüche*, Habilitationsschrift München (1846).

[Schr86]   M. R. Schroeder, "Number Theory in Science and Communication", Second Enlarged Edition, Springer Series in Information Services, Springer-Verlag, Berlin (1986).

[Ster48]   M. A. Stern, *Über die Kennzeichen der Konvergenz eines Kettenbruchs*, J. Reine u. Angew. Math. **37** (1848), 255–272.

[Wund79]   M. C. Wunderlich, *A Running Time Analysis of Brillhart's Continued Fraction Factoring Method*, "Number Theory, Carbondale 1979", Lecture Notes in Mathematics **751**, Springer-Verlag, Berlin (1979), 328–342.

[Wund84]   M. C. Wunderlich, *Factoring Numbers on the Massively Parallel Computer*, Advances in Cryptology (David Chaum, ed.) (1984), 87–102.

[Wund85]   M. C. Wunderlich, *Implementing the Continued Fraction Factoring Algorithm on Parallel Machines*, Math. of Computation, Vol. 44 (1985), 251–260.

Chapter X

Zero-free regions

About this chapter

The largest part of the present chapter deals with the problem of finding zero-free regions for sequences of polynomials, given by three-term
recurrence relations. The close connection between continued fractions
and the three-term recurrence relations makes it natural to try continued fractions as a tool in determining such zero-free regions. The main
purpose of this part is to give examples of how this can be done. In
those examples we will mainly see what can be done by *direct* use of
established results on continued fractions, i.e. we will illustrate an approach based upon "continued fraction attitude" and basic knowledge
of continued fractions.

There are, on the other hand, some continued fraction based methods
that are tailor-made (and often very fit) for certain important special
sequences. It is beyond the scope and the purpose of this little chapter
to include a discussion of those methods, let alone bring a "catalogue"
of them. We have included some good references and also a little subsection, where two such methods are briefly described.

The chapter finally contains a brief discussion of stability of polynomials
including a continued fraction test for stability.

1   Zero-free regions for certain sequences of polynomials

1.1   *Introduction*

The problem of locating zeros of polynomials is important in mathematics and applications of mathematics. Up to and including degree 4 the problem can be solved by explicit formulas (although not always very manageable). From degree 5 on, however, this can only be done in special cases. Hence numerical methods are needed. Often we use a combination of some general result on location of zeros and an algorithm for the actual determination of the zero. A reference on location of zeros is volume 1 of [Henr86].

In the present section we shall be concerned with the problem of finding zero-free regions for polynomials given by certain three-term recurrence relations. We shall briefly list some of the familiar examples of such sequences of polynomials, all given by recurrence relations of the form

$$P_n(z) = b_n(z){\cdot}P_{n-1}(z) + a_n(z){\cdot}P_{n-2}(z)$$

with some initial conditions, and where b_n and a_n are polynomials of low degrees.

Tchebycheff polynomials:
First kind:

$$
\begin{aligned}
T_n(z) &= 2zT_{n-1}(z) - T_{n-2}(z)\,, \qquad n \geq 2\,, \\
T_0(z) &= 1\,, \ T_1(z) = z\,.
\end{aligned}
$$

Second kind:

$$
\begin{aligned}
U_n(z) &= 2zU_{n-1}(z) - U_{n-2}(z)\,, \qquad n \geq 2\,, \\
U_0(z) &= 1\,, \qquad U_1(z) = 2z\,.
\end{aligned}
$$

Legendre polynomials:

$$P_n(z) = \left(2 - \frac{1}{n}\right)zP_{n-1}(z) - \left(1 - \frac{1}{n}\right)P_{n-2}(z)\,, \qquad n \geq 2\,,$$

$$P_0(z) = 1, \qquad P_1(z) = z.$$

Laguerre polynomials:

$$L_n(z) = \left(2 - \frac{1+z}{n}\right) L_{n-1}(z) - \left(1 - \frac{1}{n}\right) L_{n-2}(z), \qquad n \geq 2,$$
$$L_0(z) = 1, \qquad L_1(z) = (1 - z).$$

Hermite polynomials:

$$H_n(z) = 2z H_{n-1}(z) - 2(n - 1) H_{n-2}(z), \qquad n \geq 2,$$
$$H_0(z) = 1, \qquad H_1(z) = 2z.$$

The close connection between continued fractions and three-term recurrence relations makes it natural to try continued fraction techniques in the search for zero-free regions for such sequences of polynomials. And indeed, this has been done. We refer to the recent survey article by de Bruin, Gilewicz and Runckel [BrGR87]. There some of the most important techniques are described (also for some polynomials satisfying a k-term recurrence relation). The article contains an extensive bibliography.

Many of the papers referred to in [BrGR87] establish continued fraction like methods for the purpose of determining regions where the zeros must be located, or equivalently: zero-free regions. We shall not do this here. We shall give some examples of how direct use of established continued fraction results can lead to results on zero-free regions. See [Waad88].

The idea is to use continued fractions $\mathbf{K}(a_n(z)/b_n(z))$ where the polynomials in question are canonical denominators of the approximants. The key to the argument is then the following simple observation:

Lemma 1 *Let A_n and B_n be the canonical numerators and denominators of $\mathbf{K}(a_n/b_n)$, where all $a_n \neq 0$. Then*

$$A_n + A_{n-1}w \qquad and \qquad B_n + B_{n-1}w$$

cannot vanish simultaneously for any $w \in \mathbf{C}$.

Proof :   One way to see this is that

$$S_n(w) = \frac{A_n + A_{n-1}w}{B_n + B_{n-1}w}$$

is a well defined, non-singular linear fractional transformation when all $a_n \neq 0$. It also follows immediately from the determinant formula, since

$$(A_n + A_{n-1}w)B_{n-1} - (B_n + B_{n-1}w)A_{n-1}$$
$$= A_n B_{n-1} - B_n A_{n-1} = - \prod_{k=1}^{n}(-a_k) \neq 0 \,.$$

■

Hence, if all $a_n(z) \neq 0$ for each z in some set D, and $V(z)$ is a value set for $\mathbf{K}(a_n(z)/b_n(z))$ for each $z \in D$ with $\infty \notin V(z)$, then

$$\frac{A_n(z) + A_{n-1}(z)w(z)}{B_n(z) + B_{n-1}(z)w(z)} \neq \infty \qquad \text{for } w(z) \in V(z),$$

and thus $B_n(z) + B_{n-1}(z)w(z) \neq 0$ for $z \in D$. (Of course, one can carry out the same type of argumentation if $\{V_n(z)\}$ is a sequence of value sets for $\mathbf{K}(a_n(z)/b_n(z))$ with $\infty \notin V_0(z)$. Sometimes numerators of approximants are more convienient to use, rather than denominators. The argument is the same, only with ∞ replaced by 0.)

As an illustration of this we shall first see what we can get "almost for free" about the zeros of the polynomials above. We temporarily disregard the knowledge we may have about their zeros.

Example 1 From the recurrence relations and the initial conditions we see that the Tchebycheff polynomials of the second kind for $n \geq 1$ are the denominators of the approximants of the continued fraction

$$\frac{-1}{2z} + \frac{-1}{2z} + \frac{-1}{2z} + \cdots.$$

By using Śleszyński-Pringsheim's theorem, Theorem 1 in Chapter I, we find that for $|z| \geq 1$ all approximants $f_n(z) = A_n(z)/B_n(z)$ satisfy $|f_n(z)| < 1$. Since $A_n(z)$ and $B_n(z)$ cannot have any zero in common by

Lemma 1, we may conclude that the polynomials $U_n(z)$ have no zeros in the set given by $|z| \geq 1$. (See also Example 9, Chapter I.)

Although this is already pretty good we can do better. We may restrict the discussion to the disk $|z| < 1$. By an equivalence transformation the continued fraction changes to

$$
2z \left(\cfrac{-\dfrac{1}{4z^2}}{1} + \cfrac{-\dfrac{1}{4z^2}}{1} + \cfrac{-\dfrac{1}{4z^2}}{1} + \cdots \right).
$$

Take any point z in the disk $|z| < 1$ and *off* the real diameter, i.e.

$$
z = re^{i\theta}, \qquad 0 < r < 1, \qquad 0 < |\theta| < \pi.
$$

Then the point

$$
-\frac{1}{4z^2} = -\frac{1}{4r^2} e^{-2i\theta}
$$

is in the complement of the ray $(-\infty, -\frac{1}{4}]$ of the negative real axis and hence in some parabolic region P_α from the Parabola Theorem (Theorem 20 in Chapter III). Hence the sequence of approximants will have all its elements in (a bounded part of) the half plane V_α for this particular z. (Here V_α is the value set for P_α as described in the parabola theorem.) The classical approximants for the two equivalent continued fractions are of course the same (although their canonical forms differ). Hence all the approximants of $\mathbf{K}(-1/2z)$ are finite, and none of the denominators can have a zero at that z. We thus conclude that the complement of the interval $(-1, +1)$ of the real axis (complement w.r.t. $\mathbf{C}$) is zero-free.

For later use we also observe that for each such z the sequence of approximants is bounded away from the boundary of the half-plane V_α, in particular from $-\frac{1}{2}$.

The first part (that $|z| \geq 1$ is zero-free) can also easily be proved by using Worpitzky's theorem. This is left as an exercise (Problem 1).

Observe that we, at no point in the proof, have used the fact that the continued fraction is periodic. To use this would have given another (but more special) way of finding a zero-free region.

For Tchebycheff polynomials T_n of the first kind we find that they are denominators of the approximants of the continued fraction

$$\frac{-1}{z} + \frac{-1}{2z} + \frac{-1}{2z} + \cdots$$

which can be written

$$\frac{-\frac{1}{z}}{1+U},$$

where

$$U = 2 \left(\frac{-\dfrac{1}{4z^2}}{1} + \frac{-\dfrac{1}{4z^2}}{1} + \frac{-\dfrac{1}{4z^2}}{1} + \cdots \right).$$

Since the expression in paranthesis is bounded away from $-\frac{1}{2}$ for each $z \notin (-1, +1)$, we find again the same zero-free regions. The details are left to the reader (Problem 2).

$\diamond$

Example 2 The Legendre polynomials are easily seen to be the denominators of the approximants of the continued fraction

$$\left[z + \underset{n=2}{\overset{\infty}{\mathbf{K}}} (a_n/b_n) \right]^{-1}, \quad \text{where } a_n = -\left(1 - \frac{1}{n}\right) \text{ and } b_n = \left(2 - \frac{1}{n}\right) z.$$

From $n = 2$ on the Śleszyński-Pringsheim condition is satisfied in $|z| \geq 1$,

$$|b_n| = \left| \left(2 - \frac{1}{n}\right) z \right| \geq 1 - \frac{1}{n} + 1 = |a_n| + 1,$$

and thus all approximants of the continued fraction

$$\underset{n=2}{\overset{\infty}{\mathbf{K}}} \frac{a_n}{b_n}$$

have absolute value < 1 when $|z| \geq 1$. Hence the approximants of the original continued fraction are all finite, and we have established $|z| \geq 1$ as a zero-free region for the Legendre polynomials. (More information on zeros can be obtained by using the parabola theorem or the limit periodicty of the continued fraction.) By using the fact that the zeros are all real (Chapter VII) we may conclude, that the zeros are all located on the interval $(-1, +1)$ of the real axis.

$\diamond$

Example 3 The Laguerre polynomials are denominators of the approximants of the continued fraction

$$\left[1 - z + \mathop{\mathbf{K}}_{n=2}^{\infty} \frac{-(1 - \frac{1}{n})}{2 - \frac{1+z}{n}}\right]^{-1}. \tag{1.1.1}$$

Again by Śleszyński-Pringsheim's theorem we find that if z is such that for all $n \geq 2$

$$\left|2 - \frac{1+z}{n}\right| \geq 2 - \frac{1}{n},$$

i.e.

$$|2n - 1 - z| \geq 2n - 1,$$

which holds iff $\Re(z) \leq 0$, then the values of the approximants of

$$\mathop{\mathbf{K}}_{n=2}^{\infty} \frac{a_n}{b_n}$$

are all of absolute value strictly less than one. Since $|1 - z| \geq 1$ when $\Re(z) \leq 0$, all approximants of the continued fraction (1.1.1) are finite when $\Re(z) \leq 0$, and hence the closed left half plane is established as a zero-free region for the Laguerre polynomials. (Again, by using the fact that the zeros are real, we find now that they are all positive.)

—————————————————————————————◇

Example 4 The Hermite polynomials are the successive denominators of the approximants of the continued fraction

$$\frac{1}{2z +} \frac{-2}{2z +} \frac{-4}{2z +} \frac{-6}{2z +} \cdots.$$

This can be written

$$\cfrac{\cfrac{1}{2z}}{1 + \cfrac{\frac{-2}{4z^2}}{1 +} \cfrac{\frac{-4}{4z^2}}{1 +} \cfrac{\frac{-6}{4z^2}}{1 +} \cdots}.$$

If z is not real, the elements

$$\frac{-2}{4z^2}, \quad \frac{-4}{4z^2}, \quad \frac{-6}{4z^2}, \ldots$$

are located equidistantly along a ray $\neq$ negative real axis from the origin to infinity. From Theorem 20 in Chapter III it follows that the approximants of

$$\frac{-2}{4z^2} \quad \frac{-4}{4z^2} \quad \frac{-6}{4z^2}$$
$$1 \ + \ 1 \ + \ 1 \ + \cdots$$

all are finite and located in the half plane V_α, where

$$2\alpha = \arg\left(\frac{-1}{z^2}\right),$$

and hence bounded away from -1. The approximants of

$$\frac{1}{2z} + \frac{-2}{2z} + \frac{-4}{2z} + \frac{-6}{2z} + \cdots$$

are thus all finite, and the set consisting of the open upper and lower half-planes (i.e., $\mathbf{C}$ minus the real axis) is a zero-free set for the Hermite polynomials.

$$\diamondsuit$$

It is well known that for Tchebycheff polynomials of both kinds, as well as for Legendre polynomials all zeros are located on the interval $(-1, +1)$ of the real axis. The Laguerre polynomials have all their zeros on the positive real axis, and the Hermite polynomials on the whole real axis. Observe that in the Tchebycheff and Legendre cases the method we used gave us (as zero-free regions) the whole plane, except for the segment (or line) where all zeros are located.

1.2   An application of Van Vleck's theorem

The following theorem is due to Runckel [Runc86]. He proved it by using a continued fraction technique. We present an alternative (continued fraction based) proof.

Theorem 2 *Let $\{B_n(z)\}$ be the sequence of polynomials, given by the recurrence relation*

$$B_n(z) = (g_n z + h_n)B_{n-1}(z) + B_{n-2}(z), \qquad n \geq 1, \qquad (1.2.1)$$

with initial values

$$B_{-1}(z) = 0, \qquad B_0(z) = 1. \tag{1.2.2}$$

Let furthermore $\Re(h_n) \geq 0$ *for all* n, *and* $\alpha < \arg g_n < \beta$ *for all* n, *where* $\beta - \alpha < \pi$. *Then* $B_n(z) \neq 0$ *for all* $n \geq 0$ *when* z *is in the angular opening*

$$-\alpha - \frac{\pi}{2} < \arg z < -\beta + \frac{\pi}{2}. \tag{1.2.3}$$

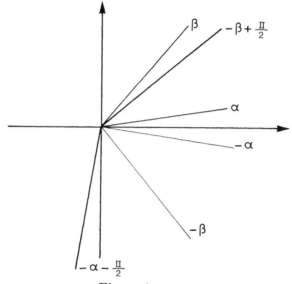

Figure 1.

Proof : $B_n(z)$ is the (canonical) denominator of the nth approximant for the continued fraction

$$\frac{1}{g_1 z + h_1 +} \, \frac{1}{g_2 z + h_2 + \cdots +} \, \frac{1}{g_n z + h_n + \cdots}.$$

From van Vleck's theorem, Theorem 2 in Chapter I, we know that if all $g_n z + h_n$ are in the angular opening

$$-\frac{\pi}{2} + \epsilon < \arg(g_n z + h_n) < \frac{\pi}{2} - \epsilon, \qquad 0 < \epsilon < \frac{\pi}{2},$$

then all approximants are finite, and located in the angular opening

$$-\frac{\pi}{2} + \epsilon < \arg \frac{A_n(z)}{B_n(z)} < \frac{\pi}{2} - \epsilon.$$

Here we are merely interested in the finiteness of $A_n(z)/B_n(z)$, from which it follows that $B_n(z) \neq 0$ for all n.

Assume now that, for all $n \geq 1$,

$$-\frac{\pi}{2} < \arg(g_n z + h_n) < \frac{\pi}{2}. \tag{1.2.4}$$

Then to each N there is an $\epsilon_N > 0$, such that the condition holds with ϵ_N for all $n \leq N$, and hence $B_n(z) \neq 0$ for $n \leq N$. Since N is arbitrary we have that (1.2.4) implies $B_n(z) \neq 0$ for all n.

Now, from the conditions of the theorem it follows that

$$-\frac{\pi}{2} < \arg(g_n z) < \frac{\pi}{2} \qquad \text{for} \quad -\alpha - \frac{\pi}{2} < \arg z < -\beta + \frac{\pi}{2},$$

and since $\Re(h_n) \geq 0$ we have

$$-\frac{\pi}{2} < \arg(g_n z + h_n) < \frac{\pi}{2},$$

and the theorem is proved.   ∎

Observe that the more we know about g_n, i.e. the more narrow the angular opening for g_n is, the more we can say about zero-free regions.

Assume that for a fixed γ we have $\arg g_n = \gamma$ for all n. Then α and β can be chosen arbitrarily close to γ ($\alpha < \gamma < \beta$). From Theorem 2 it follows that for all n, $B_n(z) \neq 0$ in the angular opening

$$\left(-\gamma - \frac{\pi}{2}\right) + \epsilon < \arg z < \left(-\gamma + \frac{\pi}{2}\right) - \epsilon$$

for any $\epsilon > 0$, and we get

Corollary 3 *If in Theorem 2 the condition on* $\arg g_n$ *is replaced by*

$$\arg g_n = \gamma \quad \textit{for all } n, \tag{1.2.5}$$

then $B_n(z) \neq 0$ *for all* $n \geq 0$ *when* z *is in the half plane*

$$-\gamma - \frac{\pi}{2} < \arg z < -\gamma + \frac{\pi}{2}. \tag{1.2.6}$$

If, in particular, all g_n *are positive, the right half plane is a zero-free region for all* $B_n(z)$. *If all* g_n *are purely imaginary with positive imaginary parts, the lower half-plane* $\Im(z) < 0$ *is zero-free.*

1.3 An application of the parabola theorem

We shall use the parabola theorem (Theorem 20 in Chapter III) to obtain results on zero-free regions. Again, we are only interested in the finiteness of the approximants, not convergence or the value set in itself. We shall here be interested in polynomials given by recurrence relations of the form

$$
\begin{align}
B_n(z) &= B_{n-1}(z) + a_n z^2 B_{n-2}(z), & n \geq 1, & \qquad (1.3.1) \\
B_{-1}(z) &= 0, \qquad B_0(z) = 1, & & \qquad (1.3.2)
\end{align}
$$

where all $a_n \neq 0$. It may seem more natural to study the problem with $a_n z$ instead of $a_n z^2$. But if the latter is solved, it is a simple transformation to obtain the solution of the problem with $a_n z$. The reason for choosing $a_n z^2$ is partly that the results are geometrically more appealing.

The polynomials $B_n(z)$ are the (canonical) denominators of the approximants of the continued fraction

$$
\frac{a_1 z^2}{1} + \frac{a_2 z^2}{1} + \frac{a_3 z^2}{1} + \cdots + \frac{a_n z^2}{1} + \cdots .
$$

From the value set part of the parabola theorem we know that if all $a_n z^2$ are in a parabolic region

$$
|w| \leq \Re(w e^{-2i\theta}) + \frac{1}{2}\cos^2\theta
$$

for some fixed $\theta \in (-\pi/2, \pi/2)$, then all approximants $A_n(z)/B_n(z)$ are finite (and located in a half plane), and hence $B_n(z) \neq 0$ for $z \neq 0$. For $z = 0$ we can see that $B_n(z) = 1 \neq 0$ for all n.

We shall, for a fixed θ and a fixed n, describe the set $S_n(\theta)$, which is such that $z \in S_n(\theta)$ iff $a_n z^2$ is in the parabolic region described above: $S_n(\theta)$ is the set of all z, such that

$$
|a_n z^2| \leq \Re(a_n z^2 e^{-2i\theta}) + \frac{1}{2}\cos^2\theta \qquad (1.3.3)
$$

holds. With $a_n = |a_n| e^{2i\psi_n}$, $-\pi/2 \leq \psi_n < \pi/2$, this transforms into

$$
|z|^2 \leq \Re\left(z^2 e^{2i(\psi_n - \theta)}\right) + \frac{\cos^2\theta}{2|a_n|}. \qquad (1.3.3')
$$

(Keep in mind that all $a_n \neq 0$.) With

$$\zeta = z e^{i(\psi_n - \theta)}$$

this can be written

$$\Re(\zeta)^2 + \Im(\zeta)^2 \leq \Re(\zeta)^2 - \Im(\zeta)^2 + \frac{\cos^2 \theta}{2|a_n|},$$

or

$$\left| \Im \left(z e^{i(\psi_n - \theta)} \right) \right| \leq \frac{\cos \theta}{2\sqrt{|a_n|}}. \qquad (1.3.4)$$

This shows that $S_n(\theta)$ is a parallel strip, as illustrated in Figure 2.

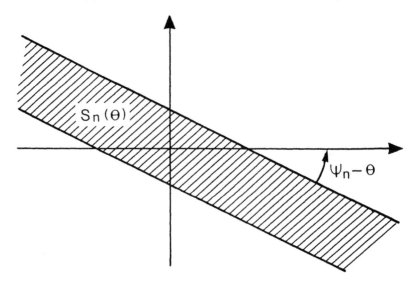

Figure 2.

Obviously all $B_n(z) \neq 0$ when

$$z \in S(\theta) = \bigcap_{n=2}^{\infty} S_n(\theta).$$

Since this statement is true for all $\theta \in (-\pi/2, \pi/2)$, the set $S = \bigcup S(\theta)$, the union taken over all $\theta \in (-\pi/2, \pi/2)$, is a zero-free region for the sequence $\{B_n(z)\}$ of polynomials.

This describes a method for determining zero-free regions for the sequences $\{B_n(z)\}$ given by recurrence relations and initial condition at the beginning of this section. But unless we know more about the coefficients, this method is hardly more than a "pre-method". In the next subsections we shall study two special cases where the determination of S can be carried out to an explicit, simple result.

1.4 The Stieltjes case

We are still studying the sequence $\{B_n\}$ from Subsection *1.3*, but now with the condition

$$a_n > 0 \quad \text{for all } n.$$

The reason for calling this the Stieltjes case is that the continued frac-

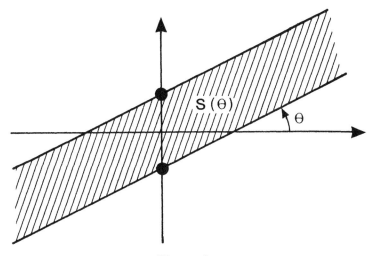

Figure 3.

tions

$$\mathop{\mathbf{K}}_{n=1}^{\infty} \frac{a_n w}{1}, \qquad a_n > 0 \tag{1.4.1}$$

are Stieltjes fractions (as in Chapter III, Subsection *4.3*). Let

$$A = \sup_{n \geq 2} a_n.$$

Then, if $A < \infty$, the set $S(\theta)$ is the parallel strip

$$|\Im(ze^{-i\theta})| \leq \frac{\cos\theta}{2\sqrt{A}}, \qquad -\frac{\pi}{2} < \theta < \frac{\pi}{2}, \qquad (1.4.2)$$

as illustrated in Figure 3.

Observe that the strip intersects the real axis at an angle of θ, and that the boundary lines intersect the imaginary axis in the points $\pm i/(2\sqrt{A})$, regardless of θ. If $A = \infty$, the strip degenerates to a line. In both cases we find that the zero-free region

$$S = \bigcup S(\theta), \qquad \theta \in \left(-\frac{\pi}{2}, \frac{\pi}{2}\right), \qquad (1.4.3)$$

is the whole plane $\mathbf{C}$ minus the two cuts from $i/(2\sqrt{A})$ to ∞ along the positive imaginary axis and from $-i/(2\sqrt{A})$ to ∞ along the negative imaginary axis. An illustration (with $A < \infty$) is shown in Figure 4.

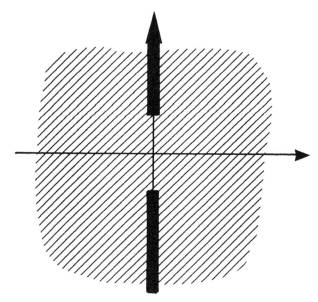

Figure 4.

Another way of phrasing this result is: All zeros of $B_n(z)$ are located on the imaginary axis at distance $> 1/(2\sqrt{A})$ from the origin. (For $A = \infty$ this means the whole imaginary axis minus the origin.)

We shall now replace the z^2 by w. It follows immediately that all $B_n(z)$ are polynomials in z^2 and hence in w. Let G_n denote those polynomials:

$$G_n(w) := B_n(\sqrt{w}).$$

(Branch of $\sqrt{w}$ is insignificant, since we never have $\sqrt{w}$ to an odd power.) The mapping

$$w = z^2$$

maps the two z-cuts onto the w-cut

$$\left(-\infty, -\frac{1}{4A}\right]$$

on the negative real axis, and the rest of the plane to the complement of the cut (or rather: two copies of it). We thus have the result:

Theorem 4 *Let $\{G_n(w)\}$ be the sequence of polynomials, given by the recurrence relation*

$$G_n(w) = G_{n-1}(w) + a_n w G_{n-2}(w), \qquad n \geq 1, \qquad (1.4.4)$$

and the initial values

$$G_{-1}(w) = 0, \qquad G_0(w) = 1. \qquad (1.4.4')$$

If $a_n > 0$ for all $n \geq 2$, then all $G_n(w) \neq 0$ in the cut plane

$$\left\{ w \in \mathbf{C}; \left| \arg\left(w + \frac{1}{4A}\right) \right| < \pi \right\}, \qquad (1.4.5)$$

where $A = \sup a_n$, $0 < A \leq \infty$.

The zero-free region is illustrated on Figure 5.

Theorem 4 is a well known result for Stieltjes fractions [HePf66]

$$\overset{\infty}{\underset{n=1}{\mathbf{K}}} \frac{a_n w}{1}, \qquad a_n > 0.$$

Moreover, it is really not much more than a restatement of Remark 2 to Theorem 20 in Chapter III, the parabola theorem.

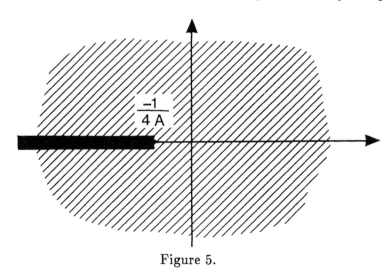

$$\frac{-1}{4\,A}$$

Figure 5.

1.5   *The case when $a_n \in \mathbf{R}$*

The "pre-method" of Subsection *1.3* can be carried out to a method under additional conditions on a_n, like e.g. in the last subsection, where all $a_n > 0$. Another natural condition is to require a fixed value, not necessarily 0 for the argument of all a_n, or, more generally: to require $\arg a_n$ to be in a given finite set. We shall here let $\arg a_n \in \{0, \pi\}$; that is, all a_n are real. We are thus interested in the sequence $\{B_n(z)\}$ of polynomials, given by the same recurrence relation and initial values as in Subsection *1.4*,

$$B_n(z) = B_{n-1}(z) + a_n z^2 B_{n-2}(z), \qquad n \ge 1, \qquad (1.5.1)$$
$$B_{-1}(z) = 0, \qquad B_0(z) = 1, \qquad\qquad\qquad (1.5.2)$$

but where we now require $a_n \in \mathbf{R}$ (instead of $a_n > 0$). If the set of negative a_n is empty, we are back to the situation discussed in Subsection *1.4*. If the set of positive a_n is empty, we get back to the situation in Subsection *1.4* by the transformation $\zeta = iz$. We shall thus, without loss of generality, assume that neither the set of positive a_n nor the set of negative a_n is empty (although we shall occasionally comment on it).

We want to determine the set $S(\theta)$ by taking the intersection of all the parallel strips $S_n(\theta)$, see Subsection *1.3*, in particular Figure 2. Let

$$\sup a_n =: A_+,$$
$$\sup(-a_n) =: A_-.$$

Since the sets of positive a_n and negative a_n are both nonempty, we have

$$0 < A_+ \leq \infty,$$
$$0 < A_- \leq \infty.$$

Let $S_+(\theta)$ be the intersection of all $S_n(\theta)$ with $a_n > 0$, and $S_-(\theta)$ the intersection of all $S_n(\theta)$ with $a_n < 0$. Then, just as in Subsection *1.4*, we find that $S_+(\theta)$ is the parallel strip, given by

$$|\Im(ze^{-i\theta})| \leq \frac{\cos\theta}{2\sqrt{A_+}}. \tag{1.5.3}$$

(See Figure 3.) Almost the same way, but with $zie^{-i\theta}$, because $\psi_n = -\frac{\pi}{2}$ when $a_n < 0$, we find that $S_-(\theta)$ is the parallel strip, given by

$$|\Im(zie^{-i\theta})| = |\Re(ze^{-i\theta})| \leq \frac{\cos\theta}{2\sqrt{A_-}}. \tag{1.5.4}$$

We have $S(\theta) = S_+(\theta) \cap S_-(\theta)$, which is a rectangle, as shown in Fig. 6.

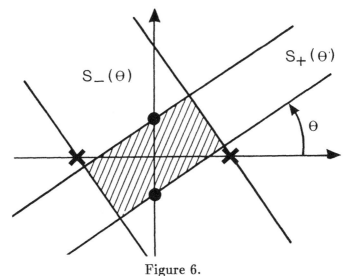

Figure 6.

The boundary lines of $S_+(\theta)$ go through $\pm i/(2\sqrt{A_+})$ (indicated by $\bullet$), and the boundary lines of $S_-(\theta)$ go through $\pm 1/(2\sqrt{A_-})$, ($\times$). Thus, by elementary geometry: When θ varies, the corners of the rectangle will describe four circles with the four line segments

$$\text{from} \quad \frac{1}{2\sqrt{A_-}} \quad \text{to} \quad \frac{i}{2\sqrt{A_+}},$$

$$\text{from} \quad \frac{i}{2\sqrt{A_+}} \quad \text{to} \quad -\frac{1}{2\sqrt{A_-}},$$

$$\text{from} \quad -\frac{1}{2\sqrt{A_-}} \quad \text{to} \quad -\frac{i}{2\sqrt{A_+}},$$

$$\text{and from} \quad -\frac{i}{2\sqrt{A_+}} \quad \text{to} \quad \frac{1}{2\sqrt{A_-}},$$

as diameters. Only the semicircles between any two neighboring points of the four indicated ones and not going through the origin are of interest to us. The zero-free set

$$S = \bigcup_\theta S(\theta), \qquad \theta \in \left(-\frac{\pi}{2}, \frac{\pi}{2}\right)$$

is the set bounded by the four semicircles described below (see Fig. 7).

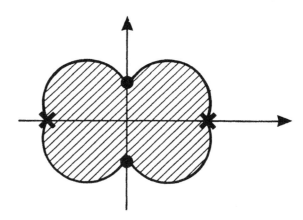

Figure 7.

The result is more simply expressed by switching from z to $1/z$, since the four circles then are transformed into straight lines. We replace $B_n(1/z)$ by $D_n(z)$:

Theorem 5 *Let $\{D_n(z)\}$ be the sequence of polynomials, given by the recurrence relation*

$$D_n(z) = D_{n-1}(z) + a_n z^{-2} D_{n-2}(z), \qquad n \geq 1, \qquad a_n \neq 0, \quad (1.5.5)$$

and the initial values

$$D_{-1}(z) = 0, \qquad D_0(z) = 1. \tag{1.5.6}$$

If $a_n \in \mathbf{R}$ for all n, and there is at least one positive and one negative a_n, then the zeros of all $D_n(z)$ are all in the closed parallelogram with corners in

$$\pm 2i\sqrt{A_+} \quad and \quad \pm 2\sqrt{A_-},$$

where $A_+ = \sup a_n$, $A_- = \sup(-a_n)$.

In Figure 8 the zeros are in the "white" region (which is bounded in this case). The indicated points correspond to the ones in Figure 7, by $z \to 1/z$.

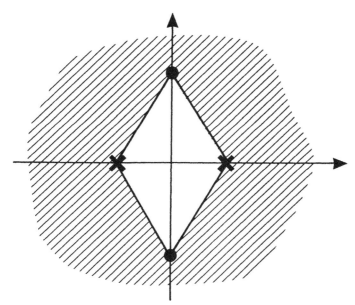

Figure 8.

Remark: So far we have used value sets in the argument, in the way described just before Example 1. Another way of doing it would be to use critical tail sequences, since $h_n(z) = B_n(z)/B_{n-1}(z)$, where we know that $B_n(z)$ and $B_{n-1}(z)$ can not have any common zeros if all $a_n(z) \neq 0$. If $\{V_n\}$ is a sequence of value sets for $\mathbf{K}(a_n/b_n)$, then

$$-h_m \notin V_m \quad \Rightarrow \quad -h_n \notin V_n \qquad \text{for all } n \geq m.$$

This follows since $s_n(V_n) \subseteq V_{n-1} \Rightarrow V_n \subseteq s_n^{-1}(V_{n-1}) \Rightarrow s_n^{-1}(\hat{\mathbf{C}} \setminus V_{n-1}) \subseteq \hat{\mathbf{C}} \setminus V_n$ and since $\{-h_n\}$ is a tail sequence so that $-h_n = s_n^{-1}(-h_{n-1})$.

1.6   A fundamental recurrence formula

The recurrence relation

$$P_n(z) = (z - c_n)P_{n-1}(z) - \lambda_n P_{n-2}(z), \qquad (1.6.1)$$

$\lambda_n \neq 0$, $P_{-1}(z) = 0$, $P_0(z) = 1$, is of great importance in the theory of orthogonal polynomials. On the one hand, if to a given quasi-definite moment functional (Chapter VII, Section 1) $P_n(z)$ are the corresponding monic orthogonal polynomials, then there exist constants c_n and $\lambda_n \neq 0$, such that (1.6.1) holds. On the other hand, by Favard's theorem (Chapter VII, Theorem 2), any sequence $\{P_n(z)\}$ satisfying some recurrence relation (1.6.1) (with all $\lambda_n \neq 0$ and including the initial conditions) is an orthogonal polynomial sequence for some linear functional.

Since the question about location of zeros of orthogonal polynomials is important, it is of interest to describe procedures leading to information about zeros of the polynomials in (1.6.1). We shall here restrict ourselves to the very simplest types of argument.

Observe first that for all $n \geq 0$ the polynomials $P_n(z)$ are the canonical denominators of the Jacobi continued fraction

$$\frac{\lambda_1}{z - c_1 +} \frac{-\lambda_2}{z - c_2 +} \cdots + \frac{-\lambda_n}{z - c_n +} \cdots . \qquad (1.6.2)$$

If we assume z to be different from all c_n this continued fraction is equivalent to

$$\cfrac{\lambda_1}{z - c_1} \cfrac{}{1} + \cfrac{\cfrac{-\lambda_2}{(z - c_1)(z - c_2)}}{1} + \cdots + \cfrac{\cfrac{-\lambda_n}{(z - c_{n-1})(z - c_n)}}{1} + \cdots . \qquad (1.6.3)$$

By using well known element/value set results we find zero-free regions by using the principle stated earlier in this chapter. From Worpitzky's theorem we know that if z is such that

$$\left| \frac{\lambda_n}{(z - c_{n-1})(z - c_n)} \right| \leq \frac{1}{4} \quad \text{for all } n \geq 2,$$

then all approximants of (1.6.3) are finite, and thus all $P_n(z)$ are $\neq 0$. A special case of this is the following result which also can be found in [Wall48, Thm. 26.2] (proved in a different way):

Proposition 6 *If in (1.6.1) $0 < |\lambda_n| \leq M^2$ for all $n \geq 2$ and $|c_n| \leq N$ for all $n \geq 1$, then all zeros of $P_n(z)$ are located in the disk $|z| < 2M + N$.*

Proof : For $|z| \geq 2M + N$ we have $z \neq c_n$ for all n and

$$\left| \frac{\lambda_n}{(z - c_{n-1})(z - c_n)} \right| \leq \frac{M^2}{(2M + N - N)(2M + N - N)} = \frac{1}{4}.$$

Hence the result follows from Worpitzky's theorem. ∎

We can also base our arguments on the parabola theorem, as we did in Subsections *1.3–1.5*. If, for some $\theta \in (-\frac{\pi}{2}, \frac{\pi}{2})$ and all $n \geq 2$

$$|d_n(z)| - \Re(d_n(z)e^{-i2\theta}) \leq \frac{1}{2} \cos^2 \theta, \qquad (1.6.4)$$

with

$$d_n(z) = \frac{-\lambda_n}{(z - c_{n-1})(z - c_n)}, \qquad (1.6.5)$$

then $P_n(z) \neq 0$ for all n. A special example of this is as follows, [Jaco89, Cor. 3.4]:

Proposition 7 *If in (1.6.1) all $c_n = 0$ and all λ_n are real, $\neq 0$ and $\lambda_n \leq M^2$, then all zeros of $P_n(z)$ are contained in the strip*

$$|\Re(z)| < 2M .$$

Proof : For $\Re(z) > 0$ take $\theta = -\arg z$, and for $\Re(z) < 0$ take $\theta = \pi - \arg z$ where the value of $\arg z$ is taken to give $-\frac{\pi}{2} < \theta < \frac{\pi}{2}$. In both cases we find in (1.6.4):

$$\text{Left hand side} \quad = \quad \frac{|\lambda_n| + \lambda_n}{|z|^2} ,$$

$$\text{Right hand side} \quad = \quad \frac{1}{2} \frac{\Re(z)^2}{|z|^2} .$$

From this the conclusion of Proposition 7 follows immediately.   ∎

1.7 Chain sequences

As in the previous subsection we shall consider monic orthogonal polynomials $\{P_n(z)\}$ satisfying the recurrence relation (1.6.1) for some $c_n \in \mathbf{C}$, $\lambda_n \in \mathbf{C}$, $\lambda_n \neq 0$. We plan to use the following part of the parabola sequence theorem, Theorem 21 in Chapter III: *Let* $-\frac{\pi}{2} < \theta < \frac{\pi}{2}$ *and* $\{g_n\}_{n=0}^{\infty}$ *be fixed numbers such that* $0 < g_0 \leq 1$ *and* $0 < g_n < 1$ *for* $n = 1, 2, 3, \ldots$. *If* $|a_n| - \Re(a_n e^{-\imath 2\theta}) \leq 2g_{n-1}(1 - g_n) \cos^2 \theta$ *for all* $n \geq 2$, *then the approximants* A_n/B_n *of* $\mathbf{K}(a_n/1)$ *are all finite.* We continue to use the notation $d_n(z)$ as in (1.6.5) for $z \neq c_k$ for all k. We get:

Proposition 8 *Let* $\{P_n(z)\}$ *be given by (1.6.1) with* $P_{-1}(z) = 0$ *and* $P_0(z) = 1$. *Then* $P_n(z) \neq 0$ *for all* $n \in \mathbf{N}$ *for all* $z \in \mathbf{C}$ *such that*

$$|d_n(z)| \leq g_{n-1}(1 - g_n) \quad for \ n = 2, 3, 4, \ldots \qquad (1.7.1)$$

where $0 < g_n < 1$ *for all* n.

Proof : According to the parabola sequence theorem with $\theta = 0$, we have that $A_n(z)/B_n(z) \neq \infty$ for all $n \geq 0$ if $d_1(z) \neq \infty$ and

$$|d_n(z)| - \Re(d_n(z)) \leq 2g_{n-1}(1 - g_n) \qquad \text{for} \ \ n \geq 2.$$

This holds in particular if $|d_n(z)| + |d_n(z)| \leq 2g_{n-1}(1 - g_n)$; i.e. if (1.7.1) holds.   ∎

Remarks

1. A sequence $\{\beta_n\} = \{(1-\gamma_{n-1})\gamma_n\}$ where $0 \le \gamma_0 < 1$ and $0 < \gamma_n < 1$ for $n = 1, 2, 3, \ldots$ is called a *chain sequence*. The sequence $\{\gamma_n\}$ is called a *parameter sequence* for $\{\beta_n\}$. (A parameter sequence for a given chain sequence is not necessarily unique.) The condition (1.7.1) can therefore be interpreted as $|d_n(z)| \le \beta_n$ for some chain sequence $\{\beta_n\}$ (with parameter sequence such that $\gamma_n = 1 - g_n$). This is a classical result in the special case where all $\lambda_n > 0$ and $c_n \in \mathbf{R}$.

2. Let $\{\beta_n\}$ be a chain sequence with parameter sequence $\{\gamma_n\}$. Then $-\beta_n = -(1-\gamma_{n-1})\gamma_n = -g_{n-1}(1-g_n)$ where $g_n = 1-\gamma_n$ as above. That is, $\{-g_n\}_{n=0}^{\infty}$ or equivalently $\{\gamma_n - 1\}_{n=0}^{\infty}$ is a tail sequence for $\mathbf{K}(-\beta_n/1)$.

The following lemma is a classical result which easily follows from the continued fraction theory:

Lemma 9 *Let $\{\beta_n\}$ be a chain sequence, and let $0 < \tilde{\beta}_n \le \beta_n$ for all n. Then $\{\tilde{\beta}_n\}$ is also a chain sequence.*

Proof : Let $\{\gamma_n\}$ be a parameter sequence for $\{\beta_n\}$ and let $g_n = 1 - \gamma_n$ for all n. We shall use the following part of the oval sequence theorem, Theorem 26 in Chapter III, with all $C_n = 0$ and $R_n = g_n$: If for all n V_n is the disk $|w| < g_n$ and E_n is the disk $|a| \le g_{n-1}(1 - g_n)$, then $\{V_n\}$ is a sequence of value sets for E_n. Since $-\tilde{\beta}_n \in E_n$ for all n, it follows that the continued fraction $\mathbf{K}(-\tilde{\beta}_n/1)$ has a tail sequence $\{-\tilde{g}_n\}$ such that $-\tilde{g}_n \in \bar{V}_n$ for all n. Hence $\tilde{\beta}_n = (1 - \tilde{\gamma}_{n-1})\tilde{\gamma}_n$ where $\tilde{\gamma}_n = 1 - \tilde{g}_n$ for all n. ∎

A reformulation of Proposition 8 is therefore: *All $P_n(z) \ne 0$ for all $z \in \mathbf{C}$ such that $|d_n(z)|$ is a chain sequence.* This result appeared from a rather rough application of the parabola sequence theorem. More careful arguments yield:

Proposition 10 *Let $\{P_n(z)\}$ be given by (1.6.1) with $\lambda_n \in \mathbf{C} \setminus \{0\}$, $c_n \in \mathbf{C}$, $P_{-1}(z) = 0$ and $P_0(z) = 1$. Then $P_n(z) \neq 0$ for all $n \in \mathbf{N}$ for all $z \in \mathbf{C}$ such that $\{\beta_n\}$, with*

$$\beta_n(z,\theta) = \frac{|d_n(z)| - \Re(d_n(z)e^{-i2\theta})}{2\cos^2\theta} \quad \text{for } n = 1,2,3,\ldots$$

is a chain sequence for some θ, $-\frac{\pi}{2} < \theta < \frac{\pi}{2}$.

Remark: An equivalent expression for $\beta_n(z,\theta)$ is

$$\beta_n(z,\theta) = \left(\frac{\Im\left(\sqrt{d_n(z)}e^{-i\theta}\right)}{\cos\theta}\right)^2$$

which is easier to check in some cases. (It does not matter which branch of $\sqrt{d_n(z)}$ we choose since the result is raised to the power 2.) For more information we refer to [Jaco89].

1.8 Two theorems on zero-free regions

So far our aim has been to give examples of how standard continued fraction results can be used directly to establish zero-free regions for polynomials satisfying certain three-term recurrence relations. But it would be strange to write a section on zero-free regions without including some of the established results in the theory. We restrict ourselves to two examples. In both cases the proofs make use of continued fraction type arguments, essentially on value sets, as we have done in the more direct approaches. Space does not allow for comparison between methods, but examples will be included to illustrate the theorems. The first theorem is the prominent Parabola Theorem by Saff and Varga [SaVa76] (not to be confused with the parabola theorem in the analytic theory of continued fractions).

Theorem 11 (Saff-Varga's Parabola Theorem) *Let the polynomials $q_n(z)$ be defined by the recurrence relation*

$$q_n(z) = (z + \beta_n)q_{n-1}(z) - \alpha_n z q_{n-2}(z), \qquad n \geq 1 \qquad (1.8.1)$$

and the initial values

$$q_{-1}(z) = 0, \qquad q_0(z) = 1, \qquad (1.8.2)$$

where $\alpha_n > 0$ for $2 \le n \le N$, $\beta_n > 0$ for $1 \le n \le N$, and

$$D_N := min\{(\beta_n - \alpha_n); 1 \le n \le N\} > 0, \qquad (1.8.3)$$

with $\alpha_1 = 0$. Then $q_n(z) \ne 0$ for $1 \le n \le N$ and all z in the parabolic region

$$\{w \in \mathbf{C}; |w| \le \Re(w) + 2D_N\}. \qquad (1.8.4)$$

Example 5 Take in (1.8.1)

$$\alpha_n = \frac{1}{n}, \quad \beta_n = 1 + \frac{1}{n}.$$

Then $D_1 = 2$ and $D_N = 1$ for all $N \ge 2$. From the Parabola Theorem by Saff and Varga we find that the parabolic region, given by

$$|w| \le \Re(w) + 2,$$

is zero-free for all $q_n(z)$.

With

$$w = u + iv, \qquad u, v \in \mathbf{R},$$

the parabolic region can also be described by the following inequality:

$$v^2 \le 4(u + 1).$$

Numerical examples:

$$n = 2 : \qquad q_2(z) = z^2 + 3z + 3$$
$$\text{Zeros:} \frac{-3 \pm \sqrt{3}i}{2}$$

In Figure 10 these zeros are indicated by $\bullet$.

$$n = 3 : \qquad q_3(z) = z^3 + 4z^2 + \tfrac{19}{3}z + 4$$
$$\text{Zeros:} - 1.61, \, -1.19 \pm 1.03i$$

In Figure 10 these zeros are indicated by $\star$.

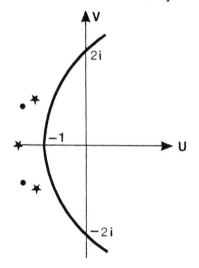

Figure 9.

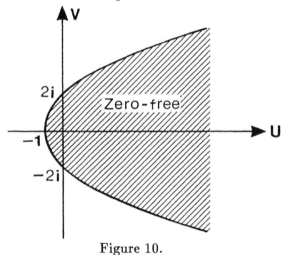

Figure 10.

We refer to the survey article [BrGR87] and to the references therein, in particular reference [32]. For applications and extensions we refer to Remark 1 at the end of the chapter and also to Problem 10 there.

The second theorem is due to Runckel, [Runc84], see for instance the survey article [BrGR87]. It gives an angular zero-free region under conditions which are the same as the ones in the Saff-Varga Parabola Theorem.

Theorem 12 *Let the conditions be as in Theorem 11. Let furthermore*

$$B_N := \max_{1 \le n \le N} \beta_n, \qquad Q_N := \max_{2 \le n \le N} \frac{\alpha_n}{\beta_n}. \qquad (1.8.5)$$

Then $q_n(z) \neq 0$ for $1 \le n \le N$ in the angular opening given by

$$z = re^{i\Phi}, \qquad r > 0, \qquad (1.8.6)$$

$$\cos \Phi \ge Q_N - \frac{D_N}{B_N}. \qquad (1.8.7)$$

As an example, we shall apply this to the sequence earlier studied in Example 5 by the Saff-Varga Parabola Theorem.

Example 6 Take in (1.8.1)

$$\alpha_n = \frac{1}{n}, \qquad \beta_n = 1 + \frac{1}{n}.$$

Then

$$\begin{aligned} B_N &= 2 \\ D_N &= 1 \qquad \text{for } N \ge 2. \\ Q_N &= \tfrac{1}{3} \end{aligned}$$

We find the following zero-free region: the angular opening $z = re^{i\Phi}$ where

$$\cos \Phi \ge \frac{1}{3} - \frac{1}{2} = -\frac{1}{6};$$

i. e. $|\arg z| \le 1.738 \ldots$. This is in some respects much better than the region obtained in Example 5 (for large z-values), in other respects it is not as good (for small z-values).

$\diamondsuit$

2   Stable polynomials

2.1   *Introductory remarks*

In the present section we shall also present a continued fraction technique for solving a problem on location of zeros. But this time it has to do with *one* polynomial, not a sequence of polynomials, and the technique is also completely different to the one used in the previous section. We shall be aiming at a necessary and sufficient condition for a polynomial $P(z)$ to have all its zeros in the left half plane $\Re(z) < 0$. Such a polynomial is called a *stable* polynomial, or a *Hurwitz* polynomial, in honor of Hurwitz, who solved it for real coefficients. Stable polynomials are important in the theory of differential equations and its application to vibration problems. The following simple example illustrates the type of problems which can be handled by using the results of the present section. The differential equation

$$\frac{d^3y}{dt^3} + 4\frac{d^2y}{dt^2} + 6\frac{dy}{dt} + 4y = 0$$

has the general solution

$$y = c_1 e^{\alpha t} + c_2 e^{\beta t} + c_3 e^{\gamma t},$$

where α, β, γ are the zeros of the polynomial

$$r^3 + 4r^2 + 6r + 4 = 0$$

(provided that they all are simple). If neither of the constants c_k is zero, we have

$$y(t) \to 0 \quad \text{as} \quad t \to \infty$$

iff the polynomial $r^3 + 4r^2 + 6r + 4 = 0$ is stable. This can of course be checked by finding the roots, but it is most useful to be able to check it without knowledge of the roots. The purpose of the present section is to present a tool for such questions, a practical test for stability, based upon continued fractions.

Let $Q(z) = a_n z^n + a_{n-1} z^{n-1} + \ldots + a_0$, $a_m > 0$ for $m = 0, 1, 2, \ldots n$ be the given polynomial to be tested. We define

$$P(z) = a_{n-1} z^{n-1} + a_{n-3} z^{n-3} + \ldots,$$

and find the continued fraction expansion of the form

$$\frac{P(z)}{Q(z)} = \frac{1}{1+d_1 z +} \frac{1}{d_2 z +} \frac{1}{d_2 z +} \cdots + \frac{1}{d_k z} ,$$

if it exists. In order to make it clear *how* this is found, by successive substitutions, we illustrate by some examples with small n:

$n = 1$:
$$Q(z) = a_1 z + a_0 ,$$
$$P(z) = a_0 ,$$
$$\frac{P(z)}{Q(z)} = \frac{a_0}{a_1 z + a_0} = \frac{1}{1 + \dfrac{a_1}{a_0} z} .$$

$n = 2$:
$$Q(z) = a_2 z^2 + a_1 z + a_0 ,$$
$$P(z) = a_1 z ,$$
$$\frac{P(z)}{Q(z)} = \frac{a_1 z}{a_2 z^2 + a_1 z + a_0} = \frac{1}{1 + \dfrac{a_2 z^2 + a_0}{a_1 z}}$$
$$= \frac{1}{1 + \dfrac{a_2}{a_1} z + \dfrac{a_0}{a_1 z}} = \frac{1}{1 + \dfrac{a_2}{a_1} z +} \frac{1}{\dfrac{a_1}{a_0} z} .$$

$n = 3$:
$$Q(z) = a_3 z^3 + a_2 z^2 + a_1 z + a_0 ,$$
$$P(z) = a_2 z^2 + a_0 ,$$
$$\frac{P(z)}{Q(z)} = \frac{a_2 z^2 + a_0}{a_3 z^3 + a_2 z^2 + a_1 z + a_0}$$
$$= \frac{1}{1 + \dfrac{a_3 z^3 + a_1 z}{a_2 z^2 + a_0}}$$
$$= \frac{1}{1 + \dfrac{a_3}{a_2} z +} \frac{1}{\dfrac{((a_1 a_2 - a_0 a_3)/a_2) z}{a_2 z^2 + a_0}} .$$

If $a_1 a_2 - a_0 a_3 = 0$ we have

$$\frac{P(z)}{Q(z)} = \frac{1}{1 + \dfrac{a_3}{a_2} z} ,$$

else we have

$$\frac{P(z)}{Q(z)} = \frac{1}{1 + \dfrac{a_3}{a_2}z +} \frac{1}{\dfrac{a_2^2}{a_1 a_2 - a_0 a_3}z +} \frac{1}{\dfrac{a_1 a_2 - a_0 a_3}{a_0 a_2}z} .$$

2.2   Polynomials with real coefficients

The following theorem uses the described type of expansion in a test for stability.

Theorem 13 *Let*

$$Q(z) = z^n + a_{n-1}z^{n-1} + \ldots + a_1 z + a_0 \qquad (2.2.1)$$

be a polynomial with real coefficients, and let

$$P(z) = a_{n-1}z^{n-1} + a_{n-3}z^{n-3} + \ldots . \qquad (2.2.2)$$

Then $Q(z)$ has all its zeros in the open left half plane if and only if the test function $t(z) = P(z)/Q(z)$ can be written as a terminating continued fraction

$$\frac{1}{1 + d_1 z +} \frac{1}{d_2 z +} \cdots \frac{1}{+ d_n z} , \qquad (2.2.3)$$

where $d_j > 0$, $1 \le j \le n$.

Proof of the "if- part":   Assume that $t(z)$ can be written in the form

$$\frac{1}{1 + d_1 z +} \frac{1}{d_2 z +} \cdots \frac{1}{+ d_n z} , \qquad \text{where } d_j > 0 \quad \text{for } 1 \le j \le n .$$

Let z be an arbitrary complex number with $\Re(z) \ge 0$. Let H denote the closed right half plane defined by $\Re(w) \ge 0$, where the closure is taken in $\hat{\mathbf{C}}$ so that $\infty \in H$, and let s_j be the linear fractional transformations, defined by

$$s_1(w) = \frac{1}{1 + d_1 z + w} , \qquad s_j = \frac{1}{d_j z + w} \quad \text{for } j \ge 2 .$$

Straightforward computation shows that $s_1(H)$ is the disk given by

$$\left| w - \frac{1}{2(1 + d_1 x)} \right| \leq \frac{1}{2(1 + d_1 x)} ,$$

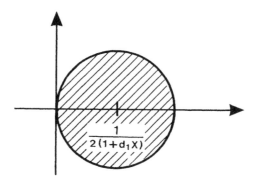

Figure 11.

where $x = \Re(z)$. Since $d_1 > 0$ and $x \geq 0$ this disk is obviously contained in the disk given by

$$\left| w - \frac{1}{2} \right| \leq \frac{1}{2} .$$

Furthermore, for $2 \leq j \leq n$, the following computation shows that $s_j(H)$ is contained in H: For $z = iy$ we have $s_j(H) = H$. For $x = \Re(z) > 0$ we find that $s_j(H)$ is the disk given by

$$\left| w - \frac{1}{2d_j x} \right| \leq \frac{1}{2d_j x} ,$$

which is contained in H. Hence $s_2 \circ \ldots \circ s_n(H) \subset H$, and thus

$$s_1 \circ s_2 \circ \ldots \circ s_n(H) \subset \left\{ w; \left| w - \frac{1}{2} \right| \leq \frac{1}{2} \right\} .$$

Since

$$\frac{P(z)}{Q(z)} = s_1 \circ s_2 \circ \ldots \circ s_n(0)$$

(the zero refers to the auxiliary variable w) we have, for $\Re(z) \geq 0$,

$$\left| \frac{P(z)}{Q(z)} - \frac{1}{2} \right| \leq \frac{1}{2} . \tag{2.2.4}$$

This shows the finiteness of $P(z)/Q(z)$. Since $Q(z)$ and $P(z)$ are the *canonical* denominator and numerator of an approximant of a continued fraction with partial numerators $\neq 0$, they can not vanish simultaneously. Therefore it follows that $Q(z) \neq 0$. This step in the argument is the same as the one used in the previous section. We have thus proved that the conditions on the continued fraction expansion imply $Q(z) \neq 0$ for $\Re(z) \geq 0$, or, in other words: $Q(z)$ is stable.

For the proof of the "only-if"-part we refer to [JoTh80, Sec. 7.4], from which also the essence of the proof above is taken.   ∎

Example 7 For the polynomial in the beginning of this section,

$$r^3 + 4r^2 + 6r + 4 \,,$$

we find the expansion

$$\frac{4r^2 + 4}{r^3 + 4r^2 + 6r + 4} = \frac{1}{1+} \frac{1}{\frac{1}{4}r+} \frac{1}{\frac{4}{5}r+} \frac{1}{\frac{5}{4}r} \,,$$

and the theorem tells us that the polynomial is stable. This agrees with the fact that the polynomial has the zeros $r_1 = -2$, $r_2 = -1 + i$, $r_3 = -1 - i$.

$\diamond$

2.3   Polynomials with complex coefficients

We shall state without proof the corresponding theorem for polynomials with complex coefficients, just to give the flavor of the more general situation. Also here we refer to [JoTh80, Sec. 7.4] for the proof.

Theorem 14 *Let*

$$Q(z) = z^n + a_{n-1}z^{n-1} + a_{n-2}z^{n-2} + \cdots + a_1 z + a_0 \qquad (2.3.1)$$

be a polynomial with complex coefficients

$$a_k = \alpha_k + i\beta_k \,, \qquad k = 0, 1, \ldots, n-1 \,. \qquad (2.3.1')$$

Let

$$P(z) = \alpha_{n-1} z^{n-1} + i\beta_{n-2} z^{n-2} + \alpha_{n-3} z^{n-3} + i\beta_{n-4} z^{n-4} + \cdots . \quad (2.3.2)$$

Then $f(z)$ is a stable polynomial if and only if the test function $t(z) = P(z)/Q(z)$ can be expressed as a terminating continued fraction of the form

$$t(z) = \cfrac{1}{1 + c_1 + d_1 z + \cfrac{1}{c_2 + d_2 z + \cfrac{1}{c_3 + d_3 z + \cdots + \cfrac{1}{c_n + d_n z}}}}, \quad (2.3.3)$$

where

$$\Re(c_j) = 0 \quad \text{and} \quad d_j > 0, \qquad j = 1, 2, \ldots, n.$$

Let it finally be mentioned that Hurwitz, who was the first one to solve the stability problem for polynomials in the real case, established a criterion for stability in terms of certain determinants of the coefficients.

Problems

(1) Use Worpitzky's theorem to prove that the set $|z| \geq 1$ is a zero-free region for the Tchebycheff polynomials of the second kind.

(2) Fill in the details in the proof that the Tchebycheff polynomials of the first kind have all their zeros in the interval $(-1, 1)$ of the real axis.

(3) Use Theorem 13 to decide which ones of the following polynomials are stable:

    (a) $z^3 + 6z^2 + 11z + 6$.

    (b) $z^3 - 7z - 6$.

    (c) $z^3 + 7z^2 + 16z + 10$.

    (d) $z^3 + 5z^2 + 4z - 10$.

In some of the "no-cases" we can see directly that the polynomial is instable. How? Give a general statement.

(4) What can be said about the zeros of the Tchebycheff polynomials of the second kind $U_n(z)$ merely by using Theorem 2. Hint: Study the polynomials $V_n(z)$, defined by

$$i^n V_n(z) = U_n(z).$$

(5) Take in Theorem 2 $g_n = i$, $h_n = 1$ for all n. Compute the zeros of $B_1(z)$ and $B_2(z)$, and check that they are in the "right" region.

(6) Let h be an arbitrary number with $\Re(h) > 0$. Take in Theorem 2 $g_n = -1/h$ and $h_n = h$ for all n. Find out as much as possible about the zeros of $B_n(z)$, and check with the theorem.

(7) Let $\{G_n(w)\}$ be the sequence in Theorem 4. Show by direct computation, that the real solutions r of the equation

$$G_5(w) = 0,$$

(i.e. *all* solutions) satisfy the inequality

$$r \leq -\frac{1}{a_2 + a_3 + a_4 + a_5},$$

and relate this to Figure 5.

(8) Let the sequence $\{D_n(z)\}$ be defined as in Theorem 5, and let a_2 be negative and a_3 be positive. Apart from that we know that

$$a_2 \geq -A_-,$$
$$a_3 \leq A_+.$$

Find by computation the set of all possible zeros of the function

$$D_3(z).$$

(9) (a) Use Saff-Varga's parabola theorem, Theorem 11, to determine a zero-free region for the polynomials $q_n(z)$ with $\beta_n = 2$, $\alpha_n = 1$. Sketch the parabola. Compute the zeros of $q_2(z)$ and $q_3(z)$, and indicate them on the same figure.

(b) Use also Theorem 12 to find a zero-free region.

(10) Let $P_n(z)$ denote the nth degree partial sum of the Taylor expansion of the exponential function e^z:

$$P_n(z) = 1 + \frac{z}{1!} + \ldots + \frac{z^n}{n!}, \qquad n \geq 0.$$

Let $q_n(z)$ be defined by

$$q_{-1}(z) = 0, \qquad q_n(z) = n!\, P_n(z) \quad \text{for } n \geq 0.$$

Prove that the sequence $\{q_n(z)\}$ satisfies the conditions in Saff-Varga's Parabola Theorem, Theorem 11, and use the theorem to prove that for $n \geq 2$

$$1 + \frac{z}{1!} + \frac{z^2}{2!} + \cdots + \frac{z^n}{n!} \neq 0$$

in the parabolic region

$$|z| \leq \Re(z) + 2.$$

See finally what you can find out about the zeros of $P_n(z)$ by using Theorem 12.

(11) Use Theorem 2 to prove that the denominator of the continued fraction from Theorem 13,

$$\cfrac{1}{1 + d_1 z + \cfrac{1}{d_2 z + \cdots + \cfrac{1}{d_n z}}}, \qquad d_j > 0, \qquad 1 \leq j \leq n,$$

has all its zeros in the *closed* left half plane.

(12) Prove by computation that any polynomial

$$z^2 + az + b, \qquad a, b \text{ real},$$

with both roots (possibly coinciding) in the open left half-plane is such that its test function has a terminating continued fraction

$$\cfrac{1}{1 + d_1 z + \cfrac{1}{d_2 z}}$$

with $d_1 > 0$, $d_2 > 0$.

(13) Prove a similar result as in Problem 12 for polynomials of degree 3.

(14) Verify by direct computation Theorem 12 for $N = 2$.

Remarks

1) The result obtained by solving Problem 10 is a very special case of a whole family of problems that can be solved by using Saff–Varga's Parabola Theorem. In fact: For a given formal power series where the coefficients satisfy certain conditions (some Toeplitz determinants $\neq 0$) the Padé numerators $U_{m,n}$ or denominators $V_{m,n}$, in both cases with a fixed $m \geq 0$ will, properly equipped with factors $C(m, n)$, $D(m, n)$ satisfy the recurrence relation (1.8.1) when α_n and β_n are chosen in the right way. With additional conditions satisfied, α_n and β_n will be positive and have a difference $\beta_n - \alpha_n$, bounded below by a positive number. In such cases the Saff-Varga Parabola Theorem permits us to conclude that the sequence of Padé numerators (or denominators) is zero-free in a parabolic region. Partial sums of power series are special cases of Padé approximants. In Problem 10 we had the partial sums of the Taylor series for e^z at 0, and the factor that made them satisfy the recurrence relations was $n!$.

   Another interesting application of Saff-Varga's theorem is for generalized Bessel polynomials, where remarkably good results have been obtained (a region bounded by a cardioid, where every point on the cardioid is a point of accumulation for the relevant zeros).

   H.-J. Runckel has generalized the Parabola Theorem substantially, to the complex case, and applied it to Bessel polynomials, Bessel functions and Lommel polynomials.

   Gilewicz has studied recurrence relations where the second coefficient has higher degree. De Bruin has studied M-term recurrence relations, in particular 4-term relations. For all these results we refer to the earlier mentioned survey article [BrGR87] and to the references mentioned there.

2) If, in Subsection *1.5*, we change the definition for A_+ and A_- slightly, the results will also cover the cases when the set of all n with $a_n < 0$ or the set of all n with $a_n > 0$ is empty. The modified definition:

$$\tilde{A}_+ := \max\{0, \sup a_n\}, \qquad \tilde{A}_- := \max\{0, \sup(-a_n)\}.$$

If the sets of positive and of negative a_n are both $\neq \emptyset$, we have $\tilde{A}_+ = A_+$, $\tilde{A}_- = A_-$. If the set of negative a_n is empty, $\tilde{A}_- = 0$, $S_-(\theta)$ is the whole plane and $S(\theta) = S_+(\theta) = $ a parallel strip. If the set of positive a_n is empty, we have $\tilde{A}_+ = 0$, and $S_+(\theta) = \mathbf{C}$, $S(\theta) = S_-(\theta)$, another parallel strip. In either case the rectangle in Figure 6 is replaced by one of the two parallel strips, and the union of the sets $S(\theta)$ will in one case be as in Subsection *1.4*, in the other case of the same form, except that the omitted slits are on the real axis. (Expand the disks on Figure 7, either with • fixed or with ⋆ fixed.) In Theorem 5, the parallelogram collapses to a segment on the imaginary axis if $\tilde{A}_- = 0$ and on the real axis if $\tilde{A}_+ = 0$.

3) Among other results on stability of polynomials we choose to refer to the recent papers [IsKi83] and [Isma85].

References

[BrGR87] M. G. De Bruin, J. Gilewicz and H.-J. Runckel, *A Survey of Bounds for the Zeros of Analytic Functions obtained by Continued Fraction Methods*, "Rational Approximation and its Applications in Mathematics and Physics, Proceedings, Lańcut 1985", (J. Gilewicz, M. Pindor and W. Siemaszko eds.), Lecture Notes in Mathematics No **1237**, Springer-Verlag, Berlin (1987), 1–23.

[CuWu87] A. Cuyt and L. Wuytack, "Nonlinear Methods in Numerical Analysis", North Holland Mathematics Studies **136**, Amsterdam (1987).

[Henr86] P. Henrici, "Applied and Computational Complex Analysis", I, II and III, J. Wiley & Sons, New York (1974, 1977, 1986).

[HePf66] P. Henrici and P. Pfluger, *Truncation Error Estimates for Stieltjes Fractions*, Numer. Math. **9** (1966), 120–138.

[Hurw95] A. Hurwitz, *Über die Bedingungen unter welchen eine Gleichung nur Wurzeln mit negativen reellen Teilen besitzt*, Math. Annalen **46** (1895), 273–284.

[IsKi83] M. Ismail and H. K. Kim, *A Simplified Stability Test for Discrete Systems Using a New z–Domain Continued Fraction Method*, IEEE Trans. Circuits Sys., Vol. CAS-**30** (July 1983), 505–507.

[Isma85] M. Ismail, *New z–Domain Continued Fraction Expansions*, IEEE Trans. Circuits Sys., Vol CAS-**32** (1985), 754–758.

479

[Jaco89]   L. Jacobsen, *Orthogonal Polynomials, Chain Sequences, Tree-term Recurrence Relations and Continued Fractions,* "Proc. of the Conference on Computational Methods and Function Theory, Valparaíso 1989", (St. Ruscheweyh, E. B. Saff, L. C. Calinas, R. S. Varga eds.), Lecture Notes in Mathematics **1435**, Springer-Verlag, Berlin (1990), 89–101.

[JoTh80]   W. Jones and W. J. Thron, "Continued Fractions: Analytic Theory and Applications", Encyclopedia of Mathematics and its Applications **11**, Addison-Wesley Publishing Company, Reading, Mass. (1980). Now distributed by Cambridge University Press, New York.

[Lange86]   L.J. Lange, *Continued Fraction Applications to Zero Location,* "Analytic Theory of Continued Fractions II", (W. J. Thron ed.), Lecture Notes in Mathematics **1199**, Springer-Verlag, Berlin (1986), 220–262.

[Runc84]   H.-J. Runckel, *Zero-free Regions for Polynomials with Applications to Padé Approximants,* "Constructive Theory of Functions, Proceedings of the International Conference on Constructive Theory of Functions, Varna 1984", Publishing House of the Bulgarian Acad. Sci., Sofia (1984), 767–771.

[Runc86]   H.-J. Runckel, *Pole- and Zero-free Regions for Analytic Continued Fractions,* Proc. Amer. Math. Soc. **97** (1986), 114–120.

[SaVa76]   E. B. Saff and R. S. Varga, *Zero-Free Parabolic Regions for Sequences of Polynomials,* SIAM J. Math. Anal. **7** (1976), 344–357.

[Wall48]   H. S. Wall, "Analytic Theory of Continued Fractions", Van Nostrand, New York (1948).

[Waad88]   H. Waadeland, *Some Recent Results in the Analytic Theory of Continued Fractions,* "Nonlinear Numerical Methods and Rational Approximation", (Annie Cuyt ed.), Reidel Publ. Co., Dordrecht, Holland (1988), 299-333.

Chapter XI

Digital filters and continued fractions

About this chapter

Linear system theory is a field where rational approximation is an important tool, and techniques from Padé theory have been quite useful in many different ways. In the present chapter we have tried to give a little taste of this. We have been rather restrictive, first of all by limiting the description and discussion to digital filters, essentially stable digital filters. And out of the many applications of Padé and continued fraction theory we have only included two: Stability test and model reduction.

Again, as in other chapters, for instance the number theory chapter, we see an example where "old mathematics" (the Schur algorithm) proves useful in "new theory". This is the case for both applications included.

1 Filters and their representation

1.1 Some introductory examples

Example 1 Let x be a real-valued, continuous function of a real variable t, defined on some interval of $\mathbf{R}$. Let furthermore x_0, x_1, x_2,...be values of the function at equally spaced, increasing values of t. Without loss of generality we may assume that $x(t)$ is defined on $[0, \infty)$, and that $x_n = x(n)$, $n = 0, 1, 2, \ldots$. Define the sequence $\{y_n\}_{n=0}^{\infty}$ by

$$y_0 = 0, \qquad y_{n+1} = y_n + \frac{1}{2}(x_n + x_{n+1}), \qquad n \geq 0.$$

Then we have

$$y_1 = \frac{1}{2}(x_0 + x_1), \qquad y_2 = \frac{1}{2}(x_0 + 2x_1 + x_2),$$

and generally for $n \geq 2$

$$y_n = \frac{1}{2}(x_0 + 2x_1 + \cdots + 2x_{n-1} + x_n),$$

i.e. y_n is the trapezoid formula approximation to the integral

$$\int_0^n x(t)dt,$$

with sub-intervals of length 1. Similarly we find, that with

$$z_0 = 0 \qquad z_{n+1} = z_n + x_{n+1}, \qquad n \geq 0,$$

z_n is the Riemann sum

$$\sum_{i=1}^{n} x(t_i^*)\Delta t,$$

with $t_i = i$, $i = 0, 1, 2, \ldots, n$, $t_i^* = t_i$, $i \geq 1$, for the same integral. ($\Delta t = 1$.)

Simpson's formula may also be described in a similar way:

$$u_0 = 0 \qquad u_{n+1} = u_n + \frac{1}{3}(x_{2n} + 4x_{2n+1} + x_{2n+2}).$$

u_n is then the Simpson approximant (with sub-intervals of length 1) to the integral

$$\int_0^{2n+2} x(t)dt \, .$$

_____ ◇

Example 2 Let x be a real-valued, continuous function of a real variable t, defined for all $t \in \mathbf{R}$. Let $\{x_n\}_{-\infty}^{\infty}$ be measured values of x, possibly containing noise (the word being used intuitively). Let $\{y_n\}_{-\infty}^{\infty}$ be defined by

$$y_n = \frac{1}{5}\left(x_{n-2} + x_{n-1} + x_n + x_{n+1} + x_{n+2}\right).$$

In this case the transformation $\{x_n\} \rightarrow \{y_n\}$ is used as a "smoothing process" for measured values of functions. Other averaging processes can also be used, generally

$$y_n = \sum_{i=-k}^{k} a_i x_{n+i} \, , \quad \text{where } a_i \geq 0 \, , \quad \sum_{i=-k}^{k} a_i = 1 \, .$$

_____ ◇

In these examples a sequence $\{x_n\}$ was given, and another sequence $\{y_n\}$ was computed from formulas

$$y_n = \sum_{k=-\infty}^{\infty} c_k x_{n-k} + \sum_{k=1}^{\infty} d_k y_{n-k} \, ,$$

where in the examples all c_k and d_k, except finitely many, are 0. Often, as in Example 1, $x_p = y_p = 0$ for $p < 0$. Such a formula is often referred to as a *digital filter*. The word *filter* comes from electrical engineering, where filters are used to transform signals from one form to another, in many cases to remove noise. The examples above merely indicated applications within mathematics itself (although strongly directed towards applications). But the scope of applications is very wide, and includes such diverse fields as astronomy, economics, medicine, radar technology, seismology and speach processing. It generally is concerned with extraction and enhancement of information contained in a sequence of measurements of continuous waveform phenomena. In many of the applications the variable t is *time* but not in all. For a short survey of applications (key words and some comments) we refer to the introductory section in the article [JoSt82]. As for books on the subject we refer to [Hamm77] and [OpSc75].

We shall not go into specific applications. The problems to be discussed here will be common to several applications of digital filters, and will illustrate how analytic theory of continued fractions can be used in this field.

We shall in the rest of the chapter use a much more restricted definition of a filter than the one given by the formula in the present section.

1.2 Digital filters

As mentioned in the previous section many different problems and situations give rise to the concept of a *digital filter*. A digital filter is a device mapping a sequence of complex numbers into a sequence of complex numbers. It is a linear mapping. Before presenting a proper definition we need to introduce some notation and basic concepts concerning sequences.

We shall let l denote the set of all sequences $\{a_n\}_{n=0}^{\infty}$ of complex numbers. With the standard operations, *addition*:

$$\{a_n\} + \{b_n\} = \{a_n + b_n\}, \tag{1.2.1}$$

and *multiplication by a scalar*:

$$c\{a_n\} = \{ca_n\}, \tag{1.2.2}$$

l is a linear space. We shall need two additional operations, *convolution*:

$$\{a_n\} * \{b_n\} := \left\{ \sum_{k=0}^{n} a_k b_{n-k} \right\}, \tag{1.2.3}$$

and *unit delay*:

$$D\{a_n\} := \{a_n'\}, \quad \text{where } a_0' = 0, \quad a_n' = a_{n-1} \quad \text{for } n \geq 1. \tag{1.2.4}$$

With each sequence $A = \{a_n\}_{n=0}^{\infty}$ we associate a formal power series $a(z)$ by

$$a(z) = \sum_{n=0}^{\infty} a_n z^{-n}. \tag{1.2.5}$$

This mapping is usually written in the following way:

$$a(z) \bullet\!\!-\!\!\circ A. \tag{1.2.6a}$$

The mapping $A \circ\!\!-\!\!\bullet a(z)$ shall be referred to as the *z-transform* and often be written

$$A \circ\!\!\overset{z}{-}\!\!\bullet a(z). \tag{1.2.6b}$$

The operations on the sequence correspond to operations on the formal power series in the following way:

$$\begin{array}{llll}
A + B & \circ\!\!\overset{z}{-}\!\!\bullet & a(z) + b(z), & cA & \circ\!\!\overset{z}{-}\!\!\bullet & ca(z), \\
DA & \circ\!\!\overset{z}{-}\!\!\bullet & z^{-1} \cdot a(z), & A * B & \circ\!\!\overset{z}{-}\!\!\bullet & a(z) \cdot b(z).
\end{array}$$

Here $a(z) \cdot b(z)$ denotes the Cauchy product of the two formal power series:

$$\left(a_0 + a_1 z^{-1} + a_2 z^{-2} + \cdots\right) \cdot \left(b_0 + b_1 z^{-1} + b_2 z^{-2} + \cdots\right) =$$

$$a_0 b_0 + (a_1 b_0 + a_0 b_1) z^{-1} + (a_2 b_0 + a_1 b_1 + a_0 b_2) z^{-2} + \cdots.$$

Observe that DA also can be written

$$A * B, \quad \text{where } B = (0, 1, 0, 0, 0, 0, \ldots).$$

In many cases it is of advantage to operate on the formal power series rather than on the sequences.

In the case of convergence $a(z)$ represents a function, holomorphic in some $|z| > r$ (also at ∞). In this case the Cauchy formula gives us the inverse transform:

$$a_n = \frac{1}{2\pi i} \oint_C a(z) z^{n-1} dz,$$

where C is a circle around the origin with radius $> r$, traversed counterclockwise.

Certain sub-spaces of l are of special importance, for instance the sub-space l^0, consisting of all $\{a_n\}_{n=0}^{\infty}$ for which

$$\limsup |a_n|^{\frac{1}{n}} < \infty .$$

(It is well known and easily proved that l^0 in fact is a linear space.) The main importance of l^0 lies in the fact that the formal power series $a(z)$ represents a function, holomorphic at $z = \infty$, if and only if $A \in l^0$.

Example 3

$$(1,1,1,\ldots) \quad \circ\!\!\xrightarrow{z}\!\!\bullet \quad 1 + z^{-1} + \cdots = \frac{z}{z-1} \quad \text{for } |z| > 1 .$$

$$(1,2,3,\ldots) \quad \circ\!\!\xrightarrow{z}\!\!\bullet \quad 1 + 2z^{-1} + 3z^{-2} + \cdots = \frac{z^2}{(z-1)^2} \quad \text{for } |z| > 1 .$$

$$\left(1,\frac{1}{2},\frac{1}{3},\ldots\right) \quad \circ\!\!\xrightarrow{z}\!\!\bullet \quad 1 + \frac{z^{-1}}{2} + \frac{z^{-2}}{3} + \cdots = z \ln\frac{z}{z-1} \quad \text{for } |z| > 1 .$$

$\diamond$

We are now ready for the definition of digital filters. In this "theoretical part" we shall use the concept for the mapping itself, not for some "device" producing the mapping. Following [JoSt82] we define:

Definition *A digital filter $F : l \to l$ is a mapping of sequences $\{x_n\}$ onto sequences $\{y_n\}$ according to formulas*

$$(F) \qquad y_n + \sum_{k=1}^{N} b_k y_{n-k} = \sum_{k=0}^{M} a_k x_{n-k} , \qquad n = 0, 1, 2, \ldots, \qquad (1.2.7)$$

where $a_0, a_1, \ldots, a_M, b_1, b_2, \ldots, b_N$ are given constants, $a_M \neq 0, b_N \neq 0$ and $x_n = y_n = 0$ for $n < 0$.

Remark: $M = 0$ or $N = 0$ is permitted. In the latter case the sum on the lefthand side is empty, and hence 0, in which case the filter is called *nonrecursive*. In all other cases it is called *recursive*. The sequence $X = \{x_n\}$ is called the *input*, and $Y = \{y_n\}$ the *output*.

Example 4 (a) $N = 0$, $M = 2$: In this case the recurrence relations are:

$$
\begin{aligned}
y_0 &= a_0 x_0 \\
y_1 &= a_0 x_1 + a_1 x_0 \\
y_2 &= a_0 x_2 + a_1 x_1 + a_2 x_0 \\
&\vdots \\
y_n &= a_0 x_n + a_1 x_{n-1} + a_2 x_{n-2} \\
&\vdots
\end{aligned}
$$

For arbitrary $n \geq 2$ the recurrence relations can be written in matrix form ($n = 4$ in the illustration):

$$
\begin{bmatrix} y_0 \\ y_1 \\ y_2 \\ y_3 \\ y_4 \end{bmatrix} = \begin{bmatrix} a_0 & 0 & 0 & 0 & 0 \\ a_1 & a_0 & 0 & 0 & 0 \\ a_2 & a_1 & a_0 & 0 & 0 \\ 0 & a_2 & a_1 & a_0 & 0 \\ 0 & 0 & a_2 & a_1 & a_0 \end{bmatrix} \begin{bmatrix} x_0 \\ x_1 \\ x_2 \\ x_3 \\ x_4 \end{bmatrix}
$$

(b) $N = 2$, $M = 2$: In this case the recurrence relations are

$$
\begin{aligned}
y_0 &= a_0 x_0 \\
y_1 + b_1 y_0 &= a_0 x_1 + a_1 x_0 \\
y_2 + b_1 y_1 + b_2 y_0 &= a_0 x_2 + a_1 x_1 + a_2 x_0 \\
&\vdots \\
y_n + b_1 y_{n-1} + b_2 y_{n-2} &= a_0 x_n + a_1 x_{n-1} + a_2 x_{n-2} \\
&\vdots
\end{aligned}
$$

For $n \geq 2$ the recurrence relations can be written in matrix form ($n = 4$ in the illustration):

$$
\begin{bmatrix} 1 & 0 & 0 & 0 & 0 \\ b_1 & 1 & 0 & 0 & 0 \\ b_2 & b_1 & 1 & 0 & 0 \\ 0 & b_2 & b_1 & 1 & 0 \\ 0 & 0 & b_2 & b_1 & 1 \end{bmatrix} \begin{bmatrix} y_0 \\ y_1 \\ y_2 \\ y_3 \\ y_4 \end{bmatrix} = \begin{bmatrix} a_0 & 0 & 0 & 0 & 0 \\ a_1 & a_0 & 0 & 0 & 0 \\ a_2 & a_1 & a_0 & 0 & 0 \\ 0 & a_2 & a_1 & a_0 & 0 \\ 0 & 0 & a_2 & a_1 & a_0 \end{bmatrix} \begin{bmatrix} x_0 \\ x_1 \\ x_2 \\ x_3 \\ x_4 \end{bmatrix}
$$

$\diamond$

Definition   *For a given digital filter (F) the rational function*

$$h(z) = \frac{\sum_{k=0}^{M} a_k z^{-k}}{\sum_{k=0}^{N} b_k z^{-k}}, \qquad b_0 = 1, \tag{1.2.8}$$

is called the transfer function of the filter. (Keep in mind that $a_M \neq 0$, $b_N \neq 0$.)

The reason for this name is that the transfer from input to output can be done by using the function h. With

$$
\begin{aligned}
A &= (a_0, a_1, \ldots, a_M, 0, 0, 0, 0, \ldots) \\
B &= (1, b_1, \ldots, b_N, 0, 0, 0, \ldots) \\
X &= (x_0, x_1, x_2, \ldots) \in l^0 \\
Y &= (y_0, y_1, y_2, \ldots) \in l^0 \qquad \text{(since } X \in l^0\text{, see remark below)}
\end{aligned}
$$

the filter can be written

$$B * Y = A * X.$$

By the z-transform we get

$$
\begin{aligned}
b(z) \cdot y(z) &= a(z) \cdot x(z) \\
y(z) &= \frac{a(z)}{b(z)} \cdot x(z),
\end{aligned}
$$

which is the same as

$$y(z) = h(z) \cdot x(z). \tag{1.2.9}$$

This gives us an algorithm for computing the output from a given input.

Remark: Observe that $X \in l^0 \Rightarrow x(z)$ is holomorphic in a neighborhood of $z = \infty \Rightarrow y(z)$ is holomorphic in a neighborhood of $z = \infty \Rightarrow Y \in l^0$.

Definition   *If $h(z)$ is the transfer function, then the inverse z-transform sequence $H = \{h_n\}$,*

$$\{h_n\} \circ\!\!\!\xrightarrow{z}\!\!\!\bullet\, h(z),$$

is called the shock response of the filter.

Observe that

$$\frac{\sum a_k z^{-k}}{\sum b_k z^{-k}} = \sum h_n z^{-n} \quad \text{(by "long division")}.$$

The reason for this name is that the input

$$X = (1, 0, 0, 0, \ldots) \quad \text{("shock")}$$

has the sequence

$$H = (h_0, h_1, h_2, \ldots)$$

as its output.

1.3   Stable filters

Let l_∞ be the linear space of bounded sequences. A filter with the property that a bounded input gives a bounded output is called a *stable filter*. This is an important property. A useful theorem for deciding the possible stability of a filter is the following, stated in [Henr86, Vol. 2] as a problem (with reference to Martin Gutknecht):

Theorem 1 *The following properties are equivalent:*

(a) *The filter is stable.*

(b) *The poles of the transfer function $h(z)$ are all located in the open unit disk $|z| < 1$.*

(c) *If $h(z) \bullet\!\!-\!\!\circ \{h_n\}$, then $\sum_{n=0}^{\infty} |h_n| < \infty$.*

We shall establish the proof in three steps, the proofs of (a) $\Rightarrow$ (b), (b) $\Rightarrow$ (c) and (c) $\Rightarrow$ (a), by operating on the z-transforms of $\{x_n\}$, $\{h_n\}$ and $\{y_n\}$. A tool in our proof is the following:

Lemma 2 *Let $g(w)$ be rational, and holomorphic in the open unit disk U, where it has the power series expansion*

$$g(w) = d_0 + d_1 w + d_2 w^2 + \cdots .$$

If g has a pole of order > 1 on the circle $|w| = 1$, the sequence $\{d_n\}$ is not bounded.

Proof of Lemma 2:   Being a rational function, $g(w)$ can be written as a linear combination of terms of the forms (where r, s are natural numbers):

$$\frac{p(w)}{(w - \alpha)^r}, \qquad |\alpha| > 1, \qquad \deg(p) < r,$$

$$\frac{q(w)}{(w - \beta)^s}, \qquad |\beta| = 1, \qquad \deg(q) < s,$$

and possibly a polynomial.

The coefficient d_n will then be the w^n-coefficient we get by expanding all the terms separately, followed by adding the expansions. The contributions of the first and third types add up to a function, holomorphic in some disk $|w| < 1 + \epsilon$, $\epsilon > 0$, and their contribution to d_n will therefore $\to 0$ when $n \to \infty$. Without loss of generality we may restrict ourself to the study of the terms of the second type. We rewrite the term:

$$\frac{q(w)}{(w - \beta)^s} = \frac{(-\bar\beta)^s q(w)}{(1 - \bar\beta w)^s} = \frac{q_0 + q_1 w + \cdots + q_n w^m}{(1 - \bar\beta w)^s}, \qquad \beta = e^{-i\theta} .$$

This has a partial fraction decomposition of the form

$$\frac{A_1}{1 - \bar\beta w} + \frac{A_2}{(1 - \bar\beta w)^2} + \cdots + \frac{A_s}{(1 - \bar\beta w)^s} ,$$

where we assume $s \geq 2$ and $A_s \neq 0$. The coefficient of w^n in the power series expansion of this expression is

$$e^{ni\theta} \left(A_s \binom{s + n - 1}{n} + A_{n-1} \binom{s + n - 2}{n} + \cdots + A_1 \binom{n}{n} \right) ,$$

which is equal to $A_1 e^{ni\theta}$ for $s = 1$. For $s \geq 2$ it is easily seen to tend to ∞ when $n \to \infty$. (No asymptotic cancelling is possible, since for $2 \leq k \leq s$

$$\frac{\binom{k-2+n}{n}}{\binom{k-1+n}{n}} = \frac{k-1}{k-1+n} \to 0 \quad \text{when } n \to \infty.)$$

The proposition is thus proved for the case when we only have one term of the second type or even in the case when we have more such terms, but only one with maximal s for the function. In case we have more, say p, terms of the second type and with s maximal, the dominant term in the coefficient for z^n is of the form

$$\left(\sum_{\nu=1}^{p} A_s^{(\nu)} \exp(in\theta_\nu)\right) \cdot \binom{s+n-1}{n},$$

where $A_s^{(\nu)} \neq 0$ for all ν and all θ_ν are distinct. Since

$$\limsup_{n \to \infty} \left|\left(\sum_{\nu=1}^{p} A_s^{(\nu)} \exp(in\theta_\nu)\right)\right| > 0,$$

(else $\lim \sum_{\nu=1}^{p} A_s^{(\nu)} \exp(in\theta_\nu) = 0$, which is not possible, see Problem 6) and

$$\binom{s+n-1}{n} \to \infty,$$

a subsequence of $\{d_n\}$ converges to infinity, and Lemma 2 is proved. ∎

In the application we have $w = 1/z$. We now proceed to the

Proof of Theorem 1:
Proof of (a) $\Rightarrow$ (b): With the input $X = (1, 0, 0, \ldots)$ the output is $H = \{h_n\}$, the shock response. This is then a bounded sequence, and hence $h(z)$ is holomorphic for $|z| > 1$ (also for $z = \infty$). Assume that $h(z)$ has a pole on $|z| = 1$, say for $z = e^{i\theta}$. Then

$$h(z) = \frac{g(z)}{z - e^{i\theta}},$$

where $g(z)$ is holomorphic for $|z| > 1$ and $g(e^{i\theta}) \neq 0$. Take $X = (1, e^{i\theta}, e^{2i\theta}, \ldots)$, which is a bounded sequence. Then

$$x(z) = \sum_{n=0}^{\infty} e^{in\theta} z^{-n} = \frac{1}{1 - e^{i\theta} z^{-1}} = \frac{z}{z - e^{i\theta}},$$

and

$$y(z) = h(z) \cdot x(z) = \frac{zg(z)}{(z - e^{i\theta})^2}.$$

$y(z)$ thus has a pole of order at least 2 at $z = e^{i\theta}$. Therefore we know from Lemma 2 that $\{y_n\}$ is not a bounded sequence, which contradicts the assumption on stability of the filter. Hence (a) $\Rightarrow$ (b) is proved.

Proof of (b) $\Rightarrow$ (c): The function $h(z)$ must be holomorphic in $|z| > 1 - \epsilon$ for some $\epsilon > 0$, which implies absolute convergence of the series expansion for $|z| \geq 1 - \epsilon/2$, in particular for z-values on the unit circle. (Remember, $h(z)$ is a rational function and has only finitely many poles.) Hence $\sum |h_n| = M < \infty$.

Proof of (c) $\Rightarrow$ (a): Let $X = \{x_n\}$ be a bounded sequence, $|x_n| \leq c$ for all n. Since

$$\{y_n\} = \{h_n\} * \{x_n\},$$

i.e.

$$y_n = \sum_{k=0}^{n} x_k h_{n-k},$$

we find

$$|y_n| \leq \sum_{k=0}^{n} |x_k||h_{n-k}| \leq c \sum_{k=0}^{\infty} |h_k| = c \cdot M < \infty.$$

Theorem 1 is thus proved. ∎

Remark: In the cases where we have the power series expansion at ∞ for the transfer function, the stability can be checked by (c). But there are cases when the transfer function is given a quite different representation, in which case other criteria may be of use. We shall return to this in Subsection 2.2.

1.4   Graph representation of filters

Thinking of the independent variable as time, the realization of a digital filter requires that past and immediate values of the input, and past values of the output be available. This requires the possibility of delay or storage of the past values. Furthermore, we need means for multiplying by coefficients and adding the results. We illustrate these operations in Figure 1. Keeping in mind that the unit delay in the sequence $\{x_n\}$

$$(x_1, x_2, x_3, \ldots) \rightarrow (0, x_1, x_2, \ldots)$$

corresponds to multiplication of the z-transform $x(z)$ by z^{-1}, we indicate the delay by using a "z^{-1}-box".

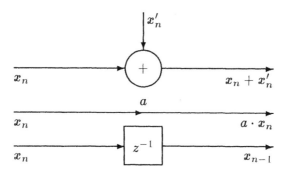

Figure 1.

This is perhaps a good place for some remarks on notation, in order to avoid confusion: It is beyond the scope of the present chapter to discuss the sampling of values from some function. From now on we shall take the sequence $\{x_n\}$ (or other sequences) for granted, and forget about the "underlying function" from which the numbers x_n are sampled. Hence, when we use the symbol $x(z)$ (or $a(z)$, $b(z)$ etc.) we shall mean the z-transform of the sequence $\{x_n\}$ (or $\{a_n\}$, $\{b_n\}$ etc.). We shall use upper case letters X, A, B as symbols for sequences: $X = \{x_n\}$ etc. In some cases we shall need (as already seen in the description of graphs of filters) symbols like X_1, $X_2, \ldots$ for different sequences, in which case the z-transform shall be denoted $x_1(z)$, $x_2(z)$, $\ldots$.

Example 5 The following difference equation fits into the definition of a recursive digital filter:

$$y_n + b_1 y_{n-1} + b_2 y_{n-2} = a_0 x_n .$$

This is an inhomogenous recurrence relation of order 2. The *block diagram* representation of this digital filter is as shown in Figure 2. In

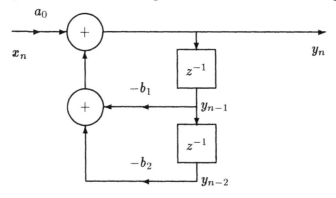

Figure 2.

order to understand this illustration we write the relation in the following form:

$$y_n = a_0 x_n + (-b_1)y_{n-1} + (-b_2)y_{n-2} .$$

─── ◇

Example 6 The digital filter

$$(F) \qquad y_n + \sum_{k=1}^{N} b_k y_{n-k} = \sum_{k=0}^{M} a_k x_{n-k}$$

has a block diagram illustrated in Figure 3.

─── ◇

This illustration is only one way of many ways to present the filter (transfer function).

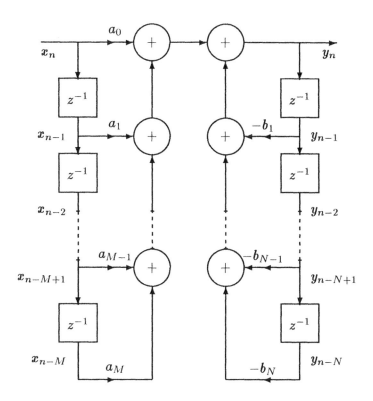

Figure 3.

A related way of implementing a digital filter is by using a *directed graph*. A directed graph consists of two types of elements, points, called *nodes*, and simple directed curves from one node to another, called *branches*. To each node there is associated a *node sequence* X_i, or equivalently, by the z-transform, a formal power series $x_i(z)$ (possibly a function). The node sequences are influenced by the other node sequences through a transmittance sequence T_{ij}, or equivalently, a transmittance function. The transmittance sequences in question are all of the form

$$(a_{ij}, b_{ij}, 0, \ldots, 0),$$

where at most one of the numbers a_{ij}, b_{ij}, is $\neq 0$. The node sequences are interrelated in the following way:

$$X_i = \sum_j T_{ij} * X_j \,,$$

where the sum is taken over the whole range of node indices, and where i also ranges over the same set. For the z-transform this can be written

$$x_i(z) = \sum_j t_{ij}(z) \cdot x_j(z) \,,$$

where $t_{ij}(z)$ is one of the three functions; 0 (no influence), $a_{ij} \neq 0$ (multiplication), $b_{ij}z^{-1}$ (delay and multiplication). We only draw the arrows (branches) between the nodes where the transition function is not 0. For a directed graph there is one particular node where no branch ends, the *source node*, and one where no branch starts, the *sink node*. In order to compare the block diagram and the directed graph we shall look at an example.

Example 7 We use as example the recurrence relation

$$y_n = ay_{n-1} + x_n + bx_{n-1} \,,$$

which is a first-order difference equation, since we only have n and $n-1$. A block diagram for this relation is shown in Figure 4.

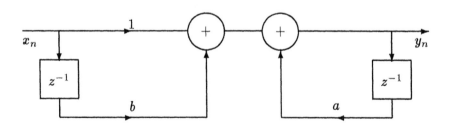

Figure 4.

With $b = 0$ and $a = 0$ respectively we would get the two block diagrams:

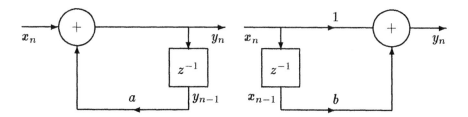

Figure 5.

By combining the two diagrams, using the delay (with z^{-1}) for X as well as Y, we get the diagram in Figure 6.

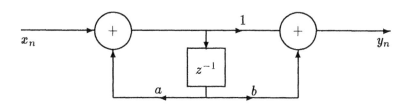

Figure 6.

A directed graph for the same recurrence relation is shown in Figure 7.

In this graph we have node sequences X_i, and their z-transforms $x_i(z)$. There is a sink node with node sequence Y, z-transform $y(z)$, and a source node with node sequence X and z-transform $x(z)$. Observe that the recurrence relation by the z-transform is turned into the following equation for formal power series $x(z)$ and $y(z)$:

$$y(z) = az^{-1}y(z) + x(z) + bz^{-1}x(z).$$

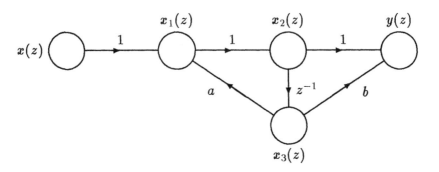

Figure 7.

This gives

$$y(z) = \frac{1 + bz^{-1}}{1 - az^{-1}} x(z) \,,$$

showing that

$$h(z) = \frac{1 + bz^{-1}}{1 - az^{-1}}$$

is the transfer function. Keep in mind that the arrows go in the direction of influence.

One way to check this is to establish that it has the right transfer function. We find:

$$
\begin{aligned}
x_1(z) &= x(z) + ax_3(z) \,,\\
x_2(z) &= x_1(z) \,,\\
x_3(z) &= z^{-1} x_2(z) \,,\\
y(z) &= x_2(z) + bx_3(z) \,.
\end{aligned}
$$

Simple computation shows that

$$
\begin{aligned}
x_2(z) &= x(z) + az^{-1} x_2(z) \,,\\
y(z) &= x_2(z) + bz^{-1} x_2(z) \,,
\end{aligned}
$$

and hence

$$y(z) = \frac{1 + bz^{-1}}{1 - az^{-1}} x(z).$$

$\diamond$

We conclude this section with an introductory example for a particular filter to be discussed later. Observe how we can work our way from node to node, starting with the source node (1), ending at the sink node (4).

Example 8 For the directed graph in Figure 8 we find:

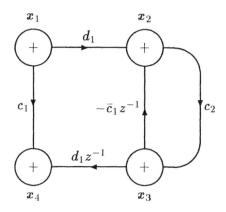

Figure 8.

$$\begin{aligned}
x_2(z) &= d_1 x_1(z) - \bar{c}_1 z^{-1} x_3(z) \\
x_3(z) &= c_2 x_2(z) \\
x_4(z) &= c_1 x_1(z) + d_1 z^{-1} x_3(z)
\end{aligned}$$

Simple computation shows that

$$x_3(z) = \frac{d_1 x_1(z)}{\bar{c}_1 z^{-1} + \frac{1}{c_2}}$$

(from the two first equations). Since the arrow from X_2 to X_3 is drawn, $c_2 \neq 0$, although the case $c_2 = 0$ is easily dealt with (trivial). If this is inserted into the last equation we get

$$x_4(z) = \left(c_1 + \frac{d_1^2 z^{-1}}{\bar{c}_1 z^{-1} + \frac{1}{c_2}} \right) x_1(z).$$

The case of particular interest is such that c_1 and c_2 are complex numbers with $|c_1| < 1$, $|c_2| \leq 1$ and d_1 is the positive number $d_1 = \sqrt{1 - |c_1|^2}$. We shall maintain these conditions in the following. In this case we have

$$x_4(z) = h_1(z) \cdot x_1(z),$$

where the transfer function is given by

$$h_1(z) = c_1 + \frac{(1 - |c_1|^2)z^{-1}}{\bar{c}_1 z^{-1}} \frac{1}{+ c_2},$$

which is a terminating continued fraction (or an approximant of some continued fraction).

$$\rule{6cm}{0.4pt}\diamond$$

In a later section we shall extend this example. We shall here only indicate the next step: From $x_2(z)$ to $x_3(z)$ we have the transfer function c_2, i.e. $x_3(z) = c_2 \cdot x_2(z)$. Assume that this transfer function is replaced by another one (possibly given by some directed graph). Let $T(z)$ be the new transfer function. Then the transfer function $h_1(z)$ is replaced by the function

$$h_1(z) = c_1 + \frac{(1 - |c_1|^2)z^{-1}}{\bar{c}_1 z^{-1}} \frac{1}{+ T(z)}.$$

In some interesting cases it is natural to write $T(z)$ in the form $T(z) = c_2 + T_0(z)$, and the transfer function

$$h_1(z) = c_1 + \frac{(1 - |c_1|^2)z^{-1}}{\bar{c}_1 z^{-1}} \frac{1}{+ c_2 + T_0(z)}$$

can be regarded as a modified approximant of some continued fraction.

2   The Schur algorithm

2.1   An old algorithm

In 1917 and 1918 a paper appeared [Schur18], in which functions holomorphic and bounded in the open unit disk were studied. The author was I. Schur, and the paper has proved to be of interest and of influence in several ways. We shall not even try to *indicate* the variety of problems upon which this paper had an impact, merely show a connection to digital filters and continued fractions. We first recall two basic facts from the theory of analytic functions of a complex variable:

Schwarz' lemma: *Let f be holomorphic in the open unit disk U, $|f(w)| \leq 1$ in U and $f(0) = 0$. Then*

$$|f(w)| \leq |w| \ in \ U$$

with equality if and only if $f(w) = e^{i\theta} w$.

(Schwarz' lemma can be found in most textbooks in complex analysis, see for instance [Ahlf53, Chapter III, Thm. 13].)

A simple mapping property.   Let α be a complex number, $|\alpha| < 1$. Then the linear fractional transformation

$$\Omega = \frac{w - \alpha}{1 - \bar{\alpha} w}$$

maps the closed unit disk U one to one and conformally onto itself.

Following Schur we shall study the family E of functions f, holomorphic in U and mapping U into $\bar{U}$, $f(U) \subseteq \bar{U}$. (Observe that the function taking only the value 1, or $e^{i\theta}$, is in E.) For a given $f_0 \in E$ we have the power series expansion

$$f_0(w) = c_0 + c_1 w + c_2 w^2 + \cdots, \quad |w| < 1. \tag{2.1.1}$$

Here $|c_0| \le 1$. If $|c_0| = 1$ the function reduces to the constant $c_0 = e^{i\theta_0}$. If $|c_0| < 1$, we put

$$c_0 =: \gamma_0 . \tag{2.1.2}$$

Then the function g_1, defined by

$$g_1(w) = \frac{f_0(w) - \gamma_0}{1 - \bar{\gamma}_0 f_0(w)}$$

is also in E (from the mapping property of Ω). Furthermore $g_1(0) = 0$, and hence the Schwarz Lemma applies, and $|g_1(w)/w| \le 1$, i.e. the function

$$f_1(w) = \frac{f_0(w) - \gamma_0}{w(1 - \bar{\gamma}_0 f_0(w))} \tag{2.1.3}$$

is again a function in E. (The removeable singularity at 0 causes no problem). The power series expansion of f_1 starts with

$$\gamma_1 := f_1(0) \ (= g_1'(0)) = \frac{c_1}{1 - |c_0|^2} .$$

If $|\gamma_1| = 1$, we have

$$f_1(w) = \gamma_1 = e^{i\theta} ,$$

which means that the function we started with was

$$f_0(w) = \frac{\gamma_0 + e^{i\theta} w}{1 + \bar{\gamma}_0 e^{i\theta} w} . \tag{2.1.4}$$

(This follows easily from (2.1.3) by putting $f_1(w) = e^{i\theta}$.)

If $|\gamma_1| < 1$, we construct a second function f_2, defined by

$$f_2(w) = \frac{f_1(w) - \gamma_1}{w(1 - \bar{\gamma}_1 f_1(w))} .$$

This is again a function in E, and we can go on. Either we come to some f_n which has a constant value of modulus 1, or we get an infinite sequence $\{f_n\}_0^\infty$ of functions in E. The forward and backward recurrence relations are

$$f_{n+1}(w) = \frac{f_n(w) - \gamma_n}{w(1 - \bar{\gamma}_n f_n(w))}, \qquad f_n(w) = \frac{\gamma_n + w f_{n+1}(w)}{1 + \bar{\gamma}_n w f_{n+1}(w)} . \tag{2.1.5}$$

Here γ_n is the constant term in the power series of $f_n(w)$. We thus have two cases:

1. The algorithm produces an infinite sequence of functions in E. In this case all γ_n have absolute value < 1. A permitted special case is that for some k the function f_k reduces to the value γ_k, in which case $f_i(z) = 0$ for all $i \geq k + 1$.

2. The algorithm produces only a finite number of functions, all in E, the last one a constant with absolute value 1. In this case $|\gamma_k| < 1$ for all $k < n$, whereas $|\gamma_n| = 1$. ($n = 0$ is permitted.)

It can be proved, that the case 2 occurs if and only if $f_0(w)$ is of the form (Blaschke product)

$$f_0(w) = \epsilon \prod_{i=1}^{n} \frac{w + w_i}{1 + \bar{w}_i w}, \qquad 0 \leq |w_i| < 1, \qquad |\epsilon| = 1, \qquad (2.1.6)$$

alternatively written

$$f_0(w) = \epsilon \frac{w^n P(w^{-1})}{\bar{P}(w)}, \qquad (2.1.6')$$

where

$$P(w) = \prod_{i=1}^{n}(1 + w_i w) = 1 + k_1 w + \cdots + k_n w^n$$

is a polynomial of at most degree n, and where $\bar{P}(w)$ is the polynomial $1 + \bar{k}_1 w + \cdots + \bar{k}_n w^n$. We shall not give the proof, although it is rather simple. We have already seen, that the statement holds for $n = 0$ and $n = 1$. See also Problem 10.

Schur calls his algorithm "kettenbruchartig", continued fraction like, and indeed it is. It must be, since we get f_n from f_{n+1} by linear fractional transformations, where $\{f_n\}$ acts like a tail sequence. We rewrite the recurrence relation

$$f_n(w) = \frac{\gamma_n + w f_{n+1}(w)}{1 + \bar{\gamma}_n w f_{n+1}(w)}$$

in the form

$$f_n(w) = \gamma_n + \frac{(1 - |\gamma_n|^2)w}{\bar{\gamma}_n w} \frac{1}{+ f_{n+1}(w)}. \qquad (2.1.7)$$

This gives

$$f_0(w) = \gamma_0 + \cfrac{(1 - |\gamma_0|^2)w}{\bar{\gamma}_0 w} + \cfrac{1}{\gamma_1 +} \cfrac{(1 - |\gamma_1|^2)w}{\bar{\gamma}_1 w} + \cfrac{1}{\gamma_2 +}$$
$$\cfrac{1}{\cdots + \gamma_{n-1} +} \cfrac{(1 - |\gamma_{n-1}|^2 w)}{\bar{\gamma}_{n-1} w} + \cfrac{1}{f_n(w)} \tag{2.1.7'}$$

for all n in case 1, and also in case 2 if $n < N$, where $N \geq 2$ is the smallest number for which $|\gamma_n| = 1$. For $n = N$ we have

$$f_0(w) = \gamma_0 \quad \text{for } N = 0,$$
$$f_0(w) = \gamma_0 + \cfrac{(1 - |\gamma_0|^2)w}{\bar{\gamma}_0 w} + \cfrac{1}{\gamma_1} \quad \text{for } N = 1,$$
$$f_0(w) = \gamma_0 + \cfrac{(1 - |\gamma_0|^2)w}{\bar{\gamma}_0 w} + \cdots + \cfrac{1}{\gamma_{N-1} +} \cfrac{(1 - |\gamma_{N-1}|^2)w}{\bar{\gamma}_{N-1} w} + \cfrac{1}{\gamma_N}$$

generally for $N \geq 2$. Hence, on one hand case 2 occurs if and only if $f_0(z)$ is of the form

$$f_0(w) = \epsilon \prod_{i=1}^{N} \frac{w + w_i}{1 + \bar{w}_i w}, \qquad 0 \leq |w_1| < 1, \qquad |\epsilon| = 1.$$

On the other hand, $f_0(w)$ is in case 2 equal to a terminating continued fraction of the type above. It is not hard to see, that if a function f_0 is primarily *given* by such a continued fraction, this function can be written in the product form above.

Continued fractions of the form

$$\gamma_0 + \cfrac{(1 - |\gamma_0|^2)w}{\bar{\gamma}_0 w} + \cfrac{1}{\gamma_1 +} \cfrac{(1 - |\gamma_1|^2)w}{\bar{\gamma}_1 w} + \cfrac{1}{\gamma_2 + \cdots}, \tag{2.1.8}$$

where $|\gamma_n| < 1$ for all n were first studied by Wall [Wall48]. They, or the terminating ones with the last γ_n being of absolute value $= 1$ are called positive *Schur fractions* in honor of I. Schur. They have many interesting properties, for instance convergence and correspondence properties. See for instance [JoNT86] and [Wall48]. They are furthermore connected to many different areas, some of which are mentioned in the remarks. We shall in the next subsection discuss the connection between terminating Schur fractions and digital filters.

2.2 Schur fractions and digital filters

We go back to Example 8 in Subsection *1.4*. Figure 8 shows the directed graph of a digital filter with transfer function

$$h_1(z) = c_1 + \cfrac{(1 - |c_1|^2)z^{-1}}{\bar{c}_1 z^{-1}} \cfrac{1}{+ c_2}.$$

In the final remark of this example the transfer function c_2 was replaced by another transfer function $T(z)$, in which case we would get the transfer function

$$c_1 + \cfrac{(1 - |c_1|^2)z^{-1}}{\bar{c}_1 z^{-1}} \cfrac{1}{+ T(z)}.$$

If $T(z)$ is given by a directed graph of exactly the same type, we would have

$$T(z) = c_2 + \cfrac{(1 - |c_2|^2)z^{-1}}{\bar{c}_2 z^{-1}} \cfrac{1}{+ c_3},$$

and the transfer function for the "combined graph" would be

$$c_1 + \cfrac{(1 - |c_1|^2)z^{-1}}{\bar{c}_1 z^{-1}} \cfrac{1}{+ c_2 +} \cfrac{(1 - |c_2|^2)z^{-1}}{\bar{c}_2 z^{-1}} \cfrac{1}{+ c_3}.$$

Again we could replace c_3 by a transfer function of the same type as the one in Example 8, and we could even go on for an arbitrary number of steps. The graph is illustrated in Figure 9, where we have renumbered the nodes for obvious reasons. Instead of merely 4 nodes, as in example 8, we have $4N$ nodes, where N is a positive integer. We maintain the condition $|c_i| < 1$, except for the last one, for which we assume $|c_{N+1}| = 1$, and $d_k = \sqrt{1 - |c_k|^2}$, $k \leq N$.

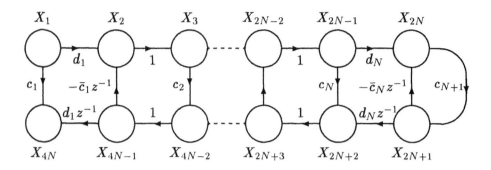

Figure 9.

It follows immediately from the above considerations that the transfer function is equal to the terminating continued fraction of the type above, only with $1/c_{N+1}$ as the last partial fraction. From the product representation in Subsection *2.1* it follows that the poles of the transfer function all are located in the open unit disk, and from Theorem 1 it follows that the filter is stable. These results deserve the status of a theorem:

Theorem 3 *Let F be a digital filter represented by a directed graph of the form shown in Figure 9 with $|c_k| < 1$ for $k = 1, 2, \ldots, N$ and $|c_{N+1}| = 1$, and let $h_N(z)$ be the transfer function. Then the filter is stable, and the transfer function is given by*

$$h_N(z) = c_1 + \cfrac{(1 - |c_1|^2)z^{-1}}{\bar{c}_1 z^{-1}} \cfrac{1}{+ c_2 + \cdots +} \cfrac{(1 - |c_N|^2)z^{-1}}{\bar{c}_N z^{-1}} \cfrac{1}{+ c_{N+1}} . \quad (2.2.1)$$

This result, due to Jones and Steinhardt [JoSt82], may seem rather special, since the graph is the special one in Figure 9. But the next theorem, also due to Jones and Steinhardt [JoSt82], and being "an almost converse" of Theorem 3, shows that it is of much more general interest that it may look like.

Theorem 4 *Let F be an arbitrary stable digital filter, and let $h(z)$ denote its transfer function. Then one of the following holds:*

a) $h(z) = a$ constant.

b) There exists a unique finite sequence $\{c_n\}_{n=1}^{N+1}$ with $|c_k| < 1$, $k = 1, 2, \ldots, N$, $|c_{N+1}| = 1$ and a positive β, such that

$$h(z) = \beta \left(c_1 + \frac{(1 - |c_1|^2)z^{-1}}{\bar{c}_1 z^{-1}} \frac{1}{+ c_2 + \cdots} \right.$$
$$\left. + \frac{(1 - |c_N|^2)z^{-1}}{\bar{c}_N z^{-1}} \frac{1}{+ c_{N+1}} \right).$$

c) There exists a unique sequence $\{c_n\}_{n=1}^{\infty}$ with $|c_k| < 1$, $k = 1, 2, \ldots$ and a positive β, such that

$$h(z) = \beta \cdot \lim_{N \to \infty} \left(c_1 + \frac{(1 - |c_1|^2)z^{-1}}{\bar{c}_1 z^{-1}} \frac{1}{+ c_2 + \cdots} \right.$$
$$\left. \frac{1}{+ c_N +} \frac{(1 - |c_N|^2)z^{-1}}{\bar{c}_N z^{-1}} \frac{1}{+ 1} \right).$$

Outline of proof: $h(z) = $ constant is obviously a (trivial) possibility. Else, since the filter is stable, $h(z)$ has all its poles in the (open) unit disk, and from Property (c) in Theorem 1 and the maximum principle it follows that $g^*(w) = h(1/w)$ is bounded in U. With

$$\beta = \sup_{w \in U} |g^*(w)|,$$

the function $g^*(w)/\beta$ satisfies the "Schur-conditions" in Subsection 2.1, and permits repeated use of the Schur algorithm. In case 2 this gives, by replacing w by z^{-1}, the b-case of the theorem. In case 1 the algorithm never stops, and one is led to a non-terminating continued fraction, whose approximants are all rational functions and the modified approximants

$$\beta \left(c_1 + \frac{(1 - |c_1|^2)w}{\bar{c}_1 w} \frac{1}{+ c_2 + \cdots +} \frac{(1 - |c_N|^2)w}{\bar{c}_N w} \frac{1}{+ t_N} \right),$$

where $|t_N| = 1$, are holomorphic in $|w| \leq 1$. From Theorem 5.11 in [Schur18] (normality of the sequence of modified approximants combined with correspondence) it follows that for any sequence $\{t_n\}$ with $|t_n| = 1$, in particular for $t_n = 1$, the sequence of modified approximants converges to one and the same function. This modification ties the c-case to the b-case of the theorem. The convergence is locally uniform on $|w| < 1$, and the limit function is holomorphic and of absolute value ≤ 1 for $|w| < 1$. Inserting $w = z^{-1}$ leads to the c-part of Theorem 4.

3 Model reduction

3.1 General remarks

What we have been discussing in this chapter belongs to (and is a very little part of) what is called *linear system theory*. One important practical problem in this theory is *the model reduction problem*. In vague terms it means to replace one system $\mathcal{S}$, having a rational transfer function of high degree, by another system $\mathcal{S}_0$, with a transfer function of lower degree, such that $\mathcal{S}_0$ in a relevant sense approximates $\mathcal{S}$. A proper definition of linear systems, as well as a survey of techniques for model reduction and a huge bibliography from the field is given by Bultheel and Van Barel in [BuBa86]. This paper, by the way, really illustrates the role of continued fractions and Padé approximants in this field.

We shall restrict ourselves to the reduction of transfer functions for stable digital filters, like the ones discussed earlier in this chapter, but first we shall discuss some general principles.

Assume that we have some kind of transfer function $h(z)$, and that this has a Taylor expansion

$$h(z) = g(w) = \sum_{k=0}^{\infty} g_k w^k \tag{3.1.1}$$

near $w = 0$. Here w may be $1/z$ in some cases, $z - \alpha$ for some constant α in other cases. One of the requirements for model reduction is often that as many as possible of the coefficients g_k be preserved, i.e. we are looking for some

$$\tilde{g}(w) = \sum_{k=1}^{\infty} \tilde{g}_k w^k \qquad (3.1.2)$$

where $\tilde{g}_k = g_k$ for $k = 0, 1, 2, \ldots, r-1$, and where r is as large as possible without violating other requirements (one of which is simplification compared to the original model). The most natural idea to think of is to produce, somehow, a sequence $\{g_r(w)\}$ of functions corresponding to the series for $g(w)$. Then raise the question: Can this be done by different kinds of continued fraction expansion, or by using some path in the Padé table for $g(w)$? One such path is the staircase formed by the successive approximants of a regular C-fraction (if a regular C-fraction exists).

3.2   Stable filters with rational transfer function

We now turn to the particular type of linear systems we are dealing with in this chapter, the digital filter. We shall furthermore assume that it is *stable*. Let $h(z)$ be the transfer function and $H = \{h_n\}$ its shock response,

$$h(z) = h_0 + h_1 z^{-1} + h_2 z^{-2} + \cdots, \qquad |z| > 1. \qquad (3.2.1)$$

In replacing this by a reduced model we want to preserve

a) stability

b) $h_0, h_1, \ldots, h_r$ for some r.

Example 9 Let

$$h(z) = \frac{40 + 28z^{-1} + 100z^{-2}}{100 + 28z^{-1} + 40z^{-2}}. \qquad (3.2.2)$$

This is, as we soon shall see, a transfer function for a stable filter. Its (terminating) regular C–fraction expansion is given by

$$h(z) = \frac{2}{5} + \frac{(84/500)z^{-1}}{1} \;-\; \frac{(472/100)z^{-1}}{1} \;+\; \frac{(300/59)z^{-1}}{1} \;-\; \frac{(5/59)z^{-1}}{1}.$$

If we try to approximate by using its 2. approximant $f_2(z)$ we get

$$f_2(z) = \frac{10 - 43z^{-1}}{25 - 118z^{-1}},$$

but this has a pole $z_0 = 118/25$ outside of the unit circle, and can therefore not be the transfer function of a stable filter. This is a disadvantage by using C–fractions, that you may lose the stability when you choose an approximant. Some methods based upon *Padé-type* approximants instead of Padé-approximants work in some cases.

———————————————————————————————————————◇

For stable filters we shall do it differently, but still based upon continued fractions and correspondence. The new idea is to use Schur fractions. Let the function $h(z)$ serve as the first example.

Example 10 Since the poles of $h(z)$ are complex conjugates, both with absolute value $\sqrt{2/5}$, $h(z)$ is the transfer function of a stable filter, and since it is not a constant, it can be represented in the form of Theorem 4b or c. Since it happens to be of the form

$$\frac{z^{-n} P_n(z)}{\bar{P}_n(1/z)}, \tag{3.2.3}$$

where P_n is a polynomial, we have the b-case, even with $\beta = 1$. (See (2.1.6').) In fact, if we expand $h(z)$ in a Schur fraction, we get

$$h(z) = \frac{2}{5} + \frac{(1 - \frac{4}{25})z^{-1}}{\frac{2}{5}z^{-1}} \; \frac{1}{+\frac{1}{5}+} \; \frac{(1 - \frac{1}{25})z^{-1}}{\frac{1}{5}z^{-1}} \; \frac{1}{+1}. \tag{3.2.4}$$

The graph is the one in Fig. 9, with

$$N = 2, \qquad c_1 = \frac{2}{5}, \qquad c_2 = \frac{1}{5}, \qquad c_3 = 1.$$

The idea is now to replace the tail

$$\frac{1}{5} + \frac{(1 - \frac{1}{25})z^{-1}}{\frac{1}{5}z^{-1}} \; \frac{1}{+1}$$

by 1, and get a function

$$\tilde{h}(z) = \frac{2}{5} + \frac{(1 - \frac{4}{25})z^{-1}}{\frac{2}{5}z^{-1}} \frac{1}{+1}. \tag{3.2.5}$$

This corresponds to replacing the rightmost part of Fig. 9 (with 4 nodes) by one single arrow $\tilde{c}_2 = 1$ between the two *now* rightmost nodes. We get a graph as the one in Fig. 9, but now with

$$N = 1, \qquad c_1 = \frac{2}{5}, \qquad \tilde{c}_2 = 1$$

representing a stable filter with (3.2.5) as its transfer function. We find

$$\tilde{h}(z) = \frac{\frac{2}{5} + z^{-1}}{1 + \frac{2}{5}z^{-1}} = \frac{2}{5} + \frac{21}{25}z^{-1} + \cdots,$$

which has a single pole $z_0 = -2/5$. Only the first coefficient coincides with the one for $h(z)$,

$$\tilde{h}_0 = h_0 = \frac{2}{5}.$$

———————————————————————————————— $\diamond$

This example illustrated how the reduction can be carried out for stable filters, assuming that we have a non-constant transfer function. Then, according to Theorem 4 the transfer function can be represented by using a Schur continued fraction. By truncating to get the $2m$th approximant (last partial fraction being $1/c_{m+1}$) we get a rational function corresponding to the transfer function up to and including the z^{-m}-term [JoNT86, Thm. 2.2]. In order to make sure it is a transfer function for a stable filter we replace $1/(c_{m+1})$ by $1/1$ (or $1/e^{i\alpha}$ for arbitrary real α), and get a terminating Schur fraction, where we know that the poles are in the unit disk. (The factor β has of course no influence on the poles.) The price we have to pay for this change, is that we lose the last coefficient in the correspondence, and hence we only preserve

$$h_0, h_1, \ldots, h_{m-1}.$$

(In Example 10 we used the approximant of order 2, i.e. $m = 1$ and replaced $1/c_2$ by $1/1$. We got stability, and the coefficient h_0.)

In order to make this a practically useful method we need to be able to transform the given transfer function into a constant times a Schur fraction, or at least a reasonably long start of such an expansion. One way is already described in Subsection 2.1, we only have to find $\beta = \sup_{|z|>1} h(z)$ first. If we have the special situation of 4b, which means that the transfer function is of the form

$$h(z) = \beta \cdot \frac{z^n \bar{Q}_n(\frac{1}{z})}{Q_n(z)} = \beta \frac{\bar{a}_n + \bar{a}_{n-1}z + \cdots + \bar{a}_0 z^n}{a_0 + a_1 z + \cdots + a_n z^n}, \qquad (3.2.3')$$

then it can be handled easier by the *Schur-Cohn algorithm*. It is rather easily proved from the Schur algorithm. We quote it without proof, and refer to [BuBa86] and [JoSt82]. The notation in (3.2.3') differs from the one in (3.2.3), in order to conform with [JoNT86]. In the latter a proof is included. The algorithm goes as follows:

From a given array

$$a_j^{(n)} = a_j, \qquad j = 0, 1, \ldots, n, \qquad (3.2.6a)$$

being the array of coefficients of the denominator polynomial $Q_n(z)$, we construct for $m = n, n-1, \ldots, 1$ new arrays $a_j^{(m-1)}$ in the following way:

$$\begin{cases} c_{n-m+1} = \dfrac{\bar{a}_0^{(m)}}{a_m^{(m)}}, \\[2ex] a_j^{(m-1)} = \dfrac{a_{j+1}^{(m)} - c_{n-m+1} a_{m-j-1}^{(m)}}{1 - |c_{n-m+1}|^2}, \end{cases} \qquad (3.2.6b)$$

with $j = 0, 1, \ldots, m-1$ and $c_{n+1} = \bar{a}_0^{(0)}/a_0^{(0)}$. Then the c_k's are the parameters of the terminating Schur fraction of $h(z)$. We shall illustrate this on an example.

Example 11

$$h(z) = \frac{z^4 - 16}{1 - 16z^4}.$$

We see that this has the form (3.2.3') with $\beta = 1$ and $n = 4$.

$$
\begin{array}{llll}
a_0^{(4)} = 1 & a_0^{(3)} = 0 & a_0^{(2)} = 0 \\
a_1^{(4)} = 0 & a_1^{(3)} = 0 & a_1^{(2)} = 0 \\
a_2^{(4)} = 0 & a_2^{(3)} = 0 & a_2^{(2)} = -16 \\
a_3^{(4)} = 0 & a_3^{(3)} = -16 \\
a_4^{(4)} = -16
\end{array}
$$

$$
\mathbf{c_1} = -\tfrac{1}{16} \qquad \mathbf{c_2} = 0 \qquad \mathbf{c_3} = 0
$$

$$
\begin{array}{ll}
a_0^{(1)} = 0 & a_0^{(0)} = -16 \\
a_1^{(1)} = -16
\end{array}
$$

$$
\mathbf{c_4} = 0 \qquad \mathbf{c_5} = 1
$$

In this case the Schur fraction is

$$
h(z) = -\frac{1}{16} + \frac{\frac{255}{256}z^{-1}}{-\frac{1}{16}z^{-1}+0+} \; \frac{1}{0} \; \frac{z^{-1}}{+0+} \; \frac{1}{0} \; \frac{z^{-1}}{+0+} \; \frac{1}{0} \; \frac{z^{-1}}{+1}.
$$

The cutting procedure described in Example 10 gives the following reductions of our model:

$$
h_1(z) = -\frac{1}{16} + \frac{(255/256)z^{-1}}{-(1/16)z^{-1}} \; \frac{1}{+1} = \frac{z-16}{1-16z},
$$

$$
h_2(z) = -\frac{1}{16} + \frac{(255/256)z^{-1}}{-(1/16)z^{-1}} \; \frac{1}{+0+} \; \frac{z^{-1}}{0} \; \frac{1}{+1} = \frac{z^2-16}{1-16z^2},
$$

and similarly

$$
h_3(z) = \frac{z^3 - 16}{1 - 16z^3}.
$$

$\diamond$

Problems

(1) Let f be a real-valued function of a real variable u, defined on an interval $[a, b]$, and let $a = u_0, u_1, \ldots, u_n = b$ be equally spaced u-values, and $f(u_0), f(u_1), \ldots$ the corresponding values of the function. Find a simple transformation $u = g(t)$, such that with $x(t) = f(g(t))$ we have $f(u_k) = x(k)$.

(2) Find the z-transform of the sequence $\{a_n\}_{n=0}^{\infty}$, where the sequence is given by

    (a) $a_n = (-1)^n$,

    (b) $a_n = \frac{1}{2}(1 + (-1)^n)$,

    (c) $a_n = 0$.

(3) Find the inverse z-transform of the formal series or functions in the following cases

    (a) $a(z) = \exp(1/z)$,

    (b) $a(z) = \ln(1 + 1/z)$,

    (c) $a(z) = 1 + z^{-1} + 2!z^{-2} + 3!z^{-3} + \cdots$.

(4) Find the transfer function and the shock response for the filters given for $n \geq 0$ by the recurrence relation below, where $x_m = y_m = 0$ for negative m.

    (a) $y_n = 2x_n + x_{n-1}$,

    (b) $y_n = y_{n-1} + 2x_n + x_{n-1}$,

    (c) $y_n = -y_{n-1} + 2x_n - x_{n-1}$.

(5) Draw the block diagram illustrations of the filters in Problem 4.

(6) Prove the following theorem: Let α and β be two distinct angular values, and A, B two complex numbers $\neq 0$. Then

$$\limsup_{n \to \infty} |Ae^{ni\alpha} + Be^{ni\beta}| > 0.$$

(7) A digital filter F has the transfer function

$$h(z) = \frac{z}{(z-\beta)^2} \, .$$

For which values of β is F a stable filter? Compute

$$\sum_{h=0}^{\infty} |h_n|$$

in all cases when F is stable.

(8) Let F be a filter with transfer function

$$h(z) = \frac{(z-1)^2}{(p+1+q)z^2 + 2(p-q)z + (p-1+q)} \, ,$$

where p and q are real numbers. Use Theorem 13 in Chapter X to determine for which p, q the filter is stable. (Hint: Transform the unit disk to the left half plane by

$$w = T(z) = \frac{z+1}{z-1} \, .$$

It then turns out that

$$[h(T^{-1}(w))]^{-1}$$

is a polynomial.)

(9) Use the result in problem 8 to establish a criterion for stability of filters F with transfer function

$$h(z) = \frac{(z-1)^2}{Az^2 + 2Bz + (A-2)} \, ,$$

where A and B are real mumbers. Illustrate in an A, B-plane.

(10) Write explicitly $f_0(w)$ in case 2 of Schur's algorithm when

(a) $|\gamma_0| < 1$, $|\gamma_1| = 1$.

(b) $|\gamma_0| < 1$, $|\gamma_1| < 1$, $|\gamma_2| = 1$.

(c) $|\gamma_0| < 1$, $\gamma_1 = \gamma_2 = \cdots = \gamma_k = 0$, $|\gamma_k| = 1$.

(11) Let $f_0(w)$ be holomorhic in the unit disk $|w| < 1$ and $|f_0(w)| \leq 1$ there. Assume that all γ_n produced by the Schur algorithm are $= 1/2$. (We have not established that the γ-values may be prescribed!) Determine the Schur fraction associated with this sequence of γ-values.

(12) Use results from Chapter III to prove that the continued fraction in Problem 11 converges for $|w| < 1$. Prove that its value is

$$f_0(w) = \frac{w - 1 + \sqrt{w^2 - w + 1}}{w} = \frac{1}{2} + \text{positive powers of } w.$$

Next prove that $f_1(w) = f_0(w)$, and hence that $f_0(w)$ gives rise to the sequence $\{\gamma_n\}$, where all $\gamma_n = 1/2$ when we use the Schur algorithm. ($f_1(w)$ is here the next function in the Schur algorithm, not to be confused with the approximants of the Schur fraction. We do not raise the question about $\sup |f_0(w)|$.)

(13) Find the continued fraction produced by the Schur algorithm in the following cases:

(a) $f_0(w) = 0$,
(b) $f_0(w) = c_1$ where $|c_1| < 1$,
(c) $f_0(w) = c_k w^k$ where $|c_k| < 1$,
(d) $f_0(w) = c_k w^k$ where $|c_k| = 1$.

(14) Let $h(z)$ be the transfer function called $h_1(z)$ in Subsection 2.2. Find the shock response $\{h_n\}$ of the filter. For which c_2-values is the filter stable, if c_1 is assumed to have absolute value < 1?

(15) Let F be a digital filter with transfer function

$$h(z) = \frac{1 + 2z^{-1}}{1 + \frac{1}{2}z^{-1}}.$$

Find the Schur fraction expansion of Theorem 3 for $h(z)$, and hence its sequence of c_n-values.

(16) Use the Schur-Cohn algorithm to find the Schur continued fraction of

(a) the function in Example 9.

(b) the function

$$h(z) = \frac{(z^2 - 4)(z - 2i)}{(1 - 4z^2)(1 + 2iz)}.$$

In problem 16b find also the first three model reductions as described in Example 11, and compare the start of their power series expansion (in powers of $1/z$) with the one for $h(z)$.

Remarks

(1) In Chapter X we have discussed stability of polynomials, i.e. the
    property of having all zeros in the open left half plane. A re-
    lated concept is the concept of disk-stability of a polynomial, i.e.
    the property of having all zeros in the open unit disk. There is
    an obvious connection between disk-stable polynomials and stable
    digital filters. Combination of Theorem 3 and Theorem 4b leads
    to the following result on disk-stability. Let

$$Q(z) = a_0 + a_1 z + \cdots + a_n z^n$$

be a given polynomial of degree n, i.e. $a_n \neq 0$. Then $Q(z)$ is
disk-stable if and only if

$$h(z) = \frac{z^n \bar{Q}(1/z)}{Q(z)}$$

has a Schur fraction expansion

$$h(z) = c_1 + \frac{(1 - |c_1|^2)}{\bar{c}_1} \frac{z}{+c_2 +} \frac{(1 - |c_2|^2)}{\bar{c}_2} +\cdots$$
$$\frac{z}{+c_n +} \frac{(1 - |c_n|^2)}{\bar{c}_n} \frac{z}{+c_{n+1}},$$

where $|c_j| < 1$ for $j = 1, 2, \ldots, n$, $|c_{n+1}| = 1$. (See also [JoSt82].)

(2) The positive Schur-fractions and the closely related PC-fractions
    (Perron-Carathéodory-fractions) are important in the study of the
    trigonometric moment problem, and, in connection with that, poly-
    nomials orthogonal on the unit circle (Szegö polynomials) and
    Gaussian quadrature on the unit circle. For references we refer
    to the remark section at the end of Chapter VII.

References

[Ahlf53]   L. Ahlfors, "Complex Analysis", McGraw-Hill, New York (1953).

[BuBa86]   A. Bultheel and M. van Barel, *Padé Techniques for Model Reduction in Linear System Theory: A Survey*, J. of Comp. and Appl. Math., **14** (1986), 401–438.

[Hamm77]   R. W. Hamming, "Digital Filters", Prentice-Hall Inc., Englewood Cliffs, New Jersey (1977).

[Henr86]   P. Henrici, "Applied and Computational Complexs Analysis", Vol. I, II, III, J. Wiley & Sons, New York (1974, 1977, 1986).

[JoNT86]   W. B. Jones, O. Njåstad and W. J. Thron, *Schur Fractions, Perron-Carathéodory Fractions and Szegö Polynomials, a Survey*, "Analytic Theory of Continued Fractions II" (W. J. Thron, ed.), Lecture Notes in Mathematics, No. **1199**, Springer-Verlag, Berlin (1986), 127–158.

[JoSt82]   W. B. Jones and A. Steinhardt, *Digital Filters and Continued Fractions*, "Analytic Theory of Continued Fractions" (W.B.Jones, W.J.Thron and H.Waadeland, eds.), Lecture Notes in Mathematics, No. **932**, Springer-Verlag, Berlin, (1982), 129–151.

[JoTh80]   W. B. Jones and W. J. Thron, "Continued Fractions: Analytic Theory and Applications", Encyclopedia of Mathematics and its Applications, Vol. **11**, Addison-Wesley,

519

Reading, Mass. (1980). Now distributed by Cambridge University Press, New York.

[OpSc75]   A. V. Oppenheim and R. W. Schafer, "Digital Signal Processing", Prentice-Hall Inc., Englewood Cliffs, New Jersey (1975).

[Schur18]   I. Schur, *Über Potenzreihen die im Innern des Einheitskreises beschränkt sind*, Journal für die reine und angewandte Mathematik **147** (1917), 205–232 and **148** (1918), 122-145.

[Wall48]   H. S. Wall, "Analytic Theory of Continued Fractions", Van Nostrand, New York (1948).

Chapter XII

Applications to some differential equations

About this chapter

A linear homogenous ODE of order 2 is a three-term linear relation between y, y' and y''. Under certain conditions differentiations and rearrangements will lead to three-term recurrence relations for the successive derivatives, and in turn to a continued fraction, where $\{y^{(n)}/y^{(n+1)}\}$ is a tail sequence (right or "wrong" tails). If the continued fraction converges to y/y' or equivalently: $\{y^{(n)}/y^{(n+1)}\}$ is a right tail sequence, then we have a continued fraction representation of the logarithmic derivative of a solution.

A Riccati equation has an important invariance property with respect to linear fractional transformations. This property has given rise to different continued fraction procedures for solving such equations.

There is a transformation from Riccati equations to second order linear equations and the converse. This opens up the possibility of treating second order linear equations the "Riccati way" or the other way around.

The topics of the present chapter are limited to the types of differential equations mentioned above. The emphasis is on the formal algorithmic part. For rigorous proofs that the expansions represent the solutions (by correspondence and convergence) we mostly refer to the sources. But in some cases, in order to present the underlying idea, we outline a proof. In some very special cases an "a posteriori verification" may be used to establish that the result of the procedure in fact represents the solution.

1 Second order linear equations

1.1 Introduction

Already in Chapter I we have seen an example of a connection between differential equations and continued fractions. In Subsection *2.2* of that chapter we saw how a linear second order differential equation gave rise to a continued fraction, which turned out to be a continued fraction expansion of $f'(x)/f(x)$, f being a particular solution of the differential equation. This can be done more generally:

Given a second order linear differential equation

$$y = P_0(x)y' + Q_0(x)y'', \qquad (1.1.1)$$

where $'$ means differentiation with respect to x, and where $P_0(x)$ and $Q_0(x)$ are infinitely differentiable. We differentiate and rearrange:

$$y' = P_1(x)y'' + Q_1(x)y''', \qquad (1.1.2)$$

where

$$P_1(x) = \frac{P_0(x) + Q_0'(x)}{1 - P_0'(x)}, \qquad Q_1(x) = \frac{Q_0(x)}{1 - P_0'(x)},$$

for all x where $P_0'(x) \neq 1$. We proceed and find generally

$$y^{(n)} = P_n(x)y^{(n+1)} + Q_n(x)y^{(n+2)}, \qquad (1.1.3)$$

where

$$P_n(x) = \frac{P_{n-1}(x) + Q_{n-1}'(x)}{1 - P_{n-1}'(x)}, \qquad Q_n(x) = \frac{Q_{n-1}(x)}{1 - P_{n-1}'(x)}. \qquad (1.1.4)$$

Since (if no denominator is 0)

$$\frac{y}{y'} = P_0 + \frac{Q_0}{y'/y''},$$

$$\frac{y'}{y''} = P_1 + \frac{Q_1}{y''/y'''},$$

$$\vdots$$

$$\frac{y^{(n)}}{y^{(n+1)}} = P_n + \frac{Q_n}{y^{(n+1)}/y^{(n+2)}},$$

we get the formal identity

$$\frac{y}{y'} = P_0 + \frac{Q_0}{P_1 +} \frac{Q_1}{P_2 +} \cdots + \frac{Q_n}{y^{(n+1)}/y^{(n+2)}} \,. \qquad (1.1.5)$$

This suggests to look at the continued fraction

$$\frac{1}{P_0 +} \frac{Q_0}{P_1 +} \frac{Q_1}{P_2 +} \cdots + \frac{Q_{n-1}}{P_n} + \cdots , \qquad (1.1.6)$$

for which we ask the questions: Does it converge (for the x-values we are interested in, e.g. an interval on the real axis, a domain in the complex plane)? In case of yes: Does it converge to $\tilde{y}'/\tilde{y}$ for some particular solution $\tilde{y}$ in that domain? If also the answer to the second question is yes, we easily find this $\tilde{y}$, and by putting $y = \tilde{y} \cdot u$ we find a linearly independent solution in the familiar standard way. This way of looking at the problem can be found in Perron [Perr57] who referred to A. Steen [Steen73].

If we have the very special case that (1.1.6) converges to a continuously differentable function f which is "manageable", e.g. we have a closed form for it, the second question can be answered by using the following lemma:

Lemma 1 *Assume that (1.1.6) converges to a differentiable function f with a continuous derivative f'. Then there is a solution $\tilde{y}$ of (1.1.1) with*

$$\frac{\tilde{y}'}{\tilde{y}} = f(x) \qquad (1.1.7)$$

if and only if

$$f \cdot f^{(1)} = Q_0(f^2 + f') , \qquad (1.1.8)$$

where

$$f^{(1)} = \frac{1}{f} - P_0 , \qquad (1.1.9)$$

is the value of the first tail of (1.1.6).

Proof of lemma: (1.1.8) is equivalent to

$$1 - P_0 f = Q_0(f^2 + f').\qquad(1.1.8')$$

Let $\tilde{y}$ be a solution of the differential equation

$$\frac{y'}{y} = f(x),$$

i.e.

$$\tilde{y} = Ce^{F(x)}, \quad \text{where } F'(x) = f(x).$$

Then (1.1.8) is equivalent to

$$1 - P_0 \frac{\tilde{y}'}{\tilde{y}} = Q_0 \left(\left(\frac{\tilde{y}'}{\tilde{y}}\right)^2 + \frac{\tilde{y}''\tilde{y} - (\tilde{y}')^2}{\tilde{y}^2} \right),$$

i.e. to

$$1 - P_0 \frac{\tilde{y}'}{\tilde{y}} = Q_0 \frac{\tilde{y}''}{\tilde{y}},$$

and hence to (1.1.1). ∎

We shall now illustrate the technique in some cases where the conditions of Lemma 1 turn out to be satisfied.

Example 1 If P_0 and Q_0 are constants, we are back to Example 3 in Subsection 2.2 of Chapter I. We shall write them as

$$\begin{aligned} P_0 &= -(\alpha + \beta), \\ Q_0 &= -\alpha \cdot \beta, \end{aligned}$$

and assume that α and β are complex numbers of different absolute values, $0 < |\alpha| < |\beta|$. The process described leads to the continued fraction (1.1.6) which now takes the form

$$\frac{1}{-(\alpha+\beta)+} \frac{-\alpha\beta}{-(\alpha+\beta)+} \frac{-\alpha\beta}{-(\alpha+\beta)+\cdots}.\qquad(1.1.6')$$

We know from Theorem 6 in Chapter III that the periodic continued fraction

$$\overset{\infty}{\underset{n=1}{\mathbf{K}}} \frac{-\alpha\beta}{-(\alpha+\beta)}$$

converges to $f^{(1)} = \alpha$, and hence that (1.1.6') converges to

$$f = \frac{1}{-(\alpha + \beta) + \alpha} = -\frac{1}{\beta}.$$

From this it follows that

$$f \cdot f^{(1)} = -\frac{\alpha}{\beta} = (-\alpha\beta) \cdot \frac{1}{\beta^2} = Q_0(f^2 + f'),$$

i.e. the relation (1.1.8) holds. Hence, one solution $\tilde{y}$ of

$$y = -(\alpha + \beta)y' - \alpha\beta y'' \qquad\qquad (1.1.1')$$

may be found by solving the equation

$$\frac{\tilde{y}'}{\tilde{y}} = -\frac{1}{\beta},$$

and we find

$$\tilde{y} = C_1 e^{-x/\beta}.$$

By putting, in the familiar way

$$y = e^{-x/\beta} \cdot u(x)$$

into (1.1.1'), we easily find a second, linearly independent solution

$$C_2 e^{-x/\alpha}$$

and hence the general solution

$$y = C_1 e^{-x/\beta} + C_2 e^{-x/\alpha}.$$

There is of course no reason to use this method for differential equations (1.1.1'), since the standard method, taught in elementary courses, is simpler. The purpose was only to illustrate the procedure and the test by the lemma in a familiar case.

$\diamond$

Remark: In Lemma 1 we have assumed that the continued fraction (1.1.6) converges to some f, and that $f^{(1)}$ is the value of the first tail. But the Lemma also holds if $\{g^{(n)}\}_{n=0}^{\infty}$ is any tail sequence for (1.1.6). For Example 1 we have for instance that

$$-\frac{1}{\alpha}, \beta, \beta, \beta, \beta, \ldots$$

is a "wrong" tail sequence (if we maintain the condition $0 < |\alpha| < |\beta|$). With $g = -1/\alpha$ and $g^{(1)} = \beta$ we have

$$g \cdot g^{(1)} = -\frac{\beta}{\alpha} = -\alpha\beta \cdot \frac{1}{\alpha^2} = Q_0(g^2 + g'),$$

and we find a particular solution of (1.1.1') by solving

$$\frac{y'}{y} = -\frac{1}{\alpha},$$

i.e. we find the particular solution

$$y = C_2 e^{-x/\alpha},$$

and together with what we did with the right tail sequence we get the general solution.

We do not even need convergence of (1.1.6). It suffices to have tail sequences with smooth elements satisfying the lemma. Taking again Example 1 as an illustration, but now with $0 < |\alpha| = |\beta|, \alpha \neq \beta$, the continued fraction (1.1.6) diverges, but it still has

$$-\frac{1}{\alpha}, \beta, \beta, \beta, \beta, \ldots \quad \text{and} \quad -\frac{1}{\beta}, \alpha, \alpha, \alpha, \alpha, \ldots$$

as tail sequences, (where both are wrong tail sequences). Since in both cases the equality (1.1.8) holds (with $(f, f^{(1)})$ replaced by $(g, g^{(1)})$), we find two linearly independent solutions just as before. (It is easy to prove, that no other tail sequences than the two mentioned above are such that the equality (1.1.8) in Lemma 1 holds.) As an example, take $\alpha = i, \beta = -i$ in Example 1. The differential equation takes the form

$$y = -y''.$$

The continued fraction (1.1.6') is in this case

$$\cfrac{1}{0+} \cfrac{-1}{0+} \cfrac{-1}{0+} \cfrac{-1}{0+} \cdots.$$

It diverges, but the two "useful" tail sequences are

$$i, -i, -i, -i, \ldots$$

and

$$-i, i, i, i, \ldots.$$

The general solution is

$$y = C_1 e^{ix} + C_2 e^{-ix} (= \tilde{C}_1 \cos x + \tilde{C}_2 \sin x).$$

The next example is also a familiar one from elementary courses. In this case we restrict ourselves to real equations and look for real solutions.

The differential equation

$$y = p_0 x y' + q_0 x^2 y'', \qquad x > 0, \qquad p_0, q_0 \in \mathbf{R}, \qquad q_0 \neq 0,$$

is an example of a differential equation with a regular singular point at the origin, actually the Euler-Cauchy equation. The first step in the standard solution procedure is to insert $y = x^r$ and solve the algebraic equation for r. Another method is to transform it to an equation with constant coefficients. But we shall here use the continued fraction technique just described.

Example 2 Given the differential equation

$$y = -xy' - 6x^2 y'', \qquad x > 0.$$

We shall see what happens if we use successive differentiations and the continued fraction method:

$$y' = -y' - xy'' - 12xy'' - 6x^2 y''',$$

i.e.

$$y' = -\frac{13}{2} x y'' - 3x^2 y'''.$$

This is again an Euler-Cauchy equation, this time for y' instead of for y, and we proceed:

$$y'' = -\frac{5}{3}xy''' - \frac{2}{5}x^2 y^{(4)}$$

$$y''' = -\frac{37}{40}xy^{(4)} - \frac{3}{20}x^2 y^{(5)}$$

$$\vdots$$

From

$$\frac{y}{y'} = -x - \frac{6x^2}{y'/y''}$$

$$\frac{y'}{y''} = -\frac{13}{2}x - \frac{3x^2}{y''/y'''}$$

$$\frac{y''}{y'''} = -\frac{5}{3}x - \frac{(2/5)x^2}{y'''/y^{(4)}}$$

$$\frac{y'''}{y^{(4)}} = -\frac{37}{40}x - \frac{(3/20)x^2}{y^{(4)}/y^{(5)}}$$

we are led to a continued fraction

$$\frac{1}{-x} + \frac{-6x^2}{(-13/2)} + \frac{-3x^2}{(-5/3)} + \frac{(-2/5)x^2}{(-37/40)} + \cdots$$

somehow associated to y'/y. A natural thing to try now is to establish general formulas for the elements, and from there on try to find out whether or not the continued fraction converges, and to what. Here we shall follow another path. With p_n and q_n as in

$$y^{(n)} = p_n x y^{(n+1)} + q_n x^2 y^{(n+2)}, \qquad x > 0, \qquad n \geq 0,$$

the continued fraction is of the form

$$\frac{1}{p_0 x} + \frac{q_0 x^2}{p_1 x} + \frac{q_1 x^2}{p_2 x} + \frac{q_2 x^2}{p_3 x} + \cdots.$$

By an equivalence transformation (divisions by x) we get the continued fraction

$$\frac{1/x}{p_0} + \frac{q_0}{p_1} + \frac{q_1}{p_2} + \frac{q_2}{p_3} + \cdots,$$

which, apart from the start, is independent of x. In case of convergence it converges to r/x for some $r \in \mathbf{R}$. In any case we try out

$$f = \frac{r}{x}, \quad f^{(1)} = \frac{x}{r} - p_0 x$$

in Lemma 1. We find

$$f \cdot f^{(1)} = \frac{r}{x} \left(\frac{x}{r} - p_0 x \right) = 1 - r p_0 = 1 + r$$

and

$$Q_0(f^2 + f') = q_0 x^2 \left(\frac{r^2}{x^2} - \frac{r}{x^2} \right) = q_0 r(r - 1) = -6r(r - 1).$$

Hence $y'/y = r/x$ for some solution of the Euler-Cauchy equation in this example, if and only if r is such that

$$-6r(r - 1) = 1 + r,$$

i.e. $r = \frac{1}{2}$ or $r = \frac{1}{3}$, and we thus find, from the equations

$$\frac{y'}{y} = \frac{\frac{1}{2}}{x} \quad \text{and} \quad \frac{y'}{y} = \frac{\frac{1}{3}}{x}$$

the two linearly independent solutions $x^{1/3}$ and $x^{1/2}$, and thus the general solution

$$y = C_1 x^{1/2} + C_2 x^{1/3}.$$

Generally we are led to the algebraic equation

$$q_0 r(r - 1) = 1 - p_0 r,$$

i.e. the same equation as in the standard method for solving Euler-Cauchy equations.

————————————————————————————◇

Example 2 shows that the continued fraction method used on Euler-Cauchy equations merely leads to the standard method of solving such equations, and that there is no need to make a "continued fraction detour". The purpose of Example 2 is on one hand to give another illustration of what the continued fraction looks like in a familiar case, on the other hand to prepare for a not so familiar case in Subsection *1.2.*

1.2 An "almost" Euler-Cauchy equation

In the present subsection we shall illustrate the use of the continued fraction method on a special case of differential equations

$$y = (p_0 + bx^s)xy' + (q_0 + \beta x^s)x^2 y'', \qquad x > 0, \qquad (1.2.1)$$

where p_0, q_0, b, β and s are real. We shall first rewrite the equation and also impose some restrictions. Observe first that if $b = \beta = 0$ we are back to the Euler-Cauchy equation. If the algebraic equation

$$1 = p_0 r + q_0 r(r - 1) \qquad (1.2.2)$$

has two distinct real roots r_1, r_2, the general solution of the Euler-Cauchy equation is $C_1 x^{r_1} + C_2 x^{r_2}$. We shall here assume this to be the case. We shall furthermore express p_0 and q_0 in terms of r_1 and r_2. We rewrite (1.2.2) (note that $q_0 \neq 0$ under our assuptions):

$$r^2 - \left(1 - \frac{p_0}{q_0}\right) r - \frac{1}{q_0} = 0. \qquad (1.2.2')$$

We have

$$r_1 + r_2 = 1 - \frac{p_0}{q_0},$$

$$r_1 r_2 = -\frac{1}{q_0},$$

from which it follows that

$$p_0 = \frac{r_1 + r_2 - 1}{r_1 r_2}, \qquad q_0 = -\frac{1}{r_1 r_2}.$$

The restrictions to be imposed are (in addition to the assumptions on r_1 and r_2)

$$s = -r_2, \qquad \beta = -\frac{b}{r_1 - 1},$$

where r_1 is required to be $\neq 1$.

The differential equations to be studied are thus of the form

$$y = \left(\frac{1 + s - r}{rs} + bx^s\right) xy' + \left(\frac{1}{rs} + \frac{b}{1 - r} x^s\right) x^2 y'', \qquad x > 0, \quad (1.2.1')$$

where r and s are real numbers, both $\neq 0$, $r \neq 1$ and $r + s \neq 0$, and where b is a real number. (We have replaced r_2 by $-s$ and r_1 by r.) Later we will put on some further restrictions on r and s. Keep in mind that for $b = 0$ the equation (1.2.1') reduces to the Euler-Cauchy equation with the general solution

$$y = C_1 x^r + C_2 x^{-s}, \qquad x > 0.$$

We find from (1.2.1') that

$$\frac{y}{xy'} = \left(\frac{1 + s - r}{rs} + bx^s \right) + \frac{\left(\dfrac{1}{rs} + \dfrac{b}{1 - r} x^s \right)}{\dfrac{y'}{xy''}}, \qquad (1.2.3)$$

provided that no dominator is 0. We differentiate (1.2.1') and rearrange:

$$\left[\frac{(r-1)(s+1)}{rs} - b(s+1)x^s \right] y' = \left[\frac{3+s-r}{rs} x + b\frac{3+s-r}{1-r} x^{s+1} \right] y''$$
$$+ \left(\frac{x^2}{rs} + \frac{b}{1-r} x^{s+2} \right) y'''.$$

By cancelling the factor

$$\left(\frac{1}{rs} + \frac{b}{1 - r} x^s \right)$$

we find

$$(r - 1)(s + 1)y' = (3 + s - r)xy'' + x^2 y''',$$

where we assume (in addition to earlier assumptions) that $s \neq -1$. This is an Euler–Cauchy equation for y'. We have

$$\frac{y'}{xy''} = \frac{3 + s - r}{(r - 1)(s + 1)} + \frac{\dfrac{1}{(r - 1)(s + 1)}}{\dfrac{y''}{xy'''}}. \qquad (1.2.4)$$

Since we have reached an Euler-Cauchy equation we could proceed as in Example 2.

But it is just as easy to go on directly: From (1.2.3) and (1.2.4) and the way it must continue, we find that the continued fraction (1.1.6) (i.e. the one associated with y'/y) in this case has the form

$$\frac{1}{\left(\frac{1+s-r}{rs}+bx^s\right)x} + \frac{\left(\frac{1}{rs}+\frac{b}{1-r}x^s\right)x}{\mathbf{K}\left(\frac{c_n}{d_n}\right)}, \qquad (1.2.5)$$

where c_n and d_n are independent of x. If it converges, it must be to

$$f = \frac{1}{\left[\left(\frac{1+s-r}{rs}+bx^s\right)+\left(\frac{1}{rs}+\frac{b}{1-r}x^s\right)c\right]x} \qquad (1.2.6)$$

for some constant $c \in \hat{\mathbf{C}}$. We assume that $c \neq \infty$ and use Lemma 1, with f as in (1.2.6) and thus with

$$f^{(1)} = \left(\frac{1}{rs}+\frac{b}{1-r}x^s\right)x \cdot c .$$

We find that

$$
\begin{aligned}
f \cdot f^{(1)} &= \frac{\left(\frac{1}{rs}+\frac{b}{1-r}x^s\right)c}{\left(\frac{1+s-r}{rs}+bx^s\right)+\left(\frac{1}{rs}+\frac{b}{1-r}x^s\right)c} \\
&= \frac{\left(\frac{1}{rs}+\frac{b}{1-r}x^s\right)c}{\frac{1+s-r+c}{rs}+b\left(1+\frac{c}{1-r}\right)x^s}, \qquad (1.2.7)
\end{aligned}
$$

and

$$Q_0(f^2+f') = \left(\frac{1}{rs}+\frac{b}{1-r}x^s\right)x^2 .$$

$$\cdot \frac{1-\frac{1+s-r+c}{rs}-b(s+1)\left(1+\frac{c}{1-r}\right)x^s}{\left[\frac{1+s-r+c}{rs}+b\left(1+\frac{c}{1-r}\right)x^s\right]^2 x^2} \qquad (1.2.8)$$

The expressions in (1.2.7) and (1.2.8) are equal if and only if

$$c = -(s+1) \quad \text{or } c = r-1 . \qquad (1.2.9)$$

From Lemma 1 we know that we get solutions of (1.2.1') by taking y'/y equal to (1.2.6) with either one of the c-values (1.2.9).

$\underline{c = -(s+1)}$ gives

$$\frac{y'}{y} = \frac{-1}{\left[\frac{1}{s} + \frac{b(r+s)}{1-r}x^s\right]x} = \frac{-s}{x\left[1 + \frac{b(r+s)s}{1-r}x^s\right]}$$

$$= -\frac{s}{x} + \frac{\frac{b(r+s)s^2}{1-r}x^{s-1}}{1 + \frac{b(r+s)s}{1-r}x^s},$$

and hence, since $x > 0$,

$$y = C_1 x^{-s}\left(1 + \frac{b(r+s)s}{1-r}x^s\right) = C_1\left(x^{-s} + \frac{b(r+s)s}{1-r}\right).$$

$\underline{c = r-1}$ gives

$$\frac{y'}{y} = \frac{r}{x}, \quad \text{and hence} \quad y = C_2 x^r.$$

By combining the two solutions we find the general solution of the differential equation (1.2.1') for $x > 0$:

$$y = C_1\left(x^{-s} + \frac{b(r+s)s}{1-r}\right) + C_2 x^r. \tag{1.2.10}$$

Example 3 We shall solve the differential equation

$$y = 2dx^{-1}y' + \left(\frac{x^2}{2} + d\right)y'',$$

where d is a real number $\neq 0$.

Simple verification shows us that this is an "almost" Euler-Cauchy equation in the meaning just described, with $r = -1$, $s = -2$ and $b = 2d$:

$$y = (0 + 2dx^{-2})xy' + (\frac{1}{2} + dx^{-2})x^2y''.$$

According to (1.2.10) the general solution is

$$y = C_1(x^2 + 6d) + C_2 x^{-1}.$$

$\diamond$

1.3   Two further examples

In the examples we have seen so far the continued fraction produced by the method led to, by convergence or by some skilled or lucky choice of the zeroth tail value, a closed form of y'/y for two linearly independent solutions of the differential equation. This is of course very special. Moreover, in many such cases, as seen by the equations with constant coefficients, by the Euler-Cauchy equation and by the "almost" Euler-Cauchy equation, the continued fraction method is in fact dispensable, except for being illustrations of how the continued fraction method works in familiar cases.

Of *practical* value is the continued fraction method only in such cases when we, on the basis of properties of the given differential equation, can come up with a statement about the solution, related to the continued fraction (1.1.6). Such a statement may for instance be that the continued fraction (1.1.6) converges to y'/y for some particular solution y of the given differential equation. (Knowledge of a possible closed form is not taken into account.) Theorems of this type would be related to theorems on power series solutions of linear second order ODE's.

The continued fraction method described in this section has been discussed and applied in many different cases, se e.g. [Khov63] and [Steen73] and the references therein. But in most cases the method has been carried out formally, with little or no attention paid to questions of justification. Several things could be done – and ought to be done. We shall not go into that here, but in the next example we shall include some remarks containing the relevant key words.

Example 4 The differential equation

$$4xy'' + 2y' + y = 0$$

has a regular singular point at $x = 0$. It can be solved by using the Frobenius method, which is to put

$$y = x^s(A_0 + A_1x + A_2x^2 + \ldots)$$

into the differential equation and determine s and the coefficients A_n. In the present case we find for s the equation

$$4s(s-1) + 2s = 0,$$

with the solution $s = 0$ and $s = \frac{1}{2}$, for which we get two linearly independent formal solutions

$$y = A_0 + A_1 x + A_2 x^2 + \ldots, \qquad A_0 \neq 0,$$
$$y = x^{\frac{1}{2}}(B_0 + B_1 x + B_2 x^2 + \ldots), \qquad B_0 \neq 0.$$

But they are more than just formal solutions. The given differential equation has the form

$$x^2 y'' + x \cdot P(x) y' + Q(x) y = 0,$$

where $P(x) = \frac{1}{2}, Q(x) = \frac{x}{4}$. If P and Q are analytic in $|x| < R$ in the complex plane, the series $\sum_{n=0}^{\infty} A_n x^n$ and $\sum_{n=0}^{\infty} B_n x^n$ both represent functions, analytic in $|x| < R$. In the present case P and Q are entire functions, and the same will be the case for the two series. The two particular integrals expressed by means of the two series, hereafter denoted y_A and y_B, are analytic in the plane and the cut plane (cut along the negative real axis) respectively. For $A_0 \neq 0$, y_A'/y_A is meromorphic in the whole plane and holomorphic in a neighborhood of the origin. For $B_0 \neq 0$, y_B'/y_B is meromorphic in the whole plane. It has a pole of order 1 at the origin.

After these preliminary remarks we switch over to the continued fraction method. Successive differentiations give

$$
\begin{array}{llll}
4xy'' & +2y' & +y & = 0 \\
4xy''' & +6y'' & +y' & = 0 \\
\cdots \\
4xy^{(n+2)} & +(4n+2)y^{(n+1)} & +y^{(n)} & = 0.
\end{array}
$$

We find

$$\frac{y}{y'} = -2 - 4x\frac{y''}{y'},$$

or

$$\frac{y'}{y} = \frac{-1/2}{1 + 2x(y''/y')},$$

and generally

$$\frac{y^{(n+1)}}{y^{(n)}} = \frac{-1/2}{2n + 1 + 2x(y^{(n+2)}/y^{(n+1)})}.$$

We are thus led to the continued fraction

$$\frac{-1/2}{1} + \frac{-x}{3} + \frac{-x}{5} + \frac{-x}{7} + \cdots.$$

Since it is equivalent to the continued fraction

$$\frac{-1/2}{1} - \frac{x/(1 \cdot 3)}{1} - \frac{x/(3 \cdot 5)}{1} - \cdots - \frac{x/(4n^2 - 1)}{1} - \cdots,$$

where $x/(4n^2 - 1) \to 0$ as $n \to \infty$, we know by Example 1 in Chapter II that it converges to a meromorphic function in the whole plane, and by Worpitzky's theorem in Chapter I that the convergence is uniform in some closed disk around the origin. Furthermore, it corresponds to a power series

$$-\frac{1}{2} - \frac{x}{6} - \cdots$$

with the same start as the Taylor expansion at 0 of y'_A/y_A. From the way the continued fraction is constructed it follows that it actually corresponds to the expansion of y'_A/y_A. From the correspondence- and convergence-properties it follows that

$$\frac{y'_A}{y_A} = \frac{-\frac{1}{2}}{1} + \frac{-x}{3} + \frac{-x}{5} + \frac{-x}{7} + \cdots$$

in the whole complex plane.

Observe that we have used the knowledge of existence and convergence of series representations of solutions, but not the series themselves. We furthermore used convergence (uniform) of the continued fraction to "something" and correspondence to the "right thing", and were able to conclude that $y'/y =$ the continued fraction, without having to depend upon a possible closed form of the continued fraction.

In some cases, for instance in using second order equations in order to solve Riccati equations, this (i.e. the logarithmic derivative of a solution) is actually what we need. In other cases we are looking for the general solution of the second order equation, in which case a continued fraction representation is not always the best starting point for further progress. (To connect it to a power series is a possibility, but in that case a very natural question is to ask if it had not been just as good to use a power series method from the very beginning.)

Another use of the method is for finding continued fraction expansions of functions. For such applications knowledge of a solution is needed. Sometimes this can be carried out by constructing a differential equation with a prescribed solution. As an illustration we shall see how the present example can produce a continued fraction expansion for the function $\tan u$. If

$$y = \cos \sqrt{x}\,, \qquad x \in \mathbf{C}\,,$$

then

$$2\sqrt{x}\,y' = -\sin\sqrt{x}\,,$$

and by differentiation

$$\frac{y'}{\sqrt{x}} + 2\sqrt{x}\,y'' = -\frac{\cos\sqrt{x}}{2\sqrt{x}} = \frac{-y}{2\sqrt{x}}\,,$$

and hence

$$4xy'' + 2y' + y = 0\,,$$

i.e. $y = \cos\sqrt{x}$ is a solution of the differential equation we just have studied. Actually we have

$$y_A = A_0 \cos\sqrt{x}\,,$$

and consequently

$$-\frac{\tan\sqrt{x}}{2\sqrt{x}} = \frac{-\tfrac{1}{2}}{1} + \frac{-x}{3} + \frac{-x}{5} + \frac{-x}{7} + \cdots$$

and we find

$$\tan u = \frac{u}{1+} \; \frac{-u^2}{3} + \frac{-u^2}{5} + \frac{-u^2}{7} + \cdots \tag{1.3.1}$$

in the whole complex plane.

◇

We conclude this section by showing another example of how a function satisfying a differential equation can be expanded in a continued fraction.

Example 5 The Bessel function $J_\nu(z)$ of the first kind of order ν is defined by

$$J_\nu(z) = \left(\frac{z}{2}\right)^\nu \sum_{k=0}^\infty \frac{(-1)^k \left(\frac{z}{2}\right)^{2k}}{k!\,\Gamma(\nu + k + 1)}, \qquad \nu \in \mathbf{C}.$$

It converges for all $z \in \mathbf{C}$, and is a solution of the Bessel equation

$$w'' + \frac{1}{z}w' + \left(1 - \frac{\nu^2}{z^2}\right)w = 0.$$

Since

$$\begin{aligned}
J_\nu'(z) &= \nu \left(\frac{z}{2}\right)^{\nu-1} \frac{1}{2} \sum_{k=0}^\infty \frac{(-1)^k (z/2)^{2k}}{k!\Gamma(\nu + k + 1)} \\
&\quad + \left(\frac{z}{2}\right)^\nu \sum_{k=0}^\infty \frac{(-1)^k 2k(z/2)^{2k-1}}{k!\Gamma(\nu + k + 1)} \cdot \frac{1}{2} \\
&= \nu \cdot \frac{2}{z} \cdot \frac{1}{2} J_\nu(z) - \left(\frac{z}{2}\right)^{\nu+1} \sum_{k=1}^\infty \frac{(-1)^{k-1}(z/2)^{2k-2}}{(k-1)!\Gamma(\nu + k + 1)} \\
&= \frac{\nu}{z} J_\nu(z) - J_{\nu+1}(z),
\end{aligned}$$

and thus

$$\begin{aligned}
J_\nu''(z) &= -\frac{\nu}{z^2} J_\nu(z) + \frac{\nu}{z} J_\nu'(z) - J_{\nu+1}'(z) \\
&= \frac{\nu^2 - \nu}{z^2} J_\nu(z) - \frac{2\nu + 1}{z} J_{\nu+1}(z) + J_{\nu+2}(z),
\end{aligned}$$

the differential equation may be written

$$J_\nu(z) = \frac{2(\nu + 1)}{z} J_{\nu+1}(z) - J_{\nu+2}(z),$$

which leads to the continued fraction expansion

$$\cfrac{z}{2(\nu+1) -} \cfrac{z^2}{2(\nu+2) -} \cfrac{z^2}{2(\nu+3) -} \cdots \qquad (1.3.2)$$

for $J_{\nu+1}(z)/J_\nu(z)$. (See Subsection *2.4* in Chapter VI.) This continued fraction was also known to Bessel himself.

◇

2 Riccati equations

2.1 General Remarks

A Riccati differential equation is a first order ODE of the form

$$y' = a_0(x) + a_1(x)y + a_2(x)y^2,\tag{2.1.1}$$

where a_0, a_1 and a_2 satisfy certain smoothness conditions on some real interval or in some domain in the complex plane. In the history of continued fractions the Riccati equation has attracted a lot of attention. Already Euler was interested in the connection between the Riccati equation and continued fractions. Later, many authors have devoted time and effort to the Riccati equation, for instance Worpitzky, to name but one example. Actually, the very first mathematical paper we know of from him [Worp62], dealt with a continued fraction expansion of a solution of a Riccati equation. There are several reasons for this great interest. First of all: The Riccati equations are among the very simplest non-linear ODE's, and they appear in applications, such as general relativity theory, system theory and acoustics. Next: They are, as we soon shall see, closely related to second order linear ODE, and finally: They have a certain invariance property with respect to linear fractional transformations, a property that makes them fit for continued fraction types of solutions.

In (2.1.1) we may assume that $a_2(x)$ is not identically 0 in the interval or domain we are interested in. Restricting ourselves to an interval or domain where $a_2(x) \neq 0$, we get, if a_2 is differentiable, that the substitution

$$y(x) = -\frac{u(x)}{a_2(x)}$$

leads to a Riccati equation for u, where the coefficient of u^2 is -1. The verification of this is left to the reader. Essentially without loss of generality we may therefore in the following assume that the Riccati equation is of the form

$$y' = a_0(x) + a_1(x)y - y^2.\tag{2.1.2}$$

The way to get from a Riccati equation (2.1.2) to a second order linear equation is to replace

$$a_0(x) \quad \text{by} \quad -\frac{f_0(x)}{f_2(x)}$$

and

$$a_1(x) \quad \text{by} \quad -\frac{f_1(x)}{f_2(x)}$$

for some "good" choices of f_0, f_1, f_2, and to replace

$$y \quad \text{by} \quad \frac{w'}{w}.$$

We then get

$$f_2(x)w'' + f_1(x)w' + f_0(x)w = 0. \tag{2.1.3}$$

We get from (2.1.3) to (2.1.2) by the opposite substitutions. We have the word "good' undefined, but the idea is of course to make a substitution leading to a differential equation we can handle.

We shall illustrate the transformation from Riccati equations to second order linear equations in two examples.

Example 6 The equation

$$y' = -1 + 2y - y^2$$

is a Riccati equation. (It is also a separable differential equation, with the general solution

$$y = 1 + \frac{1}{x + C}, \quad \text{including } y = 1.)$$

We want to transform it to a second order linear homogenous differential equation. Using the above notation, we put

$$1 = \frac{f_0(x)}{f_2(x)}, \qquad 2 = -\frac{f_1(x)}{f_2(x)}.$$

One way is to take $f_2(x) = 1$, $f_0(x) = 1$ and $f_1(x) = -2$. This gives the linear equation

$$w'' - 2w' + w = 0.$$

The general solution of this is

$$w = (C_1 + C_2 x)e^x .$$

From this it follows, when C_1 and C_2 do not both vanish:

$$y = w'/w = 1 + \frac{C_2}{C_1 + C_2 x} ,$$

which is equal to 1 for $C_2 = 0$ and to $1 + 1/(x + C_1/C_2)$ for $C_2 \neq 0$, just as expected.

$\diamond$

Example 7 Given the Riccati equation

$$y' = -1 + \frac{1}{x^2} - \frac{1}{x}y - y^2 .$$

We want this transformed to a second order homogenous ODE. With notation from above we put

$$-\frac{f_0(x)}{f_2(x)} = -1 + \frac{1}{x^2} , \qquad -\frac{f_1(x)}{f_2(x)} = -\frac{1}{x} .$$

Take $f_0(x) = 1 - 1/x^2$, $f_1 = 1/x$, $f_2(x) = 1$. We then get the differential equation

$$w'' + \frac{1}{x}w' + \left(1 - \frac{1}{x^2}\right) w = 0$$

which is the Bessel equation of index 1. (See Example 5.)

$\diamond$

We now consider the invariance property: Given a Riccati equation

$$y' = a_0(x) + a_1(x)y + a_2(x)y^2 . \tag{2.1.4}$$

By introducing the new dependent variable w given by

$$y = \frac{\alpha(x)w + \beta(x)}{\gamma(x)w + \delta(x)} , \tag{2.1.5}$$

(with obvious smoothness and non-singularity conditions on α, β, γ, δ), the equation (2.1.4) is transformed to a new Riccati equation

$$w' = \tilde{a}_0(x) + \tilde{a}_1(x)w + \tilde{a}_2(x)w^2 , \qquad (2.1.6)$$

where $\tilde{a}_0$, $\tilde{a}_1$, $\tilde{a}_2$ are easy to compute (Problem 2). The crucial thing is that the ww'-terms cancel.

For the history of continued fraction solutions to Riccati equations we refer to [Khov63], and more updated to [Coop88]. See also the remark section at the end of this chapter. Let us here merely mention some steps from the *early* history.

Lagrange had proposed the following method for solving differential equations by using continued fractions: For a given differential equation let y be "near" ξ_0 when $|x|$ is small. Write

$$y = \frac{\xi_0}{1 + y_1} \qquad (2.1.7)$$

and insert this into the differential equation. If y_1 is "near" ξ_1 we repeat the process, and if it can be repeated without stop (or terminates), we are led to a continued fraction expansion

$$\mathop{\mathbf{K}}_{a=1}^{\infty} \frac{\xi_n}{1} . \qquad (2.1.8)$$

The terms ξ_n have (in the cases studied) mostly been of the form $a_n x^{q_n}$, where $a \in \mathbf{C}$ and $q_n \geq 0$.

This idea can not be expected to work generally. But for Riccati equations we know at least that the new equation also is a Riccati equation, so in the present section we shall concentrate on this type of a differential equation. Euler and Lagrange treated differential equations of the form

$$(\alpha + \alpha' x)xy' + (\beta + \beta' x)y + \gamma y^2 = \delta x , \qquad (2.1.9)$$

where α, α', β, β', γ and δ are constants. Euler also applied Lagrange's idea to

$$\alpha x y' + \beta y + y^2 = x^k , \quad k \in \mathbf{N}, \quad y(0) = -\beta, \quad \frac{\beta}{\alpha} \notin \mathbf{N}, \qquad (2.1.10)$$

to find the solution

$$y = -\beta + \mathop{\mathbf{K}}_{n=1}^{\infty} \frac{x^k}{nk\alpha - \beta}. \tag{2.1.11}$$

Worpitzky [Worp62] was briefly mentioned in the introduction. He studied the Riccati equation

$$y' + y^2 = ax^{m-2} \tag{2.1.12}$$

and found the solution

$$xy = 1 + \mathop{\mathbf{K}}_{k=1}^{\infty} \frac{ax^m}{km + 1}. \tag{2.1.13}$$

2.2   An old example

The "technical part", i.e. the part where some formal continued fraction for the solution is created, depends upon a "good" choice of a linear fractional transformation, such that the new Riccati equation can be transformed again in a reasonably simple way. In fortunate cases one can see a pattern. The first example is closely related to equation (2.1.10), but not directly a special case of it.

Example 8 Let k be a real number. We study the Riccati equation

$$xy' + ky + y^2 = -x^2, \qquad y(0) = 0. \tag{2.2.1}$$

We try, heuristically, to find a "good" transformation: Insert the power series expansion

$$y = ax + bx^2 + \cdots.$$

The left-hand side then takes the form

$$(k+1)ax + (b(k+2) + a^2)x^2 + \cdots.$$

If k is not a negative integer, we find that $a = 0$ and $b = -1/(k+2)$. Then it is natural to try $\xi_0 = -x^2/(k+2)$ in (2.1.7), and thus (*almost* following Lagrange)

$$y = \frac{-x^2}{k+2+z} \left(= \frac{-x^2}{k+2} + \frac{x^2 \cdot z(x)}{(k+2)^2} - \frac{x^2 [z(x)]^2}{(k+2)^3} + \cdots \right),$$

where (by studying the power series expansion) we have $z(0) = 0$.

We find, by putting this into the given Riccati equation:

$$\frac{-2x^2(k+2+z) + x^3 z'}{(k+2+z)^2} - \frac{kx^2(k+2+z)}{(k+2+z)^2} + \frac{x^4}{(k+2+z)^2} = -x^2 .$$

After multiplication by $x^{-2}(k+2+z)^2$ and rearranging terms (out of which several cancel) we get

$$xz' + (k+2)z + z^2 = -x^2 , \qquad z(0) = 0 .$$

This is a Riccati equation of exactly the same type as the one we started with, only with k replaced by $k+2$. In the same manner we can use the transformation

$$z = \frac{-x^2}{k+4+u}$$

and get the Riccati equation

$$xu' + (k+4)u + u^2 = -x^2 , \qquad u(0) = 0 ,$$

and so on. If we, instead of using z, u, etc. use the notation y_1, y_2, $\ldots$, we find by repeated use of these transformations that

$$y = \frac{-x^2}{k+2+} \frac{-x^2}{k+4+} \cdots + \frac{-x^2}{k+2n+y_n} .$$

(We recall that k is assumed not to be a negative integer). From this equality one is led to the function defined by the continued fraction

$$y = \underset{n=1}{\overset{\infty}{\mathbf{K}}} \frac{-x^2}{k+2n} . \tag{2.2.2}$$

We shall outline the proof that this in fact is the solution with $y(0) = 0$:

a) The continued fraction converges in the whole plane to a meromorphic function, holomorphic in some neighborhood of $x = 0$ and with $y(0) = 0$. In order to see this one can replace $-x^2$ by ξ, and by an equivalence transformation we get the regular C-fraction

$$\frac{\frac{\xi}{k+2}}{1} + \frac{\frac{\xi}{(k+2)(k+4)}}{1} + \frac{\frac{\xi}{(k+4)(k+6)}}{1} + \cdots ,$$

whose properties can be established as for the continued fraction in Example 4.

b) The continued fraction corresponds to a formal power series solution with $y(0) = 0$ of the differential equation: The formal power series solution starts with the same term $-x^2/(k+2)$ as the series corresponding to the continued fraction. Moving from the original differential equation to the next one (for z) we have the same situation, with $-x^2/(k+4)$ as the first term, and similarly for u and all subsequent equations. (Every time we jump to the next tail of the continued fraction.) The correspondence is an immediate consequence of this.

c) The formal power series solution represents a holomorphic function in a neighborhood of the origin, since the continued fraction has this property, and is thus not only a formal solution, but a solution. Hence the continued fraction is a solution.

The outline is closely related to considerations in Example 4. Note also that a comparison between (2.2.2) and the continued fraction (1.3.2) for a ratio of Bessel functions leads to an expression for y in terms of such Bessel functions.

$\diamondsuit$

Example 9 We can use the previous example to re-establish the expansion of $\tan x$ which we found in Example 4.

The differential equation

$$y' = 1 + y^2, \qquad y(0) = 0, \tag{2.2.3}$$

is separable. It has the solution $y = \tan x$. If we insert a power series and compare coefficients, we find that it is of the form

$$y = x + \frac{x^3}{3} + \quad \text{higher powers of } x.$$

Following the procedure of Example 8 we substitute

$$y = \frac{x}{1+z}, \qquad z(0) = 0, \tag{2.2.4}$$

and get the differential equation

$$xz' + z + z^2 = -x^2, \qquad z(0) = 0,$$

which is the one in Example 8 with $k = 1$. Hence

$$z = \cfrac{-x^2}{3} + \cfrac{-x^2}{5} + \cfrac{-x^2}{7} + \cdots,$$

and finally

$$\tan x = \cfrac{x}{1+} \cfrac{-x^2}{3} + \cfrac{-x^2}{5} + \cfrac{-x^2}{7} + \cdots \qquad (2.2.5)$$

for all $x \in \mathbf{C}$. This expansion was discovered by Lambert, and later by Lagrange and Euler. The book [Khov63] contains several examples of expansions found by means of Riccati equations.

Let it briefly be mentioned, that if x is replaced by $-ix$ in (2.2.5) we get, since $i\tan(-ix) = \tanh x$,

$$\tanh x = \cfrac{x}{1+} \cfrac{x^2}{3} + \cfrac{x^2}{5} + \cfrac{x^2}{7} + \cdots \qquad (2.2.6)$$

This also converges in the whole plane.

$\diamond$

2.3 A new example

Among the newer continued fraction results on Riccati equations we have chosen to mention one particular result, namely a solution by using general T-fractions. This was presented by Sandra C. Cooper in her thesis [Coop88]. See also [CoJM88]. We shall here give a very rough and brief sketch of the idea.

We remember about the general T-fractions that they interpolate and possibly approximate simultaneously at 0 and ∞. In the method to be mentioned here two initial value problems for a Riccati equation are simultaneously solved by one and the same general T-fraction.

Here the Riccati equation is assumed to be of the form

$$zA_0(z) + B_0(z)W_0(z) + C_0(z)W_0^2(z) - zW_0'(z) = 0, \qquad (2.3.1)$$

where

$$\begin{cases} A_0(z), B_0(z), C_0(z) & \text{are analytic at } z = 0, \\ B_0(0) & \text{is not a positive integer,} \end{cases}$$

$$\begin{cases} A_0(z), B_0(z), C_0(z) & \text{are analytic at } z = \infty, \\ A_0(\infty) = C_0(\infty) = 0, \\ \lim_{z \to \infty} zC_0(z) \neq 0, & \lim_{z \to \infty} (1 - B_0(z)) \notin \mathbf{Z}_0^- . \end{cases} \qquad (2.3.2)$$

We seek possible solutions, holomorphic at $z = 0$ and satisfying the initial condition $W_0(0) = 0$, and possible solutions, holomorphic in a neighborhood of $z = \infty$, except for a pole of order 1 at $z = \infty$.

The method aims at a solution, represented by a general T-fraction of the form

$$G_0 z + \cfrac{F_1 z}{1 + G_1 z +} \cfrac{F_2 z}{1 + G_2 z + \cdots +} \cfrac{F_n z}{1 + G_n z + \cdots}, \qquad (2.3.3)$$

where $F_n \neq 0$, $G_n \neq 0$ for $n \geq 1$.

The T-fraction corresponds at $z = 0$ to a power series

$$(G_0 + F_1)z - (F_1 G_1 + F_1 F_2)z^2 + \quad \text{higher powers of } z,$$

and at $z = \infty$ to

$$G_0 z + \frac{F_1}{G_1} + \quad \text{higher powers of } z^{-1} .$$

Observe that these power series have the same form as the expansions at 0 and at ∞ of the two solutions we are looking for.

The idea is related to the one we saw in Example 8, where a formal power series solution was inserted and the initial coefficient determined, from which the transformation of the variable was decided. The main difference here is that *two* series are involved. Let the two series (formal solutions) be

$$p_{-1}^* z + p_0^* + p_1^* z^{-1} + \ldots \quad \text{and} \quad p_1 z + p_2 z^2 + \ldots .$$

We insert them into the differential equation, and use our knowledge of the expansions of A_0, B_0, C_0 at 0 and at ∞:

$$A_0(z) = a_0^{(0)} + a_1^{(0)}z + \ldots, \qquad A_0(z) = \frac{\alpha_1^{(0)}}{z} + \frac{\alpha_2^{(0)}}{z^2} + \ldots,$$

$$B_0(z) = b_0^{(0)} + b_1^{(0)}z + \ldots, \qquad B_0(z) = \beta_0^{(0)} + \frac{\beta_1^{(0)}}{z} + \ldots,$$

$$C_0(z) = c_0^{(0)} + c_1^{(0)}z + \ldots, \qquad C_0(z) = \frac{\gamma_1^{(0)}}{z} + \frac{\gamma_2^{(0)}}{z^2} + \ldots.$$

Straightforward computations (comparison of coefficients) lead to formulas for the p- and p^*-coefficients in terms of the known coefficients above, in particular for p_1, p^*_{-1} and p^*_0. Since

$$G_0 + F_1 = p_1, \qquad G_0 = p^*_{-1} \quad \text{and} \quad \frac{F_1}{G_1} = p^*_0,$$

we find G_0, F_1 and G_1 such that the two series corresponding to the general T-fraction coincide with the formal series solutions up to and including the terms $p_1 z$ and $p^*_{-1} z + p^*_0$. Rather than writing up the formulas (which, by the way, are easily found) we shall illustrate this process later in an example.

After having found G_0, F_1 and G_1 we introduce the new variable $W_1(z)$ by

$$W_0(z) = G_0 z + W_1(z).$$

$W_1(z)$ (with $W_1(0) = 0$) is the solution of a Riccati equation of exactly the same form as (2.3.1), only with subscripts 1 instead of 0. $A_1(z)$, $B_1(z)$, $C_1(z)$ are easily expressed in terms of $A_0(z)$, $B_0(z)$, $C_0(z)$ and G_0. We next introduce a new variable $W_2(z)$, defined by

$$W_1(z) = \frac{F_1 z}{1 + G_1 z + W_2(z)}. \qquad (2.3.4)$$

Here is the place where the invariance property of Riccati equations is used: Under certain mild condition $W_2(z)$ not only satisfies a Riccati equation, but a Riccati equation of exactly the same form as (2.3.1):

$$z A_2(z) + B_2(z)W_2(z) + C_2(z)W_2^2(z) - z W_2'(z) = 0. \qquad (2.3.5)$$

Continuing the process we set

$$W_2(z) = \frac{F_2 z}{1 + G_2 z + W_3(z)},$$

and so on. If we can proceed indefinitely, a general T-fraction (2.3.3) is generated, where

$$W_0(z) = G_0 z + \frac{F_1 z}{1 + G_1 z} + \frac{F_2 z}{1 + G_2 z} + \cdots + \frac{F_n z}{1 + G_n z + W_{n+1}(z)}, \quad (2.3.6)$$

and where $W_{n+1}(z)$ is the solution with $W_{n+1}(0) = 0$ of a Riccati equation (2.3.5) with subscript 2 replaced by subscript $n + 1$.

In the papers [Coop88] and [CoJM88] conditions (on A_n, B_n, C_n) are established, which ensure that the process can be carried out. Recurrence relations for A_n, B_n, C_n are proved, as well as formulas for the parameters F_n, G_n in the general T-fraction.

The main result of [Coop88] and [CoJM88] is as follows:

Theorem 2 *Let*

$$z A_0(z) + B_0(z) W_0(z) + C_0(z) W_0^2(z) - z W_0'(z) = 0 \qquad (2.3.7)$$

give rise to the general T-fraction

$$G_0 z + \mathbf{K} \frac{F_n z}{1 + G_n z} = G_0 z + \frac{F_1 z}{1 + G_1 z} + \frac{F_2 z}{1 + G_2 z} + \cdots. \qquad (2.3.8)$$

a) *If (2.3.8) converges uniformly in a neighborhood of 0 to a function $W(z)$, then $W(z)$ is the unique solution of (2.3.7), analytic in a neighborhood of $z = 0$ and with $W_0(0) = 0$.*

b) *If $(G_0 z + \mathbf{K}(F_n z/(1 + G_n z)))/z$ converges uniformly in a neighborhood of $z = \infty$ to a function $V(z)$, then $W(z) = z V(z)$ is the unique solution of (2.3.7) at $z = \infty$ with a simple pole at $z = \infty$.*

We shall illustrate the algorithm on an example, picked from [CoJM88].

Example 10 Let a, b, c be distinct real numbers, not 0 or negative integers. We shall illustrate the first two steps of the algorithm (from W_0 to W_2) on the Riccati equation

$$\frac{a(b-c)}{c}z}{1-z} + \frac{(-c + (b-a)z)}{1-z}W_0(z) - \frac{c}{1-z}W_0^2(z) - zW_0'(z) = 0 \,.$$

With notation from the text we get

$$
\begin{aligned}
A_0(z) &= \frac{a(b-c)/c}{1-z} = \frac{a(b-c)}{c}\left(1 + z + z^2 + \cdots\right) \quad \text{at } z = 0 \\
&= -\frac{a(b-c)}{c}\left(\frac{1}{z} + \frac{1}{z^2} + \frac{1}{z^3} + \cdots\right) \quad \text{at } z = \infty, \\
B_0(z) &= \frac{-c + (b-a)z}{1-z} = -c + (b-a-c)\left(z + z^2 + z^3 + \cdots\right) \\
&\qquad\qquad \text{at } z = 0 \\
&= a - b + (c + a - b)\left(\frac{1}{z} + \frac{1}{z^2} + \frac{1}{z^3} + \cdots\right) \quad \text{at } z = \infty, \\
C_0(z) &= -\frac{c}{1-z} = -c\left(1 + z + z^2 + \cdots\right) \quad \text{at } z = 0 \\
&= c\left(\frac{1}{z} + \frac{1}{z^2} + \frac{1}{z^3} + \cdots\right) \quad \text{at } z = \infty, \\
W_0(z) &= p_1 z + p_2 z^2 + \ldots \quad \text{at } z = 0, \qquad p_1 = G_0 + F_1, \\
W_0(z) &= p_{-1}^* z + p_0^* + \ldots \quad \text{at } z = \infty, \qquad p_{-1}^* = G_0, \qquad p_0^* = \frac{F_1}{G_1}.
\end{aligned}
$$

We insert this into the Riccati equation and compare coefficients to find that

$$\text{at } z = 0 \quad :$$

$$z^1 : \quad \frac{a(b-c)}{c} - cp_1 - p_1 = 0,$$

$$\text{so } p_1 = \frac{a(b-c)}{c(1+c)},$$

$$\text{at } z = \infty \quad :$$

$$z^1: \quad (a-b)p^*_{-1} + c(p^*_{-1})^2 - p^*_{-1} = 0\,,$$

$$\text{so } p^*_{-1} = \frac{1+b-a}{c}\,, \quad (p^*_{-1} = 0 \quad \text{is ruled out})$$

$$z^0: \quad -\frac{a(b-c)}{c} + (a-b)p^*_0 + (c+a-b)p^*_{-1} +$$

$$c(p^*_{-1})^2 + c\cdot 2p^*_{-1}p^*_0 = 0$$

$$\text{so } p^*_0 = \frac{(a-1-c)(1+b)}{c(2+b-a)}\,.$$

Hence

$$G_0 = \frac{1+b-a}{c}\,, \qquad F_1 = \frac{(b+1)(a-c-1)}{c(1+c)}\,, \qquad G_1 = \frac{2+b-a}{1+c}\,.$$

With $W_1(z)$ defined by

$$W_0(z) = \frac{1+b-a}{c}z + W_1(z)$$

we find that $W_1(z)$ is the unique solution, analytic at $z = 0$ and with $W_1(0) = 0$, of the Riccati equation

$$\frac{\dfrac{(b+1)(a-c-1)}{c}z}{1-z} + \frac{(-c+(a-b-2)z)}{1-z}W_1(z)$$

$$-\frac{c}{1-z}W_1^2(z) - zW_1'(z) = 0\,.$$

With $W_2(z)$ defined by

$$W_1(z) = \frac{\dfrac{(b+1)(a-c-1)}{c(1+c)}z}{1 + \dfrac{2+b-a}{1+c}z + W_2(z)}$$

we find that $W_2(z)$ is the unique solution, analytic at $z = 0$ and with $W_2(0) = 0$, of the Riccati equation

$$\frac{\dfrac{(b+2)(a-c-2)}{c+1}z}{1-z} + \frac{(-(c+1)+(a-b-3)z)}{1-z}W_2(z)$$

$$-\frac{c+1}{1-z}W_2^2(z) - zW_2'(z) = 0\,.$$

Observe that we get the differential equation for W_2 by replacing

$$b \quad \text{by } b+1 \qquad \text{and} \qquad c \quad \text{by } c+1$$

in the differential equation for W_1. We find F_2 and G_2 from F_1 and G_1 in the same way, and by repeating the argument we are led to the continued fraction

$$G_0 z + \mathop{\mathbf{K}}_{n=1}^{\infty} \frac{F_n z}{1 + G_n z} \,,$$

where

$$G_0 = \frac{1+b-a}{c}, \qquad F_n = \frac{(b+n)(a-c-n)}{(c+n-1)(c+n)},$$

$$G_n = \frac{n+1+b-a}{c+n}, \qquad n = 1,2,3,\ldots$$

The general T-fraction found here is limit periodic, $F_n \to -1$ and $G_n \to 1$ when $n \to \infty$. We may thus conclude from the theorem that the T-fraction represents both solutions (by correspondence and convergence).

$\diamond$

Problems

(1) Use the procedure of Example 1 to solve the differential equations

(a) $y = y' + 6y''$,

(b) $y = -2y' - 5y''$.

In which case does the continued fraction converge? diverge?

(2) Find the explicit expressions for $\tilde{a}_0(x)$, $\tilde{a}_1(x)$ and $\tilde{a}_2(x)$ in (2.1.6).

(3) The differential equation

$$y' = 1 + y^2, \qquad y(0) = 0$$

is separable, and can be solved as such. Do this first. But it is also a Riccati equation. Transform it into a second order linear differential equation. Solve this, and use the solution to find the solution of the given differential equation. Compare the solutions.

(4) Transform the Euler-Cauchy equation

$$x^2 w'' + x w' - \frac{1}{4} w = 0$$

to a Riccati equation, and use the solution of the first equation to find the solution of the Riccati equation.

(5) For the equation (2.1.10) with $k = 1$

$$\alpha x y' + \beta y + y^2 = x, \qquad y(0) = -\beta, \qquad \frac{\beta}{\alpha} \notin \mathbf{N}$$

we find that the power series expansion of the solution has the start

$$y = -\beta + \frac{x}{\alpha - \beta} + \quad \text{higher powers of } x.$$

This suggests the transformation

$$y = -\beta + \frac{x}{\alpha - \beta + y_1}, \qquad y_1(0) = 0.$$

Show that this leads to the differential equation

$$\alpha x y_1' + (\alpha - \beta) y_1 + y_1^2 = x, \qquad y_1(0) = 0.$$

Next try the transformation

$$y_1 = \frac{x}{2\alpha - \beta + y_2}$$

(also suggested by the power series expansion) to find a Riccati equation for y_2. Proceed in a similar way to obtain *formally* the solution (2.1.11) in the case $k = 1$. Finally, introduce

$$x = t^k \quad \text{and} \quad \frac{\alpha}{k} = \alpha_1$$

into the equation

$$\alpha x y' + \beta y + y^2 = x$$

and use what you have found to establish formally the solution of form (2.1.11) (but with t and α_1) for the equation (2.1.10) (with t and α_1).

(6) Take in the Riccati equation (2.1.12), studied by Worpitzky, $a = 1$ and $m = 2$. We then get

$$y' + y^2 = 1.$$

We are interested in the solution where $1/y \to 0$ when $x \to 0$ (actually coth x). Use Worpitzky's solution (2.1.13) to establish (again) the continued fraction expansion (2.2.6) for tanh$x$.

(7) Carry out the details of the computations in Example 10.

Remarks

(1) In her dissertation from 1988 [Coop88], Sandra Clement Cooper
has a chapter on "A history of continued fraction solutions to Riccati differential equations". We refer to this, in particular to the
part dealing with recent results, and also to the references in the
thesis.

(2) In the method of T-fraction solutions of Riccati equations, strong
conditions had to be imposed on the coefficient functions. By using
the δ-fractions of L. J. Lange instead of T-fractions a related, but
more general method has been establised [Coop89].

(3) The paper [Steen73] by A. Steen, referred to by Perron, is in dan-
ish. It is in fact a document of invitation to a celebration at the
University of Copenhagen on April 8, 1873, on the occasion of the
55th birthday of His Majesty King Christian IX. Adolph Steen was
a professor of mathematics, but was also active in many other ar-
eas. He was during a long period Rector (i.e. President) of the University of Copenhagen. The mathematical paper [Steen73], used
as an invitation document, was meant to be material for teaching
of graduate students. One may wonder how much the participants
at the party, including the King, could understand!

References

[Coop88]   S. C. Cooper, "General T–Fraction Solutions to Riccati Differential Equations", Dissertation, Colorado State University (Spring 1988).

[Coop89]   S. C. Cooper, *δ–Fraction Soluticns to Riccati Equations.* "Analytic Theory of Continued Fractions III, Proceedings, Redstone 1988", (L. Jacobsen ed.), Lecture Notes in Mathematics, Springer-Verlag, Berlin-Heidelberg (1989), 1–18.

[CoJM88]   S. C. Cooper, W. B. Jones and A. Magnus, *General T-Fraction Solutions to Riccati Differential Equations*, "Nonlinear Numerical Methods and Rational Approximation", (A. Cuyt, ed.), D. Reidel Publishing Company (1988), 409–425.

[JoTh80]   W. B. Jones and W. J. Thron, "Continued Fractions: Analytic Theory and Applications", Encyclopedia of Mathematics and its Applications **11**, Addison Wesley Publ. Co., Reading, Mass. (1980). Distributed now by Cambridge University Press, New York.

[Khov63]   A. N. Khovanskii, "The Application of Continued Fractions and their Generalizations to Problems in Approximation Theory" (translated by Peter Wynn), P. Noordhoff N.V., Groningen (1963).

[Perr57]   O. Perron, "Die Lehre von den Kettenbrüchen", Band **2**, B. G. Teubner, Stuttgart (1957).

[Steen73]   A. Steen, "Integration af lineære Differentialligninger af an-
                den Orden ved Hjælp av Kjædebrøker", Köbenhavns Uni-
                versitet, Köbenhavn (1873), 1–66.

[Worp62]   J. Worpitzky, *Beitrag zur Integration der Riccatischen Gle-
                ichung*, Greifswald (1862), 1–74.

Appendix

Some continued fraction expansions

About this section

This is a catalogue of some of the known continued fraction expansions. The list is in no way complete. Still it can be useful, both to find a continued fraction expansion of some given function and to "sum" a given continued fraction.

As for the notation in this catalogue, we write $f(z) =$ the continued fraction for $z \in D$ to mean that the continued fraction converges in the classical sense to $f(z)$ for all $z \in D$. The set D is usually an open set, such as for instance $D = \mathbf{C} \setminus i[-1, 1]$. The continued fraction will then normally diverge on the cut $i(-1, 1)$, and it may converge or diverge at the end points $z = \pm i$. We give a reference to only one of the possibly many books or papers where the expansion can be found. We have not attempted to find the origin of the various results.

1   Introduction

It is evident that not every continued fraction expansion can find room in a book like this. On the other hand, quite a number of the known continued fraction expansions can be derived from one another by simple transformations. We have for instance

$$f = b_0 + \mathbf{K}(a_n/b_n) \iff c \cdot \frac{1}{f} = \frac{c}{b_0 +} \frac{a_1}{b_1 +} \frac{a_2}{b_2 +} \cdots ; \quad c \neq 0. \qquad (1.1.1)$$

Similarly, if $f = b_0 + \mathbf{K}(a_n/b_n)$ then $g = (f-1)/(f+1) = 1 - 2/(1+f)$, that is

$$f = b_0 + \mathbf{K}(a_n/b_n) \iff \frac{f-1}{f+1} = 1 - \frac{2}{1 + b_0 +} \frac{a_1}{b_1 +} \frac{a_2}{b_2 +} \cdots . \qquad (1.1.2)$$

Another simple transformation is maybe most easily described for regular C-fractions. Assume that $f(z) = b_0 + \mathbf{K}(a_n z/1)$. Then $f(z^{-1}) = b_0 + \mathbf{K}(a_n z^{-1}/1)$. Equivalence transformations lead to

$$
\begin{aligned}
f(z) &= b_0 + \mathbf{K}(a_n z/1) \Rightarrow \\
f\left(\frac{1}{z}\right) &= b_0 + \frac{a_1}{z +} \frac{a_2}{1 +} \frac{a_3}{z +} \frac{a_4}{1 +} \frac{a_5}{z +} \cdots \\
&= b_0 + \frac{a_1/z}{1 +} \frac{a_2}{z +} \frac{a_3}{1 +} \frac{a_4}{z +} \frac{a_5}{1 +} \cdots \\
&= b_0 + \frac{a_1/\xi}{\xi +} \frac{a_2}{\xi +} \frac{a_3}{\xi +} \frac{a_4}{\xi +} \frac{a_5}{\xi +} \cdots ,
\end{aligned}
\qquad (1.1.3)
$$

where $\xi^2 = z$. We shall not list equivalent continued fractions like this separately. Another situation that often arises is the following: We have

$$f(z) = b_0 + \frac{a_1 z^2}{1 +} \frac{a_2 z^2}{1 +} \frac{a_3 z^2}{1 +} \cdots . \qquad (1.1.4a)$$

Then

$$f(iz) = b_0 - \frac{a_1 z^2}{1 -} \frac{a_2 z^2}{1 -} \frac{a_3 z^2}{1 -} \cdots . \qquad (1.1.4b)$$

Of course, every time we have a continued fraction expansion $f = b_0 + \mathbf{K}(a_n/b_n)$ with all $a_n, b_n \neq 0$, we can take its even or odd part and obtain a "new" continued fraction converging to the same value f. Some of these variations will be listed. (In particular if they turn out to be nice and simple.) We give references to other continued fractions in the appendix which have the present continued fraction as a special case, or which can be transformed into the present continued fraction by simple means.

2 Mathematical constants

$$\pi = 3 + \cfrac{1}{7+}\cfrac{1}{15+}\cfrac{1}{1+}\cfrac{1}{292+}\cfrac{1}{1+}\cfrac{1}{1+}\cfrac{1}{1+}\cfrac{1}{2+}\cfrac{1}{1+}\cfrac{1}{3+}\cfrac{1}{1+}\cfrac{1}{14+}\cdots,$$
(2.1.1)

[JoTh80, p. 23]. This is the regular continued fraction expansion of π.

$$\pi = \cfrac{4}{1+}\cfrac{1^2}{3+}\cfrac{2^2}{5+}\cfrac{3^2}{7+}\cfrac{4^2}{9+}\cdots,$$
(2.1.2)

[JoTh80, p. 25], (see also (3.6.1)). For the Riemann zeta function we have

$$\frac{1}{2}\zeta(2) = \frac{\pi^2}{12} = \cfrac{1}{1+}\cfrac{1^4}{3+}\cfrac{2^4}{5+}\cfrac{3^4}{7+}\cdots,$$
(2.1.3)

[Bern89, p.150].

$$\zeta(2) = \frac{\pi^2}{6} = 1 + \cfrac{1}{1+}\cfrac{1^2}{1+}\cfrac{1\cdot2}{1+}\cfrac{2^2}{1+}\cfrac{2\cdot3}{1+}\cfrac{3^2}{1+}\cfrac{3\cdot4}{1+}\cfrac{4^2}{1+}\cdots,$$
(2.1.4)

[Bern89, p. 153], (see also (4.7.32)).

$$\zeta(3) = 1 + \cfrac{1}{4+}\cfrac{1^3}{1+}\cfrac{1^3}{12+}\cfrac{2^3}{1+}\cfrac{2^3}{20+}\cfrac{3^3}{1+}\cfrac{3^3}{28+}\cfrac{4^3}{1+}\cfrac{4^3}{36+}\cdots,$$
(2.1.5)

[Bern89, p. 155], (see also (4.7.37)).

$$e = \frac{1}{1} - \frac{1}{1} + \frac{1}{2} - \frac{1}{3} + \frac{1}{2} - \frac{1}{5} + \frac{1}{2} - \frac{1}{7} + \cdots, \qquad (2.1.6)$$

[JoTh80, p. 25], (see also (3.2.1)).

$$e = 2 + \frac{1}{1} + \frac{1}{2} + \frac{1}{1} + \frac{1}{1} + \frac{1}{4} + \frac{1}{1} + \frac{1}{1} + \frac{1}{6} + \cdots, \qquad (2.1.7)$$

[JoTh80, p. 23].

$$e = 1 + \frac{2}{1} + \frac{1}{6} + \frac{1}{10} + \frac{1}{14} + \frac{1}{18} + \cdots, \qquad (2.1.8)$$

[Khov63, p. 114], (see also (3.2.2)).

$$e = 2 + \frac{2}{2} + \frac{3}{3} + \frac{4}{4} + \frac{5}{5} + \cdots, \qquad (2.1.9)$$

[Perr57, p. 57].

$$e = \frac{1}{1} - \frac{2}{3} + \frac{1}{6} + \frac{1}{10} + \frac{1}{14} + \frac{1}{18} + \cdots, \qquad (2.1.10)$$

[Khov63, p. 114], (see also (3.2.2) for $z = -1$).

The golden ratio:

$$\frac{\sqrt{5} - 1}{2} = \frac{1}{1} + \frac{1}{1} + \frac{1}{1} + \cdots = 1 - \frac{1}{3} - \frac{1}{3} - \frac{1}{3} - \cdots, \qquad (2.1.11)$$

[JoTh80, p. 23].

Catalan's constant $G = \sum_{k=0}^{\infty}(-1)^k/(2k+1)^2$:

$$2G = 2 - \frac{1^2}{3} + \frac{2^2}{1} + \frac{2^2}{3} + \frac{4^2}{1} + \frac{4^2}{3} + \frac{6^2}{1} + \frac{6^2}{3} + \cdots, \qquad (2.1.12)$$

[Bern89, p. 151], (4.7.30).

$$2G = 1 + \frac{1}{1/2} + \frac{1^2}{1/2} + \frac{1\cdot 2}{1/2} + \frac{2^2}{1/2} + \frac{2\cdot 3}{1/2} + \frac{3^2}{1/2} + \frac{3\cdot 4}{1/2} + \cdots, \qquad (2.1.13)$$

[Bern89, p. 153], (4.7.32).

3 Elementary functions

3.1 *Introduction*

The elementary functions listed here are all special cases of hypergeometric functions or ratios of hypergeometric functions. Their nice corresponding continued fraction expansions are special examples of expansions for the hypergeometric functions in general. Still, we prefer to list them separately in this section.

3.2 *The exponential function*

$$
\begin{aligned}
e^z &= {}_1F_1(1;1;z) = \frac{1}{1} - \frac{z}{1} + \frac{z}{2} - \frac{z}{3} + \frac{z}{2} - \frac{z}{5} + \frac{z}{2} - \frac{z}{7} + \cdots \\
&= \frac{1}{1} - \frac{z}{1} + \frac{1z}{2} - \frac{1z}{3} + \frac{2z}{4} - \frac{2z}{5} + \frac{3z}{6} - \frac{3z}{7} + \frac{4z}{8} - \frac{4z}{9} + \cdots ; \; z \in \mathbf{C},
\end{aligned}
$$
$$(3.2.1)$$

[JoTh80, p. 207]. (See also (4.1.4).) The odd part of this continued fraction is

$$e^z = 1 + \frac{2z}{2} - \frac{z^2}{z} + \frac{z^2}{6} + \frac{z^2}{10} + \frac{z^2}{14} + \frac{z^2}{18} + \cdots ; \; z \in \mathbf{C}, \qquad (3.2.2)$$

[Khov63, p. 114].

$$e^z = 1 + \frac{z}{1} \frac{z}{-1} \frac{1}{+1} \frac{z}{-1} \frac{2}{+1} \frac{z}{-1} \frac{3}{+1} \cdots ; \ z \in \mathbf{C}, \qquad (3.2.3)$$

[BoSh89, p. 32].

The even part of $(3.2.3)$ is

$$e^z = 1 + \frac{z}{1-z} \frac{1z}{+2-z} \frac{2z}{+3-z} \frac{3z}{+4-z} \cdots ; \ z \in \mathbf{C}, \qquad (3.2.4)$$

[JoTh80, p. 272]. Since $e^z = 1/e^{-z}$, we can find 4 more expansions from $(3.2.1)$–$(3.2.4)$ by use of $(1.1.1)$. For instance, $(3.2.4)$ transforms into

$$e^z = \frac{1}{1-1} \frac{z}{+z} \frac{1z}{-2+z} \frac{2z}{-3+z} \frac{3z}{-4+z} \cdots ; \ z \in \mathbf{C}, \qquad (3.2.5)$$

[Khov63, p. 113]. An unusual expansion is

$$e^{2k(\frac{\pi}{2}-\theta)} = 1 + \frac{2k}{\tan\theta - k} + \frac{k^2+1^2}{3\tan\theta} + \frac{k^2+2^2}{5\tan\theta} + \frac{k^2+3^2}{7\tan\theta} + \cdots \qquad (3.2.6)$$

for $\tan^2\theta \in \mathbf{C}\backslash[-1,0]$; i.e. $\tan\theta \in \mathbf{C}\backslash i[-1,1]$, [BoSh89, p. 50]. (See also $(3.3.8)$ with $\alpha = ik$ and $z = i\tan\theta$ and $(3.6.8)$.) Lambert's continued fraction

$$\frac{e^z - e^{-z}}{e^z + e^{-z}} = \frac{z}{1} + \frac{z^2}{3} + \frac{z^2}{5} + \frac{z^2}{7} + \cdots ; \ z \in \mathbf{C}, \qquad (3.2.7)$$

[Wall48, p. 349], is easily obtained from $(3.2.2)$ by use of $(1.1.2)$.

3.3 The general binomial function

$$
\begin{aligned}
(1+z)^\alpha &= {}_2F_1(-\alpha, 1; 1; -z) \\
&= \frac{1}{1-} \frac{\alpha z}{1+} \frac{(1+\alpha)z}{2} + \frac{(1-\alpha)z}{3} + \frac{(2+\alpha)z}{2} + \\
&\quad \frac{(2-\alpha)z}{5} + \frac{(3+\alpha)z}{2} + \frac{(3-\alpha)z}{7} + \cdots
\end{aligned}
\qquad (3.3.1)
$$

for $|\arg(z+1)| < \pi$, [JoTh80, p. 202]. (See also (4.1.6).) The odd part of this continued fraction is

$$(1+z)^\alpha = 1 + \cfrac{2\alpha z}{2 + (1-\alpha)z -} \;\; \cfrac{(1^2 - \alpha^2)z^2}{3(z+2)} \;\; \cfrac{(2^2 - \alpha^2)z^2}{5(z+2)} \;\; \cfrac{(3^2 - \alpha^2)z^2}{7(z+2)} - \cdots$$
$$(3.3.2)$$

for $|\arg(z+1)| < \pi$, [Khov63, p. 105]. (3.3.2) is also the odd part of

$$(1+z)^\alpha = \cfrac{1}{1 -} \cfrac{\alpha z}{1(1+z) -} \cfrac{(1-\alpha)z}{2} \cfrac{(1+\alpha)z}{-3(1+z) -}$$
$$\cfrac{(2-\alpha)z}{2} \cfrac{(2+\alpha)z}{-5(1+z) -} \cfrac{(3-\alpha)z}{2} - \cdots \qquad (3.3.3)$$

for $|\arg(z+1)| < \pi$, [Khov63, p. 101].

$$(1+z)^\alpha = \cfrac{1}{1 -} \cfrac{\alpha z}{1 + (1+\alpha)z -} \cfrac{1(1+\alpha)z(1+z)}{2 + (3+\alpha)z} \cfrac{2(2+\alpha)z(1+z)}{3 + (5+\alpha)z} - \cdots$$
$$(3.3.4)$$

for $\Re(z) > -1/2$, [Khov63, p. 101].

$$(1+z)^\alpha = \cfrac{1}{1 -} \cfrac{\alpha z}{1 + \alpha z +} \cfrac{1(1-\alpha)z}{2 - (1-\alpha)z +} \cfrac{2(2-\alpha)z}{3 - (2-\alpha)z +} \cfrac{3(3-\alpha)z}{4 - (3-\alpha)z +} \cdots$$
$$(3.3.5)$$

for $|z| < 1$, [Khov63, p. 102].

For the general binomial function we have

$$(1+z)^\alpha = 1/(1+z)^{-\alpha}. \qquad (3.3.6)$$

Hence the equality (1.1.1) applied to these 5 expansions gives us 5 new ones. To find a continued fraction expansion for

$$\left(\frac{z+1}{z-1}\right)^\alpha = \left(1 + \frac{2}{z-1}\right)^\alpha \qquad (3.3.7)$$

we can use any of the 5 expansions (3.3.1)–(3.3.5) with z replaced by $2/(z-1)$. For instance (3.3.2) gives Laguerre's continued fraction

$$\left(\frac{z+1}{z-1}\right)^\alpha = 1 + \cfrac{2\alpha}{z - \alpha +} \cfrac{\alpha^2 - 1^2}{3z} \cfrac{\alpha^2 - 2^2}{5z} \cfrac{\alpha^2 - 3^2}{7z} + \cdots \qquad (3.3.8)$$

for $|\arg(\frac{2}{z-1} + 1)| < \pi$; that is $z \in \mathbf{C} \setminus [-1, 1]$, [Perr57, p. 153]. An interesting special case of for instance (3.3.1) is obtained by replacing

$1 + z$ by z and α by $-1/z$:

$$z^{1/z} = 1/z^{-1/z} = 1 + \cfrac{z-1}{1z} + \cfrac{(1z-1)(z-1)}{2} + \cfrac{(1z+1)(z-1)}{3z} +$$
$$\cfrac{(2z-1)(z-1)}{2} + \cfrac{(2z+1)(z-1)}{5z} +$$
$$\cfrac{(3z-1)(z-1)}{2} + \cfrac{(3z+1)(z-1)}{7z} +\cdots \qquad (3.3.9)$$

for $|\arg z| < \pi$, [Khov63, p. 109]. The even part of (3.3.9) is

$$z^{1/z} = 1 + \cfrac{2(z-1)}{z^2+1} - \cfrac{(1^2z^2-1)(z-1)^2}{3z(z+1)} -$$
$$\cfrac{(2^2z^2-1)(z-1)^2}{5z(z+1)} - \cfrac{(3^2z^2-1)(z-1)^2}{7z(z+1)} -\cdots \qquad (3.3.10)$$

for $|\arg z| < \pi$, [Khov63, p. 110].

$$\left(\frac{1+az}{1+bz}\right)^\alpha = 1 + \cfrac{2\alpha(a-b)z}{2+(a+b-\alpha(a-b))z} - \cfrac{(a-b)^2(1^2-\alpha^2)z^2}{3(2+(a+b)z)} -$$
$$\cfrac{(a-b)^2(2^2-\alpha^2)z^2}{5(2+(a+b)z)} - \cfrac{(a-b)^2(3^2-\alpha^2)z^2}{7(2+(a+b)z)} -\cdots \qquad (3.3.11)$$

for $\frac{2+(a+b)z}{(a-b)z} \in \mathbf{C} \setminus [-1,1]$, [Perr57, p. 264].

3.4   The natural logarithm

$$\log(1+z) = z \, {}_2F_1(1,1;2;-z) = z \int_0^1 \frac{dt}{1+zt}$$

$$= \cfrac{z}{1} + \cfrac{1z}{2} + \cfrac{1z}{3} + \cfrac{2z}{2} + \cfrac{2z}{5} + \cfrac{3z}{2} + \cfrac{3z}{7} + \cfrac{4z}{2} + \cfrac{4z}{9} + \cdots$$

$$= \cfrac{z}{1} + \cfrac{1^2z}{2} + \cfrac{1^2z}{3} + \cfrac{2^2z}{4} + \cfrac{2^2z}{5} + \cfrac{3^2z}{6} + \cfrac{3^2z}{7} + \cfrac{4^2z}{8} + \cfrac{4^2z}{9} + \cdots$$
$$(3.4.1)$$

for $|\arg(1+z)| < \pi$, [JoTh80, p. 203]. (See also (4.1.6).)

$$\log(1+z) = \cfrac{z}{1+z} - \cfrac{z}{1} + \cfrac{1}{1+z} - \cfrac{z}{1} + \cfrac{1/2}{1+z} - \cfrac{z}{1} + \cfrac{1}{1+z} - \cfrac{z}{1} + \cfrac{2/3}{1+z} -\cdots$$
$$(3.4.2)$$

for $|\arg(1+z)| < \pi$, [JoTh80, p. 319 (NB! misprints)]. Here the continued fraction has the form $\mathbf{K}(a_n(z)/b_n(z))$ where all $a_{2n}(z) = -z$, $a_{4n-1}(z) = 1$ and $a_{4n+1}(z) = n/(n+1)$. (3.4.1) is the even part of (3.4.2). The odd part of (3.4.2) can be written

$$
\begin{aligned}
\log(1+z) &= \frac{z}{1+z}\left\{1 + \frac{z}{2+}\frac{2z}{3}+\frac{z}{2}+\frac{3z}{5}+\frac{2z}{2}+\frac{4z}{7}+\frac{3z}{2}+\frac{5z}{9}+\frac{4z}{2}+\cdots\right\} \\
&= \frac{z}{1+z}\left\{1 + \frac{z}{2+}\frac{1\cdot 2z}{3}+\frac{1\cdot 2z}{4}+\frac{2\cdot 3z}{5}+\frac{2\cdot 3z}{6}\right. \\
&\qquad\qquad \left.+\frac{3\cdot 4z}{7}+\frac{3\cdot 4z}{8}+\cdots\right\}
\end{aligned}
\tag{3.4.3}
$$

for $|\arg(1+z)| < \pi$. The even part of (3.4.1) is

$$
\log(1+z) = \frac{2z}{1(2+z)-}\frac{1^2 z^2}{3(2+z)-}\frac{2^2 z^2}{5(2+z)-}\frac{3^2 z^2}{7(2+z)-}\cdots
\tag{3.4.4}
$$

for $|\arg(1+z)| < \pi$, [Khov63, p. 111].

$$
\log(1+z) = \frac{z}{1+}\frac{1^2 z}{2-z+}\frac{2^2 z}{3-2z+}\frac{3^2 z}{4-3z+}\frac{4^2 z}{5-4z+}\frac{5^2 z}{6-5z+}\cdots
\tag{3.4.5}
$$

for $|z| < 1$, [Khov63, p. 111]. The connection

$$
\log(1+z) = -\log\left(\frac{1}{1+z}\right) = -\log\left(1 - \frac{z}{1+z}\right)
\tag{3.4.6}
$$

can be applied to (3.4.1)–(3.4.5) to get 5 new continued fraction expansions. For instance, from (3.4.1) we get

$$
\log(1+z) = \frac{z}{1+z-}\frac{1z}{2}-\frac{1z}{3(1+z)-}\frac{2z}{2}-\frac{2z}{5(1+z)-}\frac{3z}{2}-\frac{3z}{7(1+z)-}\cdots
\tag{3.4.7}
$$

for $|\arg(1+z)| < \pi$, [Khov63, p. 110], and from (3.4.5)

$$
\log(1+z) = \frac{z}{1+z-}\frac{1^2 z(1+z)}{2+3z}-\frac{2^2 z(1+z)}{3+5z}-\frac{3^2 z(1+z)}{4+7z}-\cdots
\tag{3.4.8}
$$

for $\Re(z) > -1/2$, [Khov63, p. 111].

$$
\begin{aligned}
\log\frac{1+z}{1-z} &= 2z\,{}_2F_1(\tfrac{1}{2},1;\tfrac{3}{2};z^2) = \log\left(1 + \frac{2z}{1-z}\right) = z\int_{-1}^{1}\frac{dt}{1+tz} \\
&= \frac{2z}{1-}\frac{1^2 z^2}{3-}\frac{2^2 z^2}{5-}\frac{3^2 z^2}{7-}\frac{4^2 z^2}{9-}\cdots
\end{aligned}
\tag{3.4.9}
$$

for $|\arg(1 - z^2)| < \pi$, [JoTh80, p. 203]. (See also (4.1.6).) Of course, also other continued fraction expansions for $\log(1 + z)$ can be used to derive expressions for $\log((1 + z)/(1 - z))$. Notice also that

$$\log \frac{z+1}{z-1} = \log \frac{1+1/z}{1-1/z} = \frac{2}{z-}\frac{1^2}{3z-}\frac{2^2}{5z-}\frac{3^2}{7z-}\frac{4^2}{9z-}\cdots \qquad (3.4.10)$$

for $z \in \mathbf{C} \setminus [-1, 1]$, [Perr57, p. 155].

3.5   Trigonometric and hyperbolic functions

$$\tan z \;=\; \frac{\sin z}{\cos z} = z \frac{{}_0F_1(3/2; -z^2/4)}{{}_0F_1(1/2; -z^2/4)}$$

$$\;=\; \frac{z}{1-}\frac{z^2}{3-}\frac{z^2}{5-}\frac{z^2}{7-}\frac{z^2}{9-}\cdots; \; z \in \mathbf{C} \qquad (3.5.1)$$

[JoTh80, p. 211], (See also (4.1.1).) The odd part of (3.5.1) is

$$\tan z \;=\; z + \frac{5z^3}{1\cdot 3\cdot 5 - 6z^2 -}\frac{1\cdot 9z^4}{5\cdot 7\cdot 9 - 14z^2 -}$$
$$\frac{5\cdot 13z^4}{9\cdot 11\cdot 13 - 22z^2 -}\frac{9\cdot 17z^4}{13\cdot 15\cdot 17 - 30z^2 -}\cdots \qquad (3.5.2)$$

for all $z \in \mathbf{C}$.

$$\tan \frac{z\pi}{4} \;=\; \frac{z}{1+}\frac{1^2 - z^2}{2+}\frac{3^2 - z^2}{2+}\frac{5^2 - z^2}{2+}\frac{7^2 - z^2}{2+}\cdots \qquad (3.5.3)$$

for all $z \in \mathbf{C}$.

[Perr57, p. 35]. (See also (4.7.7).) From these expansions one also gets continued fractions for $\cot z = 1/\tan z$, $\tanh z = -i\tan(iz)$ and $\coth z = i/\tan(iz)$ by use of (1.1.1) and (1).

Quite another type of expansion for $\tan z$ follows from

$$\tan \alpha z = -i\frac{(1 + i\tan z)^\alpha - (1 - i\tan z)^\alpha}{(1 + i\tan z)^\alpha + (1 - i\tan z)^\alpha} = -i\frac{y-1}{y+1}, \qquad (3.5.4)$$

where $y = ((1 + i\tan z)/(1 - i\tan z))^\alpha$ can be expanded according to (3.3.8). Combined with (1.1.4) we get

$$\tan \alpha z \;=\; \cfrac{\alpha \tan z}{1} \;-\; \cfrac{(\alpha^2 - 1^2)\tan^2 z}{3} \;-\; \cfrac{(\alpha^2 - 2^2)\tan^2 z}{5} \;-\; \cfrac{(\alpha^2 - 3^2)\tan^2 z}{7} \;-\cdots \qquad (3.5.5)$$

for $|\arg(1 + \tan^2 z)| < \pi$, [Khov63, p. 108]. Since

$$\frac{\pi}{2z}\left(\coth\frac{\pi z}{2} - \frac{2}{\pi z}\right) = \frac{1}{2}\sum_{k=0}^{\infty}\frac{1}{(k+1)^2 + z^2/4}, \qquad (3.5.6)$$

it follows that

$$\frac{\pi z}{2}\coth\frac{\pi z}{2} = 1 + \cfrac{z^2}{1} + \cfrac{1^2(z^2 + 1^2)}{3} + \cfrac{2^2(z^2 + 2^2)}{5} + \cfrac{3^2(z^2 + 3^2)}{7} + \cdots \qquad (3.5.7)$$

for all $z \in \mathbf{C}$, [ABJL, Entry 44].

$$\frac{a\tanh(\pi b/2) - b\tanh(\pi a/2)}{a\tanh(\pi a/2) - b\tanh(\pi b/2)} = \cfrac{ab}{1} + \cfrac{(a^2 + 1^2)(b^2 + 1^2)}{3} + \cfrac{(a^2 + 2^2)(b^2 + 2^2)}{5} + \cdots \qquad (3.5.8)$$

for all $a, b \in \mathbf{C}$, [ABJL, Entry 47].

$$\frac{\sinh(\pi z) - \sinh(\pi z)}{\sinh(\pi z) + \sinh(\pi z)} = \cfrac{2z^2}{1} + \cfrac{4z^4 + 1^4}{3} + \cfrac{4z^4 + 2^4}{5} + \cfrac{4z^4 + 3^4}{7} + \cdots \qquad (3.5.9)$$

for all $z \in \mathbf{C}$, [ABJL, Entry 49].

3.6 Inverse trigonometric and hyperbolic functions

$$\arctan z \;=\; z\,{}_2F_1(\tfrac{1}{2}, 1; \tfrac{3}{2}; -z^2) = -\frac{i}{2}\log\frac{1 + iz}{1 - iz}$$

$$=\; \cfrac{z}{1} + \cfrac{1^2 z^2}{3} + \cfrac{2^2 z^2}{5} + \cfrac{3^2 z^2}{7} + \cfrac{4^2 z^2}{9} + \cdots \qquad (3.6.1)$$

for $|\arg(1+z^2)| < \pi$; i.e. $z \in \mathbf{C} \setminus i((-\infty, -1] \cup [1, \infty))$, [JoTh80, p. 202]. (See also (4.1.6).) The C-fraction for $\arctan z$ can also be written

$$\arctan z = z - \cfrac{z^3}{3 +} \cfrac{3^2 z^2}{5 +} \cfrac{2^2 z^2}{7 +} \cfrac{5^2 z^2}{9 +} \cfrac{4^2 z^2}{11 +} \cfrac{7^2 z^2}{13 +} \cfrac{6^2 z^2}{15 +} \cdots \tag{3.6.2}$$

for $z \in \mathbf{C} \setminus i((-\infty, -1] \cup [1, \infty))$, [Khov63, p. 117].

$$\arctan z = \cfrac{z}{1(1+z^2) -} \cfrac{1 \cdot 2z^2}{3} \cfrac{1 \cdot 2z^2}{-5(1+z^2) -}$$
$$\cfrac{3 \cdot 4z^2}{7} \cfrac{3 \cdot 4z^2}{-9(1+z^2) -} \cfrac{5 \cdot 6z^2}{11} - \cdots \tag{3.6.3}$$

for $z \in \mathbf{C} \setminus i((-\infty, -1] \cup [1, \infty))$, [Khov63, p. 121]. (This follows from (3.6.6) with z replaced by $z(1 + z^2)^{-1/2}$.) Since $\operatorname{arctanh} z = i \arctan(-iz)$, we also get continued fraction expansions for $\operatorname{arctanh} z$ from (3.6.1)–(3.6.3). Also expressions for

$$\arcsin z = \arctan \frac{z}{\sqrt{1 - z^2}}, \quad \arccos z = \arctan \frac{\sqrt{1 - z^2}}{z}$$

can be obtained. For instance, from (3.6.1) we get

$$\frac{\arcsin z}{\sqrt{1 - z^2}} = \cfrac{z}{1(1 - z^2) +} \cfrac{1^2 z^2}{3} \cfrac{2^2 z^2}{+5(1 - z^2) +} \cfrac{3^2 z^2}{7} \cfrac{4^2 z^2}{+9(1 - z^2) +} \cdots \tag{3.6.4}$$

for $|\arg(1 - z^2)| < \pi$, [Khov63, p. 118] and

$$\frac{\arccos z}{\sqrt{1 - z^2}} = \cfrac{1}{z +} \cfrac{1^2(1 - z^2)}{3z} \cfrac{2^2(1 - z^2)}{+} \cfrac{3^2(1 - z^2)}{5z} + \cfrac{3^2(1 - z^2)}{7z} + \cdots \tag{3.6.5}$$

for $\Re(z) > 0$, [Khov63, p. 119]. But we also have

$$\frac{\arcsin z}{\sqrt{1 - z^2}} = z \frac{{}_2F_1(\frac{1}{2}, \frac{1}{2}; \frac{3}{2}; z^2)}{{}_2F_1(\frac{1}{2}, -\frac{1}{2}; \frac{1}{2}; z^2)}$$

$$= \cfrac{z}{1 -} \cfrac{1 \cdot 2z^2}{3} \cfrac{1 \cdot 2z^2}{-} \cfrac{3 \cdot 4z^2}{5} \cfrac{3 \cdot 4z^2}{-} \cfrac{5 \cdot 6z^2}{7} \cfrac{5 \cdot 6z^2}{-} \cfrac{}{9} \cfrac{}{-} \cfrac{}{11} \cfrac{}{-} \cfrac{}{13} - \cdots \tag{3.6.6}$$

for $|\arg(1 - z^2)| < \pi$, [JoTh80, p. 203], and thus, since $\arccos z = \arcsin \sqrt{1 - z^2}$

$$\frac{\arccos z}{\sqrt{1 - z^2}} = \cfrac{z}{1 -} \cfrac{1 \cdot 2(1 - z^2)}{3} \cfrac{1 \cdot 2(1 - z^2)}{-} \cfrac{}{5} -$$
$$\cfrac{3 \cdot 4(1 - z^2)}{7} \cfrac{3 \cdot 4(1 - z^2)}{-} \cfrac{}{9} - \cdots \tag{3.6.7}$$

for $\Re(z) > 0$, [Khov63, p. 121]. Obviously similar expressions for inverse hyperbolic functions can be derived, since $\operatorname{arcsinh} z = i \arcsin(-iz)$ and $(\operatorname{arccosh} z)/\sqrt{z^2 - 1} = (\arccos z)/\sqrt{1 - z^2}$. A neat formula can be obtained from (3.2.6) in the following way

$$\left(\frac{iz+1}{iz-1}\right)^{i\alpha} = \exp\left(i\alpha \log \frac{iz+1}{iz-1}\right) = \exp(2\alpha \arctan(1/z))$$

$$= 1 + \frac{2\alpha}{z-\alpha+} \frac{\alpha^2+1^2}{3z} \frac{\alpha^2+2^2}{5z} \frac{\alpha^2+3^2}{7z} +\cdots \quad (3.6.8)$$

for $|\arg(1 + 1/z^2)| < \pi$, i.e. $z \notin i[-1, 1]$, [Wall48, p. 346].

$$\frac{\operatorname{arcsinh} z}{(1+z^2)^{1/2}} = z\,{}_2F_1(1,1;\tfrac{3}{2};-z^2) =$$

$$\frac{z}{1+} \frac{2z^2}{1+} \frac{2(1+z^2)}{1} \frac{4z^2}{+1+} \frac{4(1+z^2)}{1} +\cdots \quad (3.6.9)$$

for $\Re(z^2) > -1/2$, [ABJL, Entry 37].

$$\arctan z = z\,{}_2F_1(\tfrac{1}{2},1;\tfrac{3}{2};-z^2) =$$

$$\frac{z}{1+} \frac{1z^2}{1+} \frac{2(1+z^2)}{1} \frac{3z^2}{+1+} \frac{4(1+z^2)}{1} +\cdots \quad (3.6.10)$$

for $\Re(z^2) > -1/2$, [ABJL, Entry 38].

3.7 Continued fractions with simple values

$$0 = -a - z + \frac{z}{1-a-z+} \frac{2z}{2-a-z+} \frac{3z}{3-a-z+}\cdots \quad (3.7.1)$$

for $z \neq 0$ if a is a non-negative integer; i. e. $a \in \mathbf{N}_0$, [Perr57, p. 279].

$$1 = \frac{z+1}{z} \frac{z+2}{+z+1+} \frac{z+3}{z+2+} \frac{z+4}{z+3+}\cdots \quad (3.7.2)$$

for $z \neq 0, -1, -2, \ldots$, [Bern89, p. 112]. (See also (4.1.5) with $z = 1$, $a = z + 1$ and $c = z + 1$.)

$$1 = \frac{z+a}{a} \frac{(z+a)^2-a^2}{a} \frac{(z+2a)^2-a^2}{a} \frac{(z+3a)^2-a^2}{a} +\cdots \quad (3.7.3)$$

for $a \neq 0$ and $z/a \neq 0, -1, -2, \ldots$, [Bern89, p. 118]. (See also (4.1.8) with $z = -1$, $a = 0$, $b = (z/a) - 2$ and $c = z/a$.)

$$a = \cfrac{ab}{a+b+d-} \; \cfrac{(a+d)(b+d)}{a+b+3d} \; - \; \cfrac{(a+2d)(b+2d)}{a+b+5d} \; -\cdots \qquad (3.7.4)$$

for $d \neq 0$, $b/d \neq 0, -1, -2, \ldots$ and $\Re((a - b)/d) > 0$ or $a = b$. It also holds for $d = 0$ if $|a| < |b|$, [Bern89, p. 119]. (See also (4.1.6) with $z = 1$, a replaced by $(a + d)/2d$, b replaced by $a/2d$ and c replaced by $(a + b + d)/2d$.) For $b = a$ replaced by $a + 1$ and $d = 1$, (3.7.4) can be transformed into

$$a = 2a + 1 - \cfrac{(a+1)^2}{2a+3} \; \cfrac{(a+2)^2}{-\;2a+5} \; \cfrac{(a+3)^2}{-\;2a+7} \; -\cdots \qquad (3.7.5)$$

for $a \neq 0, -1, -2, \ldots$, [Perr57, p. 105].

$$az = \cfrac{abz}{b-(a+1)z+} \cfrac{(a+1)(b+1)z}{b+1-(a+2)z+} \cfrac{(a+2)(b+2)z}{b+2-(a+3)z+} \cdots \qquad (3.7.6)$$

for $|z| < 1, b \neq 0, -1, -2, \ldots$, [Perr57, p. 290]. (See also (4.1.8) with z replaced by $-z$, a replaced by $b - a$, and $b = c$ replaced by $b - 1$.)

$$\frac{z+a+1}{z+1} = \cfrac{z+a}{z-1+} \cfrac{z+2a}{z+a-1+} \cfrac{z+3a}{z+2a-1+} \cdots \qquad (3.7.7)$$

for $a \neq 0$ and $z/a \neq 0, -1, -2, \ldots$. If $a = 0$, then (3.7.7) is periodic and converges to the said value (which now is 1) for $|z| > 1$, [Bern89, p. 115]. (See also (4.1.5) with z replaced by $1/a$, a replaced by $z/a + 1$ and c replaced by z/a.) If we instead let $z = 1$ and replace a by $z - 1$, c by $z - 3$ in (4.1.5) we get

$$\frac{z^2+z+1}{z^2-z+1} = \cfrac{z}{z-3+} \cfrac{z+1}{z-2+} \cfrac{z+2}{z-1+} \cfrac{z+3}{z} \cfrac{z+4}{+z+1+} \cdots \qquad (3.7.8)$$

for $z \neq 3, 2, 1, 0, -1, -2, \ldots$, [Bern89, p. 118]. For $z = 1, a = z - 1$ and $c = z - 4$ in (4.1.5) we get

$$\frac{z^3+2z+1}{(z-1)^3+2(z-1)+1} = \cfrac{z}{z-4+} \cfrac{z+1}{z-3+} \cfrac{z+2}{z-2+} \cfrac{z+3}{z-1+} \cfrac{z+4}{z} +\cdots \qquad (3.7.9)$$

for $z \neq 4, 3, 2, 1, 0, -1, \ldots$, [Bern89, p. 118]. We can continue the process.

$$\frac{\frac{\pi}{2} - \log(\sqrt{2}+1)}{2\sqrt{2}} = \int_0^1 \frac{t^2\,dt}{1+t^4} = \cfrac{1^2}{3+} \cfrac{3^2}{7+} \cfrac{4^2}{11+} \cfrac{7^2}{15+} \cfrac{8^2}{19+} \cfrac{11^2}{23} +\cdots \qquad (3.7.10)$$

[BoSh89, p. 56]. (See also (4.1.6).)

4   Hypergeometric functions

4.1   General expressions

$$c\frac{{}_0F_1(c; z)}{{}_0F_1(c+1; z)} = c + \frac{z}{c+1+}\frac{z}{c+2+}\frac{z}{c+3+}\cdots \quad (4.1.1)$$

for all $z \in \mathbf{C}$, $c \neq 0, -1, -2, \ldots$, [JoTh80, p. 210].

$$\frac{{}_2F_0(a, b; z)}{{}_2F_0(a, b+1; z)} =$$
$$1 - \frac{az}{1-}\frac{(b+1)z}{1}\frac{(a+1)z}{-1}\frac{(b+2)z}{-1}\frac{(a+2)z}{-1}\cdots \quad (4.1.2)$$

for $|\arg(-z)| < \pi$, [JoTh80, p. 213]. The even part of this one is

$$\frac{{}_2F_0(a, b; z)}{{}_2F_0(a, b+1; z)} =$$
$$1 + \frac{az}{(b+1)z-1-}\frac{(a+1)(b+1)z^2}{(a+b+3)z-1-}\frac{(a+2)(b+2)z^2}{(a+b+5)z-1-}\cdots \quad (4.1.3)$$

for $|\arg(-z)| < \pi$.

$$c\frac{{}_1F_1(a; c; z)}{{}_1F_1(a+1; c+1; z)} = c - \frac{(c-a)z}{c+1} + \frac{(a+1)z}{c+2}$$
$$- \frac{(c-a+1)z}{c+3} + \frac{(a+2)z}{c+4} - \frac{(c-a+2)z}{c+5} + \cdots \quad (4.1.4)$$

for all $z \in \mathbf{C}$, $c \neq 0, -1, -2, \ldots$, [JoTh80, p. 206].

$$\frac{{}_1F_1(a+1; c+1; z)}{{}_1F_1(a; c; z)} = \frac{c}{c-z+}\frac{(a+1)z}{c+1-z+}\frac{(a+2)z}{c+2-z+}\frac{(a+3)z}{c+3-z+}\cdots \quad (4.1.5)$$

for all $z \in \mathbf{C}$, $c \neq 0, -1, -2, \ldots$, [JoTh80, p. 278].

$$c\frac{{}_2F_1(a, b; c; z)}{{}_2F_1(a, b+1; c+1; z)} = c - \frac{a(c-b)z}{c+1} - \frac{(b+1)(c-a+1)z}{c+2}$$
$$- \frac{(a+1)(c-b+1)z}{c+3} - \frac{(b+2)(c-a+2)z}{c+4}$$
$$- \frac{(a+2)(c-b+2)z}{c+5} - \cdots \tag{4.1.6}$$

for $|\arg(1-z)| < \pi$, $c \neq 0, -1, -2, \ldots$, [JoTh80, p. 199]. The Nörlund fraction has the form

$$c\frac{{}_2F_1(a, b; c; z)}{{}_2F_1(a+1, b+1; c+1; z)} =$$
$$c - (a+b+1)z + \frac{(a+1)(b+1)(z-z^2)}{c+1 - (a+b+3)z +}$$
$$\frac{(a+2)(b+2)(z-z^2)}{c+2 - (a+b+5)z +} \; \frac{(a+3)(b+3)(z-z^2)}{c+3 - (a+b+7)z +} \cdots \tag{4.1.7}$$

for $\Re(z) < 1/2$, $c \neq 0, -1, -2, \ldots$. The Euler fraction has the form

$$c\frac{{}_2F_1(a, b; c; z)}{{}_2F_1(a, b+1; c+1; z)} =$$
$$c + (b-a+1)z - \frac{(c-a+1)(b+1)z}{c+1 + (b-a+2)z -}$$
$$\frac{(c-a+2)(b+2)z}{c+2 + (b-a+3)z -} \; \frac{(c-a+3)(b+3)z}{c+3 + (b-a+4)z -} \cdots \tag{4.1.8}$$

for $|z| < 1$, $c \neq 0, -1, -2, \ldots$.

Letting $b = 0$ in (4.1.2), (4.1.5), (4.1.6) or (4.1.7) and using (1.1.1) we get continued fraction expansions for ${}_2F_0(a, 1; z)$ and ${}_2F_1(a, 1; c+1; z)$. Similarly, $a = 0$ in (4.1.3) or (4.1.4) gives continued fraction expansions for $_1F_1(1; c + 1; z)$. A different expansion is

$$_2F_1(a, 1; c + 1; z) = \frac{\Gamma(1 - a)\Gamma(c + 1)}{\Gamma(c - a + 1)} \frac{(1 - z)^{c-a}}{(-z)^c} -$$

$$\frac{c}{1 - c + (a - 1)z +} \frac{1(1 - c)(z - 1)}{3 - c + (a - 2)z +} \frac{2(2 - c)(z - 1)}{5 - c + (a - 3)z +} \cdots \quad (4.1.9)$$

for $|z - 1| < 1$ and $c \neq -1, -2, -3, \ldots$, [Bern89, p. 164. NB! Mistake in the condition]. From this follows after some computation, [Bern89, p. 165] that

$$_1F_1(1; c + 1; z) = \frac{e^z \Gamma(c + 1)}{z^c} - \frac{c}{z +} \frac{1 - c}{1} \frac{1}{+z +} \frac{2 - c}{1} \frac{2}{+z +} \frac{3 - c}{1} + \cdots$$

$$= \frac{e^z \Gamma(c + 1)}{z^c} - \frac{c}{z + 1 - c -} \frac{1(1 - c)}{z + 3 - c -}$$

$$\frac{2(2 - c)}{z + 5 - c -} \frac{3(3 - c)}{z + 7 - c -} \cdots \quad (4.1.10)$$

for $|\arg z| < \pi$ and $c \neq -1, -2, -3, \ldots$, [Bern89, p. 165] (the second continued fraction is the even part of the first one).

4.2 Special examples with $_0F_1$

The *Bessel functions* of the first kind and order ν are

$$J_\nu(z) = \left(\frac{z}{2}\right)^\nu \sum_{k=0}^\infty \frac{(-1)^k (z/2)^{2k}}{k! \, \Gamma(\nu + k + 1)} = \frac{(z/2)^\nu}{\Gamma(\nu + 1)} \, _0F_1(\nu + 1; -\tfrac{z^2}{4}), \quad (4.2.1)$$

so that by (4.1.1)

$$\frac{J_\nu(z)}{J_{\nu+1}(z)} = \frac{2(\nu + 1) \, _0F_1(\nu + 1; -z^2/4)}{z \quad \, _0F_1(\nu + 2; -z^2/4)}$$

$$= \frac{z}{2(\nu + 1) -} \frac{z^2}{2(\nu + 2) -} \frac{z^2}{2(\nu + 3) -} \frac{z^2}{2(\nu + 4) -} \cdots \quad (4.2.2)$$

for $z \in \mathbf{C}$, $\nu \neq -1, -2, -3, \ldots$, [JoTh80, p. 211], (4.1.1).

4.3 Special examples with $_2F_0$

The connection (see for instance [Wall48, p. 352, p. 355])

$$_2F_0(a, b; -z) \sim \frac{1}{\Gamma(a)} \int_0^\infty \frac{e^{-t}t^{a-1}}{(1+tz)^b} dt = \frac{1}{\Gamma(b)} \int_0^\infty \frac{e^{-t}t^{b-1}}{(1+tz)^a} dt \quad (4.3.1)$$

implies that $(4.1.2) - (4.1.3)$ lead to continued fraction expansions for ratios of such integrals. In particular, the *incomplete gamma function* $\Gamma(a, z)$ satisfies

$$\Gamma(a, z) = \int_z^\infty e^{-t}t^{a-1}dt \sim e^{-z}z^{a-1}{}_2F_0(1-a, 1; -1/z), \quad (4.3.2)$$

[EMOT53, p. 266]. Hence, by (4.1.2)

$$\Gamma(a, z) = \frac{e^{-z}z^a}{z} + \frac{1-a}{1} + \frac{1}{z} + \frac{2-a}{1} + \frac{2}{z} + \frac{3-a}{1} + \frac{3}{z} + \cdots$$

$$= \frac{e^{-z}z^a}{1+z-a} - \frac{1(1-a)}{3+z-a} - \frac{2(2-a)}{5+z-a} - \frac{3(3-a)}{7+z-a} - \cdots \quad (4.3.3)$$

for $|\arg z| < \pi$, [AbSt65, p. 260, p. 263], [Khov63, p. 144], where the second continued fraction is the even part of the first one.

This (and the expressions to come) are to be interpreted in the following way: The integral in (4.3.2) is taken for real z. Then $\Gamma(a, z)$ is the analytic continuation of this function to the given domain. The *complementary error function* erfc z satisifies

$$\text{erfc } z = \int_z^\infty e^{-t^2}dt = \frac{1}{2}\Gamma(\tfrac{1}{2}, z^2) \sim e^{-z^2}z^{-1/2}{}_2F_0(\tfrac{1}{2}, 1; -1/z^2)(4.3.4)$$

[EMOT53, p. 266], which means that by (4.1.2)

$$\text{erfc } z = e^{-z^2}\left\{\frac{1}{2z} + \frac{2}{2z} + \frac{4}{2z} + \frac{6}{2z} + \frac{8}{2z} + \cdots\right\} =$$

$$e^{-z^2}\left\{\frac{z}{1+2z^2} - \frac{1 \cdot 2}{5+2z^2} - \frac{3 \cdot 4}{9+2z^2} - \frac{5 \cdot 6}{13+2z^2} - \frac{7 \cdot 8}{17+2z^2} - \cdots\right\} \quad (4.3.5)$$

for $\Re(z) > 0$, [JoTh80, p. 219]. (There is a slightly different notation in [JoTh80].) Again the second continued fraction is the even part of the first one. If we integrate this complementary error function we get similar expressions: Let

$$i^{-1}\operatorname{erfc}z = e^{-z^2} \,, \quad i^0\operatorname{erfc}z = \operatorname{erfc}z \,, \quad i^n\operatorname{erfc}z = \int_z^\infty i^{n-1}\operatorname{erfc}t\,dt \quad (4.3.6)$$

for $n = 1, 2, 3, \ldots$. Then

$$
\begin{aligned}
\frac{i^{n-1}\operatorname{erfc}z}{i^n\operatorname{erfc}z} &= 2z\frac{{}_2F_0\!\left(\frac{n+1}{2}, \frac{n}{2}; -1/z^2\right)}{{}_2F_0\!\left(\frac{n+1}{2}, \frac{n}{2} + 1; -1/z^2\right)} \\
&= 2z + \frac{2(n+1)}{2z} + \frac{2(n+2)}{2z} + \frac{2(n+3)}{2z} + \cdots
\end{aligned}
\quad (4.3.7)
$$

for $\Re(z) > 0$ by (4.1.2), [JoTh80, p. 219]. For the *exponential integral*

$$-\operatorname{Ei}(-z) = \int_z^\infty \frac{e^{-t}}{t}\,dt \sim \frac{e^{-z}}{z}\,{}_2F_0\!\left(1, 1; -\tfrac{1}{z}\right), \quad (4.3.8)$$

[EMOT53, p. 267], we get by (4.1.2) and its even part

$$
\begin{aligned}
\operatorname{Ei}(-z) &= -\frac{e^{-z}}{z} \ \frac{1}{+1} \ \frac{1}{+z} \ \frac{2}{+1} \ \frac{2}{+z} \ \frac{3}{+1} \ \frac{3}{+z} \ \frac{4}{+1} + \cdots \\
&= -\frac{e^{-z}}{1+z} \ \frac{1^2}{-3+z} \ \frac{2^2}{-5+z} \ \frac{3^2}{-7+z} \ \frac{4^2}{-9+z} - \cdots
\end{aligned}
\quad (4.3.9)
$$

for $|\arg z| < \pi$, [Khov63, p. 145]. Similarly for the *logarithmic integral*

$$
\begin{aligned}
\operatorname{li}z &= \int_0^z \frac{dt}{\log t} = \operatorname{Ei}(\log z) = \frac{z}{\log z} \ \frac{1}{-1} \ \frac{1}{-\log z} \ \frac{2}{-1} \ \frac{2}{-\log z} - \cdots \\
&= -\frac{z}{1 - \log z} \ \frac{1^2}{-3 - \log z} \ \frac{2^2}{-5 - \log z} \ \frac{3^2}{-7 - \log z} \ \frac{4^2}{-9 - \log z} - \cdots
\end{aligned}
\quad (4.3.10)
$$

The *plasma dispersion function* is

$$
\begin{aligned}
P(z) &= \frac{1}{\sqrt{\pi}} \int_{-\infty}^\infty \frac{e^{-t^2}}{t - z}\,dt = 2ie^{-z^2}\operatorname{erfc}(-iz) \\
&= \frac{2z}{1 - 2z^2} \ \frac{1 \cdot 2}{-5 - 2z^2} \ \frac{3 \cdot 4}{-9 - 2z^2} \ \frac{5 \cdot 6}{-13 - 2z^2} \ \frac{7 \cdot 8}{-17 - 2z^2} - \cdots
\end{aligned}
\quad (4.3.11)
$$

for $\Im(z) > 0$, [JoTh80, p. 219].

$$\frac{\int_0^\infty t^a e^{-bt-t^2/2} dt}{\int_0^\infty t^{a-1} e^{-bt-t^2/2} dt} = \frac{a}{b} \cdot \frac{{}_2F_0\left(\frac{a}{2}, \frac{a+1}{2}; -\frac{1}{b^2}\right)}{{}_2F_0(\frac{a}{2}, \frac{a-1}{2}; -\frac{1}{b^2})}$$

$$= \frac{a}{b+} \frac{a+1}{b} + \frac{a+2}{b} + \frac{a+3}{b} + \cdots \qquad (4.3.12)$$

for $\Re(b) > 0$, [Perr57, p. 297].

4.4 Special examples with ${}_1F_1$

From [EMOT53, p. 255] it follows that

$$\tag{4.4.1} {}_1F_1(a; c; z) = \frac{\Gamma(c)}{\Gamma(a)\Gamma(c-a)} \int_0^1 e^{tz} t^{a-1} (1-t)^{c-a-1} dt$$

for $\Re(c) > 0$, $\Re(a) > 0$. Hence (4.1.4), (4.1.5) and (4.1.10) lead to continued fraction expansions of ratios of such integrals. The *error function* is given by

$$\begin{aligned} \operatorname{erf}(z) = \int_0^\infty e^{-t^2} &= z \, {}_1F_1(\tfrac{1}{2}; \tfrac{3}{2}; -z^2), \quad [\text{EMOT53}, p.\, 266] \\ &= z e^{-z^2} {}_1F_1(1; \tfrac{3}{2}; z^2), [\text{JoTh80}, p.\, 282]. \end{aligned}$$

$$\tag{4.4.2}$$

Hence,

$$\begin{aligned} \operatorname{erf}(z) &= \frac{z e^{-z^2}}{1} - \frac{2z^2}{3} + \frac{4z^2}{5} - \frac{6z^2}{7} + \frac{8z^2}{9} - \cdots \\ &= \frac{z e^{-z^2}}{1 - 2z^2} + \frac{4z^2}{3 - 2z^2} + \frac{8z^2}{5 - 2z^2} + \frac{12z^2}{7 - 2z^2} + \cdots \end{aligned} \qquad (4.4.3)$$

for $z \in \mathbf{C}$, [JoTh80, p. 208, p. 282].

The error function is related to *Dawson's integral*

$$\int_0^z e^{t^2} = i\operatorname{erf}(-iz), \quad [\text{JoTh80}, p.\, 208] \qquad (4.4.4)$$

and to the *Fresnel integrals*

$$C(z) = \int_0^z \cos\left(\frac{\pi}{2}t^2\right) dt, \qquad S(z) = \int_0^z \sin\left(\frac{\pi}{2}t^2\right) dt \qquad (4.4.5)$$

by

$$
\begin{aligned}
C(z) + iS(z) &= \int_0^z e^{it^2\pi/2}\, dt = \sqrt{\frac{-2}{i\pi}} \int_0^{\sqrt{-i\pi/2}\,\cdot\, z} e^{-u^2}\, du \\
&= \frac{1+i}{\sqrt{\pi}} \operatorname{erf}\left(\frac{\sqrt{\pi}}{2}(1-i)z\right)
\end{aligned}
$$

$$(4.4.6)$$

The *incomplete gamma function*

$$
\begin{aligned}
\gamma(a,z) &= \int_0^z e^{-t} t^{a-1}\, dt = \frac{z^a}{a} e^{-z}\, {}_1F_1(1; a+1; z) \\
&= \frac{z^a e^{-z}}{a} \quad \frac{az}{-a+1+} \frac{1z}{a+2-} \frac{(a+1)z}{a+3} \frac{2z}{+a+4-} \frac{(a+2)z}{a+5} +\cdots \\
&= \frac{z^a e^{-z}}{a} \quad \frac{az}{-1+a+z-} \frac{(1+a)z}{2+a+z-} \frac{(2+a)z}{3+a+z-} \frac{(3+a)z}{4+a+z-}\cdots
\end{aligned}
$$

$$(4.4.7)$$

for all $z \in \mathbf{C}$, [JoTh80, p. 209], [Khov63, p. 149–150].

The *Coulomb wave function*

$$F_L(\eta, \rho) = \rho^{L+1} e^{-i\rho} C_L(\eta)\, {}_1F_1(L+1-i\eta; 2L+2; 2i\rho) \qquad (4.4.8)$$

where $C_L(\eta) = 2^L \exp(-\pi\eta/2)|\Gamma(L+1+i\eta)|/(2L+1)!$ for $\eta \in \mathbf{R}$, $\rho > 0$ and $L \in \mathbf{N}_0$ satisfies

$$
\frac{F_L(\eta,\rho)}{F_{L-1}(\eta,\rho)} =
$$

$$
\frac{(L+1)(L^2+\eta^2)^{1/2}}{(2L+1)(\eta+L(L+1)/\rho)-} \frac{(L+2)((L+1)^2+\eta^2)}{(2L+3)(\eta+(L+1)(L+2)/\rho)-}
$$

$$
\frac{(L+3)((L+2)^2+\eta^2)}{(2L+5)(\eta+(L+2)(L+3)/\rho)-}\cdots \qquad \text{for } L = 1,2,3,\ldots
$$

$$(4.4.9)$$

[JoTh80, p. 216]. It is well known that

$$\sum_{k=0}^{\infty} \frac{(-z)^k}{k!\,(a+k)} = e^{-z} \sum_{k=0}^{\infty} \frac{z^k}{(a)_{k+1}} = \frac{e^{-z}}{a}\,{}_1F_1(1; a+1; z), \qquad (4.4.10)$$

[Bern89, p. 166]. This means for instance that

$$\sum_{k=0}^{\infty} \frac{(-z)^k}{k!\,(a+k)} = \frac{\Gamma(a)}{z^a} - \frac{e^{-z}}{z+1-a} - \frac{1(1-a)}{z+3-a} - \frac{2(2-a)}{z+5-a} - \cdots \qquad (4.4.11)$$

for $|\arg z| < \pi$. For $a = 1/2$ and z replaced by $z/2$ this gives

$$\sum_{k=0}^{\infty} \frac{z^k}{1 \cdot 3 \cdots (2k+1)} = \sqrt{\frac{\pi}{2z}}e^{z/2} - \frac{1}{z+1} - \frac{1 \cdot 2}{z+5} - \frac{3 \cdot 4}{z+9} - \frac{5 \cdot 6}{z+13} - \cdots$$
$$(4.4.12)$$

for $|\arg z| < \pi$, [Bern89, p. 166. NB! Misprint], and

$$ze^{-z^2} \sum_{k=0}^{\infty} \frac{(2z^2)^k}{1 \cdot 3 \cdots (2k+1)} = \int_0^z e^{-t^2}\,dt$$

$$= \frac{\sqrt{\pi}}{2} - \frac{e^{-z^2}}{2z^2+1} - \frac{1 \cdot 2}{2z^2+5} - \frac{3 \cdot 4}{2z^2+9} - \frac{5 \cdot 6}{2z^2+13} - \cdots \qquad (4.4.13)$$

for $\Re(z) > 0$, [Bern89, p. 166] (even part and equivalence transformation).

4.5   Special examples with $_2F_1$

$$\int_0^z \frac{t^p\,dt}{1+t^q} = \frac{z^{p+1}}{q}\,{}_2F_1\left(\frac{p+1}{q}, 1; 1+\frac{p+1}{q}; -z^q\right)$$

$$= \frac{z^{p+1}}{0q+p+1} + \frac{(0q+p+1)^2 z^q}{1q+p+1} + \frac{(1q)^2 z^q}{2q+p+1} +$$

$$\frac{(1q+p+1)^2 z^q}{3q+p+1} + \frac{(2q)^2 z^q}{4q+p+1} + \cdots \qquad (4.5.1)$$

for $(p+1)/q \neq 0, -1, -2, -3, \ldots$ and $|\arg(1+z^q)| < \pi$, [Khov63, p. 127].

Incomplete beta functions are given by

$$B_x(p,q) = \int_0^x t^{p-1}(1-t)^{q-1}dt = \frac{x^p}{p}{}_2F_1(p, 1-q; p+1; x) \quad (4.5.2)$$

for $p > 0, q > 0$ and $0 \le x \le 1$, [EMOT53, p. 87]. Hence, by (4.1.6) and (4.1.8)

$$
\begin{aligned}
\frac{B_x(p+1,q)}{B_x(p,q)} &= \frac{px}{p+1}\frac{{}_2F_1(p+1, 1-q; p+2; x)}{{}_2F_1(p, 1-q; p+1; x)} \\
&= \frac{px}{p+1-}\frac{1(1-q)x}{p+2-}\frac{(p+1)(p+q+1)x}{p+3} - \\
&\qquad\qquad \frac{2(2-q)x}{p+4-}\frac{(p+2)(p+q+2)x}{p+5} -\cdots \\
&= \frac{px}{p+1+(p+q)x-}\frac{(p+q+1)(p+1)x}{p+2+(p+q+1)x-} \\
&\qquad\qquad \frac{(p+q+2)(p+2)x}{p+3+(p+q+2)x-}\cdots
\end{aligned}
\quad (4.5.3)
$$

for $p \ne -1, -2, -3, \ldots$ and $|\arg(1-x)| < \pi$ in the first continued fraction, $|x| < 1$ in the second, [JoTh80, p. 217]. *Legendre functions* of the first kind of degree α and order m are given by

$$
\begin{aligned}
P_\alpha^m(z) &= \frac{1}{\pi}\frac{\Gamma(\alpha+m+1)}{\Gamma(\alpha+1)}\int_0^\pi (z+(z^2-1)^{1/2}\cos t)^\alpha \cos mt\, dt \\
&= \frac{1}{\Gamma(1-m)}\left(\frac{z+1}{z-1}\right)^{m/2}{}_2F_1\left(-\alpha, \alpha+1; 1-m; \frac{1-z}{2}\right).
\end{aligned}
$$

In [Gaut70, p. 55] it is proved that

$$
\begin{aligned}
\frac{P_\alpha^m(z)}{P_\alpha^{m-1}(z)} &= \frac{(m+\alpha)(m-\alpha-1)}{-\dfrac{2mz}{(z^2-1)^{1/2}}} - \frac{(m+1+\alpha)(m-\alpha)}{-\dfrac{2(m+1)z}{(z^2-1)^{1/2}}} - \\
&\qquad\qquad \frac{(m+2+\alpha)(m+1-\alpha)}{-\dfrac{2(m+2)z}{(z^2-1)^{1/2}}} - \cdots
\end{aligned}
$$

$$\approx \quad - \frac{(m+\alpha)(m-\alpha-1)\sqrt{z^2-1}}{2mz} \quad - \quad \frac{(m+1+\alpha)(m-\alpha)(z^2-1)}{2(m+1)z} \quad -$$

$$\frac{(m+2+\alpha)(m+1-\alpha)(z^2-1)}{2(m+2)z} \quad -$$

$$\frac{(m+3+\alpha)(m+2-\alpha)(z^2-1)}{2(m+3)z} \quad - \dots \tag{4.5.4}$$

for $\Re(z) > 0$. *Legendre functions of the second kind of degree α and order m* are given by

$$Q_\alpha^m(z) \;=\; (-1)^m \frac{\Gamma(\alpha+1)}{\Gamma(\alpha-m+1)} \int_0^\infty \frac{\cosh mt}{(z+(z^2-1)^{1/2}\cosh t)^{\alpha+1}} dt$$

$$= \; \sqrt{\pi} \frac{e^{im\pi}}{(2z)^{\alpha+1}} \left(1-\frac{1}{z^2}\right)^{m/2} \frac{\Gamma(\alpha+m+1)}{\Gamma(\alpha+m+\frac{3}{2})} \cdot$$

$$_2F_1\left(\frac{\alpha+m+2}{2}, \frac{\alpha+m+1}{2}; \alpha+\frac{3}{2}; 1/z^2\right). \tag{4.5.5}$$

In [JoTh80, p. 205] it is proved that

$$\frac{Q_\alpha^m(z)}{Q_{\alpha+1}^m(z)} \;=\; \frac{1}{\alpha+m+1} \left\{ (2\alpha+3)z - \frac{(\alpha+m+2)^2}{(2\alpha+5)z} \; - \right.$$

$$\frac{(\alpha+m+3)^2}{(2\alpha+7)z} \quad \frac{(\alpha+m+4)^2}{(2\alpha+9)z} \quad -$$

$$\frac{(\alpha+m+5)^2}{(2\alpha+11)z} \quad \frac{(\alpha+m+6)^2}{(2\alpha+13)z} \quad -\dots \cdot \tag{4.5.6}$$

for $\alpha \neq -3/2, -5/2, -7/2, \dots$, and $z \notin [-1, 1]$.

4.6   Some simple integrals

Hypergeometric functions can be written in terms of integrals. This has already been used to some extent in the preceding subsections, and we refer to [AbSt65] and [EMOT53] for further details. Here we shall just list some simple examples without bringing in the hypergeometric functions themselves.

$$\int_0^\infty \frac{e^{-t}\,dt}{t+z} = \frac{1}{z+1} \frac{1^2}{-z+3} \frac{2^2}{-z+5} \frac{3^2}{-z+7-\dots}; \; \Re(z) > 0, \tag{4.6.1}$$

[BoSh89, p. 20].

$$\int_0^\infty \frac{e^{-t/z}}{(1+t)^n} dt$$

$$= \frac{z}{1+} \frac{nz}{1+} \frac{1z}{1+} \frac{(n+1)z}{1} + \frac{2z}{1+} \frac{(n+2)z}{1} + \frac{3z}{1+} \cdots$$

$$= \frac{z}{1+nz-} \frac{nz^2}{1+(n+2)z-} \frac{2(n+1)z^2}{1+(n+4)z-} \frac{3(n+2)z^2}{1+(n+6)z-} \cdots \quad (4.6.2)$$

for $|\arg z| < \pi$, $n \neq 0, -1, -2, \ldots$, [BoSh89, p. 157]. The second continued fraction is the even part of the first one. For *Jacobi's elliptic functions* sn t, cn t and dn t with modulus k we have

$$\int_0^\infty e^{-tz} \operatorname{sn} t \, dt =$$

$$\frac{1}{1^2(1+k^2)+z^2-} \frac{1 \cdot 2^2 \cdot 3k^2}{3^2(1+k^2)+z^2-} \frac{3 \cdot 4^2 \cdot 5k^2}{5^2(1+k^2)+z^2-} \cdots \quad (4.6.3)$$

for all $z \in \mathbf{C}, |k| < 1$, [Wall48, p. 374],

$$\int_0^\infty e^{-tz} \operatorname{sn}^2 t \, dt =$$

$$\frac{2}{2^2(1+k^2)+z^2-} \frac{2 \cdot 3^2 \cdot 4k^2}{4^2(1+k^2)+z^2-} \frac{4 \cdot 5^2 \cdot 6k^2}{6^2(1+k^2)+z^2-} \cdots \quad (4.6.4)$$

for all $z \in \mathbf{C}, |k| < 1$, [Wall48, p. 375],

$$\int_0^\infty e^{-tz} \operatorname{cn} t \, dt = \frac{1}{z+} \frac{1^2}{z+} \frac{2^2 k^2}{z} + \frac{3^2}{z+} \frac{4^2 k^2}{z} + \frac{5^2}{z+} \cdots \quad (4.6.5)$$

for all $z \in \mathbf{C} \setminus \{0\}, |k| < 1$, [Perr57, p. 220],

$$\int_0^\infty e^{-tz} \operatorname{dn} t \, dt = \frac{1}{z+} \frac{1^2 k^2}{z} + \frac{2^2}{z+} \frac{3^2 k^2}{z} + \frac{4^2}{z+} \frac{5^2 k^2}{z} \cdots \quad (4.6.6)$$

for all $z \in \mathbf{C} \setminus \{0\}, |k| < 1$, [Wall48, p. 374], and

$$\int_0^\infty \frac{\operatorname{sn} t \operatorname{cn} t}{\operatorname{dn} t} e^{-tz} dt =$$

$$\frac{1}{2 \cdot 1^2(2-k^2)+z^2-} \frac{1 \cdot 2^2 \cdot 3k^4}{2 \cdot 3^2(2-k^2)+z^2-} \frac{3 \cdot 4^2 \cdot 5k^4}{2 \cdot 5^2(2-k^2)+z^2-} \cdots$$

$$(4.6.7)$$

for all $z \in \mathbf{C}$, $|k| < 1$, [Wall48, p. 375]. For $k = 1$ we have

$$\operatorname{sn} t = \tanh t \quad \text{and} \quad \operatorname{cn} t = \operatorname{dn} t = \operatorname{sech} t, \quad \text{[Lawd89, p. 39]}. \quad (4.6.8)$$

This can be used in (4.6.3)–(4.6.7) to derive new expressions.

$$\int_0^\infty \left(\frac{1-c}{e^{t(1-c)} - c^b} \right)^a e^{-tz} \, dt =$$

$$\frac{r^a}{z+1} + \frac{ar}{z} + \frac{rc^b}{1} + \frac{(a+1)r}{z} + \frac{2rc^b}{1} + \frac{(a+2)r}{1} + \cdots \quad (4.6.9)$$

for $a > 0, b > 0, c > 0, |\arg z| < \pi$, where $r = (1-c)/(1-c^b)$, [Wall48, p. 359].

$$\int_0^\infty \frac{t e^{-tz}}{\sinh t} \, dt = \frac{1}{z} + \frac{1^4}{3z} + \frac{2^4}{5z} + \frac{3^4}{7z} + \cdots \quad (4.6.10)$$

for $\Re(z) > 0$, [Wall48, p. 371].

$$\int_0^\infty \frac{2t e^{-tz}}{e^t + e^{-t}} \, dt = \int_0^\infty \frac{t e^{-tz}}{\cosh t} \, dt = 2 \sum_{n=0}^\infty \frac{(-1)^n}{(z+1+2n)^2}$$

$$= \frac{1}{z^2 - 1} + \frac{4 \cdot 1^2}{1} + \frac{4 \cdot 1^2}{z^2 - 1} + \frac{4 \cdot 2^2}{1}$$

$$+ \frac{4 \cdot 2^2}{z^2 - 1} + \frac{4 \cdot 3^2}{1} + \cdots$$

$$(4.6.11)$$

for $\Re(z) > 0$, but z not real and ≤ 1, [Perr57, p. 30]. For $z = \sqrt{5}$ we get

$$\int_0^\infty \frac{4t e^{-\sqrt{5}t}}{\cosh t} \, dt = \frac{1}{1} + \frac{1^2}{1} + \frac{1^2}{1} + \frac{2^2}{1} + \frac{2^2}{1} + \frac{3^2}{1} + \frac{3^2}{1} + \cdots, \quad \text{[Perr57, p. 30]}.$$

$$(4.6.12)$$

$$\exp \int_0^\infty e^{-t} \tanh t \frac{dt}{t} = \frac{z}{z-1} + \frac{1}{2(z-1)} + \frac{3}{2(z-1)} + \frac{5}{2(z-1)} + \cdots$$

$$(4.6.13)$$

for $\Re(z) < 1$, [BoSh89, p. 157], (after a change of variable $x = 1/z$ and an equivalence transformation).

4.7   Gamma function expressions by Ramanujan

Ramanujan produced quite a number of continued fraction expansions of ratios of gamma functions. These ratios have all proved to be connected to hypergeometric functions. Let us first introduce the notation.

$$\prod_\epsilon \Gamma(a + \epsilon b + c) = \Gamma(a + b + c)\Gamma(a - b + c),$$

$$\prod_\epsilon \Gamma(a + \epsilon b + \epsilon c + d) = \Gamma(a + b + c + d)\Gamma(a - b + c + d)$$

$$\Gamma(a + b - c + d)\Gamma(a - b - c + d) \quad (4.7.1)$$

and so on. Then

$$\frac{1 - R}{1 + R} = \frac{p}{z+} \; \frac{1^2 - q^2}{z} \; + \; \frac{2^2 - p^2}{z} \; + \; \frac{3^2 - q^2}{z} \; + \; \frac{4^2 - p^2}{z} + \cdots \quad (4.7.2)$$

for $\Re(z) > 0$, [Bern89, p. 156], where

$$R = \prod_\epsilon \frac{\Gamma\left(\dfrac{z + p + \epsilon q + 1}{4}\right)}{\Gamma\left(\dfrac{z + p + \epsilon q + 3}{4}\right)} \cdot \prod_\epsilon \frac{\Gamma\left(\dfrac{z - p + \epsilon q + 3}{4}\right)}{\Gamma\left(\dfrac{z - p + \epsilon q + 1}{4}\right)}. \quad (4.7.3)$$

From this it follows that

$$\sum_{k=1}^\infty \left\{ \frac{(-1)^{k+1}}{z + q + 2k - 1} + \frac{(-1)^{k+1}}{z - q + 2k - 1} \right\}$$

$$= \lim_{p \to 0} \frac{1}{p} \frac{1 - R}{1 + R} = \int_0^\infty \frac{\cosh(qt) e^{-tz}}{\cosh t} dt$$

$$= \frac{1}{z+} \; \frac{1^2 - q^2}{z} \; + \; \frac{2^2}{z} + \; \frac{3^2 - q^2}{z} \; + \; \frac{4^2}{z} + \cdots \qquad \text{for } \Re(z) > 0, \quad (4.7.4)$$

[Bern89, p. 148], [BoSh89, p. 157] and

$$\tanh\left\{ \int_0^\infty \frac{\sinh(at) e^{-tz}}{t \cosh t} dt \right\} = \frac{a}{z+} \; \frac{1^2}{z} \; + \; \frac{2^2 - a^2}{z} \; + \; \frac{3^2}{z} + \; \frac{4^2 - a^2}{z} + \cdots \quad (4.7.5)$$

for $\Re(z) > 0$, [Wall48, p. 372], and

$$\tanh\left\{ \frac{1}{2} \int_0^\infty \frac{\sinh(2at) e^{-tz}}{t \cosh t} dt \right\} =$$

$$\frac{a}{z+} \; \frac{1^2 - a^2}{z} \; + \; \frac{2^2 - a^2}{z} \; + \; \frac{3^2 - a^2}{z} \; + \; \frac{4^2 - a^2}{z} + \cdots \quad (4.7.6)$$

for $\Re(z) > 0$, [Wall48, p. 371]. Solving (4.7.2) for $1/R$ gives

$$\frac{1}{R} = 1 + \frac{2p}{z-p+} \frac{1^2-q^2}{z} + \frac{2^2-p^2}{z} + \frac{3^2-q^2}{z} + \frac{4^2-p^2}{z} +\cdots \qquad (4.7.7)$$

for $\Re(z) > 0$, [Perr57, p. 34]. The values $p = q = 1/2$ lead to

$$\frac{z}{4} \frac{\Gamma^2\left(\dfrac{z}{4}\right)}{\Gamma^2\left(\dfrac{z+2}{4}\right)} = 1 + \frac{2}{2z-1+} \frac{1\cdot 3}{2z} + \frac{3\cdot 5}{2z} + \frac{5\cdot 7}{2z} +\cdots \qquad \text{for } \Re(z) > 0$$

$$(4.7.8)$$

and thus, for $z = 4n$ or $z = 4n - 2$ where $n \in \mathbf{N}$, we have

$$\frac{1}{n\pi} \left(\frac{2\cdot 4\cdot\cdots(2n)}{1\cdot 3\cdot\cdots(2n-1)}\right)^2 =$$

$$1 + \frac{2}{8n-1+} \frac{1\cdot 3}{8n} + \frac{3\cdot 5}{8n} + \frac{5\cdot 7}{8n} +\cdots, \qquad [\text{Perr57, p. 34}], (4.7.9)$$

$$\frac{2n^2\pi}{2n-1} \left(\frac{1\cdot 3\cdot\cdots(2n-1)}{2\cdot 4\cdot\cdots(2n)}\right)^2 = \qquad\qquad (4.7.10)$$

$$1 + \frac{2}{8n-5+} \frac{1\cdot 3}{8n-4+} \frac{3\cdot 5}{8n-4+} \frac{5\cdot 7}{8n-4+}\cdots, \qquad [\text{Perr57, p. 34}].$$

$$\frac{a+1}{a} \frac{\displaystyle\int_0^1 t^a \left(\frac{1-t}{1+t}\right)^b dt}{\displaystyle\int_0^1 t^{a-1} \left(\frac{1-t}{1+t}\right)^b dt} = \frac{a+1}{2b} + \frac{(a+1)(a+2)}{2b} + \frac{(a+2)(a+3)}{2b} +\cdots$$

$$(4.7.11)$$

for $\Re(b) > 0$, [Perr57, p. 299]. From this follows directly that also

$$\frac{\displaystyle\int_0^1 t^a \left(\frac{1-t}{1+t}\right)^b \frac{dt}{1-t}}{\displaystyle\int_0^1 t^a \left(\frac{1-t}{1+t}\right)^b \frac{dt}{1-t^2}} = 1 + \frac{a+1}{2b} + \frac{(a+1)(a+2)}{2b} + \frac{(a+2)(a+3)}{2b} +\cdots$$

$$(4.7.12)$$

for $\Re(b) > 0$, [Perr57, p. 300]. A formula of the same character as (4.7.2) is

$$\prod_\epsilon \frac{\Gamma\left(\dfrac{z + \epsilon p + \epsilon q + 1}{4}\right)}{\Gamma\left(\dfrac{z + \epsilon p + \epsilon q + 3}{4}\right)} = \tag{4.7.13}$$

$$\frac{8}{\frac{1}{2}(z^2 - p^2 + q^2 - 1)+} \quad \frac{1^2 - q^2}{1} + \frac{1^2 - p^2}{z^2 - 1} + \frac{3^2 - q^2}{1} + \frac{3^2 - p^2}{z^2 - 1} + \cdots$$

for $|\arg(z^2 - 1)| < \pi$; i.e. for $\Re(z) > 0$ with $z \notin (0, 1]$, [Bern89, p. 159]. For $p = 0$ (or $q = 0$) this reduces to

$$\left(\prod_\epsilon \frac{\Gamma\left(\dfrac{z + \epsilon q + 1}{4}\right)}{\Gamma\left(\dfrac{z + \epsilon q + 3}{4}\right)}\right)^2$$

$$= \frac{8}{\frac{1}{2}(z^2 + q^2 - 1)+} \quad \frac{1^2 - q^2}{1} + \frac{1^2}{z^2 - 1+} \quad \frac{3^2 - q^2}{1} + \frac{3^2}{z^2 - 1} + \cdots$$

$$= \frac{8}{\frac{1}{2}(z^2 - q^2 - 1)+1+} \quad \frac{1^2 - q^2}{z^2 - 1} + \frac{3^2}{1} + \frac{3^2 - q^2}{z^2 - 1} + \cdots \tag{4.7.14}$$

for $\Re(z) > 0$ with $z \notin (0, 1]$, [Bern89, p. 145]. One can also derive the formula

$$\prod_\epsilon \frac{\Gamma\left(\dfrac{z + \epsilon q + 1}{4}\right)}{\Gamma\left(\dfrac{z + \epsilon q + 3}{4}\right)} = \frac{4}{z+} \quad \frac{1^2 - q^2}{2z} + \frac{3^2 - q^2}{2z} + \frac{5^2 - q^2}{2z} + \cdots \tag{4.7.15}$$

for $\Re(z) > 0$, [Bern89, p. 140]. For $q = 0$ and $z = 4n - 1$ or $z = 4n + 1$ for an $n \in \mathbf{N}$, this reduces to

$$4\pi n^2 \left(\frac{1 \cdot 3 \cdot \cdots \cdot (2n - 1)}{2 \cdot 4 \cdot \cdots \cdot (2n)}\right)^2 = \tag{4.7.16}$$

$$4n - 1 + \frac{1^2}{8n - 2+} \frac{3^2}{8n - 2+} \frac{5^2}{8n - 2 + \cdots}, \quad \text{[Perr57, p. 36]},$$

$$\frac{1}{\pi}\left(\frac{2n + 1}{n + 1}\right)^2 \left(\frac{2 \cdot 4 \cdot \cdots \cdot (2n + 2)}{1 \cdot 3 \cdot \cdots \cdot (2n + 1)}\right)^2 = \tag{4.7.17}$$

$$4n + 1 + \frac{1^2}{8n + 2+} \frac{3^2}{8n + 2+} \frac{5^2}{8n + 2 + \cdots}, \quad \text{[Perr57, p. 36]}.$$

A formula closely related to (4.7.15) is

$$\exp\left\{\int_0^\infty \left(1 - \frac{\cosh 2at}{\cosh 2t}\right) e^{-tz}\frac{dt}{t}\right\} =$$
$$1 + \frac{2(1^2 - a^2)}{z^2} + \frac{3^2 - a^2}{1} + \frac{5^2 - a^2}{z^2} + \cdots \qquad (4.7.18)$$

for $\Re(z) > 0$, [Wall48, p. 371]. The most involved of Ramanujan's formulas of this type is

$$\frac{R - Q}{R + Q} = \frac{8abcdh}{1\{2S_4 - (S_2 - 2\cdot 0\cdot 1)^2 - 4(0^2 + 0 + 1)^2\}+}$$
$$\frac{64(a^2 - 1^2)(b^2 - 1^2)(c^2 - 1^2)(d^2 - 1^2)(h^2 - 1^2)}{3\{2S_4 - (S_2 - 2\cdot 1\cdot 2)^2 - 4(1^2 + 1 + 1)^2\}} +$$
$$\frac{64(a^2 - 2^2)(b^2 - 2^2)(c^2 - 2^2)(d^2 - 2^2)(h^2 - 2^2)}{5\{2S_4 - (S_2 - 2\cdot 2\cdot 3)^2 - 4(2^2 + 2 + 1)^2\}} + \cdots , \qquad (4.7.19)$$

where $S_4 = a^4 + b^4 + c^4 + d^4 + h^4 + 1$, $S_2 = a^2 + b^2 + c^2 + d^2 + h^2 - 1$, and

$$R = \prod_\epsilon \Gamma\left(\frac{a + \epsilon(b + c) + \epsilon(d + h) + 1}{2}\right)\cdot$$
$$\cdot\prod_\epsilon \Gamma\left(\frac{a + \epsilon(b + d) + \epsilon(c + h) + 1}{2}\right),$$
$$Q = \prod_\epsilon \Gamma\left(\frac{a + \epsilon(b - c) + \epsilon(d + h) + 1}{2}\right)\cdot$$
$$\cdot\prod_\epsilon \Gamma\left(\frac{a + \epsilon(b + c) + \epsilon(d - h) + 1}{2}\right), \qquad (4.7.20)$$

[Bern89, p. 163]. The expansion (4.7.19) only holds if the continued fraction terminates. But as a corollary one can prove

$$\frac{1 - R}{1 + R} = \frac{2abc}{z^2 - a^2 - b^2 - c^2 + 1+} \frac{4(a^2 - 1^2)(b^2 - 1^2)(c^2 - 1^2)}{3(z^2 - a^2 - b^2 - c^2 + 5)} +$$
$$\frac{4(a^2 - 2^2)(b^2 - 2^2)(c^2 - 2^2)}{5(z^2 - a^2 - b^2 - c^2 + 13)} + \cdots \qquad (4.7.21)$$

for $\Re(z) > 0$, [Bern89, p. 157], where

$$R = \prod_{\epsilon} \frac{\Gamma\left(\dfrac{z + a + \epsilon(b + c) + 1}{2}\right)}{\Gamma\left(\dfrac{z - a + \epsilon(b + c) + 1}{2}\right)} \cdot \prod_{\epsilon} \frac{\Gamma\left(\dfrac{z - a + \epsilon(b - c) + 1}{2}\right)}{\Gamma\left(\dfrac{z + a + \epsilon(b - c) + 1}{2}\right)} .$$

$$(4.7.22)$$

Replacing z by z/c and letting $c \to \infty$ in (4.7.22) leads to

$$\frac{1 - R}{1 + R} = \qquad\qquad\qquad\qquad\qquad\qquad\qquad\qquad (4.7.23)$$

$$\frac{ab}{z +} \frac{(a^2 - 1^2)(b^2 - 1^2)}{3z} + \frac{(a^2 - 2^2)(b^2 - 2^2)}{5z} + \frac{(a^2 - 3^2)(b^2 - 3^2)}{7z} + \cdots$$

for $\Re(z) > 0$, [Bern89, p. 155], where

$$R = \prod_{\epsilon} \Gamma\left(\frac{z + \epsilon(a - b) + 1}{2}\right) \bigg/ \prod_{\epsilon} \Gamma\left(\frac{z + \epsilon(a + b) + 1}{2}\right) . \qquad (4.7.24)$$

In particular

$$\sum_{k=0}^{\infty} \left\{ \frac{1}{z - a + 2k + 1} - \frac{1}{z + a + 2k + 1} \right\} = \lim_{b \to 0} \frac{1}{b} \frac{1 - R}{1 + R}$$

$$= \frac{a}{z +} \frac{1^2(1^2 - a^2)}{3z} + \frac{2^2(2^2 - a^2)}{5z} + \frac{3^2(3^2 - a^2)}{7z} + \cdots \qquad (4.7.25)$$

for $\Re(z) > 0$, [Bern89, p. 149]. Moreover,

$$\sum_{k=1}^{\infty} \frac{(-1)^{k+1}}{(a + k)(b + k)} =$$

$$\frac{1}{(a + 1)(b + 1) +} \frac{(a + 1)^2(b + 1)^2}{a + b + 3} + \frac{(a + 2)^2(b + 2)^2}{a + b + 5} + \cdots \quad (4.7.26)$$

for $a, b \neq -1, -2, -3, \ldots$, [Bern89, p. 123]. We also have

$$\frac{1 - R}{1 + R} = \frac{ab}{z^2 - 1 - a^2 +} \frac{2^2 - b^2}{1} + \frac{2^2 - a^2}{z^2 - 1 +} \frac{4^2 - b^2}{1} + \frac{4^2 - a^2}{z^2 - 1 +} \cdots$$

$$(4.7.27)$$

for $\Re(z) > 0$, [Bern89, p. 158], where

$$R = \prod_{\epsilon} \frac{\Gamma\left(\dfrac{z + \epsilon(a + b) + 3}{4}\right)}{\Gamma\left(\dfrac{z + \epsilon(a + b) + 1}{4}\right)} \bigg/ \prod_{\epsilon} \frac{\Gamma\left(\dfrac{z + \epsilon(a - b) + 3}{4}\right)}{\Gamma\left(\dfrac{z + \epsilon(a - b) + 1}{4}\right)} . \qquad (4.7.28)$$

Dividing (4.7.27) by a and letting $a \to 0$ gives

$$\sum_{k=0}^{\infty} \left\{ \frac{(-1)^k}{z - b + 2k + 1} - \frac{(-1)^k}{z + b + 2k + 1} \right\} = \int_0^{\infty} e^{-tz} \frac{\sinh(bt)}{\cosh t} dt$$

$$= \frac{b}{z^2 - 1+} \frac{2^2 - b^2}{1} \frac{2^2}{+z^2 - 1+} \frac{4^2 - b^2}{1} +\cdots \qquad (4.7.29)$$

for $\Re(z) > 0$, [Bern89, p. 150]. Of course, dividing by b and letting $b \to 0$ gives

$$2 \sum_{k=0}^{\infty} \frac{(-1)^k}{(z + 2k + 1)^2} = \frac{1}{z^2 - 1+} \frac{2^2}{1} \frac{2^2}{+z^2 - 1+} \frac{4^2}{1} \frac{4^2}{+z^2 - 1+}\cdots \qquad (4.7.30)$$

for $\Re(z) > 0$, [Bern89, p. 151].

$$1 + 2z \sum_{k=1}^{\infty} \frac{(-1)^k}{z + 2k} = \frac{1}{z+} \frac{1 \cdot 2}{z} \frac{2 \cdot 3}{+ z} \frac{3 \cdot 4}{+ z} \frac{}{+ z} +\cdots \qquad (4.7.31)$$

for $\Re(z) > 0$, [Bern89, p. 151].

$$1 + 2z^2 \sum_{k=1}^{\infty} \frac{(-1)^k}{(z + k)^2} = \frac{1}{z+} \frac{1^2}{z +} \frac{1 \cdot 2}{z} \frac{2^2}{+ z +} \frac{2 \cdot 3}{z} \frac{3^2}{+ z} +\cdots \qquad (4.7.32)$$

for $\Re(z) > 0$, [Bern89, p. 152].

$$\int_0^{\infty} \frac{\sinh at \sinh bt}{\sinh ct} e^{-tz} dt =$$

$$\frac{ab}{1(z^2 + c^2 - a^2 - b^2) -} \frac{4 \cdot 1^2 (1^2 c^2 - a^2)(1^2 c^2 - b^2)}{3(z^2 + 5c^2 - a^2 - b^2)} -$$

$$\frac{4 \cdot 2^2 (2^2 c^2 - a^2)(2^2 c^2 - b^2)}{5(z^2 + 13c^2 - a^2 - b^2)} -\cdots \,, \qquad (4.7.33)$$

where the coefficients for c^2 are $2k^2 + 2k + 1$ in the denominators [Wall48, p. 370]. (4.7.33) is valid for $|\arg(1 \pm c^2)| < \pi$.

$$c \int_0^{\infty} \frac{\sinh at}{\sinh ct} e^{-tz} dt = \frac{a}{z+} \frac{1^2 (1^2 c^2 - a^2)}{3z} \frac{2^2 (2^2 c^2 - a^2)}{+ 5z} +\cdots \qquad (4.7.34)$$

for $\Re(c/z) > 0$, [Wall48, p. 370].

$$\int_0^{\infty} \frac{e^{-tz} dt}{(\cosh t + a \sinh t)^b} = \qquad (4.7.35)$$

$$\frac{1}{z + ab+} \frac{1 \cdot b(1 - a^2)}{z + a(b + 2)+} \frac{2(b + 1)(1 - a^2)}{z + a(b + 4)} \frac{3(b + 2)(1 - a^2)}{+ z + a(b + 6)} +\cdots$$

for $\Re(a) > 0$, [Wall48, p. 369].

$$\int_0^\infty {}_2F_1\left(a, b; \frac{a+b+1}{2}; -\sinh^2 t\right) e^{-tz}\, dt =$$

$$\cfrac{1}{z+} \cfrac{4 \cdot 1ab}{(a+b+1)z+} \cfrac{4 \cdot 2(a+1)(b+1)(a+b)}{(a+b+3)z} +$$

$$\cfrac{4 \cdot 3(a+2)(b+2)(a+b+1)}{(a+b+5)z} + \cdots \qquad (4.7.36)$$

for $\Re(z) > 0$, [Wall48, p. 370].

$$\zeta(3, z+1) = \sum_{k=1}^\infty \frac{1}{(z+k)^3}$$

$$= \cfrac{1}{2(z^2+z)+} \cfrac{1^3}{1+} \cfrac{1^3}{6(z^2+z)+} \cfrac{2^3}{1+} \cfrac{2^3}{10(z^2+z)+} \cdots$$

$$= \cfrac{1}{1(2z^2+2z+1)-} \cfrac{1^6}{3(2z^2+2z+3)-}$$

$$\cfrac{2^6}{5(2z^2+2z+7)-} \cfrac{3^6}{7(2z^2+2z+13)-} \cdots \qquad (4.7.37)$$

for $\Re(z) > -1/2$, [Bern89, p. 153]. The second continued fraction in (4.7.37) is the even part of the first one.

$$\sum_{k=0}^\infty \left\{ \frac{1}{z+a+b+2k+1} + \frac{1}{z-a-b+2k+1} - \right.$$

$$\left. \frac{1}{z+a-b+2k+1} - \frac{1}{z-a+b+2k+1} \right\}$$

$$= \sum_{k=0}^\infty \frac{8ab(z+2k+1)}{\{(z+2k+1)^2 - a^2 - b^2\}^2 - 4a^2b^2}$$

$$= \cfrac{2ab}{1(z^2-1)+b^2-a^2+} \cfrac{2(1^2-b^2)}{1} \cfrac{2(1^2-a^2)}{+3(z^2-1)+b^2-a^2+}$$

$$\cfrac{4(2^2-b^2)}{1} \cfrac{4(2^2-a^2)}{+5(z^2-1)+b^2-a^2} + \cdots \qquad (4.7.38)$$

for $\Re(z) > 0$, [Bern89, p. 158]. Dividing by $2a$ and letting $a \to 0$ in (4.7.38) leads to

$$\sum_{k=0}^\infty \left\{ \frac{1}{(z-b+2k+1)^2} - \frac{1}{(z+b+2k+1)^2} \right\}$$

$$= \sum_{k=0}^{\infty} \frac{4b(z+2k+1)}{\{(z+2k+1)^2 - b^2\}^2}$$

$$= \frac{b}{1(z^2-1)+b^2+} \frac{2(1^2-b^2)}{1} \frac{2\cdot1^2}{+3(z^2-1)+b^2+}$$

$$\frac{4(2^2-b^2)}{1} \frac{4\cdot2^2}{+5(z^2-1)+b^2+\cdots}$$

$$= \frac{b}{1(z^2-b^2+1)-} \frac{4(1^2-b^2)1^4}{3(z^2-b^2+5)-}$$

$$\frac{4(2^2-b^2)2^4}{5(z^2-b^2+13)-} \frac{4(3^2-b^2)3^4}{7(z^2-b^2+25)-\cdots} \qquad (4.7.39)$$

for $\Re(z) > 0$, [Bern89, p. 158]. Let

$$u = \prod_{k=0}^{\infty}\left\{1+\left(\frac{2a}{z+2k+1}\right)^2\right\}, \quad v = \frac{\Gamma^2\left(\dfrac{z+1}{2}\right)}{\Gamma\left(\dfrac{z+2a+1}{2}\right)\Gamma\left(\dfrac{z-2a+1}{2}\right)}.$$

$$(4.7.40)$$

Then

$$\frac{u-v}{u+v} = \frac{2a^2}{1z+} \frac{4a^4+1^4}{3z} \frac{4a^4+2^4}{+} \frac{4a^4+3^4}{5z} \frac{4a^4+3^4}{7z} +\cdots \qquad (4.7.41)$$

for $\Re(z) > 0$, [ABJL, Entry 48]. Let

$$u = \prod_{k=1}^{\infty}\left\{1+\left(\frac{a}{z+k}\right)^3\right\}, \quad v = \prod_{k=1}^{\infty}\left\{1-\left(\frac{a}{z+k}\right)^3\right\}. \qquad (4.7.42)$$

Then

$$\frac{u-v}{u+v} = \frac{a^3}{1(2z^2+2z+1)+} \frac{a^6-1^6}{3(2z^2+2z+3)+} \frac{a^6-2^6}{5(2z^2+2z+7)+\cdots} \qquad (4.7.43)$$

for $\Re(z) > -1/2$, [ABJL, Entry 50]. Let

$$y = ((1+z^2)^{1/2}-1)/z \quad \text{and} \quad r = a/(1+z^2)^{1/2}, \qquad (4.7.44)$$

where $\Re\left((1+z^2)^{1/2}\right) > 0$. Then

$$2\sum_{k=0}^{\infty}\frac{(-1)^k y^{2k+1}}{r+2k+1} = \frac{z}{1+a+} \frac{1^2 z^2}{3+a+} \frac{2^2 z^2}{5+a+} \frac{3^2 z^2}{7+a+\cdots} \qquad (4.7.45)$$

for $\Re(z) > 0$, [ABJL, Entry 14],

$$y + r\left(y + \frac{1}{y}\right)\sum_{k=1}^{\infty}\frac{(-1)^k y^{2k}}{r + 2k} = \frac{z}{2 + a +}\frac{1 \cdot 2z^2}{4 + a +}\frac{2 \cdot 3z^2}{6 + a +}\frac{3 \cdot 4z^2}{8 + a +}\cdots,$$

(4.7.46)

for $\Re(z) > 0$, [ABJL, Entry 15],

$$\left(1 + \frac{1}{z^2}\right)^{(b-1)/2}(2y)^b\sum_{k=0}^{\infty}\frac{(-1)^k (b)_k y^{2k}}{k!(r + b + 2k)} =$$

$$\frac{z}{a + b +}\frac{1 \cdot bz^2}{a + b + 2 +}\frac{2(b+1)z^2}{a + b + 4 +}\frac{3(b+2)z^2}{a + b + 6 +}\cdots$$

(4.7.47)

for $\Re(z) > 0$, [ABJL, Entry 17].

5 Basic hypergeometric functions

5.1 General expressions

$$(1 - c)\frac{{}_2\varphi_1(a, b; c; q; z)}{{}_2\varphi_1(a, bq; cq; q; z)} =$$

$$1 - c + \frac{(1 - a)(c - b)z}{1 - cq} + \frac{(1 - bq)(cq - a)z}{1 - cq^2} + \frac{(1 - aq)(cq - b)qz}{1 - cq^3} +$$

$$\frac{(1 - bq^2)(cq^2 - a)qz}{1 - cq^4} + \frac{(1 - aq^2)(cq^2 - b)q^2 z}{1 - cq^5} + \cdots$$

(5.1.1)

for $|q| < 1, z \in \mathbf{C}, c \neq 1, q^{-1}, q^{-2}, \dots$, Thm 10 in Chapter VI, [ABBW85, p. 14].

$$(1 - c)\frac{{}_2\varphi_1(a, b; c; q; z)}{{}_2\varphi_1(aq, bq; cq; q; z)} = b_0 + \mathbf{K}(a_n/b_n)$$

(5.1.2)

for $|q| < 1, z \in \mathbf{C}, c \neq 1, q^{-1}, q^{-2}, \dots$ where

$$a_n = (1 - aq^n)(1 - bq^n)cq^{n-1}(1 - zabq^n/c)z$$
$$b_n = 1 - cq^n - (a + b - abq^n - abq^{n+1})q^n z,$$

Theorem 11 in Chapter VII.

$$q(1-c)\frac{{}_2\varphi_1(a,b;c;q;z)}{{}_2\varphi_1(a,bq;cq;q;z)} = (1-c)q + (a-bq)z -$$

$$\frac{(a-cq)(1-bq)qz}{(1-cq)q+(a-bq^2)z} - \frac{(a-cq^2)(1-bq^2)qz}{(1-cq^2)q+(a-bq^3)z} -$$

$$\frac{(a-cq^3)(1-bq^3)qz}{(1-cq^3)q+(a-bq^4)z} - \cdots \tag{5.1.3}$$

for $|q| < 1, c \neq 1, q^{-1}, q^{-2}, \ldots$ [ABBW85, p. 18].

If we choose $b = 1$ in (5.1.1), (5.1.2) or (5.1.3) we obtain continued fraction expansions for ${}_2\varphi_1(a, q; cq; q; z)$ or ${}_2\varphi_1(aq, q; cq; q; z)$.

5.2   *Two general results by Andrews*

$$\frac{G(a,b,c;q)}{G(aq,b,cq;q)} = 1 + \frac{aq+cq}{1} + \frac{bq+cq^2}{1} + \frac{aq^2+cq^3}{1} + \frac{bq^2+cq^4}{1} + \cdots \tag{5.2.1}$$

for $|q| < 1$, [ABJL89, p. 80] where

$$G(a,b,c;q) = \sum_{k=0}^{\infty} \frac{(-\frac{c}{a};q)_k q^{k(k+1)/2} a^k}{(q;q)_k(-bq;q)_k}. \tag{5.2.2}$$

$$\frac{H(a_1,a_2;z;q)}{H(a_1,a_2;qz;q)} = 1 + bqz + \frac{(1+aq^2z)qz}{1+bq^2z} + \frac{(1+aq^3z)q^2z}{1+bq^3z} + \cdots \tag{5.2.3}$$

for $|q| < 1, z \in \mathbf{C}, a = -1/a_1a_2$ and $b = -1/a_1 - 1/a_2$, where

$$H(a_1,a_2;z;q) = \frac{\left(\frac{qz}{a_1};q\right)_{\infty}\left(\frac{qz}{a_2};q\right)_{\infty}}{(qz;q)_{\infty}(1-z)} \cdot$$

$$\cdot \sum_{k=0}^{\infty} \frac{(1-zq^{2k})(z;q)_k(a_1;q)_k(a_2;q)_k q^{k(3k+1)/2}(az^2)^k}{(q;q)_k\left(\frac{qz}{a_1};q\right)_k\left(\frac{qz}{a_2};q\right)_k} \tag{5.2.4}$$

[ABJL89, p. 79].

5.3 *q-expressions by Ramanujan*

The formula (5.2.1) can also be found in Ramanujan's lost notebook [Andr79, p. 90]. Quite a number of Ramanujan's expressions are special cases of (5.2.1) and (5.2.3). We refer in particular to [ABJL] for more details. From (5.1.1) we find that

$$\frac{(-a;q)_\infty(b;q)_\infty - (a;q)_\infty(-b;q)_\infty}{(-a;q)_\infty(b;q)_\infty + (a;q)_\infty(-b;q)_\infty}$$

$$= \frac{a-b}{1-q} \frac{{}_2\varphi_1\left(\dfrac{bq}{a},\dfrac{bq^2}{a};q^3;q^2;a^2\right)}{{}_2\varphi_1\left(\dfrac{bq}{a},\dfrac{b}{a};q;q^2;a^2\right)}$$

$$= \frac{a-b}{1-q+} \frac{(a-bq)(aq-b)}{1-q^3} + \frac{(a-bq^2)(aq^2-b)q}{1-q^5} + \cdots \quad (5.3.1)$$

for $|q| < 1$, [ABBW85, p. 14].

$$\frac{(a^2q^3;q^4)_\infty(b^2q^3;q^4)_\infty}{(a^2q;q^4)_\infty(b^2q;q^4)_\infty} =$$

$$\frac{1}{1-ab+} \frac{(a-bq)(b-aq)}{(1-ab)(q^2+1)+} \frac{(a-bq^3)(b-aq^3)}{(1-ab)(q^4+1)} + \cdots \quad (5.3.2)$$

for $|q| < 1$, [ABBW85, Entry 12].

$$\frac{F(b;a)}{F(b;aq)} = 1 + \frac{aq}{1+bq+} \frac{aq^2}{1+bq^2+} \frac{aq^3}{1+bq^3} + \cdots ; |q| < 1, \quad (5.3.3)$$

[ABBW85, Entry 15], where

$$F(b;a) = \sum_{k=0}^{\infty} \frac{a^k q^{k^2}}{(-bq;q)_k(q;q)_k}. \quad (5.3.4)$$

If we let $a = 0$ in (5.2.1) we get

$$\frac{\varphi(c)}{\varphi(cq)} = 1 + \frac{cq}{1+} \frac{bq+cq^2}{1} \frac{cq^3}{+1+} \frac{bq^2+cq^4}{1} \frac{cq^5}{+1+} \cdots, \quad (5.3.5)$$

where

$$\varphi(c) = \sum_{k=0}^{\infty} \frac{q^{k^2}c^k}{(q;q)_k(-bq;q)_k},$$

for $|q| < 1$, [ABJL, Entry 56]. If $b = -c$ this reduces to

$$\sum_{k=0}^{\infty}(-c)^k q^{k(k+1)/2} = \frac{1}{1+}\frac{cq}{1+}\frac{c(q^2-q)}{1}\frac{cq^3}{+1+}\frac{c(q^4-q^2)}{1}+\cdots \quad (5.3.6)$$

for $|q| < 1$, [ABBW85, p. 22].

$$\frac{G(z)}{G(qz)} = 1 - \frac{qz}{1+q+}\frac{q^3z}{1+q^2-}\frac{q^2z}{1+q^3+}\frac{q^6z}{1+q^4-}\frac{q^3z}{1+q^5+}\frac{q^9z}{1+q^6-}\cdots \quad (5.3.7)$$

for $|q| < 1$, [ABJL, Formula 9.1], where

$$G(z) = \sum_{k=0}^{\infty} \frac{(-z)^k q^{k(k+1)/2}}{(q^2;q^2)_k}. \quad (5.3.8)$$

$$\frac{(q^2;q^3)_\infty}{(q;q^3)_\infty} = \frac{1}{1-}\frac{q}{1+q-}\frac{q^3}{1+q^2-}\frac{q^5}{1+q^3-}\frac{q^7}{1+q^4}-\cdots, \quad (5.3.9)$$

[ABJL, Entry 10].

$$\frac{(q^3;q^4)_\infty}{(q;q^4)_\infty} = \frac{1}{1-}\frac{q}{1+q^2-}\frac{q^3}{1+q^4-}\frac{q^5}{1+q^6}-\cdots, \quad (5.3.10)$$

[ABJL, Entry 11].

$$\frac{(-q^2;q^2)_\infty}{(-q;q^2)_\infty} = \frac{1}{1+}\frac{q}{1+}\frac{q^2+q}{1}\frac{q^3}{+1+}\frac{q^4+q^2}{1}\frac{q^5}{+1}+\cdots, \quad (5.3.11)$$

[ABJL, Entry 12].

$$\frac{(q;q^2)_\infty}{\{(q^3;q^6)_\infty\}^3} = \frac{1}{1+}\frac{q+q^2}{1}+\frac{q^2+q^4}{1}+\cdots, [\text{ABJL89, Thm 7}]. \quad (5.3.12)$$

$$\frac{(q;q^5)_\infty(q^4;q^5)_\infty}{(q^2;q^5)_\infty(q^3;q^5)_\infty} = \frac{1}{1+}\frac{q}{1+}\frac{q^2}{1}+\cdots, [\text{ABJL89, (5)}]. \quad (5.3.13)$$

$$\frac{(q;q^8)_\infty(q^7;q^8)_\infty}{(q^3;q^8)_\infty(q^5;q^8)_\infty} = \frac{1}{1+}\frac{q+q^2}{1}+\frac{q^4}{1+}\frac{q^3+q^6}{1}+\frac{q^8}{1+}\frac{q^5+q^{10}}{1}+\cdots, \quad (5.3.14)$$

[ABJL89, Thm 6].

$$\sum_{k=1}^{\infty} \frac{(a;q)_{\infty} a^k}{(q;q)_k (1+q^k z)} =$$

$$\frac{a}{1+} \quad \frac{(1-a)qz}{1} + \frac{(1-q)aqz}{1} + \frac{(1-aq)q^2 z}{1} +$$

$$\frac{(1-q^2)aq^2 z}{1} + \frac{(1-aq^2)q^3 z}{1} +\cdots, \qquad (5.3.15)$$

[Wall48, p. 376].

References

[AbSt65]   M. Abramowitz and I. A. Stegun, "Handbook of Mathematical Functions", Dover, New York (1965).

[ABBW85]   C. Adiga, B. C. Berndt, S. Bhargava and G. N. Watson, "Chapter 16 of Ramanujan's Second Notebook: Theta-Functions and q–Series", Mem. of the Amer. Math. Soc., no. **315**, Providence (1985).

[Andr79]   G.E. Andrews, *An Introduction to Ramanujan's "Lost" Notebook*, Amer. Math. Monthly, **86**, (1979), 89–108.

[ABJL89]   G. E. Andrews, B. C. Berndt, L. Jacobsen and R. L. Lamphere, *Variations on the Rogers–Ramanujan Continued Fraction in Ramanujan's Notebooks*, "Number Theory, Madras 1987" (K. Alladi ed.) Lecture Notes in Math. Springer–Verlag, **1395** (1989), 73–83.

[ABJL]   G. E. Andrews, B. C. Berndt, L. Jacobsen and R. L. Lamphere, "The Continued Fractions Found in the Unorganized Portions of Ramanujan's Notebooks". To appear in Memoirs of the Amer. Math. Soc., Providencs R.I.

[Bern89]   B. C. Berndt, "Ramanujan's Notebooks, Part II", Springer-Verlag, New York (1989).

[BBLW85]   B. C. Berndt, R. L. Lamphere and B. M. Wilson, *Chapter 12 of Ramanujan's Second Notebook: Continued Fractions*, Rocky Mountain J. Math. **15** (1985), 235–310.

[BoSh89]   K. O. Bowman and L. R. Shenton, "Continued Fractions in Statistical Applications", Marcel Dekker, Inc., New York and Basel (1989).

[EMOT53]   A. Erdélyi, W. Magnus, F. Oberhettinger and F. G. Tricomi, "Higher Transcendental Functions", Vol. 1, McGraw-Hill, New York (1953).

[Gaut70]   W. Gautschi, *Efficient Computation of the Complex Error Function*, SIAM J. Numer. Anal. **7** (1970), 187–198.

[JoTh80]   W. B. Jones and W. J. Thron, "Continued Fractions: Analytic Theory and Applications", Addison-Wesley, Encyclopedia of Mathematics and its Applications, Vol.11, London, Amsterdam, Don Mills, Ontario, Sydney, Tokyo (1980). Now distributed by Cambridge University Press.

[Khov63]   A. N. Khovanskii, "The Application of Continued Fractions and Their Generalizations to Problems in Approximation Theory", P. Noordhoff, Groningen, The Netherlands (1963).

[Lawd89]   D. F. Lawden, "Elliptic Functions and Applications", Springer-Verlag, Applied Mathematical Sciences Vol. **80**, New York (1989).

[Perr57]   Perron, O., "Die Lehre von den Kettenbrüchen", Band II, B.G. Teubner, Stuttgart (1957).

[Wall48]   H. S. Wall, "Analytic Theory of Continued Fractions", Van Nostrand, New York (1948).

Subject Index

Printed and bound by CPI Group (UK) Ltd, Croydon, CR0 4YY

03/10/2024

01040428-0013